SNOW CRYSTALS

Snow Crystals

A Case Study in Spontaneous Structure Formation

Kenneth G. Libbrecht

PRINCETON UNIVERSITY PRESS
PRINCETON & OXFORD

Published by Princeton University Press
41 William Street, Princeton, New Jersey 08540
6 Oxford Street, Woodstock, Oxfordshire OX20 1TR

press.princeton.edu

Library of Congress Cataloging-in-Publication Data

Names: Libbrecht, Kenneth, author.
Title: Snow crystals : a case study in spontaneous structure formation / Kenneth G. Libbrecht.
Description: Princeton, NJ : Princeton University Press, [2022] | Includes bibliographical references and index.
Identifiers: LCCN 2020057327 (print) | LCCN 2020057328 (ebook) | ISBN 9780691200378 (hardback) | ISBN 9780691223629 (ebook)
Subjects: LCSH: Snowflakes.
Classification: LCC QC926.32 .L5295 2022 (print) | LCC QC926.32 (ebook) | DDC 548/.5—dc23
LC record available at https://lccn.loc.gov/2020057327
LC ebook record available at https://lccn.loc.gov/2020057328

British Library Cataloging-in-Publication Data is available

Editorial: Jessica Yao, Ingrid Gnerlich, Maria Garcia
Jacket Design: Layla Mac Rory
Production: Jacqueline Poirier
Publicity: Matthew Taylor, Amy Stewart

Jacket image courtesy of author

This book has been composed in Garamond Premier Pro

Printed on acid-free paper. ∞

Printed in China

10 9 8 7 6 5 4 3 2 1

CONTENTS

SEVEN
Simple Ice Prisms

EIGHT
Electric Ice Needles

NINE
Designer Snow Crystals

TEN
Natural Snowflakes

ELEVEN
Snowflake Photography

PREFACE

Studying snow crystals is a somewhat unusual endeavor, so people often ask me what got me started on this path, and why I have kept at it for over two decades. The short answer is simply that I find the science both fascinating and entirely worthy of attention. The molecular dynamics of crystal growth is a knotty problem on many levels, and ice exhibits some especially curious behaviors. Even now, well into the twenty-first century, our fundamental understanding of why snow crystals grow into the rich variety of structures we see falling from the clouds is remarkably primitive.

Part of me feels that the humble snowflake has become a bit of an embarrassment to the scientific community. We can split the atom and sequence the human genome, but explaining the growth of a snowflake remains beyond our abilities? Every winter we see these icy works of art simply appearing, spontaneously, quite literally out of thin air. And yet we have no ready explanation as to why snowflakes look the way they do.

Another part of me feels that the general physics of crystal growth is something we ought to know better. The manufacture of semiconductor crystals underlies the entire electronics industry, yet growing crystals is a bit like growing carrots—knowing how to do it is not the same as knowing how it works. Snowflakes can be a valuable case study in this regard; if we can figure out the detailed molecular dynamics governing snow crystal formation, maybe that knowledge will be useful in other areas.

Although crystal growth is an important area in materials science and engineering, my studies have not been motivated by any specific practical applications. My focus is instead on fundamental questions regarding the molecular physics of crystal growth. Applied research can certainly be rewarding, but contemplating the overarching scientific questions has its merits as well. History clearly teaches us that the knowledge gained from basic scientific pursuits often ends up being quite beneficial, even if one cannot always imagine right now how, when, or where those future benefits might arise.

On a related note, I always make a point of telling people that I have not spent any tax dollars on this research. I have always considered my snow crystal studies to be something of a scientific hobby—interesting to me, but with no obvious financial payoff now or down the road. I figure with over seven billion people on the planet and vast resources being spent on sports, entertainment, and all manner of generally unnecessary activities, perhaps a few of us can be spared to contemplate the inner workings of a snowflake.

My foray into snow crystals began in 1995 during an idle conversation with Stephen Ross, whom I had recently hired as a postdoctoral researcher in my lab at Caltech. It occurred to us that the basic physics of crystal growth deserved more attention, and I soon realized that ice would be a good place to start. If nothing else, it was certainly an inexpensive material to work with, with no onerous safety issues, and its freezing temperature was easily accessible as well. Although our attention was quickly pulled back to ongoing projects in atomic physics, I began researching what was known about the science of snow crystals. By the fall of 1998, I had created a website devoted to the subject, which eventually morphed into what is now *SnowCrystals.com*.

Although I grew up in snowy North Dakota, I had not really seen snowflakes until I started digging through the scientific literature. This wonderful story was simply not being told outside that realm, so I set my sights on writing a popular science book on the subject, which, surprisingly, had never been done before. In search of suitable photographs, I found that Wilson Bentley's pictures were something of a standard, but they were more than 100 years old, and their quality was rather poor by modern standards. Newer snowflake photos were out there, but the quality I wanted was not to be found.

As a laboratory physicist, I was already experienced with optics and electronics, so I was soon building a better snowflake photomicroscope. This led to a collaboration with Patricia Rasmussen in Wisconsin, who put the instrument to good use during the 2001–2002 winter season, substantially raising the bar for high-resolution snowflake photography. Voyageur Press then worked with us to publish *The Snowflake: Winter's Secret Beauty* in the fall of 2003, just in time for Christmas.

When *The Snowflake* did well, I made numerous improvements to my microscope and mounted it in a rugged suitcase for traveling, with the aim of becoming a serious snowflake photographer. This led to several expeditions to northern Ontario and central Alaska, including countless hours out in the cold photographing minute ice crystals. These new photos formed the basis for *The Little Book of Snowflakes*, which came out during the 2004 holiday season.

I continued photographing snowflakes around the globe for about a decade, and the subject remained popular in the media. Voyageur Press and I produced a new book every year, including *The Art of the Snowflake*, *The Secret Life of a Snowflake*, *Ken Libbrecht's Field Guide to Snowflakes*, *The Magic of Snowflakes*, *Snowflakes*, and *The Snowflake: Winter's Frozen Artistry*. These years were something of a whirlwind experience, highlighted by a set of snowflake photos on U.S. postage stamps (over 3 billion sold!) and even an appearance on the Martha Stuart show.

With an influx of revenue from book royalties, I was able to gear up my snowflake lab to the point that I could start doing meaningful experimental research investigating the physics of snow crystal growth. This led to better measurements of the molecular attachment kinetics, studies using electric needle crystals, and making designer Plate-on-Pedestal snow crystals, all topics that are discussed at some length in the chapters that follow.

My students and I have made good progress on several scientific fronts, especially toward explaining the Nakaya diagram, as I describe in detail in this book. Why snow crystals switch back and forth between platelike and columnar forms with changing temperature had been an outstanding puzzle for 75 years, and I feel now that it is finally beginning to make sense. Still, the phenomenon of snow crystal growth is not a solved problem by any means. Like an onion, as you peel away layers, you tend to find more layers, as this is the nature of scientific research. Many interesting questions remain.

Until recently, my work on snow crystals was mostly a side project. My scientific focus has drifted over several decades from solar astrophysics to atomic/laser physics to gravitational physics and the LIGO (Laser Interferometer Gravitational-wave Observatory) project, and I dabbled with snowflakes when time permitted. Recently I began to realize that snow crystals are my new calling, so, starting around 2014, I have been focusing nearly all my research efforts in this area. It remains, at least to me, a continually fascinating scientific venture.

I am fortunate to have worked with many talented undergraduate students from Caltech and other universities on my snow crystal research, including Hannah Arnold, Robert Bell, Johanna Bible, Nina Budaeva, Timothy Crosby, Benjamin Faber, Cameron Lemon, Kevin Lui, Christopher Miller, Helen Morrison, Ryan Potter, Mark Rickerby, Molly Swanson, Victoria Tanusheva, Sarah Thomas, and Han Yu. Their determined efforts are much appreciated.

In the same vein, I have enjoyed countless rewarding interactions with fellow snow/ice enthusiasts, colleagues and collaborators, including Todd Berger, Michael Dregni, Harald Garcke, Luis González MacDowell, Janko Gravner, David Griffeath, James Kelly, Ted Kinsman, Alexey Kljatov, Don Komarechka, Heiner Müller-Krumbhaar, Nathan Myhrvold, Carol Norberg, Joseph Shaw, Matthew Sturm, Walter Tape, and Mary Anne White.

I am especially indebted to Caltech for educating me, hiring me as a young professor, and providing me gainful employment for most of my adult life. The university has provided me with ample lab space while allowing me the freedom to explore this atypical line of scientific research. Without Caltech's support, none of this work would have been possible.

Finally, my wife, Rachel Wing, and our two children, Maxwell and Alanna, have been enthusiastic participants throughout this snowflake adventure, especially on our numerous snowflake-related vacations to such far-flung venues as northern Japan, Vermont, northern Ontario, northern Sweden, Alaska, and the mountains of California, all during the cold of winter. Thanks for the memories!

Kenneth Libbrecht
Pasadena, California
July 30, 2020

SNOW CRYSTALS

FIGURE 1.1. An exceptionally large and symmetrical stellar snow crystal, measuring nearly 4 mm from tip to tip, photographed by the author in Kiruna, Sweden.

ONE

Snow Crystal Science

How full of the creative genius is the air in which these are generated!
I should hardly admire more if real stars fell and lodged on my coat.

—HENRY DAVID THOREAU, JOURNAL, 1856

This book is about the science of snowflakes. Its overarching objective is to explain why snowflakes grow into those remarkable crystalline structures that can be found floating down from the winter clouds. In these pages, I answer some of the basic scientific questions one might ask while scrutinizing a newly fallen snowflake: Where do snowflakes come from? How does formless water vapor manage to arrange itself, spontaneously, into such a variety of amazingly ornate shapes? What physical processes guide the development of these elaborate, yet symmetrical, patterns? Why does all this happen the way it does?

Comprehending the humble snowflake is a surprisingly challenging task. The seemingly simple phenomenon of water vapor freezing into ice involves a veritable symphony of subtle molecular processes, from diffusive mixing in the air to the complex attachment kinetics that govern how water molecules assimilate into a rigid crystalline lattice. Explaining this intricate act of meteorological morphogenesis requires a rather deep dive into areas of mathematical physics, statistical mechanics, materials science, and the many-body molecular dynamics of crystal growth. Even now, well into the twenty-first century, snowflake science is very much a work in progress, as several rather basic aspects of the surface structure and dynamics of ice at the molecular level remain quite mysterious.

When I first began reading about this subject in the 1990s, I was immediately struck by just how little was really understood about snowflake formation. Many different morphological types of snowflakes had been observed and cataloged over the years, yet there was no comprehensive explanation for why these different shapes appeared under different growth conditions. As illustrated in Figure 1.2, thin plates and ornate stellar crystals appear when the temperature is around −15°C, while slender needles and columns form when the temperature is near −5°C. Intermediate temperatures generally yield blocky shapes. Small platelike crystals are the norm above −3°C, while sharply faceted columnal crystals form below −30°C. Why does snow crystal morphology depend so strongly on temperature, and why specifically in this manner? I expound at some length on this topic in Chapter 4, as this has been a long-standing

FIGURE 1.2. Natural snow crystals exhibit a remarkable degree of morphological diversity. Platelike examples shown here include (a) a basic hexagonal crystal with symmetrical surface markings; (b) a stellar plate with six broad-branched extensions; and (c) a "fernlike" stellar dendrite, which is an exceptionally thin, flat crystal with copious sidebranching. These types of crystals typically appear when the temperature is in a narrow range around −15°C, although smaller platelike crystals can also be found near −2°C. Other natural specimens include (d) a simple hexagonal prism viewed from the side; (e) a pair of hollow columns, each exhibiting conical voids that almost touch at the crystal centers; and (f) a capped column, resulting when two platelike crystals grow out from the ends of a stout column, like two wheels on an axle. Columnar and needle forms are common near −5°C, while a capped column results when the temperature changes as the crystal grows.

scientific puzzle, and I have developed a few new ideas aimed at solving it. But a complete understanding of even this straightforward observation remains elusive.

At first glance, the snowflake appears to be a somewhat basic natural phenomenon. It is made of little more than pure ice, and it assembles itself, quite literally, out of thin air (Figure 1.3). Nevertheless, trying to understand snowflake formation in detail will take us to the cutting edge of contemporary science. The journey will be neither short nor simple, so let us begin with the basics.

COMPLEX SYMMETRY

I often use the term *snowflake* synonymously with *snow crystal.* The latter is a single crystal of ice, in which water molecules are all lined up in a precise hexagonal array. Whenever you see that characteristic sixfold symmetry associated with snowflakes, you are actually looking at a snow crystal. A snowflake is a more general meteorological term that can mean an individual snow crystal, a cluster of snow crystals that form together, or even a large aggregate of snow crystals that collide and

FIGURE 1.3. A gray winter scene looking out over Lake Superior near Houghton, Michigan. Water vapor evaporating from the warm lake quickly condenses into a mist of water droplets, because the air is substantially colder than the water. But the mist soon evaporates back to water vapor as it rises above the lake. The vapor condenses once again into droplets at higher altitudes, forming thick clouds. Should the clouds cool down sufficiently, most of the liquid droplets will evaporate to feed the formation of snowflakes that fall back into the lake, completing the water cycle.

stick together mid-flight. Those large puff-balls you see floating down in warmer snowfalls are called "snowflakes," and each is made from hundreds or even thousands of individual snow crystals. Snow crystals are commonly called "snowflakes," and this is fine, like calling a tulip a "flower."

A snow crystal is not a frozen raindrop; that type of precipitation is called "sleet." Instead, a snow crystal forms out of water vapor in the atmosphere, as water molecules in the gaseous state transition directly to the solid state. Complex structures emerge as the crystal grows, driven mainly by how water vapor molecules are transported to the developing crystal via diffusion, together with how readily impinging molecules stick to different ice surfaces.

From Clouds to Crystals

To begin our study of snow crystal formation, consider the life of a large, well-formed snowflake that falls from the winter clouds. The story begins as weather patterns transport and cool a parcel of moist air until its temperature drops below the *dew point*, meaning the relative humidity rises above 100 percent and the air becomes *supersaturated* with water vapor. When this happens, the gaseous water vapor in the air tends to condense out as liquid water. Near the ground, the water vapor might condense as dew on the grass (which is why this temperature is called the "dew point"). At higher altitudes, however, water vapor condenses into countless cloud droplets. Each liquid droplet forms around a microscopic particle of dust, and these are typically abundant in the atmosphere. Cloud droplets are so small—about 10–20 microns in diameter—that they can remain suspended in the air almost indefinitely.

If the cloud continues cooling and its temperature drops significantly below 0°C, then the liquid water droplets will start freezing into ice. Not all the droplets freeze at once, and none will freeze right at 0°C. Instead, the droplets become *supercooled* as their temperature drops, often remaining in a metastable liquid state for long periods of time. Some droplets will freeze when the temperature is as high −5°C, but most will freeze somewhere in the vicinity of −10°C. A hearty few may survive unfrozen at −20°C or below, but all will become solid ice before the temperature reaches −40°C.

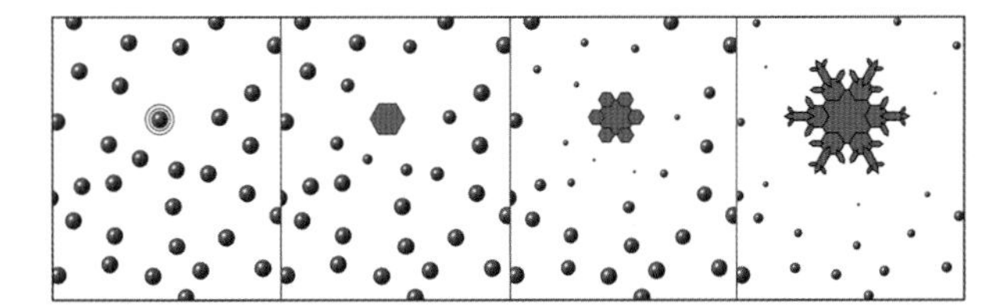

FIGURE 1.4. A snowflake is born when a liquid cloud droplet freezes into ice (left sketch). The ice particle initially grows into a faceted prism, as the growth is limited by anisotropic attachment kinetics on the ice surface. After the crystal grows larger, the diffusion of water molecules through the air causes branches to sprout from the six corners of the prism, which continue growing to become elaborate dendritic structures. The growing crystal removes water vapor from the air, which is replenished by the evaporation of nearby water droplets. About 100,000 cloud droplets evaporate to provide enough material to make one large stellar snow crystal. The flake continues growing inside the cloud until it becomes so heavy that it falls to earth.

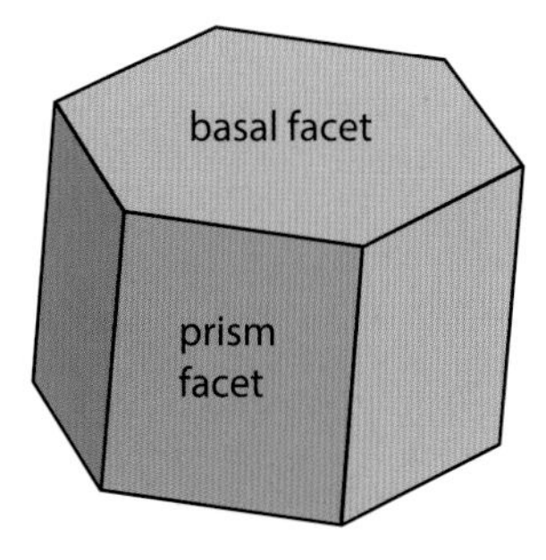

FIGURE 1.5. The most basic shape of a snow crystal is a *hexagonal prism* with two basal facets and six prism facets. This shape arises because of the underlying hexagonal structure of the ice crystal lattice.

The freezing temperature of a specific cloud droplet is determined in large part by the speck of dust it contains. Pure water can be cooled to nearly −40°C before freezing, while some materials (silver iodide in particular) will nucleate freezing as high as −4°C. Certain bacterial proteins can even promote freezing at temperatures as high as −2°C. These exotic materials are not much present in the atmosphere, however, so your average speck of dust will nucleate freezing around −10°C. Note that the character of an included dust particle usually has little effect on the final snow crystal shape, because the dust is soon buried within the ice.

Once a cloud droplet freezes, it becomes an embryonic snow crystal that commences its growth by absorbing water vapor from the air around it. Because the vapor pressure of liquid water is higher than that of solid ice (see Chapter 2), the cloud droplets surrounding the nascent snowflake begin to evaporate away, as illustrated in Figure 1.4. During this process, there is a net transfer of water molecules from liquid water droplets to water vapor, and then from vapor to ice. About 100,000 cloud droplets will evaporate away to provide enough material to make one good-sized snowflake. This roundabout route is how the liquid water in a cloud freezes into solid ice.

As the temperature inside a cooling cloud falls below −10°C, cloud droplets will begin freezing in large numbers, thus initiating a full-fledged snowfall. By the time the cloud has cooled to around −20°C, however, most of the liquid droplets will be gone, as some will have frozen and many will have evaporated away to form snow crystals. At temperatures below −20°C, it is sometimes said to be "too cold to snow," because nearly all the liquid cloud droplets will have already disappeared before the cloud cools to that temperature. And when no liquid water remains to feed growing snowflakes, there will not be a lot of falling snow.

Faceting and Branching

Soon after a cloud droplet freezes, it initially grows into the shape of a small, faceted hexagonal prism, as illustrated in Figure 1.5. The prism shape is defined by two *basal facets* and six *prism facets* that arise from the underlying sixfold symmetry of the ice crystal lattice (see Chapter 2). The molecular mechanism that creates this faceted shape is one of the key physical processes that guide snow crystal formation, and Figure 1.6 illustrates how it works. Water vapor molecules strike the ice crystal everywhere on its surface, but they are more likely to stick when the surface is molecularly "rough," meaning it has a lot of dangling chemical bonds. The facet surfaces are special, because they are aligned with the lattice

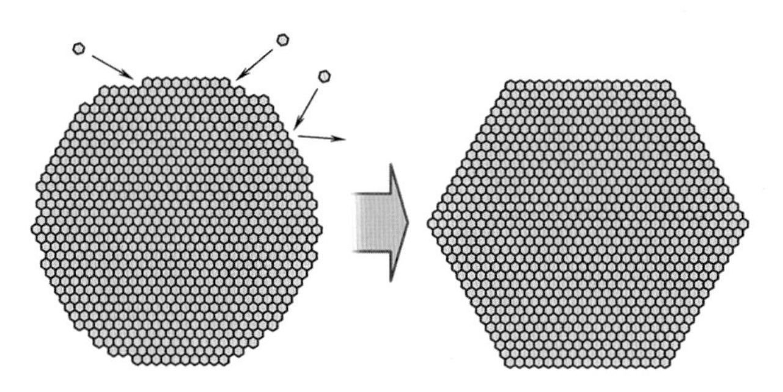

FIGURE 1.6. When water vapor molecules strike a molecularly rough ice surface, they tend to stick and become incorporated into the ice lattice. But when they strike a molecularly smooth facet surface, they are less likely to stick. As the crystal grows, the rough areas soon fill in to yield a fully faceted ice prism.

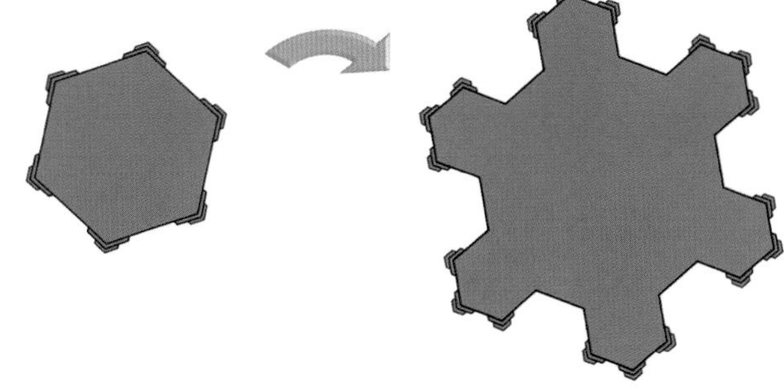

FIGURE 1.7. As a thin hexagonal plate grows larger, its six corners stick out into the humid air around it. Water vapor condenses preferentially on the corners as a result, making them stick out even farther. This leads to a *growth instability* that causes six branches to sprout from the corners of the hexagon. The same process later yields sidebranches on the main branches.

structure of the crystal, so these surfaces exhibit fewer open molecular bonds. Thus, the facet surfaces accumulate water vapor at a lower rate than do the rough surfaces, and this process soon yields a faceted ice prism. The rate at which impinging water molecules stick to various surfaces is called the *attachment kinetics,* and I discuss this subject in considerable detail in Chapter 4.

If the cloud temperature drops to near −15°C, which is often the case during an ample snowfall, then the basal surfaces will accumulate material especially slowly, while water vapor will condense on the prism facets much more readily. As a result, a frozen droplet in those conditions will soon develop into a thin, flat, hexagonal plate, which is an early stage of what will eventually become a large stellar snow crystal. As the small hexagonal plate is growing, its six corners stick out slightly into the surrounding humid air, causing the tips of the hexagon to absorb water vapor a bit more quickly than other parts of the crystal. The faster growth makes the corners stick out farther still, causing them to grow even faster. This positive-feedback effect causes a set of six branches to sprout from the hexagonal plate, as illustrated in Figure 1.7. I describe this *branching instability* (also known as the *Mullins-Sekerka instability*) at length in Chapter 3, as it is responsible for most of the complex structure seen in snow crystals.

Once the six branches begin to develop, most of the subsequent growth occurs near the branch tips, where the supply of water vapor is greatest. Moreover, the growth behavior of each branch is quite sensitive to the temperature and humidity of the air surrounding it. As the crystal travels through the inhomogeneous clouds, it experiences ever-changing conditions that modify how the crystal grows. Sometimes the branch tips become faceted, while at other times, they may sprout additional sidebranches. It all depends on the growth conditions seen by the crystal at any given time. The final shape of the branch, therefore, reflects the entire history of its growth, which was determined by the meandering path the flake took through the atmosphere.

The six branches of a snow crystal develop in near synchrony, simply because they all travel together through the cloud. All six branches experience essentially the same growth conditions at the same times, so all six develop into the same elaborate shape, as illustrated in Figure 1.8. Note that the growth of the separate branches is not synchronized by any communication between them, but rather by their common history. And because no two snowflakes follow exactly the same path through the turbulent atmosphere, no two look exactly alike. (The full story of snowflake uniqueness is a bit more involved, as I describe later in this chapter.) The formation of a large stellar snow crystal takes about 30–45 minutes, and we can reproduce the process in the lab

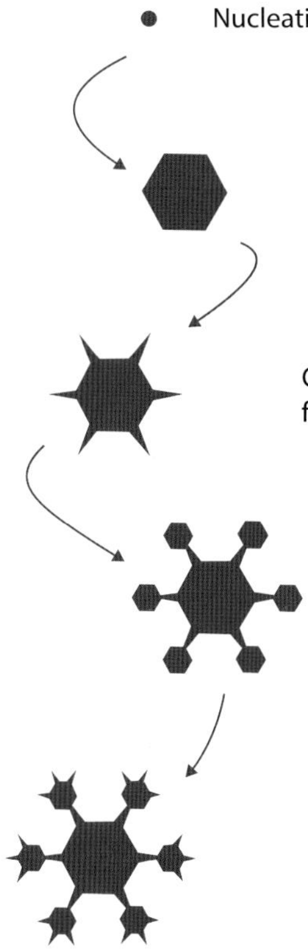

Nucleation of ice particle

Develops into hexagonal prism because smooth facets growth most slowly

Corners stick out farther, accumulate ice faster ⟶ corners sprout branches

Crystal moves to new position ⟶ plates grow on ends of branches

Crystal travels through clouds, experiences many changes in growth behavior ⟶ a complex stellar snow crystal

FIGURE 1.8. The final shape of a complex stellar snow crystal depends on the path it traveled through the clouds. Sudden changes in the temperature and humidity around a crystal can cause abrupt changes in its growth behavior, perhaps stimulating the formation of sidebranches or plates. However, because the six arms see the same changes at the same times, they grow in near synchrony. The final snow crystal thus exhibits a complex structure with an overall sixfold symmetry.

with relative ease, studying how environmental changes induce symmetrical dendritic structures.

Although this narrative reasonably describes the origin of complex symmetry in stellar snow crystals, the story deepens when we also consider the variety of morphological types illustrated in Figure 1.2. The full menagerie of natural snow crystals is presented in Chapter 10, and laboratory studies have found that these can be organized according to the *Nakaya diagram* shown in Figure 1.9, which is also called the *snow crystal morphology diagram*. This empirical chart summarizes the various snow crystal forms that appear at different temperatures and humidity levels, including platelike and columnar forms with varying amounts of structural complexity. Explaining this morphological organization is no easy task, requiring a methodical characterization of the relevant physical processes involved. The quest to fully comprehend the science of snow crystal formation is the primary focus of this book, continuing an endeavor that has been ongoing for some 400 years.

A BRIEF HISTORY OF SNOW CRYSTAL SCIENCE

I like to think about the snow crystal as something of a case study of scientific reasoning. Science is fundamentally about understanding the natural world, so snowflakes, being part of that world, deserve an explanation. Richard Feynman commented that "nature uses only the longest threads to weave her patterns, so each small piece of her fabric reveals the organization of the entire tapestry" [1964Fey, p. 34]. There is hardly a more fitting example of this truism than the intricate patterns of common snowflakes, as the entire panoply of modern scientific knowledge is still not quite enough to fully explain their origin.

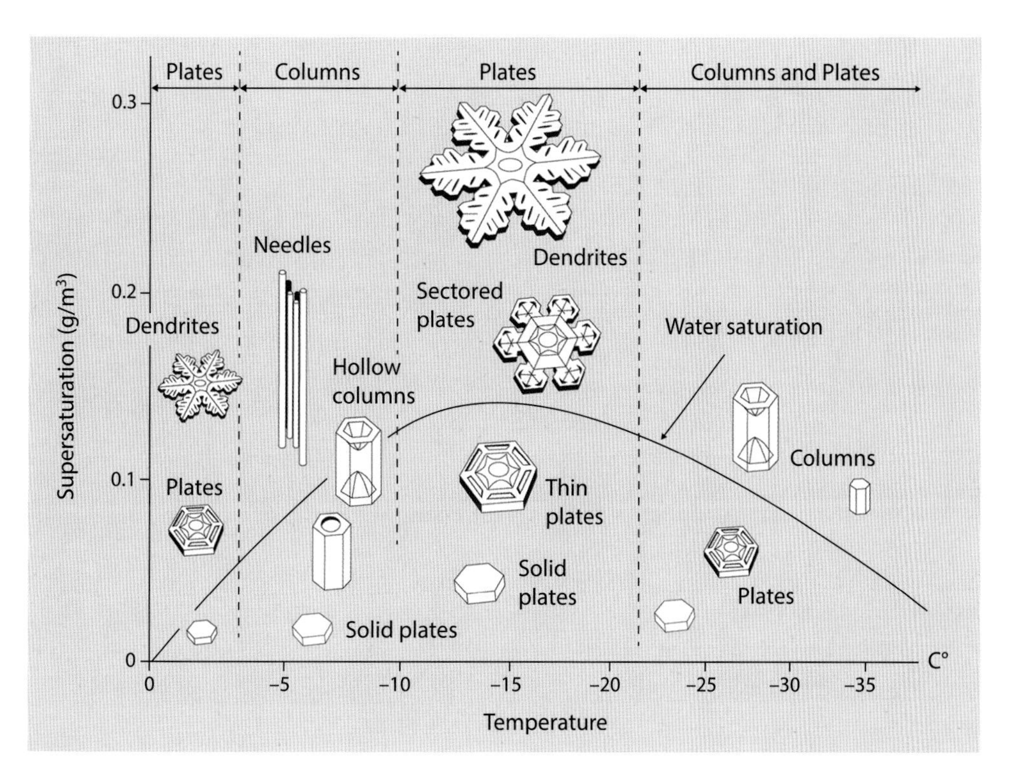

FIGURE 1.9. The *Nakaya diagram* plots the morphological types of snow crystals that develop at different temperatures and humidity levels. Here the supersaturation is shown as the "excess" water vapor density in the air, above the value for saturated air, and the *water saturation* line shows the supersaturation in a dense winter cloud made of liquid water droplets. Explaining why snow crystals experience such varied growth behaviors at different temperatures and supersaturations is a remarkably challenging scientific puzzle, with many parts still unsolved.

The study of snowflake science began when the distinctive sixfold symmetry of individual snow crystals was first recognized as something that could be investigated and possibly understood. Over time, this led to a greater scrutiny of what fell from the clouds, yielding early sketches that began to document the remarkable variety of different morphological types. With advances in technology, snow crystals were examined in greater detail by using optical microscopy and were further documented in extensive photographic studies. And as sophisticated scientific tools became available, researchers progressed from observations of natural snowfalls to scrutinizing laboratory-grown snow crystals, eventually leading to precision measurements of ice growth behaviors, molecular dynamics investigations, and studies using computer-generated snowflake simulations.

In many ways, the snowflake story mirrors the historical development of science itself. As mathematics became intertwined with natural philosophy, the precise symmetry of snow crystal facets suggested an underlying geometrical basis, foreshadowing the molecular order found in crystal lattices. As laboratory-based science emerged, synthetic snowflakes revealed an underlying organization for the observed diversity of natural snow crystal morphologies. And as the nanoscale structure of crystalline materials has become better characterized in the modern era, our understanding of the molecular attachment kinetics governing snow crystal growth has improved as well. We can only guess as to what future scientific tools will be brought to bear in our effort to comprehend the inner workings of the common snowflake.

Early Observations

The earliest account (of which I am aware) describing the sixfold symmetry of individual snow crystals was written in 135 BCE by Chinese philosopher Han Yin [2002Wan, p. 3], who commented: "Flowers of plants and trees are generally five-pointed, but those of snow, which are called ying, are always six-pointed." Subsequent Chinese authors mentioned snow crystal symmetry as well, an example being the sixth-century poet Hsiao Tung, who penned, "The ruddy clouds float in the four quarters of the cerulean sky. And the white snowflakes show forth their six-petaled flowers" [2002Wan, p. 3].

European authors began documenting snowflakes many centuries after the first Asian accounts, and one oft-quoted reference is the woodcut shown in Figure 1.10, created by Olaus Magnus in 1555 [1982Fra]. It can be seen, however, that the clergyman depicted snowflakes as having a curious assortment of odd shapes, including crescents, arrows, and even one that looked like a human hand, so perhaps this account does not quite warrant being called a historical first. It appears that English astronomer Thomas Harriot was the first in Europe to clearly identify and document the snowflake's sixfold symmetry in 1591 [1982Fra].

French philosopher and mathematician René Descartes recorded the first detailed account of snow crystal structures in his famous *Les Météores* in 1637, including the sketches shown in Figure 1.11. In his essay, Descartes described some remarkably thorough naked-eye observations of snow crystals, which included several uncommon forms [1982Fra, p. 5]:

> After this storm cloud, there came another, which produced only little roses or wheels with six rounded semicircular teeth . . . which were quite transparent and quite flat . . . and formed as perfectly and symmetrically as one could possibly imagine. There followed, after this, a further quantity of such wheels joined two by two by an axle, or rather, since at the beginning these axles were quite thick, one could as well have described them as little crystal columns, decorated at each end with a six-petaled rose a little larger than their base. But after that there fell more delicate ones, and often the roses or stars at their ends were unequal. But then there fell shorter and progressively shorter ones until finally these stars completely joined, and fell as double stars with twelve points or rays, rather long and perfectly symmetrical, in some all equal, in others alternately unequal.

FIGURE 1.10. This 1555 woodcut by Olaus Magnus was perhaps the first European illustration depicting a sixfold symmetrical snow crystal, although the artist was perhaps a bit overzealous in his depiction of morphological diversity. Adapted from [1982Fra].

In this passage, we can see snowflakes influencing—in their own small way—the early development of modern science. Descartes was clearly impressed with the geometrical perfection he saw in snow crystal forms, with their flat facets and hexagonal symmetry. Pondering this and other observations, he went on to reason about how the principles of geometry and mathematics play a central role in describing the natural world. Although we take this for granted now, using mathematics to explain ordinary phenomena was still something of an unconventional idea at the time, and a major step forward in science.

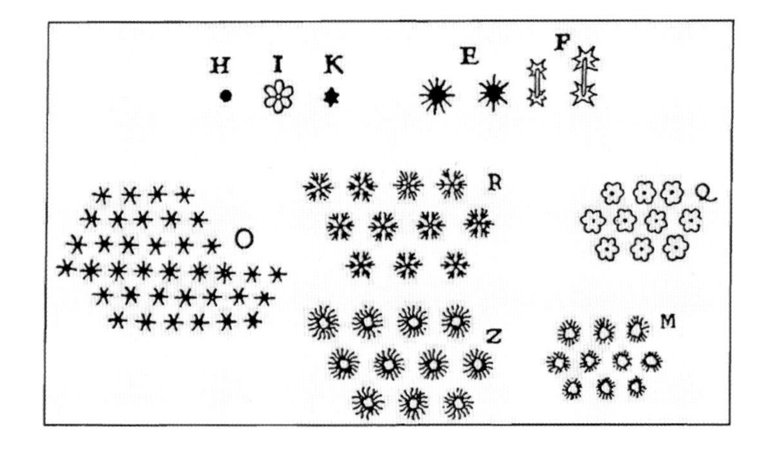

FIGURE 1.11. René Descartes made some of the first accurate sketches of different snow crystal morphologies in 1637, including observations of capped columns (group F in this sketch) [1637Des].

Emerging Science

The first scientist to speculate on a theoretical explanation of the sixfold symmetry of snow crystals was German astronomer and mathematician Johannes Kepler. In 1611, Kepler presented a small treatise titled *The Six-Cornered Snowflake* to his patron, Holy Roman Emperor Rudolf II, as a New Year's Day gift [1611Kep, p. 35]. In his treatise, Kepler contrasted the sixfold symmetry of snowflakes with similar symmetries found in flowers, deducing that the similarities must be in appearance only, because flowers are alive, while snowflakes clearly are not:

> Each single plant has a single animating principle of its own, since each instance of a plant exists separately, and there is no cause to wonder that each should be equipped with its own peculiar shape. But to imagine an individual soul for each and any starlet of snow is utterly absurd, and therefore the shapes of snowflakes are by no means to be deduced from the operation of soul in the same way as with plants.

Kepler saw that a snowflake is a relatively simple thing, made only from ice, compared to the utterly baffling complexity of living things. He offered, therefore, that there might be some relatively simple organizing principle that was responsible for snow crystal symmetry. Drawing on correspondence with Thomas Harriot, Kepler noted that stacking cannonballs also yielded geometric structures with sixfold symmetry, and he further surmised that there might be a mathematical connection between these two phenomena. There was certainly a germ of truth in this reasoning, as the geometry of stacking water molecules lies at the heart of snow crystal symmetry. But this was long before the atomistic view of matter was accepted canon, so Kepler could not carry the cannonball analogy very far.

Kepler realized that the genesis of crystalline symmetry was a worthy scientific question, and he also recognized the similarity between snow crystals and mineral crystals, as they both exhibited symmetrical faceted structures. At the end of his treatise, however, Kepler accepted that the science of his day was not advanced enough to explain any of it. He was certainly correct in this conclusion: Three centuries would pass before scientists knew enough about atoms, molecules, and their arrangement in solid materials to finally answer Kepler's 1611 query.

Microscopic Observations

The invention of the microscope in the mid-seventeenth century quickly led to more and better snowflake observations. English scientist and early microscopist Robert

FIGURE 1.12. Robert Hooke sketched these observations of snowflakes 1665, enabled by his newly invented microscope [1665Hoo].

FIGURE 1.13. English explorer William Scoresby made these sketches during a winter voyage through the Arctic, which he recounted in 1820 [1820Sco]. These are the first drawings that accurately depicted many features of snow crystal structure, as well as several rare forms, including triangular crystals and 12-branched snowflakes. Scoresby also noted that the cold arctic climate produced more highly symmetrical crystals than were typically seen in Britain.

Hooke sketched snowflakes (Figure 1.12) and practically everything else he could find for his book *Micrographia*, published in 1665 [1665Hoo]. Although his microscope was crude by modern standards, Hooke's drawings nevertheless began to reveal the complexity and intricate symmetry of snow crystal structures, details that could not be detected with the unaided eye.

As the quality and availability of optical magnifiers improved, so did the accuracy of snow crystal drawings. By the mid-nineteenth century, several observers around the globe had recorded the diverse character of snow crystal forms, and one notable example is shown in Figure 1.13. Given the ephemeral nature of a snowflake, however, observers inevitably had to rely on memory to complete their sketches. As a result, even the best snow crystal drawings lacked detail and were not completely faithful to their original subjects.

Snowflake Photography

It took Wilson Bentley, a farmer from the small town of Jericho, Vermont, to create the first photographic album of falling snow, thus awakening the world to the hidden wonders of snowflakes. Bentley became interested in the microscopic structure of snow crystals as a teenager in the 1880s, and he soon began experimenting with the new medium of photography as a means of recording what he observed. He constructed an ingenious mechanism for attaching a camera to his microscope for this purpose, and he succeeded in photographing his first snow crystal in 1885 when he was 19 years old.

To say Bentley was dedicated to this task is an understatement. Snowflake photography became his lifelong passion, and over the course of 46 years, he captured more than 5,000 snow crystal images, all on 4-inch glass photographic plates. He resided his entire life in the same Jericho farmhouse, photographing snowflakes each winter, using the same equipment he constructed as a teenager. Figure 1.14 shows Wilson Bentley demonstrating his apparatus, although the grass at his feet suggests there were no snowflakes to be found that day.

Bentley usually presented his photographs as white snowflakes on a black background, as shown in Figure 1.15, but the original photos had a bright background. A snow crystal is made of pure ice, which is clear, not white (see Chapter 11). When illuminating a snow crystal from behind, as Bentley did, the resulting photo exhibits a somewhat low-contrast "bright-on-bright" appearance. To increase the contrast, Bentley made a copy of each photographic negative and painstakingly scraped away the emulsion from the background areas. A print made from the modified negative then yielded a white snowflake on a black background, as illustrated in the figure. Bentley preferred this high-contrast look, so he modified most of his photos using this technique. Some have accused Bentley of altering his photos to augment what nature had provided, but he did not hide the fact that he processed his photos this way. And he was always quick to point out that he never changed the snow crystal images themselves during this process.

One aspect of his work that Bentley rarely emphasized is that large, symmetrical stellar snow crystals are not the norm, and near-perfect specimens are quite rare (see Chapter 10). Over the course of an entire winter season, he only photographed about 100 crystals on average, reserving his expensive emulsions for only the most photogenic snow crystals he could find. Modern automated cameras that photograph falling snow without any selection bias confirm that well-formed symmetrical snow crystals are exceedingly rare [2012Gar].

FIGURE 1.14. Vermont farmer Wilson Bentley first developed the art of snowflake photography in the 1880s, eventually producing a large album of images. He is shown here with his specially built snow crystal photomicroscope [1931Ben].

Bentley's photographs appeared in numerous publications over several decades, providing for many their first look at the inner structure and symmetry of snow crystals. And with thousands of snowflakes, each unique, the world was exposed to their incredible variety as well. The now-familiar old chestnut that no two snowflakes are exactly alike appears to have had its origin in Bentley's photographs.

In the late 1920s, Bentley teamed with W. J. Humphreys, chief physicist for the United States Weather Bureau, to publish his magnum opus, containing more than 2,000 snow crystal photographs [1931Ben]. The book appeared in November 1931, and the 66-year-old Vermont farmer died of pneumonia just a few weeks later. In the decades following this seminal work, many others have taken up the challenge of capturing the structure and beauty of snow crystals using photography, and I describe some modern techniques and recent results in Chapter 11.

FIGURE 1.15. These are just a few of the thousands of snowflake photographs taken by Wilson Bentley between 1865 and 1931. The original photos showed bright crystals against a bright background, as the clear snowflakes were illuminated from behind. The photos were subsequently modified by essentially cutting each crystal out and placing it on a black background. Adapted from [1931Ben].

Crystallography

The word "crystal" derives from the Ancient Greek *krystallos*, meaning "ice" or "rock ice." Contrary to what the definition implies, *krystallos* was not originally used to describe ice, but rather the mineral quartz. The early Roman naturalist Pliny the Elder described clear quartz *krystallos* as a form of ice, frozen so hard that it could not melt. Pliny was certainly mistaken on this point, as quartz is not a form of ice, nor is it even made of water. Nevertheless, after nearly 2,000 years, Pliny's misunderstanding is still seen in the language of the present day. If you look in your dictionary, you may find that one of the definitions for crystal is simply "quartz."

While mineral collectors have admired beautiful crystalline specimens for millennia, understanding the origin of their faceted structures required a bona fide scientific breakthrough. In 1912, German physicist Max von Laue and coworkers discovered that when X-rays were shone through a crystal of copper sulfate, the crystal acted like a grating and produced a diffraction pattern that could be recorded on photographic film. Australian-born British physicists William Henry Bragg and William Lawrence Bragg (father and son) soon developed a mathematical theory showing how the atomic structures of crystalline materials could be ascertained from these diffraction patterns, thus creating the field of crystallography.

Working with the Braggs in their Cambridge laboratory, William Barnes used X-ray crystallography to determine the structure of ice for the first time in 1929 [1929Bar], discovering the now-familiar hexagonal lattice of normal ice Ih (see Chapter 2), and Figure 1.16 shows Barnes's discovery photograph. In subsequent studies over many decades, numerous additional solid phases of water have been discovered and characterized, mostly at extremely high pressures. Three hundred years after Kepler's initial musings, scientists had finally proven that the geometry of stacking was indeed the underlying source of the snowflake's sixfold symmetry.

In the decades that followed these early crystallographic discoveries, the development of quantum mechanics and quantum chemistry have allowed precise *ab initio* calculations of the water molecule electronic and atomic structure, including two-body and higher order interactions between water molecules. From these fundamental quantum-mechanical calculations, researchers have been able to reproduce the known structures of water in many of its solid phases. As a result, the lattice structure of ice Ih, from which snow crystals are made, is now well understood at a fundamental physical level.

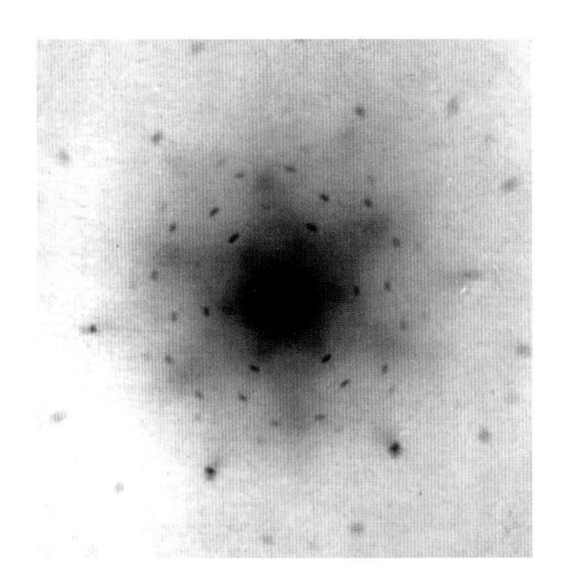

FIGURE 1.16. This X-ray diffraction pattern was made by a crystal of normal ice Ih, allowing William Barnes to first determine the hexagonal lattice structure of the ice crystal. Adapted from [1929Bar].

Attachment Kinetics

While the sixfold symmetry of a snowflake ultimately derives from the symmetry of the ice crystal lattice, how the nanoscale structure of the molecular matrix translates into the large-scale morphology of a growing crystal is a separate matter. For example, quartz and copper are both crystalline minerals, but quartz often exhibits striking faceted features that reveal its lattice structure, while copper rarely does. Why? A big part of the answer lies in the physical processes that govern the formation of faceted surfaces, collectively called the *surface attachment kinetics.*

Around the beginning of the twentieth century, scientists began examining the physics of solidification using the newly discovered laws of statistical mechanics and thermodynamics, which had been developed by James Clerk Maxwell, Ludwig Boltzmann, J. Willard Gibbs, Amedeo Avogadro, Lord Kelvin, and other scientific luminaries throughout the nineteenth century. An early result came from German physicist Heinrich Hertz [1882Her] and independently from Danish physicist Martin Knudsen [1915Knu], who calculated the growth rate of a solid from its vapor phase (like ice from water vapor) from the net flux of vapor molecules striking the solid surface. The resulting Hertz-Knudsen law provides the starting point for the attachment kinetics theory I describe in Chapter 4.

Some decades later, however, it had become clear that the Hertz-Knudsen law did not provide a complete description of the growth of faceted crystalline surfaces. The net flux of molecules striking the surface was only one factor determining the growth rate; another was the probability that an impinging molecule would permanently attach to the surface and become part of the bulk crystal lattice. This probability, ranging from 0 to 1, is now called the *attachment coefficient,* also discussed at length in Chapter 4. Figure 1.6 shows how an anisotropic attachment coefficient produces faceted crystal growth, and this mechanism is also essentially responsible for the appearance of faceted minerals like those shown in Figure 1.17.

Beginning around the 1930s, physicists I. N. Stranski [1928Str], R. Kaischew [1934Str], R. Becker and W. Döring [1935Bec], M. Volmer [1939Vol], and others pushed the field forward by developing a detailed statistical-mechanical theory describing the nucleation and subsequent growth of one-molecule-high *terraces* on flat faceted surfaces. Many researchers fortified this theory in the following decades, notably W. K. Burton, N. Cabrera, and F. C. Frank [1951Bur], building it into the modern theory of crystal growth and surface attachment kinetics that is described in modern textbooks [1996Sai, 1999Pim, 2002Mut, 2004Mar]. This theory provides the starting point for understanding snow crystal growth.

The attachment kinetics are a major factor in determining the growth rates and resulting morphologies in different environmental conditions. For example, the principal difference between a thin platelike snow crystal and a slender columnar form (see Figure 1.2) lies in how readily the impinging water vapor molecules attach to the basal and prism surfaces. As a result, the large-scale

FIGURE 1.17. Besides ice, many other mineral crystals grow into faceted morphologies under the right conditions, as seen in these examples (from upper left to lower right): gypsum, quartz, pyrite, and synthetic bismuth. The lattice structure determines the overall symmetry of each crystalline form, but the attachment kinetics are largely responsible for the appearance of faceted surfaces.

morphology of nearly every snowflake, from platelike to columnar, is determined to a large degree by how the surface attachment kinetics changes with temperature, supersaturation, surface orientation, and other factors.

It is a common misconception that crystallography explains crystal growth, but this is far from the truth. Crystallography refers to the lattice structure of crystalline materials, and this is entirely a statics problem describing the lowest-energy molecular configuration in equilibrium. Crystal growth, by contrast, is a dynamical problem involving many-particle interactions in systems far from equilibrium. Modern science is quite adept at solving statics problems, but less so with many-body dynamics problems. For this reason, the crystallography of ice has been essentially solved for nearly a century, while many important aspects the ice attachment kinetics remain quite puzzling. Terrace nucleation theory nicely explains some aspects of snow crystal attachment kinetics, but certainly not all. Creating a comprehensive model of this many-body molecular process is very much a work in progress, with many unsolved problems still outstanding.

The Nakaya Diagram

Japanese physicist Ukichiro Nakaya conducted the first true scientific investigation of snow crystals at Hokkaido University in the 1930s. Motivated by the abundant snowfalls in Hokkaido, and inspired by Wilson Bentley's photographs, Nakaya began his investigations by cataloging the different types of falling snow. Unlike Bentley, however, Nakaya looked beyond stellar crystals and focused his attention on describing the full range of different snowflake types, including columns, needles, capped columns, and other less-common forms. Nakaya thus produced the first photographic documentation of the broader menagerie of falling snow.

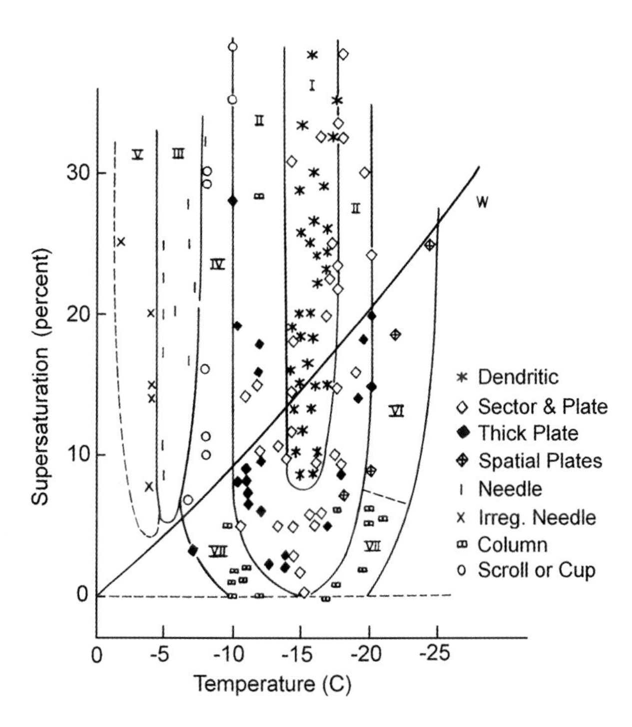

FIGURE 1.18. The original Nakaya diagram [1954Nak, 1958Nak] illustrates snow crystal morphology plotted versus growth temperature and water vapor supersaturation. Nakaya's observations of laboratory-grown snow crystals revealed, for example, that large stellar dendrites only form in a narrow temperature range around −15°C, while slender needle crystals only appear near −5°C. Adapted from [1954Nak].

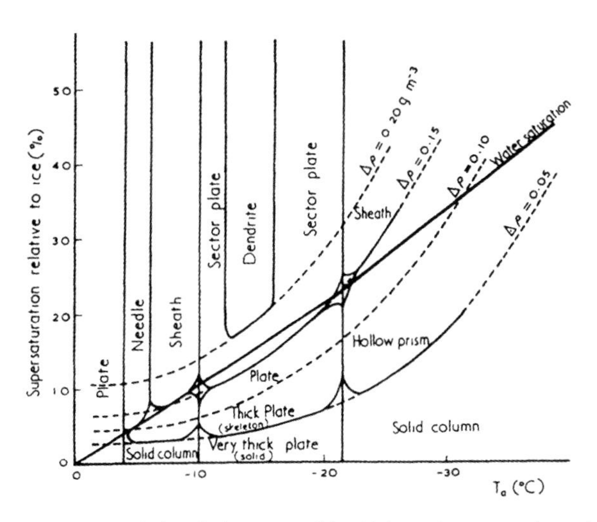

FIGURE 1.19. A detailed version of the Nakaya diagram, adapted from [1990Yok]. Additional data further indicate that plates grow at all temperatures above −20°C when the supersaturation is sufficiently low [2019Lib1] while columns are common down to −70°C [2009Bai, 2012Bai].

While learning a great deal from observations of natural snow crystals, Nakaya quickly realized that laboratory experiments would be essential for better understanding the origin of what he saw falling from the clouds. To this end, he constructed a walk-in freezer laboratory at Hokkaido, using it for a variety of experimental investigations of ice crystal growth. Prominent among them, Nakaya created the world's first laboratory-grown snowflakes in his lab in 1936 [1954Nak].

Nakaya and his collaborators spent years examining how synthetic snow crystals grew and developed at different temperatures and supersaturations in their growth chamber, soon combining all these observations into what is now called the *Nakaya diagram,* shown in Figure 1.18. Subsequent researchers further refined and expanded the Nakaya diagram [1958Hal, 1961Kob, 1990Yok], and the most recent of these is shown in Figure 1.19. The more stylized version in Figure 1.9 includes newer observations indicating that simple plates form in low supersaturations at all temperatures above −20°C [2019Lib1]. Bailey and Hallett further extended these results with additional observations, exploring temperatures down to −70°C and finding an abundance of columnar forms in these frigid conditions [2009Bai, 2012Bai].

The Nakaya diagram was immediately recognized as being like a Rosetta Stone for snowflakes. With it, one can translate the shape of a falling snow crystal into a description of its growth history. On seeing a slender needle crystal, for example, one can deduce that it must have grown in high humidity at a temperature near −5°C. A large stellar crystal indicates growth near −15°C, and the amount of sidebranching provides an indication of the level of supersaturation it experienced. The formation of a capped column (see Figure 1.2) arises from an initial period of growth near −5°C (columnar) followed by subsequent growth near −15°C (yielding plates on both ends of the column).

Nakaya liked to remark that snowflakes are like "hieroglyphs from the sky." With the Nakaya diagram, a spectator on the ground can decipher the observed crystal morphology to ascertain the conditions of the clouds in which it formed, like a kind of meteorological hieroglyphics. The Nakaya diagram also tells us that snow crystal growth is remarkably sensitive to temperature. Even a change of a few degrees can dramatically alter a crystal's growth behavior, which helps explain why snowflakes have such a remarkable diversity of shapes. We will come back to the Nakaya diagram many times in this book, as it has become an essential tool for understanding the variable nature of snow crystal formation.

Crystal Dendrites

In 1917, Scottish zoologist D'Arcy Wentworth Thompson published *On Growth and Form*, in which he pondered on the physical, biological, and mathematical origins of complex structures in nature [1917Tho, 1961Tho]. While confessing that crystal growth was somewhat outside the province of his book, Thompson commented:

> Yet snow-crystals . . . have much to teach us about the variety, the beauty and the very nature of form. To begin with, the snow-crystal is a regular hexagonal plate or thin prism; that is to say, it shows hexagonal faces above and below, with edges set at co-equal angles of 120°. Ringing her changes on this fundamental form, Nature superadds to the primary hexagon endless combinations of similar plates or prisms, all with identical angles but varying lengths of side; and she repeats, with an exquisite symmetry, about all three axes of the hexagon, whatsoever she may have done for the adornment and elaboration of one.

In his celebrated treatise, Thompson used extensive examples to focus scientific attention on the central question of how complex structures arise spontaneously in natural systems. Humans tend to create intricate objects via a subtractive process, beginning with bulk material and carving it into a final desired form, following a preconceived design. At the opposite end of the fabrication spectrum, living things develop into amazingly sophisticated organisms quite spontaneously, using the additive process of growth. Thompson strove to comprehend the underlying physical and chemical principles that guide the development of living organisms, thus pioneering what has become the field of developmental biology.

Like Kepler 300 years before him, however, Thompson found that the whole of biological structure formation presented a challenging problem, to say the least. An easier approach, therefore, might be to consider something like the snowflake, which exhibits an interesting degree of spontaneous structure formation, but in a far simpler physical system. Over time, physicists also began to appreciate that the patterns arising during solidification offered a worthy phenomenon to investigate. Just as the hydrogen atom was a first step toward understanding the complex chemistry of large biomolecules, perhaps the physical origin of structure formation during solidification can provide insights into systems having far greater complexity.

A significant step forward in this direction was made in 1964, when American physicists William W. Mullins and Robert F. Sekerka realized that *growth instabilities* are often associated with pattern-forming systems, with solidification being a specific example. In their seminal paper [1964Mul], the authors showed that many of the simplest solutions to the equations describing diffusion-limited growth were mathematically unstable to small perturbations that developed into complex dendritic structures. This spontaneous branching process that arises during solidification—the *Mullins-Sekerka instability*—plays a central role whenever diffusion limits the solidification of materials, and as described in Chapter 3, growth instabilities are necessary for producing essentially all the complex morphological features seen in snow crystals.

Although dendritic structures had been documented in a broad range of physical and biological systems by D'Arcy Thompson and others for many decades, the underlying causes of these forms was beyond the reach of early scientific knowledge. Counting the petals on a flower was one thing; explaining their existence was another matter entirely. Indeed, comprehending even quite simple biological structures remains largely an intractable problem to this day. Mullins and Sekerka showed, however, at least for simple physical systems, that it was possible to make some progress toward understanding how complex structures arise spontaneously in nonequilibrium systems.

A systematic study of growth instabilities in laboratory solidification was undertaken in the 1970s by American materials scientist Martin Glicksman and others, who examined the growth of dendritic structures when liquids cooled and solidified [1976Gli, 1981Hua]. In an extensive series of influential experiments, these researchers made detailed measurements of structure formation during the freezing of liquid succinonitrile, choosing this material because it is transparent with a freezing temperature near room temperature, while its growth behavior is similar to most common metals. When unconstrained by container walls, Glicksman found that freezing often yielded branched structures like that shown in Figure 1.20, with growth characteristics that depended mainly on crystal symmetry and the degree of supercooling of the liquid. Similar branching is seen in some stellar snow crystals, like the fernlike stellar dendrite shown in Figure 1.2, and this same kind of dendritic growth behavior has been observed to be quite ubiquitous during solidification from both liquids and vapors over a broad range of materials.

The work of Glicksman and others soon called attention to the Mullins-Sekerka instability and its consequences for structure formation during crystal growth. There followed a concerted push by physicists, material scientists, and applied mathematicians to form a self-consistent theory describing the characteristics of the diffusion-limited growth of dendritic structures. Efforts in the 1980s, led by James Langer [1978Lan, 1980Lan, 1989Lan], Hans Müller-Krumbhaar, Efim Brener, Herbert Levine, and others eventually yielded what has become known as *solvability theory*, which explains many of the defining characteristics of dendritic crystal growth. As I describe in Chapter 3, this theory provides an overarching explanation for why there is a marked increase in dendritic structure with supersaturation, as seen in the Nakaya diagram.

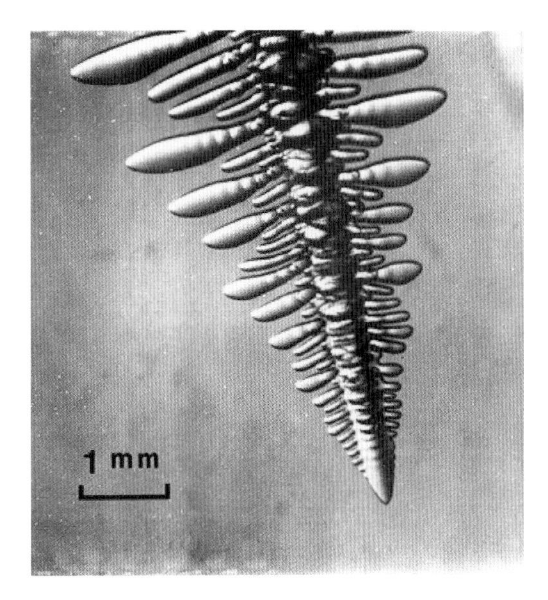

FIGURE 1.20. This photo shows a dendritic crystal of succinonitrile growing into a supercooled melt of the same material. In a vessel with a fixed temperature, the tip advances at a constant growth velocity, while the radius of curvature and overall shape of the tip do not change with time. Adapted from [1981Hua].

TWENTY-FIRST-CENTURY SNOWFLAKES

In the decades following Nakaya's seminal work, many scientists have conducted specific investigations into snow crystal structures and growth, including Matthew Bailey, Norihiko Fukuta, Yoshinori Furukawa, Takehiko Gonda, John Hallett, Jerry Harrington, Katsuhiro Kikuchi, Charles Knight, Teisaku Kobayashi, Toshio Kuroda, R. Lacmann, Dennis Lamb, Basil Mason, Jon Nelson, and Vincent Schaefer. Some of their work is cited throughout the chapters that follow, so I do not review it separately here. Suffice it to say at this point that much effort has been expended by numerous researchers on this topic, including aspects related to meteorology, material science, chemical physics, and the mathematics of pattern formation.

My primary goal in this book is to help carry the torch forward as snow crystal science advances through the twenty-first century. Review papers are useful in this regard [1987Kob, 2001Nel, 2005Lib, 2017Lib], but their inevitable page limitations make it difficult to give the subject a proper treatment. I found much inspiration in Nakaya's book when I first began studying snowflakes, and it is still a fascinating read [1954Nak]. But there has been no comparable treatise on the subject for nearly 70 years, so clearly an update is long overdue. When written by a single author, books inevitably give a somewhat biased view of a subject, and this book is no exception in that regard. I find certain topics especially intriguing, so I dwell perhaps too long in those areas. And I may skip quickly over related topics that deserve more attention, especially when I do not feel proficient in those subjects. I have made some attempt to provide a broad overview of snow crystal science, but knowledge is a limited commodity, and, like most people, I tend to write about what I know (or what I think I know, as the case may be). I have been studying the physics of snow crystal growth for more than 20 years already, trying to understand the detailed physical processes that govern growth rates and morphological development. During that time, I have photographed more than 10,000 natural snowflakes, measured the growth of countless small ice prisms, and developed techniques for growing complex dendritic structures under well-controlled laboratory conditions. My interest is rooted in the fundamental physics of snow crystal formation, so that is the primary focus in this volume.

The Big Picture

Like most technical subjects, it is impossible to present snow crystal science in an entirely linear fashion. Different topics are invariably interconnected to some extent, so one cannot fully appreciate any individual chapter in this book without having at least some understanding of the material presented in all the other chapters. I believe it is useful, therefore, begin with a brief synopsis of some key areas that form the backbone of this book.

Ice Crystal Structure: Chapter 2

Chapter 2 examines the properties of ice in equilibrium, including lattice structures, crystallography, general material properties, and thermodynamic quantities like latent heats and vapor pressures. Special attention is given to *terrace step energies* on the basal and prism facets, as these play an important role in snow crystal growth. Equilibrium physics is generally quite well understood, so these topics form a foundational basis for studying snow crystal growth. I have some fondness for lattice projections and crystal twinning in natural snow crystals, so these topics are presented in this chapter as well.

Although the lattice structure of crystalline ice is well characterized in the bulk, the molecular structure and chemical physics of the ice surface remains an area of active research [1970Fle, 2004Ike, 2010Pfa]. For example, *surface premelting* has been investigated using a variety of surface probes over a wide range of temperatures, but this phenomenon remains somewhat enigmatic overall [2007Li, 2018Qui]. Surface premelting appears to play a major role in snow crystal growth, and I touch on the topic in this chapter. But there is no comprehensive theory of surface premelting, so its detailed effects on the dynamics of ice crystal growth are not yet understood at even a qualitative level, a fact that has hindered progress for many decades.

This chapter also connects to current research in *molecular dynamics simulations* of the ice crystal surface, which have provided many insights into surface premelting and other ice properties [2002Mat, 2009Pae]. Molecular dynamics (MD) simulations are just beginning to calculate terrace step energies in ice [2020Llo], which have been measured as a function of temperature from ice-growth experiments. These developments suggest that computational chemistry will soon become an essential tool for better understanding the detailed molecular dynamics governing snow crystal growth.

Diffusion-Limited Growth: Chapter 3

The formation of elaborately branched snow crystal structures results from the *Mullins-Sekerka instability*, which arises from the slow diffusion of water vapor molecules through air. Analytical solutions presented in Chapter 3 indicate that *particle diffusion* and *surface attachment kinetics* are the primary physical processes guiding snow crystal growth. The complex interplay of these two effects yields the full menagerie of observed morphologies that are simultaneously branched and faceted. These same analytical solutions show that *heat diffusion* and *surface energies* play relatively minor roles in snow crystal dynamics, although they can become important considerations in special circumstances.

The physics of particle diffusion is well described by the statistical mechanics of ideal gases, so this aspect of snow crystal science is essentially a solved problem. Applying this theory to the growth of complex structures continues to be a nontrivial challenge, however. Analytic solutions are suitable for especially simple examples, like growing spheres and parabolic needlelike forms, and these are extremely useful for examining scaling relations and revealing the relative importance of competing factors in overall growth behaviors. But numerical modeling (Chapter 5) is needed to reproduce the complexity seen in all but the simplest snow crystals.

This chapter connects to a large body of more-general scientific work examining the growth of dendritic structures during solidification, including work on *solvability theory* [1992Mus, 1997Kar, 2019Liu]. Much of this work pertains to solidification from the melt, however, where heat diffusion and surface energies are dominant forces, while particle diffusion and attachment kinetics are often negligible factors. As a result, the connections are not as strong as one might expect, because solidification from vapor and from liquid can have quite different growth behaviors. This chapter ends with qualitative discussions of numerous specific snow crystal growth behaviors that arise from the interplay of the oft-competing processes of particle diffusion and surface attachment kinetics.

Attachment Kinetics: Chapter 4

In this chapter, I present a comprehensive attachment kinetics (CAK) model that can explain most of the morphological transitions seen in the Nakaya diagram. The model begins with *terrace nucleation theory*, as this physical process nicely explains the growth of large facet surfaces. The theory incorporates *terrace step energies* on the basal and prism facets, which have been measured over a broad range of temperatures. This foundational element of the model relies heavily on a several precision ice-growth measurements that are presented separately in Chapter 7.

The CAK model goes on to incorporate the concept of *structure-dependent attachment kinetics* (SDAK), which stipulates that the molecular attachment kinetics on small faceted terraces can be dramatically different from that on large terraces. Enhanced surface diffusion factors into this phenomenon, facilitated by temperature-dependent surface premelting that differs on the basal and prism facets. The SDAK effect further yields an *edge-sharpening instability* (ESI), which can explain the rather abrupt temperature transitions seen in the Nakaya diagram, including the formation of thin plates at −15°C and hollow columns at −5°C. The CAK model extends and improves on earlier models of the snow crystal attachment kinetics, and it makes numerous predictions that are (so far) holding up to experimental scrutiny. The chapter ends with a detailed look at snow crystal measurements near −5°C, as this temperature presents an especially interesting nexus of growth behaviors.

The general topic of surface attachment kinetics connects with a great deal of work associated with crystal growth theory and mesoscale molecular dynamics across many areas of scientific research [2007Mic,

2010Zha]. However, while terrace nucleation theory has deep roots in these areas, the SDAK and ESI phenomena appear to be unique to snow crystal growth. The reason is simply that ice is a special material in the world of materials science, as it has a high vapor pressure and is often investigated near the solid/liquid/vapor triple point. The phenomenology of ice crystal growth thus stands apart from that found with the low-vapor-pressure, high-melting-point materials commonly studied for technology applications.

A somewhat different connection is with studies of antifreeze proteins (AFPs) in water, which can strongly inhibit attachment kinetics at the ice/water interface [2003Du, 2009Pet]. Molecular dynamics simulations have become an important tool in these studies, which aim to understand the underlying molecular mechanisms and to develop AFPs for applications in the food industry and in cryobiology. These topics are all related at the molecular level, where they can be grouped into the general area of chemically mediated crystal growth. Notably, the effects of chemical additives on snow crystal growth have been documented by numerous researchers, but there is essentially no theoretical framework to describe the observations at present.

Computational Snow Crystals: Chapter 5

Numerical models of solidification have been extensively studied since the 1980s, but only around 2005 did researchers begin demonstrating realistic model structures that exhibited both branching and faceting. This area has attracted considerable attention from applied mathematicians and metallurgists seeking to better understand the solidification process [2002Bra, 2011Mil]. The snow crystal case involves highly anisotropic attachment kinetics exhibiting deep cusps at the facet angles, requiring specialized computational modeling techniques that are not needed in most metallurgical applications. At the time of this writing, several existing models have reproduced reasonable-looking structures, but only when some nonphysical assumptions are adopted. Resolving these issues is the subject of current research described in this chapter. Several numerical strategies have been investigated, including *front-tracking methods*, *phase-field techniques*, and *cellular automata*. The cellular automaton models seem to be especially adept at dealing with strongly anisotropic attachment kinetics, and these appear to be winning the race to develop physically realistic computational snow crystals. This area is evolving rapidly, however, and it is difficult to predict how future advances in numerical algorithms will impact the different modeling strategies.

Laboratory Snow Crystals: Chapter 6

Quantitative experimental observations are an essential part of snow crystal science, so this chapter explores some laboratory techniques that have been applied in this area. I examine a variety of ice growth chambers that allow in situ observations of growing snow crystals over a wide range of environmental conditions, along with nucleation methods, sample support mechanisms, and techniques for creating clean ice samples with oriented basal and prism facets. Controlling, measuring, and modeling supersaturation are discussed, as are unwanted systematic errors from substrate interactions, chemical impurities, and other experimental factors. Imaging techniques at all scales are discussed briefly, along with other types of surface probes that can yield information about premelting and other aspects of the ice surface structure.

The technology associated with ice crystal investigations connects with a broad range of studies in atmospheric science [1992Don, 2019Har, 2019Nel2], environmental science [2004Czi, 2012Hoo], atmospheric chemistry [2016Sei], materials science [1999Pet, 2013Dev], planetary science [2002Pou], cryobiology [1987Ban], and ice technology [2009Pat]. Ice has a ubiquitous presence in a broad range of scientific disciplines, linking these areas like no other single material can. Thus, although I treat snow crystal formation as its own microcosm of scientific investigation, the subject is clearly woven into a much larger tapestry.

Simple Ice Prisms: Chapter 7

The surface attachment kinetics comprise the crux of the snow crystal problem—necessary to understand even basic observations like the Nakaya diagram, yet remarkably difficult to comprehend at even a qualitative level. Progress toward developing a comprehensive physical model of the attachment kinetics begins with accurate measurements of the growth of simple ice prisms, a core topic that is examined in this chapter. I examine two key investigations in considerable detail, focusing on experimental techniques, growth modeling, and managing systematic errors. Although this topic is quite specialized, the resulting data provide the empirical foundation for the CAK model presented in Chapter 4, so precision measurements like these are central to our understanding of even the most basic elements of snow crystal formation.

Electric Ice Needles: Chapter 8

Because of the SDAK and ESI phenomena, the growth of simple ice prisms cannot reveal all aspects of the attachment kinetics. Further studies of complex growth morphologies are needed, which must be performed at high supersaturations and ultimately compared with full three-dimensional computational models of diffusion-limited snow crystal growth. This chapter presents what is perhaps the best overall experimental technique so far developed for examining the SDAK and ESI effects under controlled conditions. By growing snow crystals on the ends of slender "electric" ice needles in a dual diffusion chamber, a wide range of growth behaviors can be examined with in situ observations and a well-defined initial seed-crystal geometry.

Designer Snow Crystals: Chapter 9

This chapter examines an especially artistic laboratory technique that exploits the ESI to create thin platelike snow crystals perched atop small, blocky, ice-prism "pedestals." While not as flexible or scientifically valuable as the electric-needle method, these *Plate-on-Pedestal (PoP) snow crystals* are nearly ideal for recording high-resolution images of growing stellar-plate crystals. This apparatus has yielded the first photographs of stellar snow crystals that exhibit qualities that are overall superior to the best natural specimens, including better symmetry and sharper faceted features. The PoP technique has also yielded the highest-quality videos of growing stellar snowflakes thus far produced, along with the first observations of "identical-twin" snow crystals.

Natural Snowflakes: Chapter 10

This chapter examines the full menagerie of natural snow crystal types with illustrative sketches and abundant photographic examples. Classification schemes are discussed along with descriptions of a variety of common snow crystal structural features. This chapter presents something of a naturalist's guide to falling snow, suitable for the convenient examination and identification of different types of falling snow crystals.

Snowflake Photography: Chapter 11

Capturing quality imagery of natural snow crystals in cold conditions presents some unusual challenges for aspiring snowflake photographers. Finding especially photogenic specimens is nontrivial; handling them can be challenging; attaining suitable magnification requires special lenses; and lighting is problematic, because single-crystal ice is quite transparent. This chapter looks at each of these issues in detail and presents examples of a variety of innovative techniques that have been pioneered by the community of snowflake photographers.

Reductionism and Holism

My overarching goal in snow crystal science is a combination of reductionism and computational holism. The reductionism side aims to break down the physics of snow crystal growth into its constituent parts and processes, including crystal structure, attachment kinetics, diffusion-limited growth, and other factors. Each of these can be isolated and examined separately, perhaps right down to the molecular level, with the hope of developing precise mathematical models of all the relevant physics.

Some of the reductionist pieces are already well understood, while others remain quite puzzling. For example, the statistical mechanics of diffusion is well known for ideal gases, and the ideal-gas approximation is more than adequate for describing snow crystal growth. The ice lattice structure and various thermodynamic properties of water and its phase transitions are also quite well understood. However, the structure and dynamics of surface premelting are rather poorly known, and how premelting affects the attachment kinetics is a remarkably difficult problem. Nevertheless, at least in principle, it is possible to isolate, investigate, and ultimately comprehend all the relevant physical process involved in snow crystal growth.

Reductionism, however, is not sufficient to describe all of snow crystal science. Characterizing all the pieces of a puzzle and assembling the puzzle are two different endeavors. Learning the fundamental laws of quantum physics does not immediately explain everything in the field of chemistry, because understanding how atoms assemble into molecules is a separate problem from understanding individual atoms alone. Similarly, comprehending the formation of a complete snow crystal, as illustrated in Figures 1.21 and 1.22, is not the same as characterizing the separate physical processes involved in its growth. Holism in this case is not so much that the whole is greater than the sum of the parts. Rather, seeing the whole requires that you be able to assemble the parts.

Holism thus compels us to create a computational model that incorporates all the known physical processes involved in ice crystal growth, with an accuracy sufficient to yield realistic snow crystal simulations. In principle, using a large enough computer, it would be straightforward to create the necessary algorithms. But the devil is in the details, and computational models involve a lot of details. Numerical inaccuracies and instabilities can be problematic, and even the fastest supercomputers cannot come close to realizing full molecular resolution in large-scale phenomena. Thus, reductionism and holism tend to separate into distinct aspects of understanding snow crystal growth, and neither can be accomplished with absolute precision. The science of complex systems is always an activity of successive approximations.

In any area of scientific investigation, progress is measured by quantitative comparisons between theory and experiment, and this certainly applies to snow crystals. Creating numerical models that generate snowflakelike structures is a start, but meaningful progress requires direct comparison with laboratory observations. Chapters 5 and 8 point toward this eventual nexus of theory and experiment, when comparisons between laboratory and computational snow crystals are sufficient to determine whether our physical description of snow crystal formation is correct. We are certainly not at that point yet.

Some might argue that creating accurate computer simulations of all types of growing snow crystals would not constitute a true understanding of the underlying phenomenon. Debating this point would require a precise definition of the word "understanding," which is itself a nontrivial philosophical point. Snow crystal formation involves a multitude of complex physical processes acting over a broad range of length and time scales. It may indeed be the case that a single brain cannot simultaneously comprehend all aspects of what is happening. If that is true, then we have little choice but to let our machines do the heavy lifting for us. I would argue that a detailed computer model that reproduces laboratory snow crystals with high fidelity is as close to a true understanding as we are likely to achieve. It is a good place to start, at any rate.

As of this writing, many aspects of this overarching scientific strategy are coming together. Our understanding of the attachment kinetics is beginning to make sense (Chapter 4), suitable computational models are rapidly becoming feasible (Chapter 5), and detailed comparisons between experiments and theoretical models of complex snow crystal morphologies are becoming pos-

FIGURE 1.21. (Above) This "pond crystal" froze from liquid water on the surface of a still pond, and basal faceting yielded a thin platelike sheet. Note how the angular branching exhibits the six-fold symmetry that characterizes the ice crystal. (Below) Frost forms from water vapor near ground level, and the resulting structures can look much like snow crystals. Large specimens like these (about 2 cm in length) are called "hoarfrost" crystals.

sible (Chapter 8). There appears to be no serious roadblocks impeding the path forward, and steady progress is being made on all fronts.

Curiosity-Driven Science

The quest to understand how snow crystals grow and develop is not motivated by any practical applications or societal needs, but rather simply by a desire to explain a common natural phenomenon. The rainbow provides a good analogy (see Figure 1.23), as considerable intellectual determination and scientific advancement was required to fully comprehend its colorful arcs. After centuries of musings and slow progress toward understanding these vibrant meteorological displays, René Descartes described a detailed physical model of the rainbow in his 1637 treatise, *Discourse on the Method* [1637Des], explaining how the primary and secondary rainbows arise

FIGURE 1.22. This series of images shows the growth and development of a laboratory-grown snow crystal. The temperature and supersaturation were changed as the crystal grew, alternating between periods of faceted and branched growth. After the initial seed crystal had grown into a larger hexagonal plate, the conditions were changed to sprout branches from the six corners of the plate. Additional changes were applied to stimulate the growth of sidebranches at various times. Because the six branches experienced the same changing growth conditions, they all developed symmetrically. It took about 45 minutes to grow the full crystal, which measured about 2 mm from tip to tip.

from single and double reflections inside raindrops. Isaac Newton discovered color dispersion three decades later, thereby explaining the rainbow's characteristic colors. And around 1800, Thomas Young further identified the fainter supernumerary bows as a manifestation of the wavelike nature of light. The rainbow only became a truly solved problem in 1908, when Gustav Mie developed a comprehensive mathematical theory for light scattering from spherical water droplets. Does this knowledge substantially improve the human condition? In 1880, Mark Twain wrote that "we have not the reverent feeling for the rainbow that the savage has, because we know how it is made. We have lost as much as we gained by prying into that matter" [1880Twa, p. 466]. Perhaps, but seeking truth seems to have worked for us in the past, especially in the natural sciences.

The snowflake is following a similar path through scientific history, except it is about 400 years behind the rainbow. We are just now piecing together a rudimentary picture of the attachment kinetics, slowly deciphering the detailed molecular forces that guide growth rates and structure formation. The snowflake and the rainbow are two small parts of nature's tapestry, and that is sufficient reason to justify our curiosity regarding their physical nature. The aim of this book is to push forward this icy microcosm of science just a bit more.

NO TWO ALIKE?

Is it really true that no two snowflakes are alike? I hear this question frequently, perhaps because our elementary-school teachers planted the notion into all our heads as we practiced cutting paper snowflakes, and there it has remained throughout our lives. It is a funny question, almost like a Zen *koan*—if two identical snowflakes fell in the vast winter forests, my inquisitive friend, would anyone ever know? And can you truly be sure that no two are alike, as you cannot possibly check them all to find out? There is indeed a certain level of unknowability to the question of snowflake alikeness, but as a physicist, I feel that I can shed some light on this issue. As I will demonstrate, the answer depends to a large extent on what you mean by the question. (Physics does occasionally have its Zenlike qualities.)

The short answer to the question is yes—it is indeed extremely unlikely that any two complex snowflakes will look exactly alike. It is so unlikely, in fact, that even if

FIGURE 1.23. Explaining the rainbow required a synthesis of many advances in physics, optics, and mathematics, so these colorful arcs were not fully understood until early in the twentieth century. Snow crystal formation is a substantially tougher problem than the rainbow, involving subtle many-body molecular processes at the ice surface. Comprehending the snowflake remains very much a work in progress. Rainbow photo by Steven E. Nelson.

you looked at every one ever made, over all of Earth's history, you would almost certainly not find any exact duplicates. The long answer is a bit more involved, however, as it depends on just what you mean by "alike" and on how you define a "snowflake." For example, I will claim that it is possible that two nano-snowflakes could be exactly alike. When developing the theory of quantum mechanics, physicists discovered that some things in nature are exactly, precisely, perfectly alike—*indistinguishable* is the proper technical term. For example, our understanding of elementary particles indicates that all electrons are fundamentally indistinguishable from one another. This is one of the cornerstones of quantum physics, and alikeness in this arena is a profound concept. Indistinguishability is part of what defines a truly elementary particle.

A water molecule is considerably more complex than an electron, and not all water molecules are exactly alike. If we restrict ourselves to water molecules that contain two ordinary hydrogen atoms and one ordinary ^{16}O atom, then again, physics tells us that all such water molecules are exactly alike. However, about one molecule out of every 5,000 naturally occurring water molecules will contain an atom of deuterium in place of one of the hydrogens, and about one in 500 will contain an atom of ^{18}O instead of the more common ^{16}O. And these rogue atoms can be distinguished from their common cousins.

Because a typical small snow crystal might contain 10^{18} water molecules, we see that about 10^{15} of these will be isotopically different from the rest. These unusual molecules will be randomly scattered throughout the snow crystal lattice, giving it a unique design. The probability that two snow crystals would have identical placements of these isotopic anomalies is essentially zero. Even with 10^{24} snow crystals being made per year on Earth, the probability that any two would be exactly identical within the entire lifetime of the Universe is zero in any practical sense.

Thus, at this extreme level of atomic precision, no two snow crystals can be exactly alike because of these isotopic differences. However, an exception (there are few absolute statements in science) would be a snow crystal with only a handful of molecules. If we assemble an ice crystal of only six molecules, for example, then it could easily happen that each of the six will contain two ordinary hydrogen atoms and one ordinary ^{16}O atom. Furthermore, a cluster of six molecules will only have a few stable configurations. Therefore, there is a reasonable probability that two six-molecule snow crystals would be exactly alike, quantum mechanically indistinguishable from one another. However, perhaps an assembly of just six molecules does not a snowflake make.

If we restrict ourselves to isotopically pure water molecules, it is still extremely improbable that two macroscopic snow crystals would be exactly alike. When a crystal grows, its molecules do not always stack together with perfect regularity, so a typical snow crystal contains a large number of minor crystal dislocations, which again are scattered throughout the crystal in a random fashion. One can then argue, as with the isotopes, that the probability of two crystals growing with exactly the same pattern of dislocations is vanishingly small. And again, one has the exception of few-molecule crystals, which can easily be free of dislocations.

Another part of this tale is that small snow crystals can at least look alike, even if they are not precisely identical down to the last molecule. So, let us relax our definition of alikeness and say that two snow crystals are alike if they just look alike in an optical microscope. The smallest features one can see in an optical microscope are about 1 micrometer in size, which is 10,000 times larger than an atom. With this relaxed definition, you could probably find two identical-looking hexagonal prisms falling from the sky if you conducted a diligent search, and it is certainly easy to make such simple crystals in the laboratory. Figure 1.24 shows a pair of identical-looking snowflakes that happened to fall next to one another in my lab. When the overall crystal morphology is little more than a simple hexagon, one looks much like any other.

As the morphology of a snow crystal becomes more complex, however, the number of possible ways to make it soon becomes staggeringly large. To see just how rapidly the possibilities increase, consider a simpler question—how many ways can you arrange books on your bookshelf? With three books, there are six possible arrangements, and you can easily sketch all of them for yourself. Increasing to 15 books, there are 15 choices when you place the first book on the shelf, then 14 for the second, 13 for the third, and so on. Multiply it out and there are more than a trillion ways to arrange just 15 books. With 100 books, the number of possible arrangements goes up to just under 10^{158}, which is about 10^{70} times larger than the total number of atoms in the entire known universe!

If you gaze at a complex snow crystal under a microscope, you can often pick out a hundred separate features if you look closely. Because all those features could have developed differently, or could have appeared in slightly different places, the math ends up being like that with the books, and it applies to fingerprints as well. The exact calculation would depend on the details, along with how you define individual features and their locations. But the details are not important, because the total number of possible ways to make a complex snow crystal soon becomes unfathomably large. Thus, it is essentially impossible that any two complex snow crystals, out of all those made over the entire history of the planet, have ever looked exactly alike.

The story of snowflake alikeness takes another amusing turn when you start looking at complex laboratory-grown crystals. As I described earlier in this chapter, the

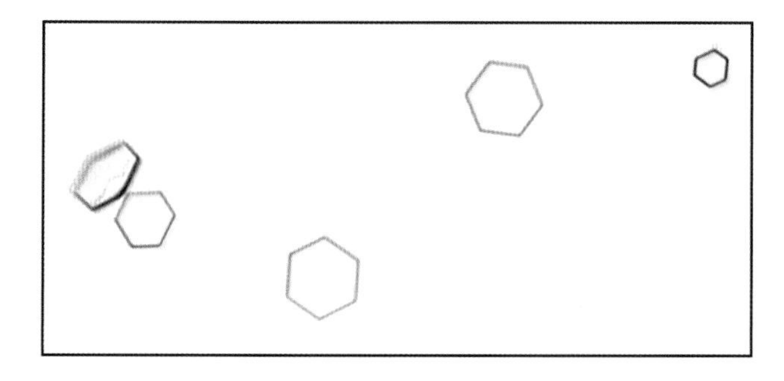

FIGURE 1.24. This laboratory photo shows several small, thin-plate snow crystals that grew while falling freely in air and then landed on a transparent substrate (see Chapter 6). Because the crystals have a simple hexagonal shape, one can easily find a pair of nearly identical specimens next to each other, like the two centered here.

FIGURE 1.25. A pair of laboratory-grown "identical-twin" snow crystals, surrounded by a field of water droplets. These grew side-by-side on a fixed transparent substrate using the Plate-on-Pedestal technique described in Chapter 10. Because both crystals experienced the same growth conditions as a function of time, they grew into nearly identical shapes. Analogous to identical-twin people, these similar snowflakes are clearly related, even though they are not precisely equal in every detail.

final shape of a large stellar snow crystal is determined by the path it traveled through the clouds as it formed. Because the air is usually turbulent to some extent, even under calm conditions, the paths of different snow crystals are typically quite meandering and chaotic. Trajectories that bring two crystals close to one another will soon diverge, separating them again by large distances. In the laboratory, however, it is possible to place two seed crystals near one another on a fixed substrate and then subject them both to the same growth conditions as a function of time. Doing this with some care yields results like that shown in Figure 1.25. As these crystals were developing, I occasionally subjected them to abrupt changes in temperature and/or supersaturation. Because both crystals saw the same changes at the same times, they responded with synchronized growth behaviors. I like to call these "identical-twin" snowflakes, in analogy to identical-twin people. They are clearly so alike that there must have been some underlying connection between them, yet they are not identical in an absolute sense.

FIGURE 2.1. This synthetic snow crystal, measuring 2.3 mm from tip to tip, grows while supported above a transparent sapphire substrate, surrounded by a field of small supercooled water droplets. The precise alignment of its faceted surfaces reveals that the entire complex structure is a single crystal of ice with its constituent molecules arranged on an ordered lattice. Chapter 9 describes how snow crystals like this one can be created in the laboratory.

TWO

Ice Crystal Structure

The chief forms of beauty
are order and symmetry and definiteness.
—ARISTOTLE, *METAPHYSICS*, BOOK XIII (350 BCE)

The lattice structure and physical properties of ice are foundational elements needed to understand the formation of snow crystals. Several books about the materials science of ice have appeared over the years [1970Fle, 1974Hob, 1999Pet], so I will not delve into all the multifarious mechanical, thermal, chemical, electrical, and optical properties of this fascinating substance here. Instead I examine the specific qualities of ice that factor directly or indirectly into snow crystal growth. One focus will be on the basic molecular arrangement of the ice lattice, as this underlies the well-known sixfold symmetry seen in snow crystals, plus it explains several varieties of crystal "twins" that are observed in natural snowflakes. Throughout this discussion, two-dimensional (2D) projections of the molecular lattice will be useful for visualizing different aspects of the ice surface structure, terrace steps, and crystal twinning.

The molecular organization of the ice/vapor interface is another important theme in this chapter, because that interface is where growth happens. The basal and prism facets are identified with specific lattice planes for which the ice surface is particularly "smooth" on a molecular scale. This feature gives each facet a lower-than-average surface energy, plus it hinders the attachment kinetics of impinging molecules to the surface. This latter property, much more than the former, controls the formation of snow crystal facets and other growth behaviors. The structural characteristics of these faceted surfaces change substantially with temperature near 0°C, and this fact is ultimately responsible for many of the morphological transitions seen in the Nakaya diagram, as I discuss at length in Chapter 4.

Thermal equilibrium will be a shared aspect of most of the topics discussed in this chapter, distinguishing it from the chapters that follow. Understanding a system in equilibrium is a statics problem, involving energy minimization, basic statistical mechanics, and related topics where the underlying physics is well in hand, at least in principle. In contrast, crystal growth is a dynamics problem, involving nonstatic molecular interactions, energy flow, particle fluxes, and nonequilibrium statistical mechanics, where often the governing physical processes are difficult to comprehend and quantify. For this reason, we

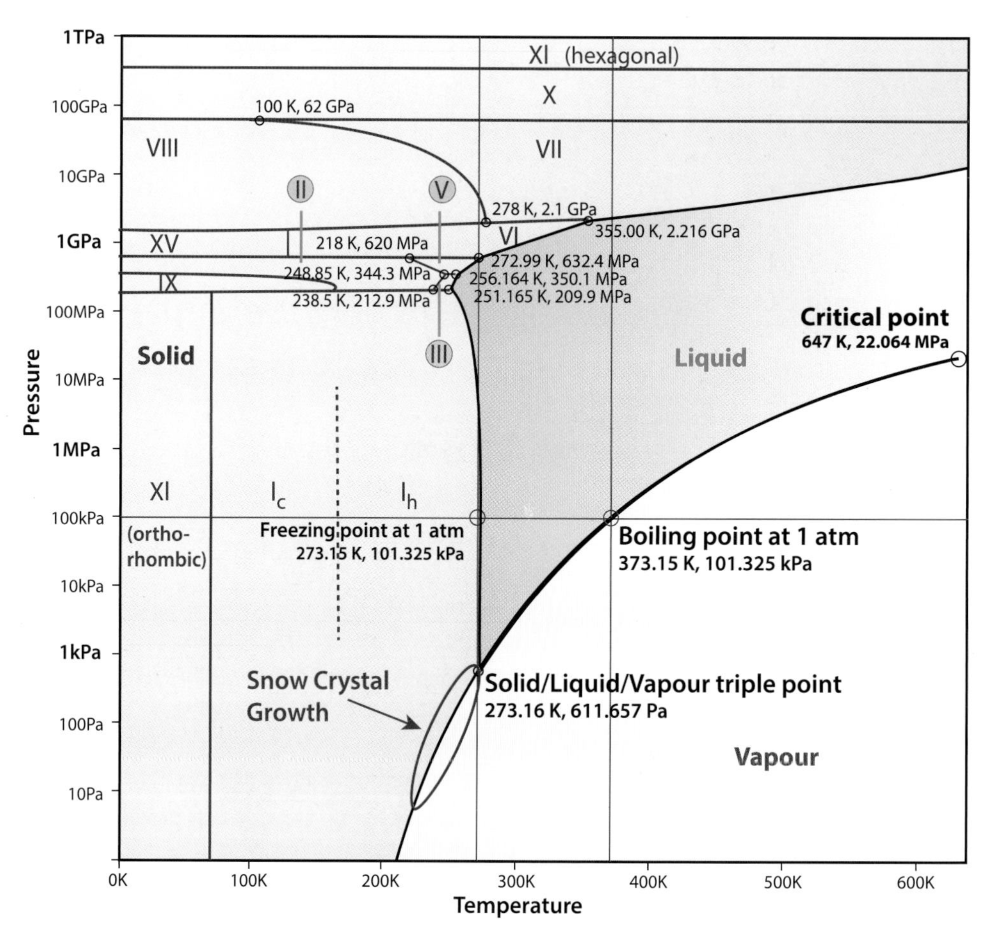

FIGURE 2.2. The phase diagram of water shows its equilibrium form as a function of temperature and pressure. Snow crystal growth mostly occurs on the ice/vapor boundary below the triple point at temperatures between 0° and −40°C, where pressure refers to the partial pressure of water vapor. Note that most of the known phases of ice exist only at extremely high pressures. Image adapted from [2018Cmg].

begin our scientific discussion of snow crystal formation by examining the most relevant properties of the ice crystal in equilibrium.

The full phase diagram of water is a complicated beast, as shown in Figure 2.2, including some 15 known forms of ice, with perhaps more to come as ever-higher pressures are being experimentally explored. Although there is considerable scientific interest in the crystallography and stability of the various exotic states of ice, these topics lie beyond the scope of this book. Our concern will be the basic, no-frills ice/vapor transition, circled in red in the figure. Although this just a small slice of the water phase diagram, the complex dynamical behavior that takes place along this phase boundary is already more than enough to fill a book.

The water phase diagram only delineates the various equilibrium phases of ice, water, and water vapor, depicting the lowest-energy state at each temperature and pressure. As such, the chart conceals the phenomenological richness inherent in transitions between the different states. Each line in the phase diagram tells us under what equilibrium conditions the adjacent phases can co-exist, but it says little about the nonequilibrium physical processes that define the character of each phase transition. It is likely that there are many fascinating stories to tell about the growth dynamics that must occur along every line segment in the phase diagram. We restrict our attention to one specific phase transition, because it has proven itself worthy of study, plus the others remain almost completely unexplored [2005Mar].

ICE CRYSTALLOGRAPHY

The usual scientific definition of crystal is any material in which the atoms and molecules are arranged in an ordered lattice (although liquid crystals, quasicrystals, and other uses of the word can be found in the scientific literature). Figure 2.3 shows a molecular model of ice Ih, the only form of ice commonly found on the earth's surface, including in snow crystals. Note how two hydrogen atoms closely flank each oxygen atom, so the trio forms an essentially intact H_2O molecule. Strong covalent bonds tightly bind the two hydrogens in each water molecule, while weaker hydrogen bonds connect adjacent molecules. The ice crystal, therefore, can be considered as a collection of intact water molecules arranged in a lattice structure.

While the arrangement of oxygen atoms is fully described by the hexagonal lattice structure in ice Ih, the relative orientations of the molecules are described by the Bernal–Fowler rules [1933Ber], placing exactly one hydrogen atom between each adjacent pair of oxygen atoms, as shown in Figure 2.4. It is possible to twist the individual H_2O molecules around into many possible arrange-

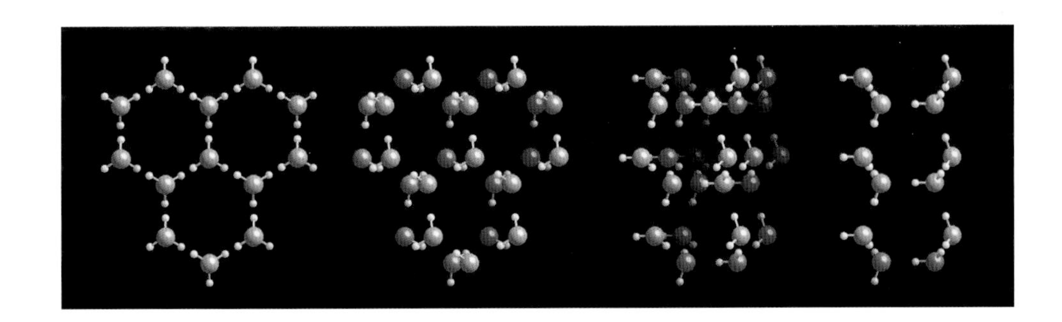

FIGURE 2.3. A molecular model of the ice Ih crystal rotated by 0, 18, 72, and 90 degrees about the $[01\bar{1}0]$ axis. The large blue spheres represent oxygen atoms, and the smaller spheres represent hydrogen atoms. The first image is looking down the c-axis, revealing the hexagonal lattice structure.

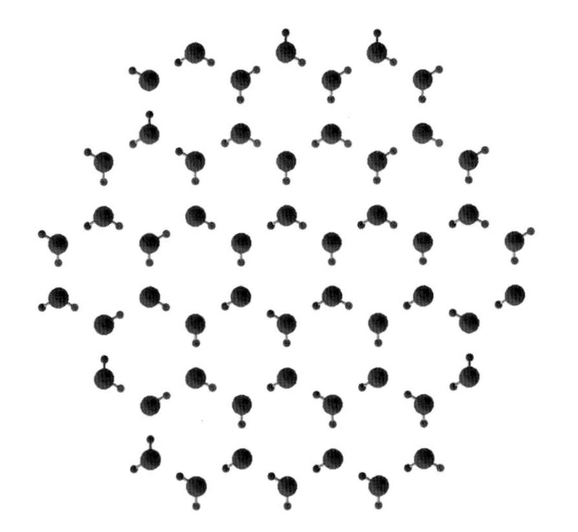

FIGURE 2.4. A 2D toy-model example of the Bernal–Fowler rules in ice Ih (although the full 3D rules do not visualize especially well in two dimensions). The basic idea sketched here is that there are many possible ways to orient the individual H_2O molecules in the crystal while keeping exactly one hydrogen atom between each adjacent pair of oxygen atoms, with each hydrogen forming one strong and one weak bond.

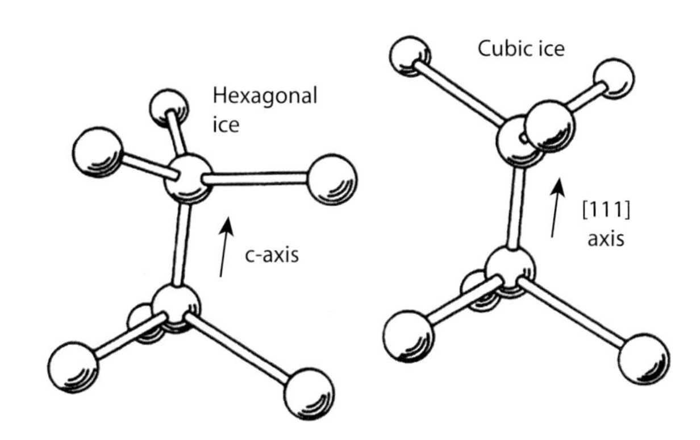

FIGURE 2.5. The essential lattice difference between hexagonal ice Ih (left) and cubic ice Ic (right) is a twist in some of the bonds, specifically, those between the basal layers in ice Ih. Here the balls and sticks represent oxygen and hydrogen atoms, respectively.

ments while maintaining two strong and two weak O-H bonds on each oxygen atom. This orientational ambiguity appears to play no important role in snow crystal growth, but it is a basic feature of the ice Ih lattice. Of course, this is all better visualized in three dimensions, and ice molecular model kits are readily available for this purpose.

Hexagonal and Cubic Ice

The two O-H bonds in a free H_2O molecule meet at an angle of 104.5 degrees, which is close to the tetrahedral angle of 109.5 degrees. Because of this near match, the four O-H bonds emanating from each oxygen atom in the ice crystal (two tight bonds, two weaker bonds) are essentially in a tetrahedral arrangement. Given this bond structure, there are two ways to form a crystal lattice: the hexagonal (icelike) structure and a cubic (diamondlike) structure, as shown in Figure 2.5. The difference between these two structures comes down to a twist in the bonding of adjacent tetrahedra separating the basal planes. Adopting the ice Ih bonding gives the normal ice lattice with hexagonal symmetry. Choosing the alternate bonding throughout the lattice yields a structure with a cubic symmetry called ice Ic. Both crystal structures can be found in the phase diagram in Figure 2.2, but only ice Ih is stable under ordinary environmental conditions. Nevertheless, as we will see below, cubic ice bonding plays a small role in some snow crystal structures.

Stacking spheres is another way to see the difference between the Ih and Ic lattices. Beginning with a flat surface, the first layer of spheres is optimally stacked in the usual hexagonal close-packed arrangement. The next layer goes on top of the first with no ambiguity; there is only one obvious way to place a second layer of spheres on top of the first. With the third layer, however, there are two choices in its placement. If the first sphere is placed directly over a sphere in the first layer, then the Ih lattice results (assuming this same choice is made for all subsequent layers). Shifting the third layer slightly, so the third-layer spheres are not directly over spheres in the first layer, results in the Ic structure (again, assuming this choice is made consistently). If some layers use the Ih placement and others use the Ic placement, this is called a "stacking disordered" structure.

Although cubic ice Ic has been created in the laboratory at low temperatures, no examples of faceted "cubic" snow crystals have been made, although this would certainly be an interesting sight. More significantly, no pure samples of cubic ice Ic, faceted or not, have ever been definitively observed in the laboratory or in nature. Laboratory samples are generally stacking disordered to a substantial degree and are thus essentially molecular jumbles of ice Ic and Ih. It appears that ice Ic has a slightly lower surface energy than ice Ih, causing it to nucleate more readily from liquid water under certain conditions [1995Hua, 2005Joh]. There is even some evidence that ice Ic can be found as minute ice grains in high-altitude clouds, although this is not known with certainty. The line between ice Ih and ice Ic in Figure 2.2 is dotted, not solid, representing that this is not a real

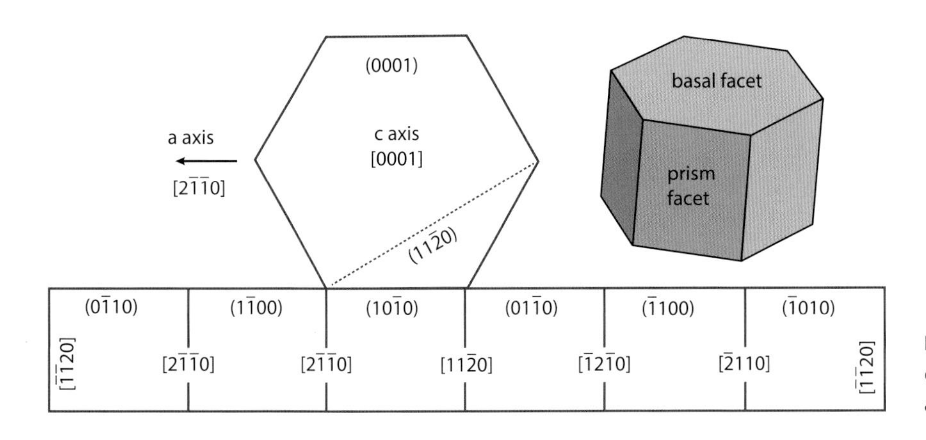

FIGURE 2.6. The Miller-Bravais indices displayed on an "unfolded" hexagonal prism of ice Ih.

phase boundary in the usual thermodynamic sense of a first-order phase transition. In general, the overall equilibrium and nonequilibrium relationships between ices Ih and Ic are not completely understood.

Lattice Projections

When referring to the various planes and axes in crystal lattices, it is customary to use either the 3-axis Miller indices or (more usually) the 4-axis Miller-Bravais indices, and the latter are shown in Figure 2.6 for ice Ih. Curly brackets refer to families of planes, including the $\{0001\}$ basal facets and the $\{\bar{1}010\}$ family of prism facets. Parentheses refer to specific planes, including the basal facet (0001) and the six prism facets $(1\bar{1}00)$, $(10\bar{1}0)$, $(01\bar{1}0)$, $(\bar{1}100)$, $(\bar{1}010)$, and $(0\bar{1}10)$. Square brackets denote directions, for example, the c-axis [0001] that is perpendicular to the basal face, or the a-axis $[11\bar{2}0]$ that is perpendicular to the $(11\bar{2}0)$ face. All six a-axes point away from corners of the hexagonal prism, as shown in the figure. The $\{11\bar{2}0\}$ planes are sometimes called the "secondary prism faces," although growing ice crystals do not form facets on these planes.

While the $\{0001\}$ basal facets and the $\{\bar{1}010\}$ prism facets are by far the most common faceted surfaces seen in snow crystals, the $\{10\bar{1}1\}$ pyramidal facets can also be observed in nature, as illustrated in Figure 2.7. Little is known about the growth of pyramidal facets, but the evidence suggests that they form only rarely, typically at temperatures below −20°C.

To relate the Miller-Bravais indices to the ice crystal facets and the molecular structures of these surfaces, it is useful to examine several 2D projections of the three-dimensional (3D) lattice structure of ice Ih. For example, Figure 2.8 shows the lattice structure of the $\{\bar{1}010\}$ prism facets and the $\{11\bar{2}0\}$ surfaces, both looking down along the c-axis. Since the early days of X-ray crystallography, it had been assumed that the $\{\bar{1}010\}$ surfaces coincided with the well-known prism facets seen in snow crystals, as the $\{\bar{1}010\}$ surfaces have

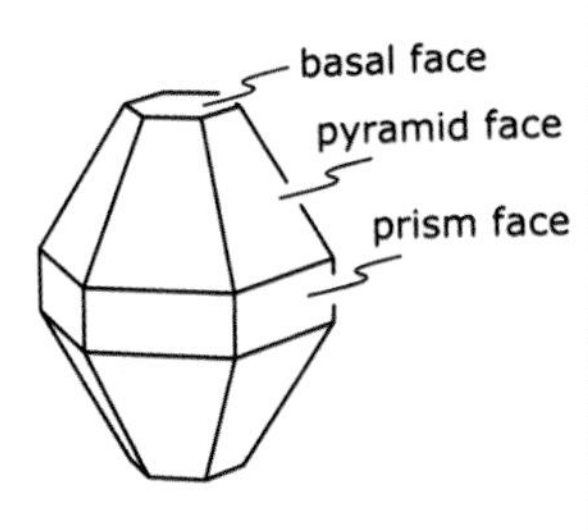

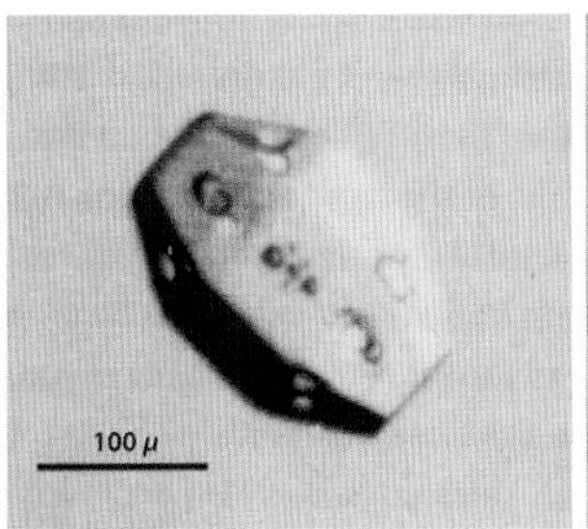

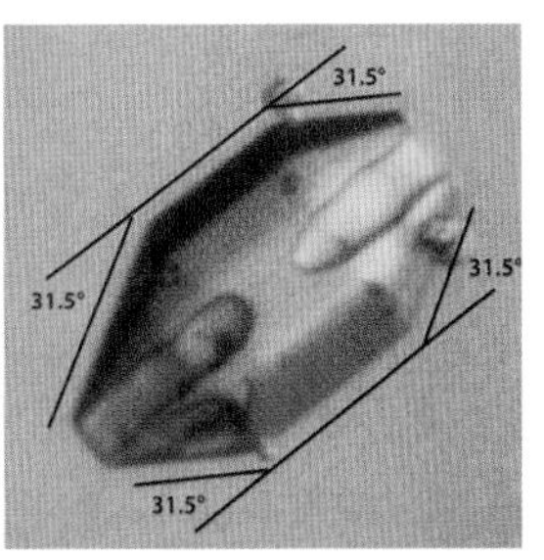

FIGURE 2.7. These south-pole snow crystals display pyramidal facets [2006Tap]. The scale bar in the first photo is 100 μm long, and the angles shown in the second are 31.5 degrees. Images courtesy of Walter Tape and Jarmo Moilanen.

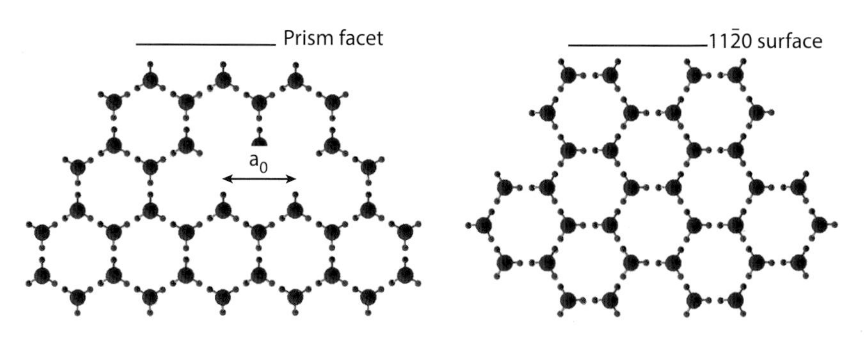

FIGURE 2.8. Two projections of the ice Ih lattice looking down along the [0001] direction (perpendicular to a basal facet). A prism $\{\bar{1}010\}$ facet and a $\{11\bar{2}0\}$ surface are labeled. The latter is sometimes called a "secondary prism surface," although it does not exhibit any known faceting.

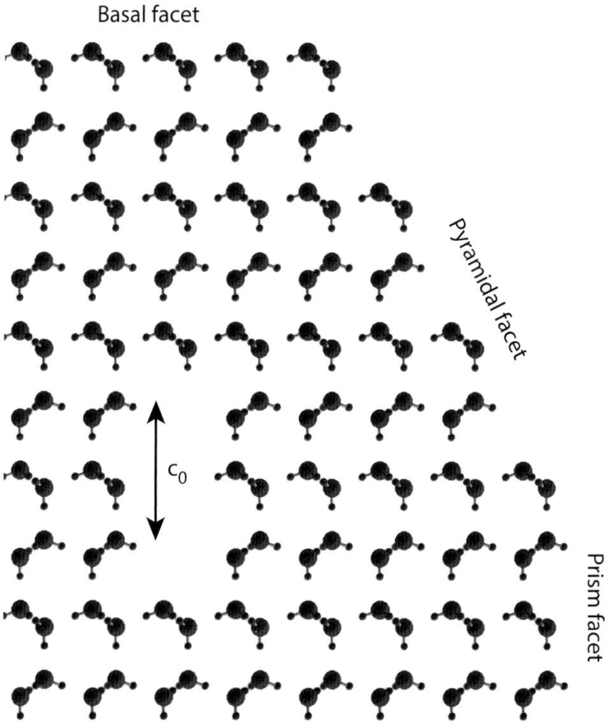

FIGURE 2.9. A projection of the ice Ih lattice looking along the $[11\bar{2}0]$ direction, showing the basal, prism, and pyramidal facets. The angle between the prism and pyramidal facets is $\tan^{-1}(\sqrt{3}a_0/2c_0) = 28.0$ degrees. (The 31.5-degree angle seen in Figure 2.7 is from looking along the $[\bar{1}010]$ direction, which is not parallel to a pyramidal facet.)

a more tightly bound lattice structure compared to the $\{11\bar{2}0\}$ surfaces. Although this early assessment was indeed correct, it was only definitively confirmed by direct observation quite recently [2017Bru].

Figure 2.9 shows another lattice projection that includes the basal, prism, and pyramidal facets. The ice lattice parameters a_0 and c_0 are defined in Figures 2.8 and 2.9, with crystallographic measurements near 0°C giving

$$\begin{aligned} a_0 &= 0.452 \text{ nm}, \\ c_0 &= 0.736 \text{ nm}, \end{aligned} \tag{2.1}$$

and the respective spacings between basal and prism layers are then

$$\begin{aligned} x_{basal} &= \frac{c_0}{2} = 0.37 \text{ nm} \\ x_{prism} &= \left(\frac{\sqrt{3}}{2}\right) a_0 = 0.39 \text{ nm}, \end{aligned} \tag{2.2}$$

which is comparable to the effective size of a water molecule in ice, $a \approx c_{ice}^{-1/3} \approx 0.32$ nm, where c_{ice} is the number density of water molecules in ice.

Terrace Steps

As we will see in Chapter 4, the nucleation of new molecular terraces is a key factor in the formation of snow crystal facets, and terrace nucleation is governed by the terrace step energies. It is useful, therefore, to examine the molecular structure of terrace steps, and several lattice projections that do so are shown in Figures 2.10 and 2.11. While the surface energies of the facet planes are generally lower than nonfaceted surfaces, this anisotropy in the surface energy seems to play only a minor role in snow crystal faceting. The anisotropy in the attachment kinetics is much stronger, and this is the more important factor governing snow crystal growth rates and faceting (Chapter 4).

The fact that the terrace steps are relatively shallow on the $\{11\bar{2}0\}$ surfaces, as seen in Figure 2.11, likely explains the absence of faceting on these surfaces. In contrast, the larger terrace steps on the basal and prism surfaces creates a substantial nucleation barrier that promotes strong faceting on both these surfaces. This line of reasoning further predicts that the pyramidal step energy becomes substantial only at low temperatures, as pyramidal facets apparently only form in quite cold conditions.

When the temperature is sufficiently low, water molecules on some metallic surfaces assemble into 2D hexagonal lattice structures that can be imaged using atomic force microscopy [2016Mai]. Edge structures resemble the lattice models in Figure 2.11, and it is even becoming possible to investigate growth processes in this regime [2020Ma].

As a general word of caution, it should be remembered that these sketches of lattice projections are oversimplified representations of the molecular surface structures of ice, approximately valid only at extremely low temperatures. At temperatures typically associated with snow crystal growth, thermal fluctuations can easily distort (or even completely rearrange) the lattice structure at the surface, and the high vapor pressure of ice means that molecules are continuously leaving the surface and reattaching at a prodigious rate, of order one monolayer per microsecond at −5°C. Although these lattice sketches can be quite useful for visualization purposes, real ice surfaces are neither rigid nor static.

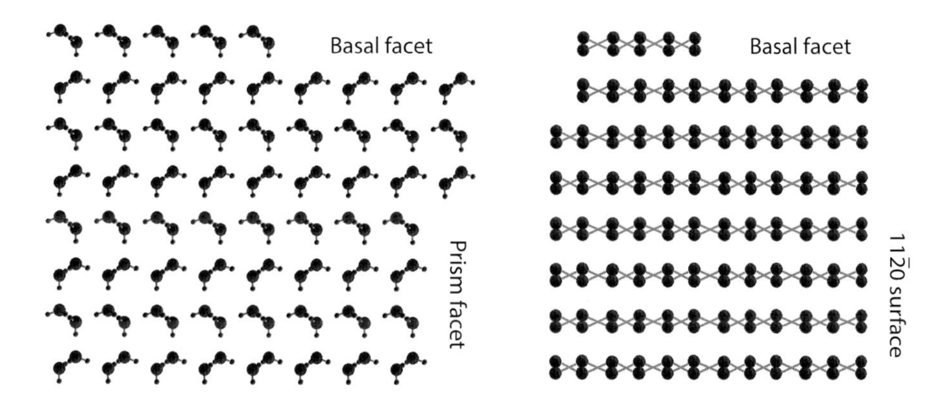

FIGURE 2.10. (Left) A projection of the ice Ih lattice looking along the $[11\bar{2}0]$ direction, showing basal and prism facets and terrace steps. The spacing between basal layers is $c_0/2 = 0.37$ nm, while the spacing between prism layers is $(\sqrt{3}/2)a_0 = 0.39$ nm. (Right) A projection of the ice Ih lattice looking along the $[\bar{1}010]$ direction, showing basal and $(11\bar{2}0)$ surfaces and terrace steps. From this perspective, it is easy to see why the basal terraces are sometimes called molecular "bilayers."

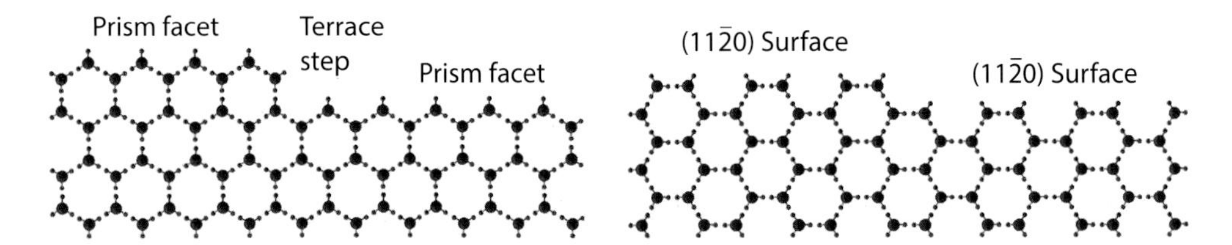

FIGURE 2.11. Terrace steps on a prism $\{\bar{1}010\}$ facet and a $\{11\bar{2}0\}$ surface, as seen looking along the [0001] direction. Note that the prism step is more distinct, and admolecules would likely bind much more strongly at a terrace step than on the prism facet. In contrast, the $\{11\bar{2}0\}$ step is less distinct, and binding at the step edge is likely not much stronger than elsewhere on the surface. This difference in terrace steps explains why faceting is strong on prism surfaces and absent on $\{11\bar{2}0\}$ surfaces.

SURFACE PREMELTING

Because water molecules at an ice surface are less tightly bound compared to those in the bulk, a disordered *quasiliquid layer* (QLL) appears near the melting point, and this phenomenon is known as *surface premelting* [1999Pet, 2006Das, 2007Li, 2014Lim]. First predicted by Michael Faraday in 1859 [1859Far], much recent research has sought to better understand this enigmatic feature of the ice/vapor interface, which is illustrated in Figure 2.12. One intuitive way of thinking about surface premelting is through the *Lindemann criterion* [1910Lin], which states that a solid will melt if thermal fluctuations of the intermolecular distance become larger than approximately 10–15 percent of the average distance. This empirical rule applies (roughly) to a broad range of materials, and one expects that thermal fluctuations will be larger near the surface, owing to the reduced binding there. One can turn this into a pedagogically useful toy model [2005Lib], but more sophisticated theoretical treatments of the phenomenon have been developed [2006Das, 2014Lim, 2019Sla]. Molecular dynamics simulations are an especially powerful tool for understanding the dynamics of surface premelting and the temperature-dependent structure of the quasiliquid layer [2007Li, 2019Ben, 2020Llo].

Surface premelting is an equilibrium phenomenon demonstrating that the phase boundary between the liquid and solid phases of a material is only precisely defined when the sample has infinite extent (called "bulk" material). For nanometer-scale molecular clusters, the melting point can be substantially lower than the bulk melting temperature [2005Bre, 2011Agu], and this shift is related to surface premelting. There is considerable current theoretical and experimental interest in the topic of cluster premelting and finite-sample thermodynamics more generally. Unfortunately, there is no comprehensive theory at present that provides an accurate quantitative description of either cluster premelting or surface premelting in detail.

The experimental evidence to date suggests that the first ice bilayer becomes at least partially disordered at quite low temperatures, roughly −70°C [2002Wei], and a second bilayer joins the QLL at around −15°C [2017San]. The QLL thickness increases rapidly at still higher temperatures, diverging logarithmically as the temperature approaches the melting point. As illustrated in Figure 2.12, the many conflicting measurements of premelting in ice make it difficult to determine the QLL characteristics with much confidence. The phenomenon has been investigated using a variety of surface probes, but different experiments measure different surface properties, and it is not always clear how to compare various results [2019Sla]. Sum frequency generation spectroscopy has proven to be an especially useful tool, as this probe is noncontact and quite surface specific [2002Wei, 2017San]. However,

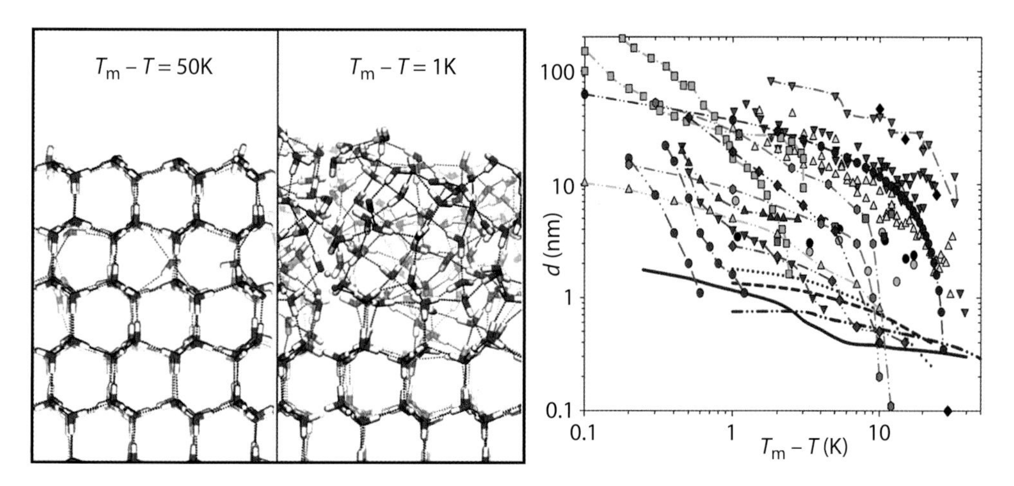

FIGURE 2.12. (Left) A molecular-dynamics simulation demonstrating surface premelting [2014Lim]. At temperatures far below the melting point, the bulk lattice structure persists up to the ice/vapor interface. Near the melting point, however, the top molecular layers lose their ordered structure, forming an amorphous premelted layer, also called a "quasiliquid layer" (QLL). (Right) The QLL thickness $d(T)$ generally increases at higher temperatures, but different experiments and MD simulations yield quite different results [2018Con]. Surface premelting likely plays an important role in snow crystal growth, particularly in how the attachment kinetics vary with temperature (see Chapter 4). However, understanding this role remains a challenging problem. Images courtesy of David Limmer and Julián Constantin.

producing accurate, quantitative comparisons between experiments and theoretical models remains a topic of much current research.

It has long been speculated that surface premelting plays a major role in snow crystal growth, especially in relation to the temperature-dependent morphological transitions seen in the Nakaya diagram [1984Kur1, 1987Kob]. While such a dramatic restructuring of the ice surface is almost certainly an important factor, establishing a clear physical connection between surface premelting and snow crystal growth is a nontrivial challenge, as discussed in Chapter 4. Our current understanding of ice premelting, based on a combination of experiments and molecular dynamics (MD) simulations, is simply not yet advanced enough to make reliable predictions regarding the ice/vapor attachment kinetics.

Surface premelting becomes especially pronounced at temperatures above −1°C, when the QLL becomes quite thick and perhaps subject to instabilities that result in a nonuniform QLL thickness [2015Asa, 2019Nag]. This topic relates to a long-debated question of whether water completely "wets" ice at 0°C. The evidence suggests that the ice/water contact angle near the triple point is small but nonzero, although the subject has not been fully settled. Snow crystal growth at temperatures above −1°C could be phenomenologically quite interesting, although few observations have been made in this high-temperature regime.

ICE ENERGETICS

Many ice properties factor into the physics of snow crystal growth, notably the bulk, surface, and step energies of the ice crystal. The terrace step energies are particularly important for determining growth rates, so these will receive special attention throughout this book. For future reference, I document here a list of particularly relevant ice energetics.

Bulk Energies

Bulk material properties refer to sample sizes that are large enough that all surface effects are negligible. The bulk energies include the latent heats of the water/vapor transition (called "evaporation," "condensation," or "vaporization," depending on conventions), the ice/vapor transition (typically called "sublimation" or "deposition"), and the ice/water transition (melting, freezing, or fusion). Each of these is the amount of energy needed to cross a line in the water phase diagram (see Figure 2.2), as all of these are basic first-order phase transitions. Because water is an extremely well-studied material, all these quantities have been accurately measured over a broad range of temperatures and pressures. Near the triple point of water (using the notations s/l/v = solid/liquid/vapor = ice/water/wv),

$$\begin{aligned} L_{sv} &\approx 2.8 \times 10^6 \text{ J/kg} \\ L_{lv} &\approx 2.5 \times 10^6 \text{ J/kg} \\ L_{sl} &\approx 0.33 \times 10^6 \text{ J/kg}, \end{aligned} \quad (2.3)$$

and we see $L_{sv} \approx L_{sl} + L_{lv}$, as we would expect at the triple point. The latent heats vary somewhat with position on the respective phase boundaries, but these differences are not important for our focus on snow crystal growth.

Related useful quantities include the specific heat capacities (here at constant pressure) of water vapor, liquid water, and ice:

$$\begin{aligned} c_{p,wv} &\approx 2.0 \text{ kJ/kg} \cdot \text{K} \\ c_{p,water} &\approx 4.2 \text{ kJ/kg} \cdot \text{K} \\ c_{p,ice} &\approx 2.1 \text{ kJ/kg} \cdot \text{K}; \end{aligned} \quad (2.4)$$

thermal conductivities

$$\begin{aligned} \kappa_{wv} &\approx 0.02 \text{ W/m} \cdot \text{K} \\ \kappa_{water} &\approx 0.6 \text{ W/m} \cdot \text{K} \\ \kappa_{ice} &\approx 2.3 \text{ W/m} \cdot \text{K}; \end{aligned} \quad (2.5)$$

and material densities

$$\begin{aligned} \rho_{ice} &\approx 917 \text{ kg/m}^3 \\ \rho_{water} &\approx 1000 \text{ kg/m}^3. \end{aligned} \quad (2.6)$$

Both the ice and water densities can be assumed constant for our purposes, as both change only slightly with temperature. The equilibrium vapor density, in contrast, depends strongly on temperature and so requires special treatment.

Vapor Pressure and Supersaturation

The equilibrium (saturated) vapor pressure of ice and water can be written in the Arrhenius form

$$c_{sat} \approx C(T) \exp\left(-\frac{\ell}{kT}\right), \quad (2.7)$$

where $\ell \approx L m_{mol}$ is the latent heat per molecule, and $C(T)$ is a weak function of temperature. Table 2.1 gives measurements of the vapor pressure of water and ice [1971Mas] along with other useful quantities, and Figure 2.13 shows a fit to the vapor-pressure data, expressed as the Arrhenius curves:

$$\begin{aligned} p_{ice} &\approx 3.7e10 \cdot \exp\left(-\frac{6150}{T_K}\right), \\ p_{water} &\approx (2.8e9 + 1700 T_C^3) \cdot \exp\left(-\frac{5450}{T_K}\right), \end{aligned} \quad (2.8)$$

where T_K is the temperature in Kelvin and T_C is the temperature in degrees Celsius. Figure 2.13 also shows the supersaturation of supercooled liquid water relative to ice,

$$\sigma_{water} = \frac{c_{sat,water} - c_{sat,ice}}{c_{sat,ice}}, \quad (2.9)$$

along with the "excess" water vapor mass density plotted as a function of temperature,

$$[c_{sat,water}(T) - c_{sat,ice}(T)]\, m_{water}. \quad (2.10)$$

The data in the table and in these two plots are often useful for understanding the physics underlying snow crystal growth and for estimating experimental quantities.

TABLE 2.1
Several handy physical quantities that commonly appear in the study of snow crystal growth, listed as a function of temperature

Temp (°C)	P_{water} (mbar)	P_{ice} (mbar)	c_{sat} (#/m^3)	c_{sat}/c_{ice}	σ_{water}	v_{kin} (μm/sec)	η (K^{-1})	χ_0	C_{diff} (K^{-2})
−45	0.111	0.072	2.29×10^{21}	7.46×10^{-8}	0.5431	9.6	0.1109	0.017	6.48×10^{-3}
−44	0.123	0.081	2.55×10^{21}	8.31×10^{-8}	0.5280	10.8	0.1136	0.019	6.37×10^{-3}
−43	0.138	0.091	2.87×10^{21}	9.35×10^{-8}	0.5154	12.1	0.1143	0.022	6.26×10^{-3}
−42	0.153	0.102	3.20×10^{21}	1.04×10^{-7}	0.5015	13.6	0.1107	0.024	6.15×10^{-3}
−41	0.170	0.115	3.57×10^{21}	1.17×10^{-7}	0.4882	15.2	0.1101	0.026	6.04×10^{-3}
−40	0.189	0.128	3.99×10^{21}	1.30×10^{-7}	0.4739	17.0	0.1091	0.029	5.93×10^{-3}
−39	0.210	0.144	4.44×10^{21}	1.45×10^{-7}	0.4603	19.0	0.1082	0.032	5.83×10^{-3}
−38	0.232	0.161	4.95×10^{21}	1.61×10^{-7}	0.4458	21.2	0.1072	0.036	5.73×10^{-3}
−37	0.257	0.179	5.50×10^{21}	1.80×10^{-7}	0.4326	23.6	0.1061	0.039	5.63×10^{-3}
−36	0.284	0.200	6.12×10^{21}	2.00×10^{-7}	0.4191	26.3	0.1052	0.043	5.54×10^{-3}
−35	0.314	0.223	6.79×10^{21}	2.22×10^{-7}	0.4059	29.3	0.1044	0.048	5.44×10^{-3}
−34	0.346	0.249	7.54×10^{21}	2.46×10^{-7}	0.3924	32.5	0.1036	0.052	5.35×10^{-3}
−33	0.382	0.277	8.35×10^{21}	2.72×10^{-7}	0.3790	36.1	0.1026	0.057	5.26×10^{-3}
−32	0.420	0.308	9.25×10^{21}	3.02×10^{-7}	0.3658	40.1	0.1017	0.063	5.17×10^{-3}
−31	0.463	0.342	1.02×10^{21}	3.34×10^{-7}	0.3529	44.5	0.1010	0.069	5.08×10^{-3}
−30	0.509	0.380	1.13×10^{22}	3.69×10^{-7}	0.3397	49.3	0.1003	0.076	5.00×10^{-3}
−29	0.559	0.421	1.25×10^{22}	4.08×10^{-7}	0.3264	54.5	0.0993	0.083	4.91×10^{-3}
−28	0.613	0.467	1.38×10^{22}	4.50×10^{-7}	0.3138	60.3	0.0983	0.091	4.83×10^{-3}
−27	0.673	0.517	1.52×10^{22}	4.96×10^{-7}	0.3012	66.6	0.0976	0.100	4.75×10^{-3}
−26	0.737	0.572	1.68×10^{22}	5.47×10^{-7}	0.2885	73.6	0.0967	0.109	4.68×10^{-3}
−25	0.807	0.632	1.85×10^{22}	6.02×10^{-7}	0.2762	81.2	0.0957	0.118	4.60×10^{-3}
−24	0.883	0.698	2.03×10^{22}	6.62×10^{-7}	0.2645	89.5	0.0953	0.130	4.52×10^{-3}
−23	0.965	0.771	2.23×10^{22}	7.28×10^{-7}	0.2516	98.6	0.0947	0.142	4.45×10^{-3}
−22	1.054	0.850	2.45×10^{22}	8.00×10^{-7}	0.2394	108.5	0.0935	0.154	4.38×10^{-3}
−21	1.150	0.937	2.69×10^{22}	8.78×10^{-7}	0.2276	119.3	0.0931	0.168	4.31×10^{-3}
−20	1.254	1.032	2.95×10^{22}	9.64×10^{-7}	0.2149	131.2	0.0922	0.182	4.24×10^{-3}
−19	1.366	1.135	3.24×10^{22}	1.06×10^{-6}	0.2035	144.0	0.0912	0.198	4.17×10^{-3}
−18	1.487	1.248	3.54×10^{22}	1.16×10^{-6}	0.1915	158.0	0.0906	0.215	4.10×10^{-3}
−17	1.618	1.371	3.88×10^{22}	1.27×10^{-6}	0.1802	173.3	0.0898	0.233	4.04×10^{-3}
−16	1.759	1.505	4.24×10^{22}	1.38×10^{-6}	0.1688	189.9	0.0891	0.253	3.97×10^{-3}
−15	1.911	1.651	4.63×10^{22}	1.51×10^{-6}	0.1575	207.9	0.0885	0.275	3.91×10^{-3}
−14	2.075	1.810	5.06×10^{22}	1.65×10^{-6}	0.1464	227.4	0.0878	0.298	3.85×10^{-3}
−13	2.251	1.983	5.52×10^{22}	1.80×10^{-6}	0.1351	248.7	0.0872	0.323	3.79×10^{-3}

(continued)

TABLE 2.1 (Continued)

Temp (°C)	P_{water} (mbar)	P_{ice} (mbar)	c_{sat} (#/m³)	c_{sat}/c_{ice}	σ_{water}	v_{kin} (μm/sec)	η (K⁻¹)	χ_0	C_{diff} (K⁻²)
−12	2.440	2.171	6.02×10^{22}	1.97×10^{-6}	0.1239	271.8	0.0864	0.349	3.73×10^{-3}
−11	2.644	2.375	6.56×10^{22}	2.14×10^{-6}	0.1133	296.7	0.0859	0.378	3.67×10^{-3}
−10	2.862	2.597	7.15×10^{22}	2.33×10^{-6}	0.1020	323.8	0.0851	0.408	3.62×10^{-3}
−9	3.096	2.837	7.78×10^{22}	2.54×10^{-6}	0.0913	353.1	0.0843	0.440	3.56×10^{-3}
−8	3.348	3.097	8.46×10^{22}	2.76×10^{-6}	0.0810	384.7	0.0837	0.475	3.51×10^{-3}
−7	3.617	3.379	9.20×10^{22}	3.00×10^{-6}	0.0704	419.0	0.0831	0.512	3.45×10^{-3}
−6	3.906	3.684	9.99×10^{22}	3.26×10^{-6}	0.0603	455.9	0.0824	0.552	3.40×10^{-3}
−5	4.214	4.014	1.08×10^{23}	3.54×10^{-6}	0.0498	495.9	0.0818	0.595	3.35×10^{-3}
−4	4.544	4.371	1.18×10^{23}	3.84×10^{-6}	0.0396	539.0	0.0812	0.640	3.30×10^{-3}
−3	4.897	4.756	1.28×10^{23}	4.16×10^{-6}	0.0296	585.3	0.0806	0.689	3.25×10^{-3}
−2	5.274	5.173	1.38×10^{23}	4.5×10^{-6}	0.0195	635.5	0.0800	0.741	3.20×10^{-3}
−1	5.677	5.622	1.50×10^{23}	4.88×10^{-6}	0.0098	689.4	0.0793	0.795	3.15×10^{-3}
0	6.107	6.106	1.62×10^{23}	5.28×10^{-6}	0.0000	747.4	0.0786	0.824	3.10×10^{-3}

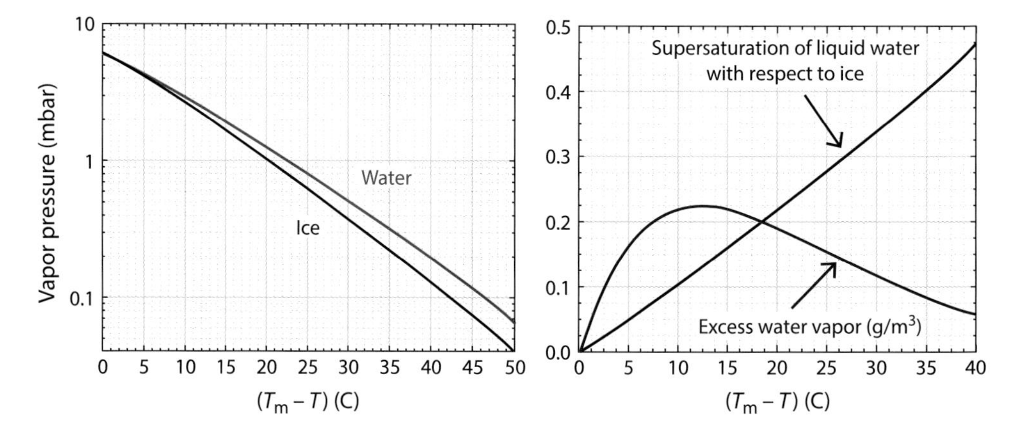

FIGURE 2.13. (Left) The vapor pressure of ice and supercooled water as a function of temperature below 0°C. When the water vapor pressure is above the Ice line, ice will grow; when the pressure is above the Water line, supercooled water droplets will condense and grow. (Right) The supersaturation of supercooled water with respect to ice, given by $\sigma_{water}(T)=[c_{water}(T)-c_{sat}(T)]/c_{sat}(T)$. This is also plotted as the "excess" water vapor, equal to $[c_{water}(T)-c_{sat}(T)]m_{water}$.

When the water vapor pressure exceeds $c_{sat,ice}$ at the surface of an existing ice crystal, the crystal will generally absorb water vapor and grow. But this vapor pressure is not sufficient to nucleate the formation of a new ice crystal. On most chemically inactive surfaces below 0°C, water vapor will not condense until the pressure is above $c_{sat,water}$, at which point liquid water droplets will appear. A separate nucleation step is then required to turn a liquid droplet into ice. This second step will occur almost instantaneously when the substrate temperature is below $-40°C$, but liquid droplets can remain in a supercooled state for long periods at higher temperatures, easily an hour or more at $-15°C$. The necessity of this two-step nucleation process is an example of *Ostwald's step rule*, which has numerous manifestations in physical chemistry.

Surface Energies

The surface energy, simply put, is the amount of energy needed to create additional interface between two material phases. For example, cutting a piece of ice in two requires breaking the chemical bonds holding the pieces together. The amount of energy needed to do this is proportional to the area of the new surface created during the break, and this defines the surface energy. If the material is intrinsically anisotropic, like a crystal lattice, then the surface energy could be anisotropic as well, depending on the angle of the surface relative to the crystal axes.

Because liquid water is an amorphous material, the water/vapor surface energy (also known as the surface tension of water) is isotropic, equal to

$$\gamma_{lv} \approx 76 \text{ mJ/m}^2 \tag{2.11}$$

near the triple point. This decreases with increasing temperature, dropping to $\gamma_{lv} \approx 72$ mJ/m^2 at 25°C and $\gamma_{lv} \approx 59$ mJ/m^2 at 100°C. The water/vapor surface energy is known to quite high accuracy, being determined from observations of the oscillation frequencies of liquid droplets.

The ice/water surface energy is best measured from the homogeneous nucleation of ice from supercooled water droplets as a function of temperature, yielding

$$\gamma_{sl} \approx 30 \pm 5 \text{ mJ/m}^2 \tag{2.12}$$

near the triple point. Measurements of this quantity as a function of temperature are somewhat model dependent, as they assume a good understanding of nucleation theory together with extrapolations of measured properties of liquid water down to temperatures as low as $-40°C$. The uncertainty in γ_{sl} given above is a rough estimate based on the data presented in [2015Ick]. The value of γ_{sl} decreases with decreasing temperature, down to roughly $\gamma_{sl} \approx 20$ mJ/m^2 at temperatures near $-40°C$, again with considerable uncertainty [2015Ick].

The ice/vapor surface energy γ_{sv} is more difficult to measure than either γ_{lv} or γ_{sl}, plus it has not received as much experimental or theoretical attention. Surface wetting measurements have produced the best measurements of γ_{sv} [1974Hob, 1999Pet], but these can be quite susceptible to surface contamination, so I tend to be conservative and give them a high uncertainty. Nevertheless, the measurements appear to be consistent with Antonow's relation, which states that

$$\gamma_{sv} \approx \gamma_{sl} + \gamma_{lv} \approx 106 \pm 15 \text{ mJ/m}^2 \tag{2.13}$$

near the triple point. The measurement uncertainty is somewhat subjective, based here on an examination of existing data from various sources. The ice/vapor surface energy γ_{sv} is also likely nearly isotropic, as I discuss below. Surface energy anisotropies are best determined from measurements of the equilibrium crystal shape, which has not yet been definitively observed for either the ice/water or ice/vapor systems.

In a naïve chemical bond picture, one can approximate of the ice/vapor surface energy as

$$\gamma_{0,sv} \approx \frac{1}{6} a \rho_{ice} L_{sv} \approx 135 \text{ mJ/m}^2, \tag{2.14}$$

because vaporizing bulk ice creates a total surface area of (roughly) $6a^2$ per molecule. This crude estimate does not account for surface relaxation or the specific ice lattice structure, so $\gamma_{0,sv}$ is mainly useful for pedagogical purposes However, it does illustrate that there are numerous connections between bulk, surface, and step energies, along with other equilibrium quantities, and these are further explored in materials science textbooks.

Terrace Step Energies

Terrace step energies factor into the rate of nucleation of new molecular terraces on faceted ice surfaces, which is one of the most important processes in snow crystal growth. In analogy to the surface energy, the step energy is the amount of energy needed to create the edge, or step, of a molecular terrace on a faceted surface. For example, separating a single island terrace into two smaller islands requires energy to break the molecular bonds holding the unseparated terrace together. The amount of energy needed is proportional to the length of new terrace step created, and this defines the terrace step energy. As with bulk energies and surface energies, the step energies on different faceted surfaces are equilibrium properties of a material.

For ice surfaces in water, the basal step energy has been accurately measured from nucleation-limited ice growth measurements [1958Hil, 1966Mic, 2014Lib], yielding $\beta_{sl,basal} \approx 5.6 \pm 0.7 \times 10^{-13}$ J/m near the triple point. This step energy provides a substantial nucleation barrier that results in basal faceting and the formation of thin, platelike ice crystals when liquid water freezes at low supercooling. Prism faceting is not observed in ice grown from liquid water near the triple point, and $\beta_{sl,prism}$ has not yet been measured, because it is much smaller than $\beta_{sl,basal}$. However, prism faceting at the ice/water interface has been observed at very high pressures [2005Mar], when the ice/water phase transition occurs near −20°C (see Figure 2.2), suggesting that $\beta_{sl,prism}$ becomes relatively high in that region of the phase diagram.

The ice/vapor step energies on both the basal and prism facets have been measured with good accuracy at temperatures ranging from −2°C to −40°C, giving the results shown in Figure 2.14. A closer examination of the ice growth data yielding these measurements is given in Chapter 7. As with γ_{sl}, the step energies were determined using nucleation-limited ice growth measurements, inferring the step energies using classical nucleation theory as described in Chapter 4.

Limiting Cases

Because the ice/vapor terrace step energies are so central to the snow crystal story, it is worthwhile to examine the physics underlying the measured values shown in Figure 2.14. At the low-temperature end of this graph, surface premelting is essentially absent on the ice surface, so the terrace step structure tends toward that of a rigid lattice. In this regime, the lattice projections above illustrate specific terrace steps, and low-temperature MD simulations (e.g., Figure 2.12) confirm the rigid lattice limit. We can make a toy model of these terrace steps by assuming a simple geometrical step of height a, and introducing such an idealized step increases the total surface area of the crystal by al, where l is the step length. Assuming the simplest possible case of a continuous, isotropic surface energy γ_{sv}, this toy model indicates that the step energy can be approximated as

$$\beta_{sv} \approx \beta_0 = a\gamma_{sv} \approx 3\times 10^{-11}\ \mathrm{J/m}. \qquad (2.15)$$

This is a crude estimate, much like the surface energy estimate in Equation 2.14, but it should give a reasonable upper limit. Moreover, we might expect the measured β_{sv} to approach β_0 at low temperatures, when the ice lattice structure becomes especially rigid. And indeed, both the prism and basal step energies seem to be trending toward β_0 at the lowest temperatures in Figure 2.14, lending credence to this basic physical picture.

At the other extreme, when the temperature approaches the melting point, surface premelting dramatically changes the structure of the ice surface, yielding a substantial quasiliquid layer that diverges in thickness as the temperature approaches 0°C. In this case, there will be a correspondence between the ice/QLL step energy and the ice/water step energy at sufficiently high temperature, as illustrated in Figure 2.15. With this overall picture of surface premelting, we see that the terrace nucleation process must take place at the ice/QLL interface, as this is where the ice lattice structure begins. Inasmuch as the disordered QLL resembles bulk liquid water, we

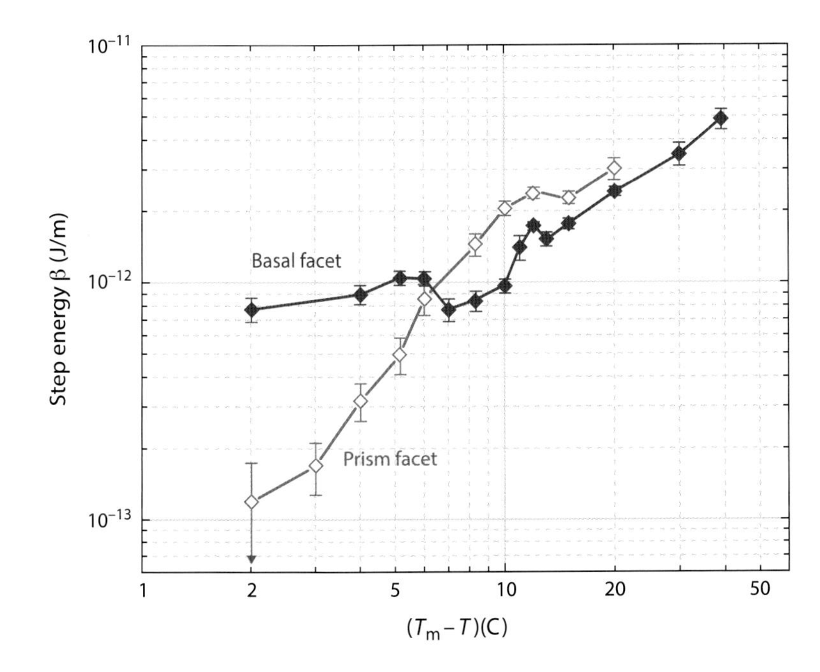

FIGURE 2.14. Measured ice/vapor terrace step energies β_{sv} for the basal and prism facets as a function of temperature [2013Lib].

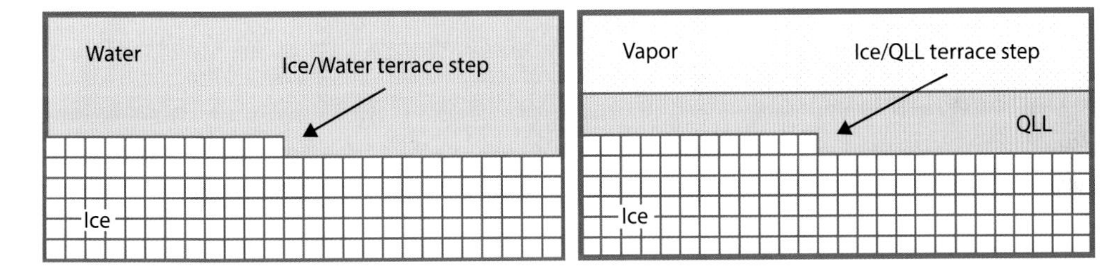

FIGURE 2.15. Sketches of terraces steps at (left) the ice/water interface and (right) the ice/vapor interface in the presence of significant surface premelting. At temperatures just below the melting point, surface premelting turns the ice/vapor interface into an ice/QLL/vapor interface, where crystal terrace steps are localized essentially at the ice/QLL interface. Inasmuch as the QLL resembles bulk water in its properties, the ice/QLL step energy should nearly equal the ice/water step energy. As discussed in the text, step-energy measurements at the ice/water and ice/vapor interfaces do show a good correspondence as the temperature approaches 0°C.

thus expect that terrace nucleation at the ice/QLL interface will begin to resemble that at the ice/water interface, as shown in the figure. In turn, this suggests that the ice/QLL step energy should tend toward the ice/water step energy at high temperatures. In other words, we should expect $\beta_{sv} \rightarrow \beta_{sl}$ as $T \rightarrow 0$°C for both the basal and prism facets.

This correspondence of step energies can be seen quite clearly in the data. In the basal case, we see that $\beta_{sv,basal}$ in Figure 2.14 levels off and is consistent with $\beta_{sv,basal} \rightarrow \beta_{sl,basal} \approx 5.6 \pm 0.7 \times 10^{-13}$ J/m at the highest temperatures measured. In the prism case, $\beta_{sv,prism}$ drops to quite low values at high temperatures, which is again consistent with the fact that $\beta_{sl,prism} \ll \beta_{sl,basal}$ in the ice/water system. Although a lot can happen between −2°C and the melting point, the data in Figure 2.14 do seem to support this correspondence between the ice/vapor and ice/water step energies near 0°C.

Putting all this together, we see that the ice/vapor step energies are reasonably well explained at both temperature extremes. The simple geometrical model applies at the lowest temperatures when the ice lattice assumes a rigid structure, allowing a satisfactory approximation for both the basal and prism step energies in that limit. And the correspondence between the ice/water and ice/QLL interfaces constrains the ice/vapor step energies as the temperature approaches 0°C. Therefore, the measured step energies in Figure 2.14 provide the transition between these low- and high-temperature behaviors. All in all, this provides a reasonably self-consistent picture of the step energies for a combination of the ice/vapor and ice/water systems on both the basal and prism facets.

MOLECULAR DYNAMICS SIMULATIONS

Many interfacial properties of the ice/vapor and ice/liquid systems are ultimately determined by complex many-body molecular interactions, and theoretical advances are likely best obtained using molecular dynamics (MD) simulations. The basic idea is to create a detailed mathematical model for the electronic potential between individual water molecules, assemble a computational system of many thousands of molecules, and then evolve and/or relax the system by calculating the many-body dynamics. In principle, with a sufficiently accurate model of the relevant molecular interactions, one can examine the detailed structure and dynamics of quite complex systems using MD simulations. Beyond their intrinsic usefulness for understanding water and ice, there is considerable motivation to develop accurate water models for studying the molecular biology of proteins and other large biomolecules in solution. This highly technical area of computational science is far outside my expertise, so I will comment little on specific molecular models or computational methodologies. Instead, I briefly summarize some results from MD simulations that have examined the structure and growth of ice surfaces.

Much work has focused on developing models of water molecular interactions that reproduce bulk equilibrium properties like the freezing temperature, latent heats, diffusion coefficients, and the peak density of liquid water near 4°C. Equilibrium properties are generally the most amenable to MD simulations, and considerable progress has been made over the past several decades [2005Car, 2006Fer, 2008Con]. In particular, surface premelting has been seen in numerous simulations [1987Kar, 1988Kar, 1996Nad, 1997Fur, 2004Ike, 2016Ben, 2020Llo], allowing investigation of the QLL thickness as a function of temperature for both the basal and prism facets, along with estimates of molecular diffusion rates and other material characteristics in the QLL. These investigations clearly confirm the existence of surface premelting, which has proven to be a robust computational phenomenon that is insensitive to details of the water model used [2008Con].

Observing actual ice growth in MD simulations remains a substantial challenge, so there is little chance that direct simulations of attachment kinetics will be practical soon. While simulations now routinely observe the freezing process [2005Car, 2012Roz, 2012Seo], this has only been possible in rather extreme conditions, when growth rates are of order 0.1 m/sec or higher. Experiments rarely access this region of parameter space [2005Shi, 2017Lib], and accurate measurements of such rapid growth rates are difficult. Snow crystal growth occurs in far more benign conditions, where the growth rates are orders of magnitude slower than what is typically observed in MD simulations.

I am especially optimistic that MD simulations of ice/vapor step energies will soon provide a direct link between fundamental molecular physics and snow crystal attachment kinetics. On the theory side, the step energies $\beta_{basal}(T)$ and $\beta_{prism}(T)$ are basic equilibrium properties of the ice crystal, so they should be amenable to direct calculation using MD simulations. Computational methods for computing terrace step energies have only recently been developed for silicon [2012Fro] and ice [2019Ben, 2020Llo], demonstrating that numerical methods are

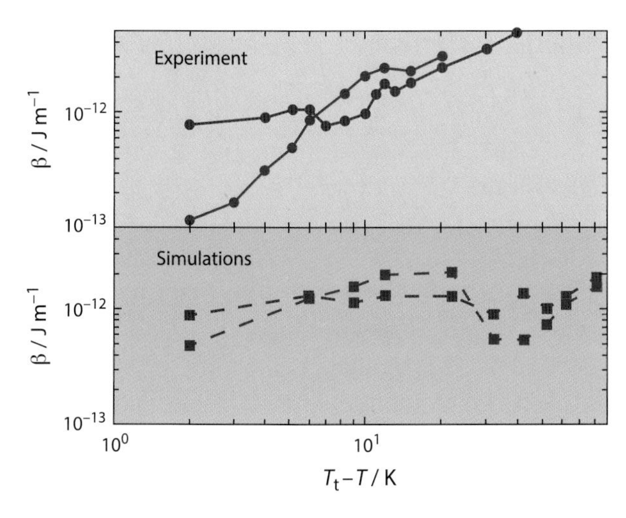

FIGURE 2.16. Ice/vapor terrace step energies on basal (blue) and prism (red) faceted surfaces as a function of temperature, comparing experimental measurements (top) with MD simulations (bottom) [2020Llo]. That it is now possible to make such comparisons bodes well for future improvements in our understanding of the ice/vapor attachment kinetics. Image courtesy of Luis Gonzalez Mac-Dowell.

suitable for this task. Figure 2.16 shows some first results comparing experiments and simulations.

Further investigation of step energies should be an especially fruitful research direction for several reasons:

1) On the experimental side, we already have ice/vapor step energy measurements over a broad range of temperatures for both the basal and prism facets (Figure 2.14), and there is considerable room for improvement in these measurements at both higher and lower temperatures [2019Har]. Measurements of ice/water step energies at 0°C are also well established.
2) Exploring the correspondence between ice/water and ice/vapor step energies near 0°C should be especially interesting. To my knowledge, there have not yet been any MD simulations of the ice/water step energy, although the ice/water system is quite important in many areas of science.
3) The ice lattice structure is especially simple at low temperatures when surface premelting is largely absent, and the step energy is especially large in that regime. This suggests that step energies in the low-temperature limit should be especially amenable to calculation using MD simulation, although the initial results in Figure 2.16 have not shown this to be the case so far.

SURFACE ENERGY CONSIDERATIONS

Although the ice/vapor surface energy is not a dominant factor in snow crystal growth compared to particle diffusion and the attachment kinetics, it does play a role and needs to be included in computational models of diffusion-limited growth. Of particular importance is the *Gibbs-Thomson effect*, which describes how the vapor pressure of a convex ice surface is slightly higher than that of a flat surface [1871Tho]. In terms of molecular interactions, a molecule on a convex surface is slightly more distant from its collective neighbors than one on a flat surface, simply from geometrical considerations. The convex-surface molecule is thus less tightly bound, resulting in a higher vapor pressure compared to a flat surface. This picture helps visualize the Gibbs-Thomson phenomenon, but it is difficult to quantify without precisely adding up all the binding energies.

The Gibbs-Thomson effect can be calculated by considering how the surface energy connects to the vapor pressure. For an ice sphere of radius R, pulling one molecule off the sphere reduces its surface area by an amount $\delta A = 2/c_{ice}R$, as the size of the sphere is reduced slightly with the loss of one molecule. This, in turn, results in a reduction in the surface energy by an amount $\delta E = 2\gamma_{sv}/c_{ice}R$, where γ_{sv} is the ice/vapor surface energy. Including this additional energy term in the Arrhenius equation gives the modified equilibrium vapor pressure

$$\begin{aligned} c_{eq}(R) &\approx C(T)\exp\left(-\frac{\ell - \delta E}{kT}\right) \\ &\approx c_{sat}(1 + d_{sv}\kappa), \end{aligned} \tag{2.16}$$

where $d_{sv} = \gamma_{sv} / c_{ice} kT \approx 1$ nm and $\kappa = 2/R$ is the curvature of the spherical surface. For a smooth but nonspherical surface, the curvature is defined as $\kappa = 1/R_1 + 1/R_2$, where R_1 and R_2 are the two principal radii of curvature of the surface. Note that this expression reduces to the normal flat-surface vapor pressure c_{sat} when $R \rightarrow \infty$, as it must. The analysis is more complicated when γ_{sv} is anisotropic, but the functional form will be like the above expression if the surface energy anisotropy is small, as it is for ice. For a small ice prism of size L, replacing $\kappa = 2/R$ with $\kappa = 2G/L$, where G is a geometrical factor of order unity, gives a reasonable approximation for most simple shapes.

This additional vapor pressure can have a significant effect on snow crystal growth if the supersaturation is especially low while the surface curvature is quite high. For example, taking $R = 1$ μm gives a change in the effective supersaturation of $\Delta\sigma = 2d_{sv} / R \approx 0.1\%$, which is often negligible under many normal circumstances in snow crystal growth. Such small feature sizes appear mostly in growth morphologies that occur at quite high supersaturations. However, detailed modeling reveals that the Gibbs-Thomson effect plays a large role in preventing the growth of thin plates at especially low supersaturations. I examine the Gibbs-Thomson effect in more detail in Chapter 3 and discuss how it can be incorporated into snow crystal modeling in Chapter 5.

Equilibrium Crystal Shape

The equilibrium crystal shape (ECS) of an isolated snow crystal is the shape that minimizes the total surface energy at constant volume. For a perfectly isotropic surface energy, the ECS is a perfect sphere, as this shape minimizes the surface area and thus also minimizes the surface energy. If $\gamma_{facet} < \gamma_{unfaceted}$, however, then it becomes energetically favorable to increase the facet surface areas with a nonspherical shape. If γ_{facet} is only slightly smaller than $\gamma_{unfaceted}$, then the ECS becomes essentially a spherical shape decorated with small faceted "dimples." The dimples become larger with increasing anisotropy, and, when the surface energy anisotropy is sufficiently high, the ECS becomes a fully faceted prism.

To see how much surface energy anisotropy is needed to produce a fully faceted ECS, consider the case of a cubic crystal with γ_{facet} on the facets and $\gamma_{unfaceted}$ on all other surfaces. Comparing the surface energy of a cube with that of a sphere of the same total volume, basic geometry shows that the cube will have a lower overall surface energy if $\gamma_{facet} < 0.806\ \gamma_{unfaceted}$, independent of the crystal size. Although this result is too simple for direct comparison with more complex crystal symmetries, it does tell us that we can expect an ECS that is fully faceted if the facet surface energy is of order 10–20 percent lower than the surface energy of unfaceted surfaces. If the surface energy anisotropy is less than 5 percent, the ECS will likely be a faceted prism with rounded corners, and the degree of rounding will depend on the amount of the anisotropy. More precise ECS calculations can be done using the Wolff construction, and there is much discussion of this in the scientific literature and in textbooks on crystal growth [1996Sai, 2004Mar].

The appearance of faceted surfaces on the ECS is related to the terrace step energies on those surfaces. If β_{facet} goes to zero, then there is no energy penalty for creating terrace steps, and therefore no energy gain in having faceted surfaces. In the limit of small β_{facet}, the half-angle of a faceted ECS dimple is approximately [2001Bon]

$$\theta \approx \frac{\beta_{facet}}{a\gamma_{facet}}. \qquad (2.17)$$

Taking $\beta_{facet} \approx 10^{-12}$ J/m and $\gamma_{facet} \approx 0.1$ J/m^2 gives $\theta \approx 0.03$ radians, suggesting that the ice ECS is nearly spherical at temperatures above -10°C [2012Libl]. The ice/vapor ECS could become quite faceted at lower temperatures, however, when the step energies are larger.

Direct measurement of the ice ECS is challenging because of the timescales involved. It takes a substantial amount of time for an ice crystal in an initially arbitrary shape to relax to its equilibrium shape, as this involves the transport of molecules between different parts of the crystal surface. Observations of bubble migration in ice [2010Dad] indicate that vapor transport (and not surface migration) is the dominant path to equilibration, so the equilibration time can be estimated using the Gibbs-Thomson effect. Beginning with a slightly nonspherical shape of overall radius R, the equilibration time becomes approximately

$$\tau_{eq} \approx \frac{c_{ice} R^3}{2 c_{sat} d_{sv} D}, \tag{2.18}$$

and inserting the diffusion constant for normal air (Chapter 3) gives $\tau_{eq} \approx 8$ days for $R = 50$ μm. This casts doubt on the one reported ECS measurement for ice [1985Col, 1997Pru], suggesting that slow growth may have yielded the observed faceted surfaces, given the known highly anisotropic attachment kinetics for ice. Lowering the background air pressure P increases the diffusion constant as $D \sim P^{-1}$, until the vapor transport becomes limited by attachment kinetics instead of diffusion. In the kinetics-limited case, the equilibration time becomes

$$\tau_{eq} \approx \frac{R^2}{2 \alpha v_{kin} d_{sv}}, \tag{2.19}$$

which becomes a more favorable 2 hours for $R = 50$ μm and $\alpha = 1$ at -15°C. However, the fact that $\alpha \to 0$ on faceted surfaces at low supersaturations can increase this time dramatically.

Overall, the evidence suggests that the ice ECS is nearly spherical at higher temperatures [2012Libl], but better experiments could settle this question by directly measuring the ECS as a function of temperature. However, in nearly all snow crystal growth scenarios, the equilibration times mentioned above are much longer than typical growth times $\tau_{growth} \approx R/v$. This fact suggests that surface energy effects are less important than effects from attachment kinetics in most snow crystal

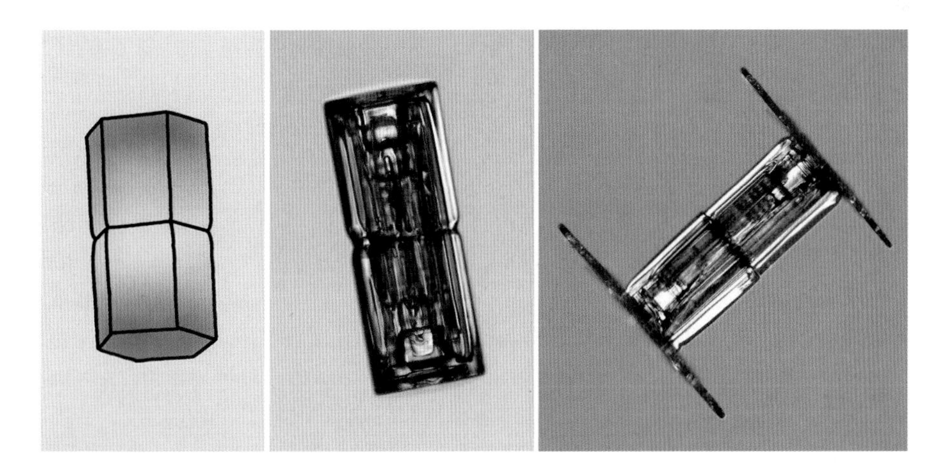

FIGURE 2.17. *Columnar twins.* (Left) A sketch of a columnar twin snow crystal, illustrating that these are essentially two ordinary columnar crystals connected by a *twin plane* between them. The crystal structure is relatively weaker in the twin plane, so sublimation often produces an *evaporation groove* between the two columns. (Center) The presence of an evaporation groove identifies this as a twinned column, which otherwise looks just like a normal single-crystal column. (Right) This capped column began as a simple twinned column, as evidenced by the evaporation groove.

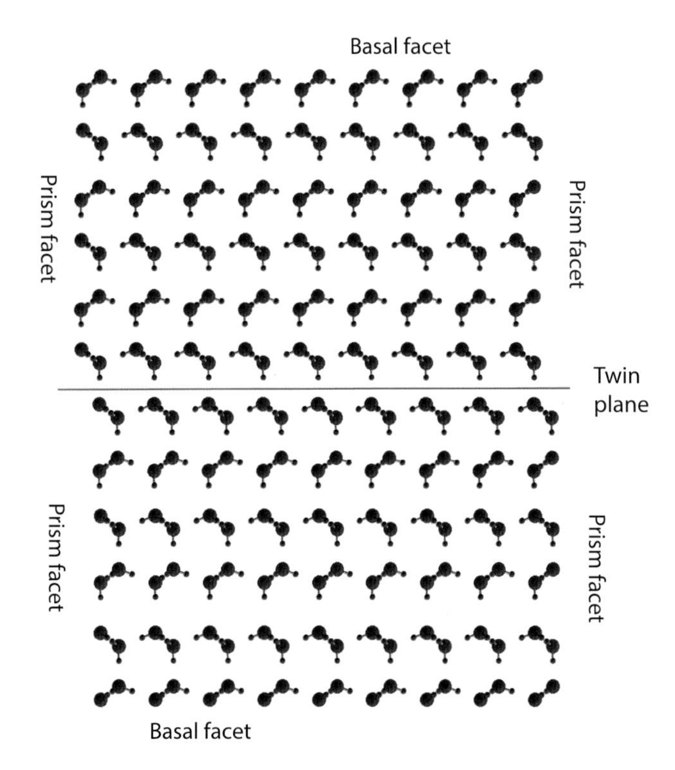

FIGURE 2.18. *Columnar twins.* The crystal lattice structure of a columnar twin crystal. Comparing this with Figure 2.5, it becomes apparent that the twin plane consists of a layer cubic bonds between water molecules, whereas the rest of the structure is made from the typical hexagonal bonds. Also note the small offset between the columns, necessary to accommodate the cubic bond structure. The twin plane could consist of multiple, randomly arranged layers of cubic/hexagonal bonds (i.e., a stacking-disordered region), rather than the single cubic layer shown here.

growth circumstances, except at extremely low supersaturations and reduced pressures. Thus, while surface energy effects are not always completely negligible in snow crystal growth, they are usually overwhelmed by other factors.

SNOW CRYSTAL TWINNING

When two single crystals grow together with a specific orientation between their respective lattices, this is called crystal *twinning*. Many mineral crystals exhibit twinning in various forms, and ice is no exception. Two questions immediately arise with crystal twinning: 1) What defines the orientation between the twin crystals? and 2) What circumstances bring about the different twinned states? The first of these questions is usually the easier to answer, as it involves statics and energetics. A twinned state is typically a metastable state: a local energy minimum that the crystal structure fell into during the early phases of its growth, from which it cannot reach the lowest-energy configuration. The second question is one of dynamics, and therefore more difficult to answer, involving how the crystal's nucleation and growth history happened to produce a twinned state. Estimating the probability that a twinned state will form under different circumstances is an extremely challenging task. Our goal here is relatively modest: simply to report on some examples of snow crystal twins and try to explain their structures as best we can. Although certainly not a well-studied topic, snow crystal twinning has been discussed in the scientific literature over many decades [1971Iwa, 1978Fur, 1987Kob, 2011Kik, 2013Kik].

Columnar Twins

Figure 2.17 shows examples of columnar twins, illustrating the most common form of twinning found in natural snow crystals [1974Kob]. If you look carefully when columnar crystals are falling from the clouds near −5°C, you are likely to find some twin columns in the mix. This form of twinning is also easily explained by ice crystallography, as shown in Figure 2.18. The two halves of a twin column fit perfectly together at the twin plane if a single plane of hexagonal bonds is replaced with a plane of cubic bonds. This can be nicely demonstrated using a 3D molecular model as well. A columnar twin plane likely originates when the initial crystal nucleation (typically from a liquid water droplet) results in some stacking disorder. Nucleation can be a somewhat violent event, as the supercooling just prior to nucleation is often quite high. Thus, it is perhaps not surprising that some lattice disorder can arise during this process. The initial ice growth soon warms the liquid and thus reduces its supercooling, so subsequent growth quickly settles into the energetically favorable ice Ih configuration. Once the initial droplet is completely frozen and growth from water vapor commences, the stacking disorder can result in a pair of twinned columns growing out from the nucleation site. The initial disorder may be more complex than a single, clean cubic plane, but subsequent growth can bury many dislocations. Polycrystalline forms are also common via this same mechanism, including bul-

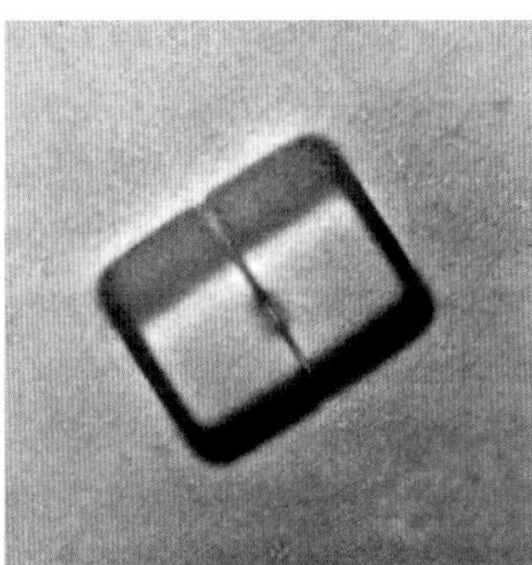

FIGURE 2.19. *Columnar twins.* When a columnar twin is growing (left), the facet corners are sharp, so the twin plane is not readily apparent. When it begins to sublimate (right), the corners become rounded, and the evaporation groove deepens. Image adapted from [1987Kob].

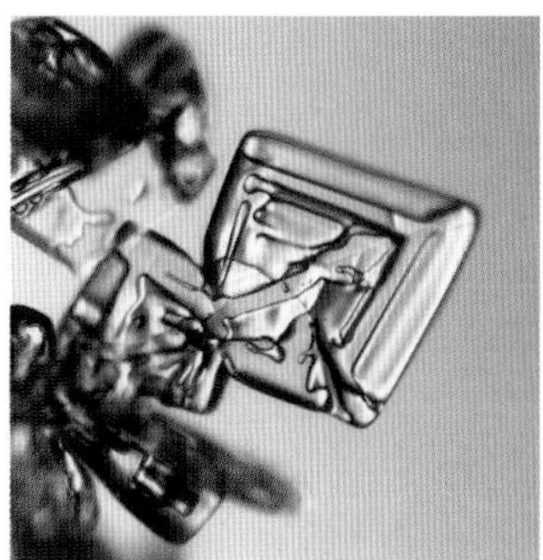

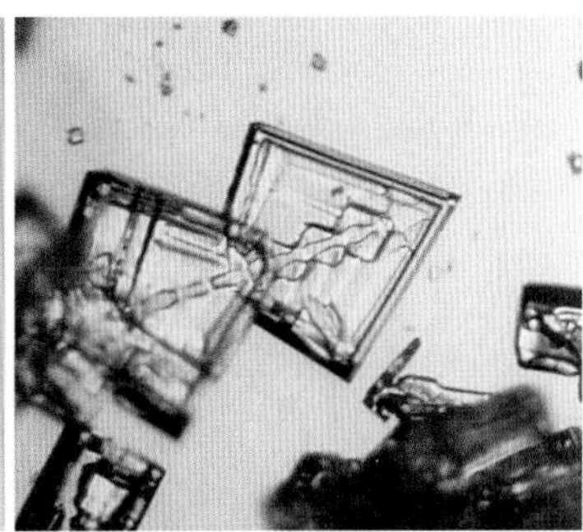

FIGURE 2.20. *Arrowhead twins variant I.* These photos show examples of arrowhead twin snow crystals with 78-degree apex angles. The first two crystals (left, center) fell at a temperature near −5°C [2003Lib2], while the third (right) fell at a temperature near −40 C [2006Tap]. Photos by Patricia Rasmussen [2003Lib2] and courtesy of Walter Tape.

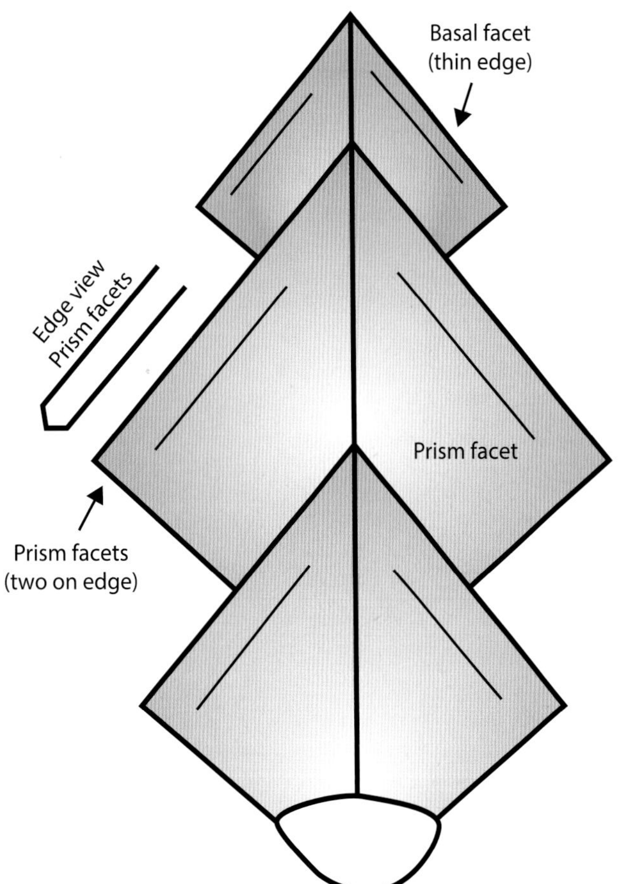

FIGURE 2.21. *Arrowhead twins variant I.* This sketch illustrates the peculiar faceted structure of a 78-degree arrowhead twin. It is a thin platelike crystal (seen face-on in this sketch and in the photos), but the two faces of the plate are prism facets, not the usual basal facets seen in thin hexagonal plates. A pair of narrow prism facets make up each of the lower-facing edges, while the top-facing edges are narrow basal facets. An arrowhead twin typically emerges from some unknown initial structure, here shown as an ill-formed blob at the bottom of the sketch.

let rosettes, which are especially common at low temperatures (Chapter 10).

Because ice Ih is energetically favorable over ice Ic, the cubic bond plane comes with an energy cost, which slightly increases the equilibrium vapor pressure at the twin plane. As a result, when the crystal begins slowly sublimating away (as usually happens after it falls from the clouds, or when it is being photographed), the twin plane sublimates faster than the surrounding prism facets, yielding an *evaporation groove* that appears like a belt around the columns at the twin plane, as illustrated in Figure 2.19. An evaporation groove is a characteristic marking that identifies a columnar twin; otherwise, the twinned column is essentially indistinguishable from a normal columnar snow crystal.

During nucleation and growth of a columnar crystal, the most likely scenario is that no twin planes form. Thus, most columns are untwinned single crystals. The next most likely scenario is that there is some stacking disorder during nucleation, resulting in a single twin plane. Perhaps this plane contains a single, clean, cubic layer, as shown in Figure 2.18, but it may contain several randomly stacked layers. In any case, the situation eventually sorts itself out, often yielding two normal columns on either side of the twin plane. Because both columns grow outward at about the same rate, the evaporation groove is usually near the midpoint of the twinned column. However, one column may grow a bit faster than the other, so the twin symmetry need not be perfect in every case. Creating

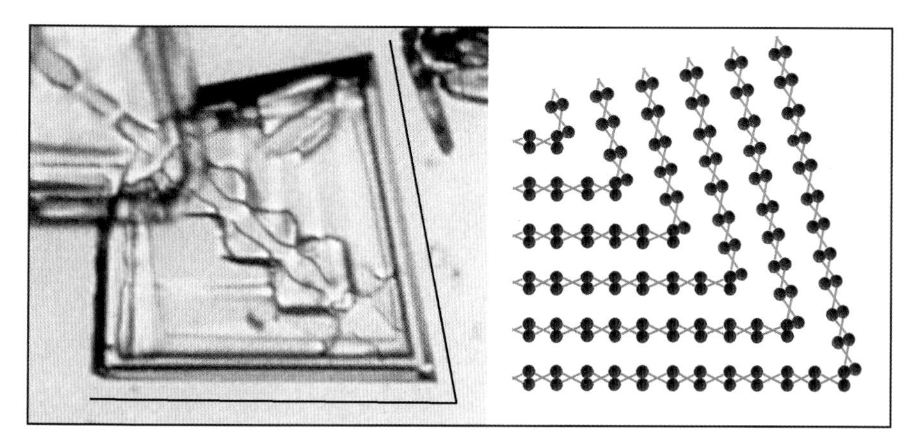

FIGURE 2.22. *Arrowhead twins variant I.* The crystal lattice projection in the sketch is looking down along the $[\bar{1}010]$ axis, which is perpendicular to a prism facet. This is called a *contact twin*, because the twins meet at a single plane, and further a *reflection twin*, because the contact plane is also a reflection plane. The model gives a theoretical apex angle of $2\tan^{-1}(c_0/2a_0) = 78.3$ degrees. A pair of black lines separated by 78 degrees has been superimposed on the photograph [2006Tap], showing good agreement with this lattice model. Image courtesy of Walter Tape.

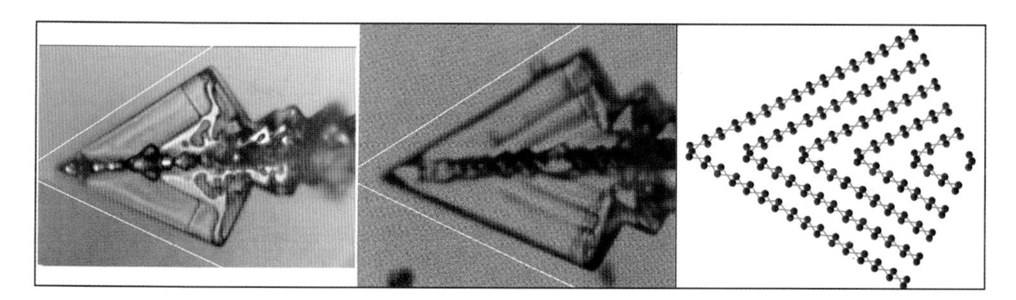

FIGURE 2.23. *Arrowhead twins variant II.* The left and middle photos above show two examples of this arrowhead-twin variant. The lattice projection in the sketch is similar to that in Figure 2.22, giving a theoretical apex angle of $2\tan^{-1}(c_0/3a_0) = 57.0$ degrees. The pairs of white lines in the photos subtend this angle, showing good agreement with theory. Images adapted from [2011Kik].

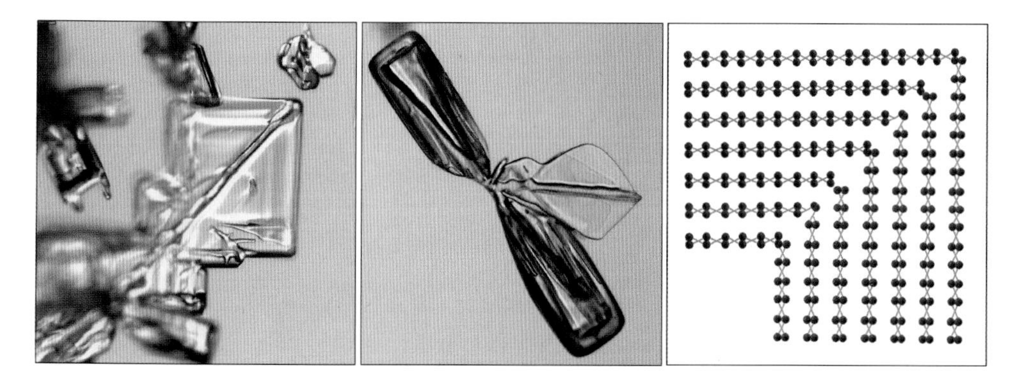

FIGURE 2.24. *Arrowhead twins variant III.* This variant of an arrowhead twin displays an apex angle that is close to 90 degrees, and the model structure shown in the sketch has a theoretical value of $2\tan^{-1}(3c_0/5a_0) = 88.7$ degrees. Images by Patricia Rasmussen [2003Lib2].

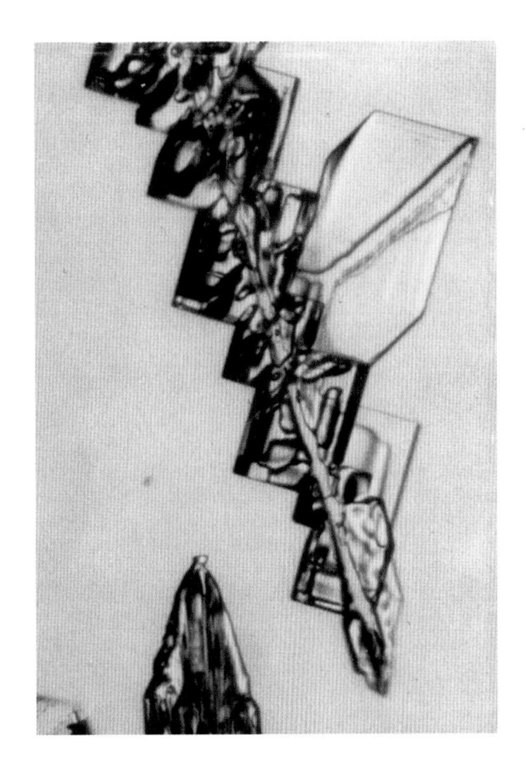

FIGURE 2.25. *Arrowhead variant III.* Walter Tape photographed this sharply faceted example of an arrowhead twin near the South Pole [2006Tap]. A careful measurement of the crystal (by the author) yielded an apex angle of 88.5±0.5 degrees, in good agreement with the model shown in Figure 2.24. Image courtesy of Walter Tape.

more than one twin plane in a single structure would be quite unlikely, and I know of no photographic examples of a simple hexagonal column exhibiting multiple evaporation grooves.

Arrowhead Twins

Figure 2.20 shows several photographs of one variant of *arrowhead* snow crystal twins, also known as *Gohei twins* [2004Bai]. The sketch in Figure 2.21 illustrates their overall faceted structure. Note that the platelike faces of an arrowhead twin are prism facets, as opposed to the basal facets that one normally associates with platelike snow crystals. This morphology results because the thin basal edges grow most rapidly, like an unrolled, sheathlike hollow column. With this geometry, the basal edges flanking the apex experience the fastest growth, extending the apex forward. By comparison, the broad prism surfaces accumulate water vapor very slowly.

Figure 2.22 shows a crystal lattice model that explains the structure of this arrowhead twin variant, giving a theoretical apex angle of $2\tan^{-1}(c_0/2a_0) = 78.3$ degrees, in good agreement with observations. Arrowhead twins are generally quite small (seldom more than a millimeter in size), typically appearing at temperatures near −5°C or −40°C, often mixed with hollow columnar crystals. Arrowhead twins are uncommon, but they have an easily recognizable shape, so they can be found if you go searching for them.

Figure 2.23 shows a second variant of arrowhead twinning, like Variant I but with a different apex angle. Comparing Figures 2.22 and 2.23, it becomes apparent that the twinning mechanisms are quite similar, and the geometry of Variant II gives it an apex angle of $2\tan^{-1}(c_0/3a_0) = 57.0$ degrees. Figure 2.24 shows yet a third arrowhead variant, this time with an apex angle just slightly below 90 degrees. The model does not give as tight a lattice match as the previous two arrowhead variants, but the theoretical apex angle of $2\tan^{-1}(3c_0/5a_0) = 88.7$ degrees agrees well with one of the best photographic specimens, as shown in Figure 2.25.

Crossed Plates

Figure 2.26 shows several examples of another class of snow crystal twinning, this time in the form of crossed platelike crystals. These can be found in nature, although specimens are relatively uncommon, quite small, and most are rather poorly formed, as shown in the figure. They seem to appear mostly at temperatures near −2°C. Crossed plates have been seen in the laboratory as well, but there are few good photographs. Figure 2.27 shows

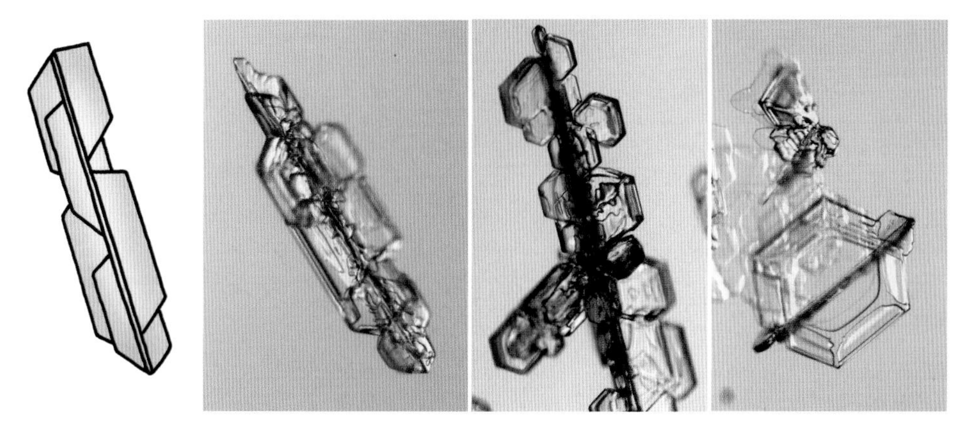

FIGURE 2.26. *Crossed plates variant I.* These photos show examples of natural crossed-plate crystals, along with a sketch of their overall structure. Note that one prism facet edge is parallel to the intersection axis of the twinned crystals. Images by the author and Patricia Rasmussen [2003Lib2].

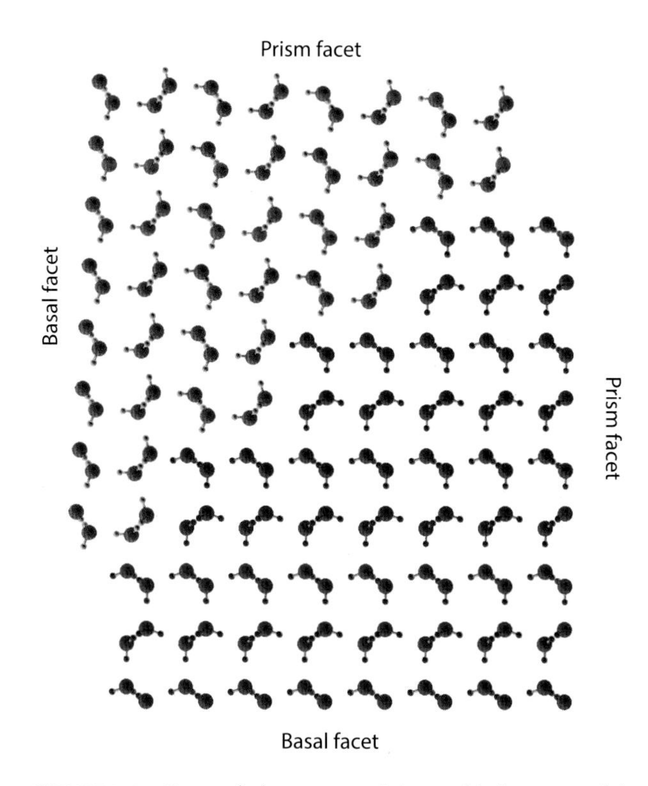

FIGURE 2.27. *Crossed plates variant I.* A possible lattice model for variant-I crossed-plate twinning. This is a contact twin where the theoretical angle between basal facets is $2\tan^{-1}(c_0/\sqrt{3}a_0)=86.5$ degrees.

a possible lattice structure for this first crossed-plate variant. The twin angle is $2\tan^{-1}\left(c_0/\sqrt{3}a_0\right)=86.5$ degrees, which is consistent with the observations showing plates crossing at roughly 90 degrees. Unfortunately, I have not found any photographic examples that have allowed a precise measurement of the angle between the crossed plates.

Figure 2.28 shows a second crossed-plate variant that is observed quite readily in free-falling laboratory crystals grown in air near $-10°$C, this time exhibiting a basal-plane angle of about 70 degrees. This form was first documented by Kunimoto Iwai [1971Iwa], who proposed the lattice model shown in Figure 2.29 with a theoretical angle between the plates of $\tan^{-1}\left(\sqrt{3}c_0/a_0\right)=70.5$ degrees. However, this is not a simple contact-twin model, so the lattice connection about the twin plane does not extend to large distances. The connection points (circled in Figure 2.29) have theoretical spacings of $\sqrt{3a_0^2+9c_0^2}$ in the top crystal and $3\sqrt{3}a_0$ in the bottom crystal. Although these values only differ by about 0.2 percent, this model geometry leads to an inevitable lattice mismatch as the twin plane propagates outward.

FIGURE 2.28. *Crossed plates variant II.* Several snow crystal twins grown by the author in air in a free-fall chamber near –10°C. Top/bottom pairs of images show the same crystal with a different microscope focus, one image focusing on the flat plate resting on the substrate and the other image focusing on the top edge of the twin plate.

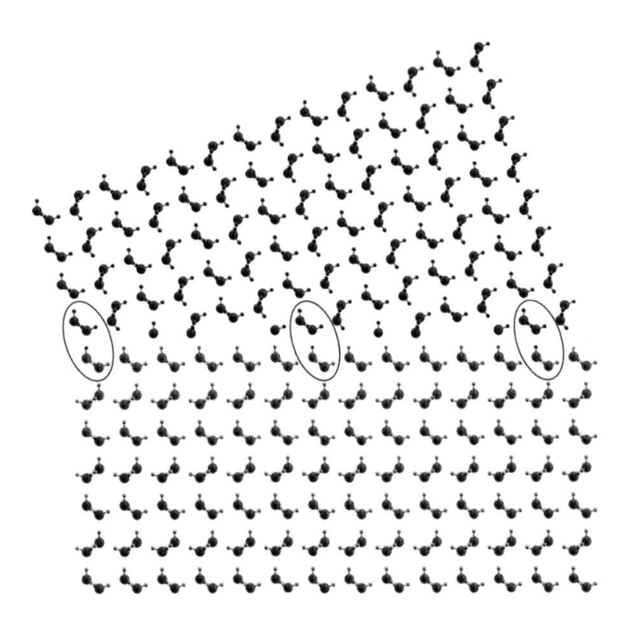

FIGURE 2.29. *Crossed plates variant II.* This diagram shows the likely crystal structure for variant-II crossed-plate twinning [1971Iwa, 1978Fur].

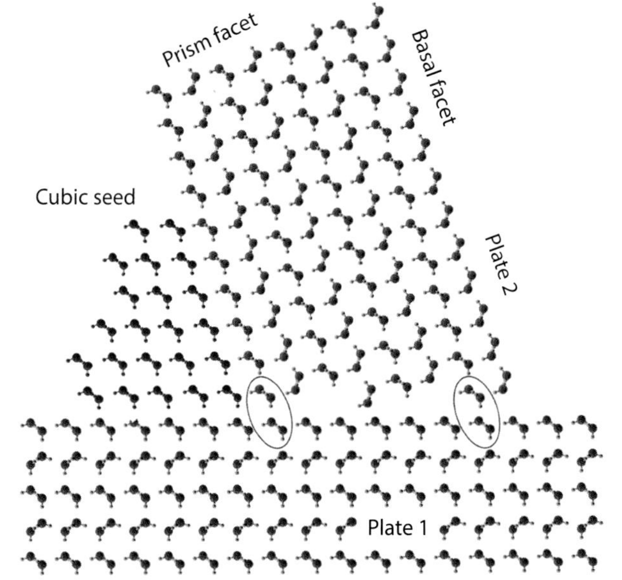

FIGURE 2.30. *Crossed plates variant II.* This diagram shows how a cubic seed crystal can give rise to a variant-II crossed plate [1987Kob].

Kobayashi and Kuroda [1987Kob] pointed out that this crossed-plate variant could originate from an ice Ic seed crystal, as shown in Figure 2.30. In this physical picture, the cubic seed appears during the initial nucleation process, stimulating the twin ice Ih crystals shown in the figure. The subsequent vapor growth soon produces two crossed plates, burying the minute seed in the process. This cubic-nucleation model nicely explains why such a high-order twinning would occur so readily, and it supports the hypothesis that stacking disordered crystals, containing a mix of Ih and Ic bonding, can play a significant role in snow crystal nucleation.

There are several additional observations of snow crystals with rather odd geometries that have been reported in the literature, and these might be explained as variants of twinning beyond those described here. To date, however, these observations are rather poor, so it is perhaps premature to extend our discussion of twinning much further. As with many other aspects of snow crystal science, better observations may yield additional surprises in the future.

FIGURE 3.1. This stellar snow crystal displays complex sidebranching brought about by diffusion-limited growth. It also experienced a major induced-sidebranching event when the primary branches were about half their final length. Photo taken by the author in Kiruna, Sweden.

THREE

Diffusion-Limited Growth

If the Lord Almighty had consulted me
before embarking on Creation,
I should have recommended
something simpler.
—ALPHONSO THE WISE, ATTRIBUTED, CA. 1250

When a snow crystal grows in the atmosphere, it does so by removing water vapor molecules from the air in its vicinity. To continue growing, more water molecules must diffuse through the surrounding air, making their way into the depleted region near the crystal. Because diffusion is a slow process, it can substantially limit the development of the crystal, so we say its growth is *diffusion limited*. As I discuss in this chapter, diffusion-limited growth is responsible for the creation of branches and other structures, making this physical phenomenon especially important in the formation of complex snow crystal patterns.

The word "diffusion" derives from the Latin *diffundere*, meaning to spread out over time. The diffusion of water molecules in air results from the normal thermal jostling of air and water molecules, which tends to mix the two species together. If the water vapor density is not spatially uniform, then the random molecular motions will, on average, transport water molecules from higher-density to lower-density regions. Therefore, as a growing snow crystal consumes water vapor molecules in its vicinity, more will flow inward toward the crystal from afar, providing additional material for continued growth. Diffusion is a common phenomenon in everyday life, although we may not readily notice diffusion in action, especially when it involves invisible gases like air and water vapor. Diffusion of colored liquids can be more easily visualized, as shown in Figure 3.2. Most people are familiar with material dispersing away from a central source, but diffusion toward a central sink is more relevant to snow crystal growth, as shown in the illustration. In either case, the net diffusive transport is always from high to low densities, and the net flow at any point is along the local density gradient, with a flow rate proportional to the magnitude of the gradient.

There are two types of diffusion involved in snow crystal growth: particle diffusion and heat diffusion. The latter arises when latent heat is generated by vapor deposition, as this heat must diffuse away into the surrounding the air. As I show later in this chapter, the effects of heating and heat diffusion are relatively small compared to particle diffusion. Except in a few special circumstances, to a good approximation one can describe the

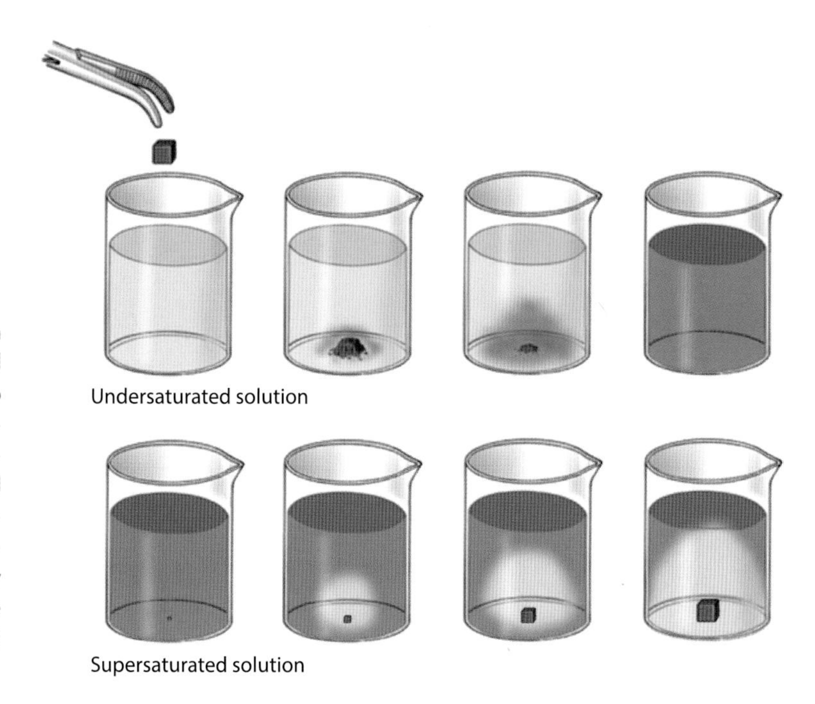

FIGURE 3.2. The process of diffusion, shown operating in two directions. A crystal dropped into an undersaturated solution will dissolve (top row). Diffusion will then slowly spread the dissolved material throughout the solution. In contrast, a seed crystal placed into a supersaturated solution (bottom row) will grow as diffusion carries material to the crystal and depletes the solution nearby. The first case is analogous to a snow crystal sublimating in undersaturated air, and the second case is analogous to a snow crystal growing in supersaturated air.

growth of snow crystals by neglecting heat diffusion entirely, and that will be my default assumption unless otherwise indicated.

Figure 3.3 illustrates how a growing ice crystal depletes the water vapor density around it, creating a supersaturation gradient. As seen in this computer simulation, the gradient is highest near the tips of a growing crystal, resulting in a high flow of water vapor at these points, and thus fast growth at the tips. Around the interior parts of the crystal, the supersaturation gradients are lower, resulting in lower water vapor flow and slower growth. The result can be seen in movies of growing snow crystals (both in computer models and in laboratory observations), as the outer regions grow outward quickly, while the interior structures evolve more slowly.

Figure 3.4 shows another illustration of the depletion of water vapor around a growing snow crystal, this time in a laboratory setting. Water droplets condense on non-ice surfaces when the supersaturation is above σ_{water}, meaning that the humidity is above the dew point (Chapter 2). When humid air is blown down

FIGURE 3.3. This 2D numerical simulation demonstrates the depletion of water vapor around a snow crystal [2008Gra]. The supersaturation is constant (gray) far away from the crystal, but it drops to near zero (white) at the growing crystal surface. The supersaturation gradient produces a diffusion-driven inward flow of water vapor that continuously supplies material to the growing crystal. Image courtesy of Janko Gravner.

FIGURE 3.4. This laboratory-grown Plate-on-Pedestal (PoP) snowflake (see Chapter 9) indirectly shows the depletion of water vapor around a growing crystal. As moist air blows down onto the substrate supporting the crystal, a fog of tiny water droplets condenses onto its surface. But no droplets condense near the ice crystal, because the water vapor density is lower in that region. The boundary between these two regions shows where the humidity passes through the dew point (or, equivalently, the supersaturation passes through σ_{water}), which is when water droplets begin to condense.

on the growing crystal, a fog of water droplets condenses on the substrate around it, because $\sigma > \sigma_{water}$ in that outer region. But the supersaturation is depleted near the growing crystal, giving $\sigma < \sigma_{water}$ and no condensed droplets in the inner region. The boundary between the inner and outer regions reveals where $\sigma \approx \sigma_{water}$ near the substrate surface. While this laboratory image nicely illustrates water vapor depletion around a growing snow crystal, it would require a rather sophisticated 3D numerical simulation to accurately reproduce the $\sigma \approx \sigma_{water}$ contour.

Large-scale air flow can also transport and mix water vapor in air, and these flows operate in addition to diffusion. Wind and turbulence thus affect snow crystal growth, sometimes substantially, and I discuss this topic later in the chapter. These effects tend to be insignificant for small atmospheric snow crystals, however, as these simply travel with the wind as it blows, like specks of dust. As a result, ordinary particle diffusion is by far the most important transport process, as it operates even in still air. To a first approximation, therefore, we can ignore large-scale air flow and focus our attention on understanding basic particle diffusion and how it affects snow crystal growth.

The process of diffusion is defined mathematically by the classical diffusion equation. Unlike attachment kinetics, where our comprehension of the underlying molecular dynamics is still an area of active scientific research (see Chapter 4), the essential physics of diffusion is extremely well understood, and has been for over a century. Calculating the effects of diffusion in complex geometries (for example, surrounding a branched snow crystal) remains a challenging computational problem, but at least we have a firm grasp of the underlying physics.

Our overarching goals in this chapter are both qualitative and quantitative. The first is to describe the phenomenology of diffusion-limited growth as it pertains to the specific case of snow crystal formation. Phenomenological descriptions are not always the best way to understand the underlying science, as they can involve rough approximations and empirical descriptions. Our brains, however, are very much attuned to visual inputs, such as graphs, sketches, and photographs, and less so to mathematical formulas. Therefore, a well-crafted phenomenological description is not without pedagogical value. Moreover, I have always found it useful to develop a basic mental picture of a physical phenomenon under study, unfettered by words or equations, as this often provides a helpful intuitive grasp of the subject.

The second goal is to develop the quantitative side of diffusion-limited growth, writing down the relevant equations and then outlining the techniques used to solve them. This mathematical background will be required when we examine computational modeling, plus careful diffusion modeling can be quite important in designing and interpreting laboratory observations of snow crystal growth. Throughout this discussion, our general game plan is a straightforward application of modern reductionist science: 1) break down a complex phenomenon (snow crystal formation) into its simplest component pieces, including the physics of diffusion; 2) study and understand those pieces as best we can by using realistic physical models informed by precise laboratory measurements; and 3) reassemble the pieces into a computer simulation that recreates the original phenomenon and (hopefully) agrees with quantitative observations. As I have been striving to execute this plan over the years, I have found that both the quantitative and qualitative perspectives are valuable to fully understand and appreciate the science of snow crystal formation, so I try to address both in this chapter.

FACETING AND BRANCHING

A good starting point for the discussion is to examine how particle diffusion affects the transition from a faceted to a branched snow crystal. Faceting is driven primarily by attachment kinetics, which I describe in detail in Chapter 4, while branching is driven by diffusion-limited growth. We can begin to see the interplay of faceting and branching by examining the growth of the simple faceted crystal shown in Figure 3.5. Because of the depletion effect outlined above, the supersaturation will generally decrease as one approaches the crystal surface. It will equal some constant value σ_∞ far away, reducing to lower values σ_{surf} at the ice surface, where σ_{surf} depends on position around the crystal. Because the six corners of the hexagonal crystal stick out farther into the humid air, σ_{surf} is slightly higher at the corners compared to σ_{surf} near the facet centers. This phenomenon is a basic outcome of the particle diffusion process and is sometimes called the *Berg effect* [1938Ber].

Although the faceted surfaces appear to be perfectly flat, consider what happens if this is initially the case. By virtue of the Berg effect, the higher supersaturation at the hexagonal corners causes the corners to grow more rapidly than the facet centers. Soon the faceted surfaces are no longer perfectly flat but slightly concave at the molecular level. New molecular terraces nucleate preferentially near the corners, where the supersaturation is highest, and the steps subsequently grow inward toward the facet centers. Moreover, because the supersaturation is highest near the corners, the terrace steps move fastest there, slowing down as they approach the facet centers. This change in step velocity causes the terrace steps to bunch up near the facet centers, as shown in the sketch.

As the slightly concave facet shape develops, however, it changes how readily water molecules attach to the surface. Terrace steps typically absorb water vapor molecules more readily than do faceted surfaces with no steps, so the increased step density near the facet centers results in faster growth there. This is quantified by the attachment coefficient α, which is essentially the probability that an impinging water vapor molecule will attach to a surface (see Chapter 4). The value of α_{center} increases as the surface becomes more concave, while α_{corner} remains essentially unchanged at the faceted corners. Eventually this process of nucleation and motion of mo-

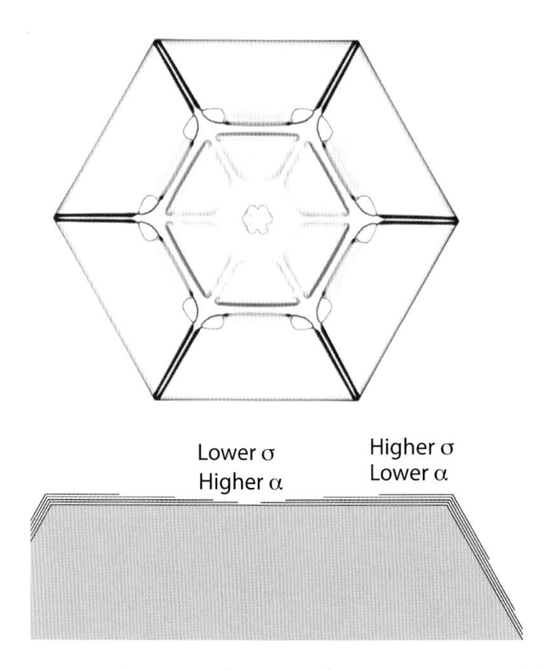

FIGURE 3.5. The prism facets on this PoP snow crystal (top photo) look straight, but they must be slightly concave at the molecular scale (bottom sketch). The hexagonal corners stick out farther into the humid air, so σ is slightly higher at the corners than at the facet centers. At the same time, the density of molecular terrace steps is higher nearer the facet centers, making the attachment coefficient α slightly higher there (the surface lines in the sketch represent molecular layers). Because the growth rate is proportional to $\alpha\sigma$, the entire facet surface grows outward at a constant rate, maintaining its flat appearance.

lecular steps results in stable, self-regulating facet growth. There is a negative feedback that maintains the precise concave shape needed to keep the growth velocity constant across the entire facet surface. If the surface became too flat, the corners grow a bit faster and increase the surface curvature. If the surface became too concave, the facet center regions grow a bit faster and again restore the correct concave shape. There is a stable point in the cur vature that is determined by the supersaturation, crystal size, attachment kinetics, and perhaps other parameters. This process is one of the simplest examples of spontaneous structure formation, as the rules of crystal growth bring about the sustained, stable growth of a faceted crystal with slightly concave faces.

Because molecular steps are extremely small, only a small amount of surface curvature is usually needed to produce a stably growing facet. Thus, the concave facet shape is normally imperceptible. In a similar vein, one cannot easily observe the supersaturation variations around the crystal, as water vapor is an invisible gas. When you look at faceted snow crystals with the naked eye, with a magnifier, or even with a high-powered microscope, you often see that the faceted surfaces appear as smooth and flat as a pane of glass.

Transition to Branching

This picture of facet formation takes on a new twist as the growth rate increases, or as the crystal grows larger. In either of these cases, the faceting mechanism continues working only until $\alpha \approx 1$ at the facet centers, meaning that all available water vapor molecules are immediately assimilated onto the surface. Once this happens, α can no longer increase, which causes the self-regulating process described above to break down. At some point, the facet centers will no longer be able to keep pace with the corners, and the facet will no longer maintain its flat appearance. When this happens, branches sprout from the hexagonal corners, as illustrated in Figure 3.6.

This transition from faceted to branched growth tends to be rather abrupt. Once the hexagonal tips sprout branches, they quickly grow outward and leave the regions between the branches far behind. Exactly when the transition occurs depends on several factors. A general rule of thumb is that faceting dominates when 1) crystals are small, 2) the degree of anisotropy in the attachment kinetics is large, and 3) the growth is slow. When these three criteria no longer describe a growing crystal, branching becomes more likely. For example, if α_{facet} is just below unity, then the facet stability is weak, as there can be little difference between the value of α at the corners and at the facet centers. In this case, branches form readily and will sprout from quite small crystals even when they are growing slowly. In contrast, if $\alpha_{facet} \ll 1$, then faceted growth with be highly stable, and a crystal

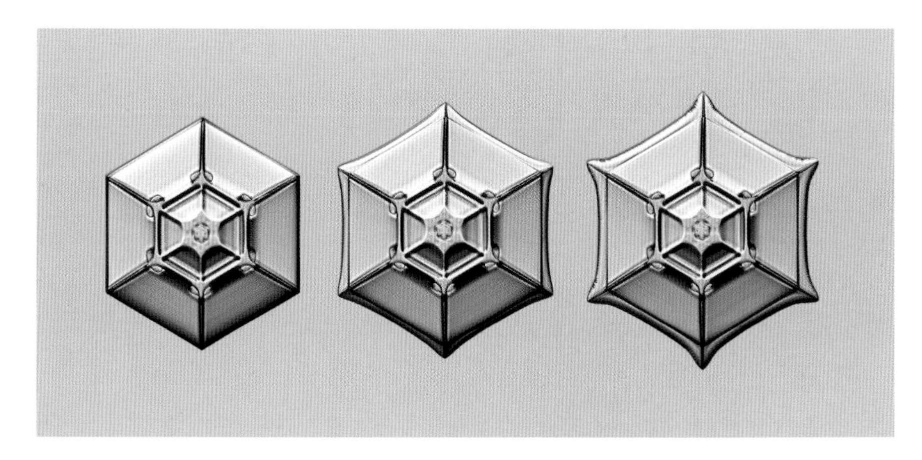

FIGURE 3.6. This series of photographs shows branches sprouting from the six corners of a hexagonal snow crystal, which happened when the applied supersaturation was increased. Branching like this becomes more likely when a hexagonal crystal is large and/or its growth rate is fast.

may grow quite large before branches sprout. Just to make things interesting, the value of α_{facet} often depends strongly on the surface supersaturation σ_{surf}, which in turn depends on the size of the crystal, the functional form of $\alpha_{facet}(\sigma_{surf})$, and the far-away supersaturation σ_{∞}. Because of all these contingencies, determining the exact point at which branches appear becomes a nontrivial problem best left to computational modeling.

The formation of a sixfold symmetrical branched snow crystal results partly because faceting is more stable on smaller crystals. When a nascent snow crystal begins its existence, it is small and tends to grow into a faceted hexagonal prism. As it grows larger, diffusion becomes a larger factor, until at some point a set of six branches will sprout simultaneously from the six corners of the hexagon, as seen in Figure 3.6. When you see a large branched snow crystal with sixfold symmetry, it is almost certain that the primary branches first sprouted in unison from an initially faceted ice crystal.

The Mullins-Sekerka Instability

The transition from stable faceted growth to branching is just one example of a more general phenomenon in diffusion-limited growth called the *Mullins-Sekerka instability*, named after its discoverers William Mullins and Robert Sekerka [1963Mul, 1964Mul]. In some contexts, I also call it the *branching instability*, because this is a simpler and more descriptive moniker. The hallmark of any instability is positive feedback, and Figure 3.7 illustrates this for the special case of an initially flat ice surface with $\alpha \approx 1$. If a small bump randomly appears on the surface, then the top of the bump sticks out slightly into the humid air above it. As a result, slightly more water vapor in the air diffuses to the top of the bump than diffuses to the flat surface around it. With this slight enhancement in material transported to it, the top of the bump grows slightly faster than its surroundings, and so it grows taller. Soon the bump sticks out even farther than it did before, causing it to grow even faster, which makes it stick out still farther, and so on. In this way, positive feedback yields a *growth instability*. The Mullins-Sekerka instability applies to many physical systems when growth is limited by diffusion.

Interestingly, the uniform surface growth shown in the top panel in Figure 3.7 is a perfectly valid solution to the diffusion equation; it is just not a stable solution. Any deviation from perfect flatness, no matter how small, will grow larger. Thus, the Mullins-Sekerka instability will eventually turn a simple growing structure into a complex, branched structure with an ever-changing morphology. When you get to the heart of the matter, this growth instability is one of the main reasons that the simple process of water vapor freezing into ice

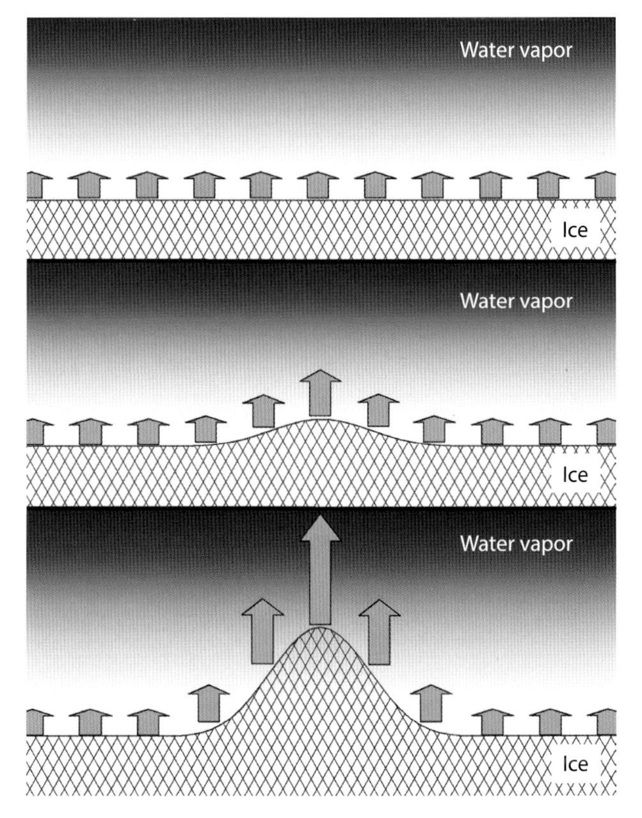

FIGURE 3.7. The diffusion-limited growth of an initially flat surface with $\alpha \approx 1$ (top sketch) is susceptible to the Mullins-Sekerka instability, also known as the branching instability. If a small bump appears on the surface (center), it will stick out farther into the supersaturated medium, so the top of the bump will grow slightly faster than the surrounding flat surface. This initiates a positive feedback effect, causing the bump to become larger, increasing its relative growth rate even more (bottom). This illustration assumes zero anisotropy in the attachment kinetics ($\alpha \approx 1$), which eliminates the possibility of faceting and thus emphasizes the Mullins-Sekerka instability.

creates the beautiful, complex snow crystal patterns we observe falling from winter clouds.

When anisotropic attachment kinetics are present in addition to diffusion-limited growth, we see that faceting provides a stabilizing influence that initially inhibits the Mullins-Sekerka instability. The faceting process described above provides a negative feedback that reduces perturbations and maintains the ever-so-slightly concave shape of the surface. If there is no anisotropy in the attachment kinetics (for example, when $\alpha \approx 1$), then the surface is always susceptible to the Mullins-Sekerka instability. In the opposite extreme, for a large anisotropy in the attachment kinetics ($\alpha_{facet} \ll 1$), faceted growth will continue for quite some time. In general, the growth behavior of a snow crystal is determined by a combination of branching and faceting, with the details depending on the crystal size, growth rate, attachment kinetics, and other factors. Faceting dominates in some regions of parameter space, while branching dominates in others. The complex interplay of the processes of faceting and branching is what gives snow crystal growth its especially rich phenomenology.

The Mullins-Sekerka instability is also well known for its repeated application in the formation of elaborately branched dendritic structures. Once a branch sprouts and grows outward, perturbations on its surfaces may again become amplified, thereby sprouting additional sidebranches, as illustrated in Figure 3.8. In principle, this could lead to sidebranches on the sidebranches, and so on, yielding quite intricate structures. Dendritic snow crystals forming near −15°C are often characterized by a set of six primary branches that are decorated with copious sidebranches, as shown in Figure 3.9. Side-sidebranches are sometimes seen, although they are somewhat rare. More common is a mixture of both faceting and branching behaviors on a single crystal, producing the endless morphological variations we associate with snowflakes.

The branching instability can become quite chaotic when it is driven hard, and Figure 3.9 illustrates the resulting unruly behavior in the snow crystal world. When this crystal was small, the six primary branches must have simultaneously sprouted from the corners of a tiny hexagonal prism, as the symmetry of these branches reflects the initial faceted order of the crystal. In contrast, the sidebranches arose from random perturbations that occurred near the primary branch tips. Because the supersaturation was quite high around this crystal, the sidebranching events were somewhat random in nature, yielding a

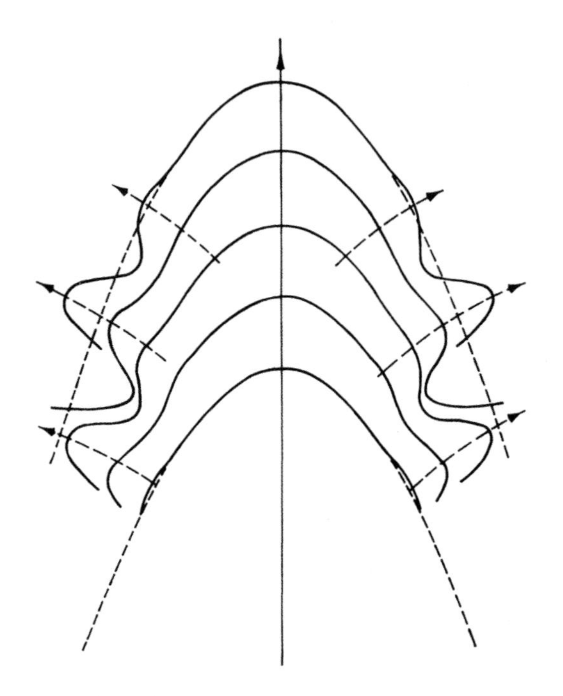

FIGURE 3.8. A schematic diagram showing the creation of dendritic sidebranches resulting from repeated application of the Mullins-Sekerka instability. Random perturbations on the sides of the growing tip typically yield somewhat erratically placed sidebranches. However, a characteristic length scale is often involved in the formation of these perturbations, resulting in a minimum spacing between adjacent sidebranches. Image adapted from [1980Lan].

disorganized arrangement of sidebranches, such that even the two sides of a single primary branch exhibit quite different structures. Thus, the detailed sixfold symmetry in this crystal is quite poor, even though the six branches have similar overall shapes. While such chaotic sidebranching is common, it is also possible to stimulate symmetrical sidebranches with not-so-random perturbations, and I discuss this phenomenon later in the chapter.

FREE DENDRITES

An individual branch like the one in Figure 3.10 is often called a "free" dendrite, because it is a self-assembling structure that can be considered free from the constraints of container walls or competing crystals. The branch grows out into open space with a uniform far-away supersaturation σ_∞, and the overall tip morphology of a free dendrite is essentially independent of time. If you photographed the near-tip structure at different times, you would find that the photos all looked about the same. The detailed placement of the sidebranches with respect to the tip is always changing, but the overall morphology remains constant. Also, the initial origin of the dendritic branch is largely unimportant; once it becomes fully developed, the branch automatically assumes its characteristic shape and properties. Fernlike stellar dendrites, like the one shown in Figure 3.9, are basically made of six free-dendrite branches that grew out from the six corners of a small hexagonal plate.

Dendritic structures are quite ubiquitous in solidification systems, reflecting the universal nature of the Mullins-Sekerka instability. When diffusion-limited growth is driven hard enough, branching invariably results. Particle diffusion is the main limiting factor in snow crystal growth, while heat diffusion can yield free dendrites when ice freezes from liquid water (as shown later in the chapter). Countless other solidification systems exhibit similar behaviors, and there is a large scientific literature on the general physics of dendrite formation during solidification [1989Lan, 1993Cro]. Besides diffusion, surface energy effects and the attachment kinetics both guide the overall development of these structures, as I describe further below. Examining the universal nature of this rich field of study is beyond the scope of this book, however, so my focus here will be to apply what is known about dendrite formation to the specific problem at hand.

In the world of snow crystals, fernlike stellar crystals are a common and easily recognizable form of dendritic growth. They grow readily at temperatures near −15°C when the supersaturation is high (as indicated in the Nakaya diagram), and they exhibit several characteristic traits:

1) The branched morphology is mainly confined to a flat plane because of strong basal faceting.

FIGURE 3.9. Both branching and faceting play large roles in fernlike stellar dendrites like this one. The branching instability is clearly responsible for the copious sidebranching and for the largely random sidebranch placement. However, this crystal began its life as a faceted hexagonal prism, because the six primary branches must have spouted from the small prism's six corners. Moreover, the crystal is thin and flat, indicating strong basal faceting even in the presence of highly developed dendritic branching. The complex interplay of faceting and branching (attachment kinetics and diffusion-limited growth) causes the morphological richness seen in snow crystal formation.

FIGURE 3.10. A fernlike dendrite growing out from the tip of a wire substrate at a temperature of –15°C. The complex branched structure is mostly confined to a thin plane by slow basal growth.

2) The tip of each dendritic branch grows outward with a constant tip velocity v_{tip} that increases approximately linearly with the far-away supersaturation, so $v_{tip} \sim \sigma_\infty$.
3) The radius of curvature of the tip is equal to about $R_{tip} \approx 1$ micron, a value that does not change with time and is roughly independent of supersaturation.
4) Each distinct sidebranch grows out at an angle of 60 degrees relative to the primary branch.
5) New sidebranches typically spout at a characteristic distance from the tip that is roughly several times R_{tip}.
6) The sidebranch spacing is generally erratic, with little correlation between branches or on either side of a primary branch.

I delve a bit more into the mathematical aspects of dendrite formation in the section on solvability theory later in this chapter, as this theory explains some aspects of dendrite growth near the tip. However, computational models have not yet advanced to the point where they can reproduce fernlike stellar dendrite snow crystals, so many of their detailed properties remain unexplained, including the sidebranch spacing, the distinctive "backbone" construction, and how the attachment kinetics guide the overall morphology. Although fernlike stellar dendrites are easy to find in nature and straightforward to create in the laboratory, developing a comprehensive theory of their construction remains a significant challenge.

Fishbone Dendrites Near –5°C

While fernlike dendrites near −15°C are something of a canonical snow crystal form, other dendritic structures appearing at different temperatures are also worthy of attention. The "fishbone" dendrites shown in Figure 3.11 are especially pronounced, as they grow rapidly near −5°C and make up the "fishbone peak" often seen in snow crystal diffusion chambers (see Chapters 6 and 8). While they look quite different from fernlike dendrites, fishbones are also free dendrites with many of the same characteristic traits just described. However, the sidebranches are not conveniently confined to a nearly flat plane, and thus their structure is not so easy to convey using a single photograph. Moreover, the formation of fishbone dendrites requires high supersaturation levels that do not occur in nature, so they are entirely a laboratory creation [2009Lib1].

Fernlike dendrites near −15°C and fishbone dendrites near −5°C are the fastest growing and most distinctive cases of snow crystal dendritic structures. However, these are just two examples from the full spectrum of free dendrites that appear in the snow crystal morphology diagram, and additional photos of morphologies as a function of growth temperature are presented in Chapter 8. These show that the direction of dendrite tip growth depends on both temperature and supersaturation. Flat fernlike dendrites have v_{tip} aligned with the crystal a-axis, but typically the growth direction is not aligned with any specific axis, being determined by details of the attachment kinetics. In particular, the growth direction depends on the ratio of α_{basal} to α_{prism}, which depends on both temperature and supersaturation at the growing tip. Even at −15°C, fernlike dendrites exhibit growth out of the basal plane at high supersaturations. Also notable is that no snow crystal free dendrites grow along the c-axis, although e-needles (see Chapter 8) can be coaxed to grow in that direction using chemical vapor additives.

Sidebranch Competition

Another manifestation of the Mullins-Sekerka instability can be seen in the development of sidebranches after they sprout near a dendrite tip, as illustrated in Figures 3.9 and 3.10. As the individual sidebranches grow longer, each competes with its neighbors for available water vapor. If one branch becomes slightly

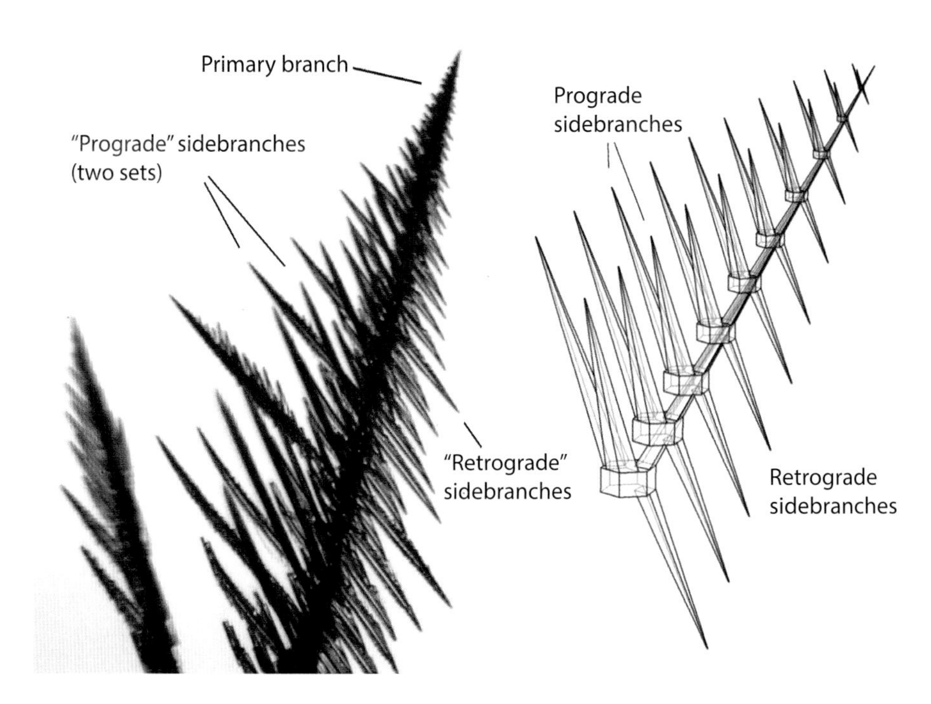

FIGURE 3.11. The sidebranch structure of a fishbone dendrite is more three dimensional than a fernlike dendrite. This figure compares a photograph with a sketch that shows the orientations of the different sidebranches with respect to the ice crystal axes (defined by the hexagonal prisms in the sketch). Unlike fernlike dendrites, fishbone dendrites are clearly not confined to a flat plane.

longer than others nearby, then it sticks out farther into the humid air and shields its neighbors. Diffusion brings the longer branch a greater supply of water vapor, so soon it shoots ahead, while its immediate neighbors are greatly stunted. The underlying physical effect is essentially the same as with the Mullins-Sekerka instability described above. Over time, this competition plays out on many length scales, so the spacing between the fastest growing sidebranches becomes ever larger, as a few players become dominant by appropriating available resources at the expense of the others. Diffusion-driven competition between neighboring structures is a common feature in snow crystal growth dynamics, and many examples can be found scattered throughout this book. In socioeconomic circles, a similar phenomenon is sometimes called the *Matthew effect*, from the biblical quote: "For unto every one that hath shall be given, and he shall have abundance: but from him that hath not shall be taken away even that which he hath" (Matthew 25:29).

Fractal Structure

Although the topic comes up frequently, I have not found that the concepts of fractal mathematics add much to our understanding of snow crystal formation and structure. Perhaps this view results from my perspective as an experimental physicist focusing on the materials science and crystal growth aspects of this problem. But my bias is reinforced by the failure of fractal mathematics to provide much predictive power when it comes to understanding snow crystal formation. Nevertheless, snow crystals do exhibit some fractal characteristics. The most apparent of these is a degree of self-similarity in the formation of dendritic structures, as illustrated in Figures 3.12 and 3.13. Primary branches yield sidebranches, and these can yield side-sidebranches, and so forth. If a dendrite sidebranch develops sufficiently, its overall structure will be indistinguishable from the central branch from which it arose. And the same would be true of side-sidebranches if they mature to the same extent.

Observations generally reveal that the degree of self-similarity seen in snow crystal structure is relatively minor. Moreover, the concept of self-similarity does not provide a physics-based explanation of the formation of the dendritic structure in the first place; that requires the Mullins-Sekerka instability. Explaining snow crystal structure necessitates a broad understanding of diffusion-limited growth, attachment kinetics, and ultimately the molecular dynamics of the ice crystal surface. Saying that a snowflake has some self-similar fractal characteristics is an accurate description, but, by itself, this description does not provide many useful insights that allow one to comprehend the underlying physical phenomena.

A Brief Classification of Solidification Systems

When surveying the scientific literature, one soon finds that there is a large body of work on dendrite structure formation and on the physics of solidification more generally. Some of this work applies readily to the problem of snow crystal growth, but much of it does not. Some aspects of the underlying physics are quite universal and can be described by a general mathematical formalism. But other aspects can be quite dependent on the material being studied, requiring different mathematical approaches for different systems. To avoid confusion regarding which physical effects are important and which can be neglected in snow crystal growth, it is useful to list the different types of systems in which dendritic structures arise in diffusion-limited crystal growth.

Unfaceted Solidification from the Melt. Most of the scientific literature on dendritic solidification can be found in this category, as it has important applications in metallurgical and semiconductor systems. Succinonitrile and pivalic acid are two oft-studied materials [2004Gli , 2012Mel], popular because they are easy to work with and are considered good proxies for simple metals; Figure 3.14 shows a typical example of an unfaceted free dendrite. The dominant physical effects that need to be considered in this category include, in order of importance:

1) Thermal diffusion. Removing latent heat is a major consideration in solidification from the melt, so the thermal diffusion equation is of paramount importance. Dendritic structures in this system arise mainly from thermal-diffusion-limited growth.
2) Surface energy. Although less important than thermal diffusion, this physical effect sets the scale for

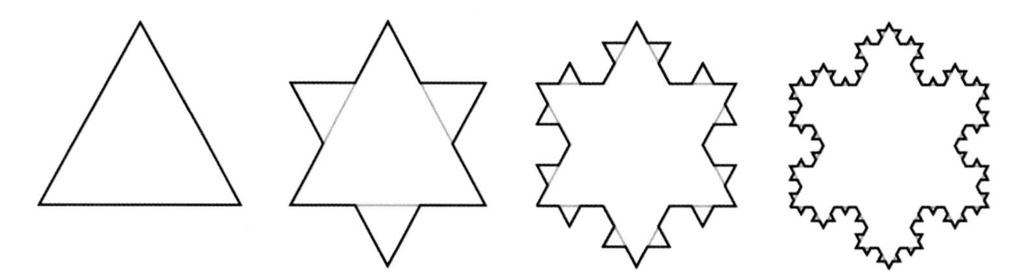

FIGURE 3.12. The construction of geometrical curve known as the *Koch snowflake*, first described by Swedish mathematician Helge von Koch in 1904. It is one of the earliest known examples of a *fractal structure*. As ever-smaller triangular sidebranches are attached ad infinitum, the area of the Koch snowflake converges to 8/5 times the area of the original triangle, while its perimeter length diverges to infinity. Consequently, the Koch snowflake has a finite area bounded by an infinitely long perimeter. Although this fractal structure bears some resemblance to a stellar dendrite snow crystal, there is little real connection between fractal mathematics and the physics of snow crystal formation.

FIGURE 3.13. A close-up of one section of an exceptionally large fernlike stellar dendrite. The crystal exhibits a somewhat self-similar fractal structure with side-branches begetting side-sidebranches, and even a few side-side-sidebranches.

R_{tip} and therefore the overall structure of a free dendrite. Together, thermal diffusion and surface energy effects define the primary features seen in metallurgical solidification.

3) Anisotropic surface energy. As I discuss later in this chapter, stable dendrites require some anisotropy in the surface boundary conditions, and here that is provided by an anisotropic surface energy.
4) Attachment kinetics. This is often neglected entirely, as the attachment kinetics are so fast that they do not limit growth significantly. This fast-kinetics approximation appears to provide a good description of solidification systems that exhibit no faceting, which is true for many metals and metal analogs.
5) Particle diffusion. This process is also usually neglected in solidification from the melt, because liquid is always present at the solidification interface. Particle diffusion may be a factor in multi-component systems, but not in growth from pure melt.

Faceted Solidification from Vapor. Snow crystal formation stands out as perhaps the most studied example of solidification of a high-vapor-pressure system. Materials science, usually classified as a branch of engineering, has little interest in materials that readily evaporate away, so over the years, the solidification of high-vapor-pressure materials has received less attention than metals. Listing the dominant physical effects for the snow crystal case, in order of importance, we obtain:

1) Particle diffusion. In air, particle diffusion transports water vapor molecules to the ice surface, and this slow process greatly limits growth. Particle diffusion is responsible for branching and essentially all the complex structure seen in snow crystals.
2) Attachment kinetics. In the formation of snow crystal free dendrites, attachment kinetics set the scale for R_{tip} and the overall branched structure, as I describe in detail later in this chapter. Together, particle diffusion and attachment kinetics define the primary features seen in snow crystal growth.

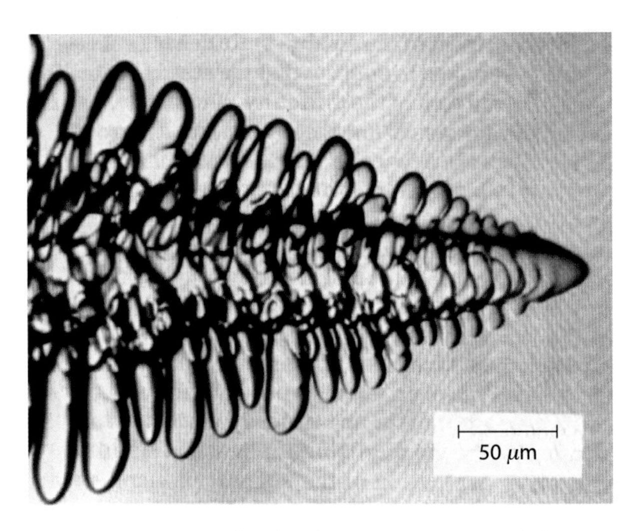

FIGURE 3.14. A free dendrite forming in succinonitrile from its melt. This transparent material is often used as a metal analog in studies of solidification, as it forms dendritic structures that are typical for solidification from the melt when attachment kinetics is not an important factor. Image adapted from [1976Gli].

3) Anisotropic attachment kinetics. Snow crystal attachment kinetics are generally highly anisotropic, which tends to yield strongly faceted structures.
4) Surface diffusion. This is nominally part of the attachment kinetics, but surface diffusion introduces nonlocal effects that are not included with a simple attachment coefficient. As discussed in Chapter 4, surface diffusion effects appear to be quite important in snow crystal growth.
5) Thermal diffusion. A minor effect compared to particle diffusion, thermal diffusion is often neglected. Its role increases close to 0°C, but this can often by approximated by a simple rescaling of σ_∞ (as discussed later in this chapter).
6) Surface energy. Almost negligible, because surface energy effects are dwarfed by similar effects arising from attachment kinetics. However, the surface energy is necessary in modeling to avoid unphysical results at exceptionally low supersaturations.
7) Anisotropic surface energy. This effect is negligible. Surface energy effects are small to begin with, plus the surface energy anisotropy in ice is quite small. Any residual effects are dwarfed by similar effects from anisotropic attachment kinetics.

Chemical Vapor Deposition. So much work has been done with chemical vapor deposition (CVD) that it deserves a separate listing, although in principle it could be included in other categories. The primary focus in CVD systems has been on technological applications, so these materials almost always exhibit low vapor pressures. Theoretical descriptions of CVD often make an implicit assumption of zero vapor pressure from the outset, which greatly simplifies the theory but also immediately changes the underlying physics compared to high-vapor-pressure systems. Thus, the vast literature on CVD systems often has remarkably little (although certainly not zero) direct application to snow crystal growth.

Unfaceted Solidification from Vapor. To my knowledge (quite limited in this case), the formation of unfaceted free dendrites from vapor has received little scientific attention. I performed a few experiments using carbon tetrabromide, as this seemed to be a convenient test system, but little came out of those observations. Important physical effects could include all the items mentioned above (particle diffusion, heat diffusion, surface energy, and attachment kinetics), all to varying degrees depending on the specific material under consideration. With few practical applications, substantial experimental challenges, and especially complex input physics, it is perhaps little surprise that vapor solidification of unfaceted free dendrites has not been a popular research topic.

Faceted Solidification from the Melt. Ice growth from liquid water falls into this category, as is clearly indicated by the presence of strong basal faceting (Figure 3.15). The dominant physical processes are the same as with unfaceted solidification from the melt, except now one must include effects from both anisotropic attachment kinetics and anisotropic surface energy, as possibly neither is

FIGURE 3.15. This ice crystal grew from liquid water on the surface of a quiet pond when the overnight temperature dipped slightly below 0°C. Attachment kinetics limited the basal growth, resulting in a thin plate of ice, while thermal diffusion brought about dendritic branching. The underlying sixfold symmetry of the ice crystal lattice guided the orientations of the branches and sidebranches, as it does with snow crystal growth. Photo courtesy of Bathsheba Grossman.

negligible compared with the other. Unfortunately, this complicates matters substantially and creates a full plate on the theory side, so again this has not been a popular research topic. The ice/water system is a good example of the current situation in this category; basal faceting clearly plays an important role, but surprisingly little is known about the attachment kinetics at the ice/water interface. Several ice/water solidification studies ignore attachment kinetics entirely, which seems unwise considering the clear occurrence of faceting. Given the ease of creating and studying ice structures from liquid water, and the many recent advances in numerical modeling, this topic seems ripe for additional experimental research.

DIFFUSION IN SNOW CRYSTAL GROWTH

Now that we have examined a few of the more prominent morphological effects of diffusion on snow crystal growth, it is time to delve into the underlying mathematics. Most textbooks on mathematical physics derive the diffusion equation and examine its solution, and I will assume that the reader already has a basic familiarity with diffusion physics. Therefore, my focus in the following discussion is on describing how to apply the diffusion equation to the specific problem at hand, examining physical concepts and specific model systems.

Particle Diffusion

We begin with the particle diffusion equation that describes the transport of water molecules through the air

$$\frac{\partial c}{\partial t} = D_{air} \nabla^2 c, \tag{3.1}$$

where $c(\vec{x})$ is the number density of water molecules, D_{air} is the diffusion constant for water molecules in air, and $\vec{x}$ is the position vector. For typical atmospheric conditions, $D_{air} \approx 2 \times 10^{-5}$ m^2/sec.

If the temperature is equal to a fixed value everywhere (the isothermal approximation), then Equation 3.1 can be rewritten in terms of the supersaturation as

$$\frac{\partial \sigma}{\partial t} = D_{air} \nabla^2 \sigma, \tag{3.2}$$

where $\sigma(\vec{x})$ is defined by

$$\sigma(\vec{x}) = \frac{c(\vec{x}) - c_{sat}}{c_{sat}} \tag{3.3}$$

and c_{sat} is the saturated water vapor density, equal to c above a flat ice surface in equilibrium with the vapor phase. Here we assumed that c_{sat} is a constant independent of $\vec{x}$, which is true in the isothermal approximation. Because the values of both c and σ vary with position around a growing crystal, I often refer to the *fields* $c(\vec{x})$ and $\sigma(\vec{x})$.

In addition to the isothermal approximation, we can also employ a quasi-static approximation that reduces Equation 3.2 to Laplace's equation

$$\nabla^2 \sigma = 0. \tag{3.4}$$

To see why this is a good approximation, consider suddenly placing a snow crystal into a uniform body of preexisting supersaturated air. The crystal will begin growing immediately and thereby create around it a region somewhat depleted of water vapor. The size of this depleted "hole" in the water vapor density will be a few times larger than R, the size of the crystal, and its creation will take a time roughly equal to $\tau_{diffusion} \approx R^2/D$. (This is a well-known result from diffusion physics. To make the notation more compact, I often use D in place of D_{air}.) If we put in some typical numbers, taking $D = D_{air} \approx 2 \times 10^{-5}$ m^2/sec and $R \approx 1$mm, we obtain $\tau \approx 50$ msec.

Meanwhile, it takes a time $\tau_{growth} \approx 2R/v_n$ for a snow crystal to grow appreciably, where v_n is the growth velocity. The ratio of the relaxation time to the growth time is called the *Peclet number*, defined as

$$p_{Peclet} = Rv_n/2D, \tag{3.5}$$

and its value is typically less than 10^{-4} for a snow crystal growing in air. (In contrast, the Peclet number is usually *not* small for solidification from the melt, so the quasi-static approximation is not valid in those systems. Solidification from the vapor generally yields much smaller Peclet numbers than solidification from the melt.) The Laplace approximation works because the depleted region around a snow crystal adjusts itself almost instantaneously to changes in the crystal shape, and this is equivalent to a quasi-static approximation. Adopting Equation 3.4 from the outset affords a substantial simplification in the mathematics, allowing us to assume that the water vapor field surrounding a growing snow crystal is always in its completely relaxed state. This state changes as the crystal grows, but we need not worry about the relaxation process itself.

Boundary Conditions

To solve the diffusion equation, we also need to supply appropriate boundary conditions. These are nontrivial for snow crystal growth, so we need to consider them with some care.

Faraway Boundary. One commonly used boundary condition is to assume that the supersaturation is equal to some fixed value σ_∞ far from the growing crystal. The term "far" in this context usually means at a distance much larger than the size of the growing crystal in question. This boundary condition works well in three dimensions if the growing crystal is small in all three dimensions. It is possible, however, to apply this boundary condition incorrectly. Assuming a simple faraway boundary may not work with infinitely long cylinders, infinitely large walls, large dendritic structures, or other system geometries. We will encounter examples of such cases later in this chapter.

Mass Flux. To look at additional boundary conditions, we need to understand the flow of material in a diffusing system. Particle diffusion always has a net particle flux associated with a density gradient, in our case given by

$$F = D(\hat{n}\cdot\vec{\nabla}c) = Dc_{sat}(\hat{n}\cdot\vec{\nabla}\sigma). \quad (3.6)$$

(This is also a standard result from basic diffusion physics found in textbooks.) Note that the equals sign goes both ways. If there is a gradient in the water vapor density in air, it necessarily results in a flow of water vapor molecules given by F. Likewise, any net diffusive flow of water vapor molecules through air must be accompanied by a density gradient $\vec{\nabla}c$.

Mass Conservation. If a snow crystal is growing in air, then there must be a particle flux into the surface of the ice, as the flow of particles is what supplies the growth. Doing the math yields a surface boundary condition

$$v_n = \frac{c_{sat}D}{c_{ice}}\left(\frac{\partial\sigma}{\partial n}\right)_{surf}, \quad (3.7)$$

where v_n is the growth velocity of the crystal normal to the surface, and $(\partial\sigma/\partial n)_{surf}$ is the normal gradient of the supersaturation just above the ice surface. Combining this with the Hertz-Knudsen relation (see Chapter 4),

$$v_n = \alpha v_{kin}\sigma_{surf}, \quad (3.8)$$

then gives the surface boundary condition as

$$X_0\left(\frac{\partial\sigma}{\partial n}\right)_{surf} = \alpha\sigma_{surf}, \quad (3.9)$$

where

$$X_0 = \frac{c_{sat}}{c_{ice}}\frac{D}{v_{kin}}. \quad (3.10)$$

This is called a *mixed* boundary condition, because it involves both the value and gradient of σ at the surface.

In some circumstances, it is reasonable to just assume $\sigma_{surf} \approx 0$ at the surface of a growing snow crystal, and one occasionally sees this assumption in the literature. But it is an oversimplification that often obscures interesting aspects of snow crystal growth. Assuming $\sigma_{surf} = 0$ on a growing crystal can never be absolutely accurate, as Equation 3.8 would then imply zero growth. Quantifying this discussion is important, and I defer that topic to the section below on spherical solutions. Equation 3.9 is the usual boundary condition needed at the surface of a growing snow crystal.

Ice-Free Walls. If particles cannot flow into or out of an ice-free wall in an experimental chamber, then zero net particle flux implies zero density gradient at the wall. Thus the boundary condition at an ice-free wall is given by

$$\left(\frac{\partial\sigma}{\partial n}\right)_{wall} = 0, \quad (3.11)$$

where $(\partial\sigma/\partial n)_{wall}$ is the gradient of the supersaturation in the direction of the surface normal.

Ice-Covered Walls. In many experimental situations, a boundary might consist of an ice-covered surface at some temperature T. Assuming the ice is neither growing nor sublimating appreciably, the vapor pressure will equal the equilibrium value, $c \approx c_{sat}(T)$, at the ice-covered surface. For an isothermal environment, this means $\sigma \approx 0$ at the ice surface. Even here, however, we must be a bit careful with this boundary condition. For the isothermal case, the surface boundary condition is given more accurately by

$$\sigma_{surf} \approx \frac{v_n}{\alpha v_{kin}}, \quad (3.12)$$

where v is the growth velocity of the ice on the surface. This can be close to zero for a large ice-covered reservoir wall, but it is often *not* a good assumption to take $\sigma \approx 0$ at the surface of a small, isolated ice crystal. In any case, the only time σ_{surf} is exactly zero is in equilibrium, when the growth velocity is also zero, as shown in Equation 3.8. If the temperature varies in an experimental system, then we must be careful about the definition of σ itself, as c_{sat} is temperature dependent. In this case, the boundary condition is best left as $c \approx c_{sat}(T)$ at the surface of a large, ice-covered wall, as written above.

Heat Diffusion

The solidification of water molecules at a growing snow crystal surface releases latent heat that increases the surface temperature and thus slows growth. The temperature rise is countered by the diffusion of heat away from the surface through the surrounding air, producing another type of diffusion-limited growth. Heat diffusion is less important than particle diffusion in snow crystal growth, so it is rightfully ignored in most numerical models, at least for the time being. Nevertheless, the separate contributions of heat and particle diffusion have been observed at least once [2016Lib], so researchers will have to face the full dual-diffusion problem (particle plus heat diffusion) at some point in the future.

Heat diffusion is described by the thermal diffusion equation

$$\frac{\partial T}{\partial t} = D_{therm} \nabla^2 T, \tag{3.13}$$

where $T(\vec{x})$ is the temperature field surrounding the crystal, with

$$D_{therm} = \frac{\kappa_{air}}{\rho_{air} c_{p,air}}, \tag{3.14}$$

where κ_{air} is the thermal conductivity of air, ρ_{air} is the air density, and $c_{p,air}$ is the heat capacity of air. For typical atmospheric conditions, $D_{therm} \approx 2 \times 10^{-5}$ m^2/sec, and the fact that $D_{therm} \approx D_{air}$ reflects the universal nature of diffusion through ideal gases.

The quasi-static approximation applies for heat diffusion as it does for particle diffusion, giving

$$\nabla^2 T = 0 \tag{3.15}$$

to quite high accuracy. Equating the heat flux away from the crystal surface to the heat generated gives the surface boundary condition

$$\kappa_{air} \left(\frac{\partial T}{\partial n} \right)_{surf} = v_n \rho_{ice} L_{sv}. \tag{3.16}$$

With all the relevant diffusion equations and boundary conditions now in hand, we can proceed to examine their simplest analytic solutions.

THE SPHERICAL SOLUTION

The case of a spherical snowflake is remarkably useful for understanding the relevant physics of snow crystal growth. The problem is simple enough to be solvable analytically, yet its solution provides a great deal of intuition that can be applied to more challenging scenarios. The spherical solution is especially useful for examining the relative importance of different physical effects and for making suitable simplifying approximations in other analyses. Moreover, analytical solutions like the spherical case play an important role in testing quantitative computation models of snow crystal growth, verifying that the simulations reproduce known analytical results with acceptable accuracy. If you really want to understand the growth of snow crystals, with all their branching, faceting, and other complex structures and growth behaviors, I recommend starting your quest with the simplest possible example—the growth of a spherical ball of ice.

Kinetics Plus Diffusion

The spherical problem can be solved analytically and exactly, and I like to start with the minimum physics needed to describe the basic problem. Thus, let us begin by including particle-diffusion-limited growth and attachment kinetics with a constant α on the surface of the sphere. This addresses the heart of the problem without a lot of unnecessary complications. The solution of the diffusion equation gives

$$\sigma(r) = \sigma_\infty - \frac{R}{r}(\sigma_\infty - \sigma_{surf}), \qquad (3.17)$$

where

$$\sigma_{surf} = \frac{\alpha_{diff}}{\alpha + \alpha_{diff}} \sigma_\infty \qquad (3.18)$$

and

$$\alpha_{diff} = \frac{c_{sat}}{c_{ice}} \frac{D}{v_{kin}} \frac{1}{R} = \frac{X_0}{R}. \qquad (3.19)$$

The crystal growth velocity is then

$$v_n = \left(\frac{\alpha \alpha_{diff}}{\alpha + \alpha_{diff}} \right) v_{kin} \sigma_\infty. \qquad (3.20)$$

There are two limiting cases that deserve special attention.

Kinetics-Limited Growth

This limit applies when

$$\alpha \ll \alpha_{diff} \qquad (3.21)$$

and gives

$$v_n \approx \alpha v_{kin} \sigma_\infty \qquad (3.22)$$

$$\sigma_{surf} \approx \sigma_\infty \qquad (3.23)$$

As the name implies, kinetics-limited growth depends on α but is independent of D and X_0. Because X_0 is typically about 0.15 microns in normal air, α_{diff} is quite small for even a small natural snow crystal. For this reason, kinetics-limited growth usually applies in air only when α is extremely small. At low pressure, however, X_0 can be substantially larger, so kinetics-limited growth is more likely to apply in near-vacuum conditions.

Diffusion-Limited Growth

This limit applies when

$$\alpha_{diff} \ll \alpha, \qquad (3.24)$$

and gives

$$v_n \approx \frac{c_{sat} D}{c_{ice} R} \sigma_\infty \approx \frac{X_0}{R} v_{kin} \sigma_\infty, \qquad (3.25)$$

with

$$\sigma_{surf} \approx \frac{\alpha_{diff}}{\alpha} \sigma_\infty \approx \frac{X_0}{\alpha R} \sigma_\infty. \qquad (3.26)$$

Figure 3.16 shows some example solutions for $\sigma(r)$ using a constant R and several different values of α.

Looking beyond the spherical solution, the spherical analysis tells us that faceting becomes a dominant growth characteristic in the kinetics-limited regime, whereas branching tends to dominate in the diffusion-limited regime. Thus, tiny snow crystals (small R) tend to grow into simple faceted prisms, as do slow-growing crystals (small α). Crystals grown at low pressures (large D) often

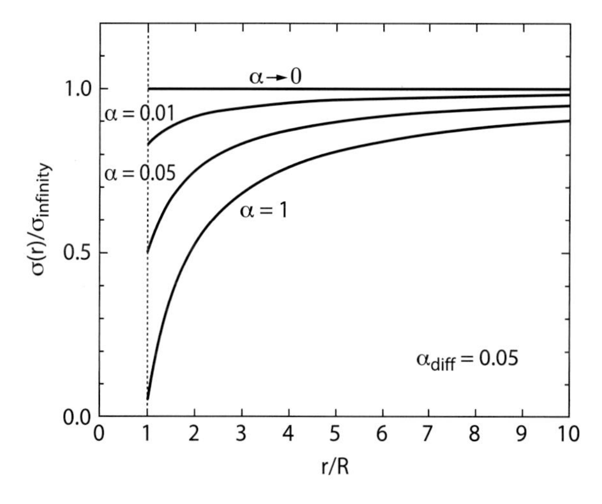

FIGURE 3.16. The solution to the diffusion equation for the growth of a spherical snow crystal with $\alpha_{diff} = 0.05$. When $\alpha \to 0$, the growth is kinetics limited, and $\sigma_{surf} \approx \sigma_\infty$. As α increases, the growth becomes more diffusion limited, and σ_{surf} decreases.

grow as simple prisms for the same reason. Conversely, branching tends to dominate over faceting in the diffusion-limited regime. Although the spherical solution is of little use for describing the detailed formation of complex snow crystals, it is invaluable for understanding different limiting behaviors.

The spherical solution also tells us that that v_n is independent of α in the diffusion-limited regime, but σ_{surf} is not. We also see that σ_{surf} generally becomes smaller as R becomes larger. However, σ_{surf} never reduces fully to zero for a growing crystal, as zero supersaturation would be equivalent to a zero-growth equilibrium state. It is also worth noting that even though the spherical solution is a perfectly correct and accurate solution to the diffusion equation, in real life it is not a stable solution. Diffusion-limited spherical growth is subject to the Mullins-Sekerka instability, eventually producing dendritic structures.

Kinetics, Diffusion, and Heating

When latent heating is included in the spherical problem, we must then simultaneously solve both the heat and particle diffusion equations, which is a substantially more difficult problem. Notably, the isothermal approximation clearly no longer holds, so c_{sat} is not a simple constant, and one must be quite careful with the definition of the supersaturation field $\sigma(\vec{x})$. The mathematics is straightforward but a bit tedious [2005Lib], yielding a result that can be written in the same basic form as Equation 3.20:

$$v_n = \left(\frac{\alpha \alpha_{diff,heat}}{\alpha + \alpha_{diff,heat}} \right) v_{kin} \sigma_\infty, \tag{3.27}$$

where

$$\alpha_{diff,heat} = \frac{X_0}{R} \frac{1}{1+\chi_0} \tag{3.28}$$

and χ_0 is a dimensionless parameter

$$\chi_0 = \frac{\eta D L_{sv} \rho_{ice}}{\kappa_{air}} \frac{c_{sat}}{c_{ice}}, \tag{3.29}$$

with all the variables in Equation 3.29 evaluated at T_∞. Values of χ_0 as a function of temperature are given in Table 2.1 in Chapter 2. In addition, the surface temperature of the growing spherical crystal is given by

$$\Delta T = \frac{1}{\eta} \frac{\alpha}{\alpha + \alpha_{diff,heat}} \frac{\chi_0}{1+\chi_0} \sigma_\infty, \tag{3.30}$$

where $\Delta T = T_{surf} - T_\infty$. If $\alpha_{diff} \ll \alpha$, so the growth is purely diffusion limited (which gives the maximum ΔT), then Equation 3.30 reduces to

$$\Delta T \approx \frac{1}{\eta} \frac{\chi_0}{1+\chi_0} \sigma_\infty, \tag{3.31}$$

which, interestingly, is independent of the crystal radius R.

A first takeaway message from this analysis is that heat diffusion plays a somewhat minor role in snow crystal growth compared to particle diffusion. The relevant variable χ_0 equals about 0.8 at −1°C, drops to about 0.4 at −10°C, and it continues falling with colder temperatures. Moreover, if the growth is mainly diffusion limited, then Equation 3.27 becomes

$$v_n \approx \alpha_{diff} v_{kin} \frac{\sigma_\infty}{1+\chi_0}, \tag{3.32}$$

which means that the main effect of heating can be incorporated into a simple rescaling of σ_∞.

This is a significant result; once we can create realistic computer models of snow crystal growth incorporating only particle diffusion and attachment kinetics, then adding heat diffusion can be done to a reasonably good approximation simply by taking $\sigma_\infty \to \sigma_\infty/(1+\chi_0)$ in the same models. The takeaway message is that we should probably ignore heating (in atmospheric snow crystal growth) until we first solve the problem including just particle diffusion and attachment kinetics. One step at a time.

Qualitatively, we can understand the heating effects from the underlying physics. Deposition generates latent heat, which warms the growing snow crystal until a balance is reached, when the heat carried away by diffusion equals that generated. The increased crystal temperature then lowers the effective supersaturation by changing c_{sat} at the surface. Moreover, the heat conductivity of ice is much higher than that of air, so the whole snow crystal heats nearly uniformly. When all this is considered in the spherical solution, we see that thermal diffusion effects, while not always negligible, are not nearly as important as particle diffusion and attachment kinetics for snow crystal growth in air.

Experimental Verification

Although diffusion theory is well understood, it is nevertheless good to see an experimental verification, if for no other reason than to obtain a "reality check" to make sure one is on the correct theoretical track. Producing a suitable experiment is nontrivial, however, as spherical growth is generally unstable to the Mullins-Sekerka instability, plus just getting to an interesting region in parameter space is not a simple task.

I was able to validate the particle + heat diffusion model using measurements of the growth of long ice needles [2016Lib], and the results are shown in Figure 3.17. Although needles are certainly not spheres, the mathematics of cylindrical growth is nearly identical to that of spherical growth, as I describe later in this chapter. Moreover, slightly tapered cylinders have the desirable property that α is large enough to make $\alpha_{diff} \ll \alpha$ valid, and so the growth is mainly diffusion limited. Thus, with (almost) no adjustable parameters, experiment and theory were found to agree nicely. These data confirm the relative roles of particle and heat diffusion, demonstrating that the net effect of heating is greater near the melting point, reflecting the dependence of χ_0 on temperature. To my knowledge, this is the first and only experimental result demonstrating that snow crystal growth in air is indeed limited by a combination of particle and heat diffusion.

Although this experiment nicely demonstrates that latent heating plays a role in snow crystal growth in air, I reiterate that it is a relatively modest perturbation when

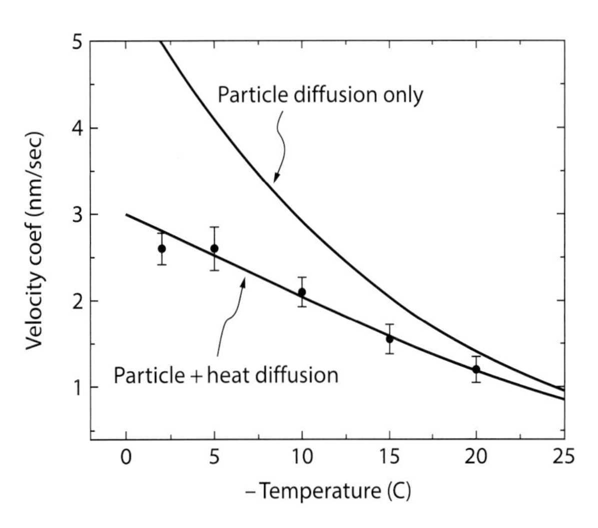

FIGURE 3.17. Measurements of the radial growth of thin ice needles, together with an analytical model that includes only particle diffusion (top line), plus a similar model that includes both particle and heat diffusion (lower line) [2016Lib]. The plotted velocity coefficient is equal to the cylinder growth rate at a fixed faraway supersaturation and a fixed cylinder radius of 5 microns. The measurements show good agreement with the particle + heat diffusion model, confirming the temperature-dependent reduction in growth rate caused by latent heating.

compared with the dominant effects of particle diffusion and attachment kinetics. Someday, we will need to solve the combined particle + heat double-diffusion problem in three dimensions to explain all the subtleties of snow crystal growth, but that day is not yet upon us. Throughout most of this book, therefore, I have largely ignored latent heating and heat diffusion for growth in air (except when dealing with possible systematic errors in precision growth experiments). Until we have a better understanding of the attachment kinetics over the full range of growth conditions, it is reasonable to mostly ignore heating effects, at least for the immediate future.

Kinetics, Diffusion, Heating, and Surface Energy

Rounding out our analysis of spherical ice growth, we add surface energy by including the Gibbs-Thomson effect, which gives the new Hertz-Knudsen relation

$v_n = \alpha v_{kin}(\sigma_{surf} - d_{sv}\kappa)$, where $\kappa = 2/R$ is the surface curvature for a sphere (see Chapter 2). Plugging this in gives

$$\sigma_{surf} = \left(\frac{\alpha_{diff,heat}}{\alpha + \alpha_{diff,heat}}\right)_{v_{kin}}\left(\sigma_\infty + \frac{\alpha}{\alpha_{diff,heat}} d_{sv}\kappa\right) \quad (3.33)$$

and

$$v_n = \left(\frac{\alpha\alpha_{diff,heat}}{\alpha + \alpha_{diff,heat}}\right) v_{kin}\left(\sigma_\infty - d_{sv}\kappa\right). \quad (3.34)$$

If one wishes to ignore heating effects, $\alpha_{diff,heat}$ can be replaced with α_{diff}.

In many snow crystal growth scenarios, the added Gibbs-Thomson term is a minor effect, especially with large crystals or fast growth rates. With a fernlike stellar dendrite, for example, the tip radius is $R \approx 1\ \mu$m, giving $d_{sv}\kappa \approx 0.2$ percent, while the supersaturation is typically $\sigma_\infty > 20$ percent. In contrast, during the growth of exceptionally thin plates at low supersaturations, the Gibbs-Thomson effect does limit the growth and prevent the formation of even thinner plates than what is observed.

Low-Pressure Growth

While heating effects are generally small in normal air, the situation changes at low pressures. To first order, D is inversely proportional to background gas pressure P, so particle diffusion speeds up considerably at low pressures. But κ_{air} is roughly independent of P down to quite low pressures (until the molecular mean free path becomes larger than other scales in the problem). For this reason, $\chi_0 \sim P^{-1}$, and heating effects can become important at lower pressures.

For example, in the case of an ice crystal levitated in near-vacuum conditions, the diffusion constant increases substantially, making $\chi_0 \gg 1$. In this limit, particle diffusion no longer limits growth in comparison to heat diffusion. Then $\alpha_{diff,heat}$ in Equation 3.27 should be replaced with

$$\alpha_{heat} = \frac{\kappa_{air}}{\eta L_{sv}\rho_{ice}v_{kin}}\frac{1}{R} \approx 0.03\left(\frac{300\ \frac{\mu m}{sec}}{v_{kin}}\right)\left(\frac{10\ \mu m}{R}\right). \quad (3.35)$$

At temperatures near 0°C, α_{heat} is low enough that heat diffusion becomes the dominant factor limiting growth in many situations, making it difficult to extract information about the attachment kinetics from levitation experiments. This problem is reduced at lower temperatures, as v_{kin} is strongly temperature dependent (see Table 2.1).

Heating effects are reduced for ice growth on a substrate, and this can be estimated by considering the plane-parallel problem of a thin sheet of ice resting a substrate. The solution is like those above, except with $\alpha_{diff,heat}$ being replaced with

$$\alpha_{therm} \approx \frac{\kappa_{ice}}{\eta L_{sv}\rho_{ice}v_{kin}}\frac{G}{H}$$
$$\approx 3G\left(\frac{300\ \frac{\mu m}{sec}}{v_{kin}}\right)\left(\frac{10\ \mu m}{H}\right), \quad (3.36)$$

where H is the crystal thickness, and $G = 1$. For the case of a small ice prism on a substrate, G can be replaced by a dimensionless geometrical factor of order unity. Here we see that heating effects are reduced by about a factor of $\kappa_{ice}/\kappa_{air} \approx 100$ compared to the levitated-crystal case, owing to the higher thermal conductivity of ice. From this analysis, we see that both heating and diffusion effects are reduced for ice growth on a substrate in near vacuum, making this experimental system well suited for learning about the attachment kinetics (see Chapter 7).

Finite Outer Boundary

Bringing the outer boundary in from infinity complicates the analysis, but the finite-boundary case is useful for validating numerical models to make sure they obtain correct quantitative results. Including both particle diffusion and attachment kinetics, the solution becomes

$$\sigma(r)=\sigma_{out}-\left(\frac{R'}{r}-\frac{R'}{R_{far}}\right)\sigma_{out}, \qquad (3.37)$$

where

$$R'=\left[\frac{\gamma}{R}-\frac{1}{R_{far}}\right]^{-1} \qquad (3.38)$$

and

$$\gamma=\frac{\alpha+\alpha_{diff}}{\alpha_{diff}}. \qquad (3.39)$$

This then gives the crystal growth velocity

$$v_n=\left(\frac{\alpha\alpha_{diff}}{\alpha+\alpha_{diff}}\right)v_{kin}\sigma_{\infty}\left[1-\frac{R}{\gamma R_{far}}\right]^{-1}, \qquad (3.40)$$

and we see that this reduces to Equation 3.20 when $R_{far}\to\infty$, as it must.

ADDITIONAL ANALYTIC SOLUTIONS

While the spherical solution is the best starting point for any quantitative discussion of diffusion-limited growth, several other analytic solutions are known. In this section, I examine some of these additional solutions and their application.

Cylindrical Growth

The analytic solution for an infinitely long growing cylinder is useful when examining the growth of electric needle crystals (see Chapter 8), so I mention the results here. The diffusion analysis is analogous to the spherical case, the main change being to work in a cylindrical coordinate system. Once again, the solution can be written in a form like Equation 3.20, giving the radial growth velocity

$$v_n=\left(\frac{\alpha\alpha_{diffcyl}}{\alpha+\alpha_{diffcyl}}\right)v_{kin}\sigma_{far}, \qquad (3.41)$$

where $\sigma_{far}=\sigma(R_{out})$ and

$$\alpha_{diffcyl}=\frac{1}{B}\frac{X_0}{R_{in}}, \qquad (3.42)$$

with $B=\log(R_{out}/R_{in})$, where R_{in} is the radius of the cylinder and R_{out} is the radius of the faraway boundary. Note that one cannot assume $R_{out}\to\infty$ in this solution without encountering a logarithmic divergence, a feature that is well known from cylindrical electrostatics problems. It is straightforward to extend the analysis to include latent heating and heat diffusion, and the resulting model is the one compared with experimental data in Figure 3.17. The cylindrical solution is also useful for validating numerical models, as described in [2013Lib1].

The Ivantsov Solution

While the spherical case is quite useful for a general examination of different physical processes, solving the diffusion equation in parabolic coordinates yields many insights into the growth of free dendrites, including the snow crystal dendrites described earlier in the chapter. The parabolic solution was discovered in 1947 by Russian physicist G. P. Ivantsov [1947Iva], and in three dimensions, it takes the form of a needlelike paraboloid of revolution that is parameterized solely by its tip radius R_{tip}, as shown in Figure 3.18.

For purely diffusion-limited growth, the Ivantsov solution (for either particle or heat diffusion separately) shows that the entire paraboloid grows at a constant velocity v_{tip} in the direction of the needle axis, while R_{tip} and the full parabolic shape of the crystal remain unchanged in time. I will not describe the derivation of the Ivantsov solution here but present the primary result

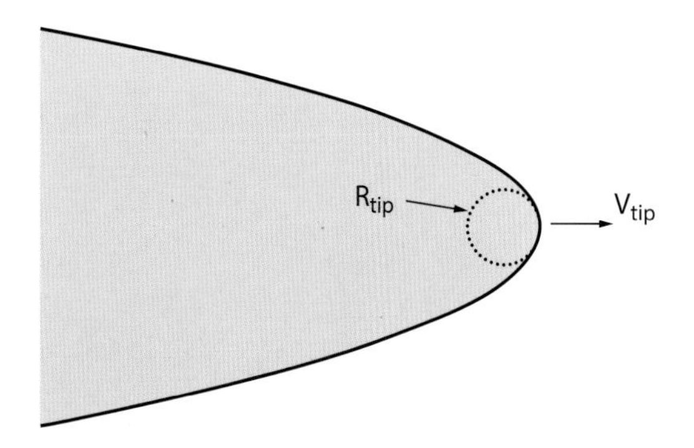

FIGURE 3.18. The Ivantsov solution to the diffusion equation describes a crystalline paraboloid of revolution with a constant parabolic shape and tip radius R_{tip}, growing forward with a constant velocity v_{tip}. If viewed from a frame of reference that moves in the growth direction with velocity v_{tip}, the system would appear completely static.

for the case of ice growing from water vapor in air, where the Laplace approximation applies to high accuracy. Neglecting surface energy and attachment kinetics ($\alpha_{diff} \ll \alpha \approx 1$), the tip velocity is given by [1996Sai, 2002Lib]

$$v_{tip} \approx \frac{2D}{BR_{tip}} \frac{c_{sat}}{c_{ice}} \sigma_{far} \approx \frac{2}{B} \frac{X_0}{R_{tip}} v_{kin} \sigma_{far}, \quad (3.43)$$

where $B = \log(\eta_{far}/R_{tip})$, η_{far} is the position of the faraway boundary (using a parabolic coordinate system with standard variables (ξ, η, φ), and $\sigma_{far} = \sigma(\eta_{far})$. Like the cylindrical case, one cannot assume $\eta_{far} \to \infty$ without encountering a logarithmic divergence.

The functional form of this equation is analogous to Equation 3.25, but its overall behavior is quite different. In the spherical case, R increases as the crystal grows, while $v_n \sim 1/R$ decreases. In the parabolic case, however, both R_{tip} and v_{tip} remain unchanged as the crystal grows. Note that the Ivantsov solution is actually a family of solutions, because it does not specify a unique R_{tip}. The diffusion equation alone only specifies the Ivantsov relationship between v_{tip} and R_{tip}.

As with the spherical and cylindrical cases, the Ivantsov solution is subject to the Mullins-Sekerka instability, resulting in complex branched structures. Nevertheless, the parabolic shape often provides a reasonable description of the overall envelope of free dendrite growth, and it does especially well near the tip, which is often nearly parabolic in form. The Mullins-Sekerka instability can create copious sidebranching away from the tip, but the tip behavior is often well represented by the Ivantsov parabola.

Although sidebranches clearly complicate the picture, the Ivantsov form is a remarkably robust solution to the diffusion equation. The large-scale outline of a typical free dendrite is roughly parabolic, and usually the structure near the tip is relatively smooth with a nearly parabolic form. Put another way, the Ivantsov solution creates a self-assembling free-dendrite morphology with constant v_{tip} that is generally insensitive to perturbations from other growth effects. This property helps explain why free-dendrite growth in diffusion-limited solidification is such a commonly observed phenomenon.

Looking closely at the fernlike and fishbone dendrite morphologies in Figures 3.10 and 3.11, respectively, we see that both have roughly parabolic envelopes and tip shapes, but the appearance of strong faceting means that the Ivantsov solution does not tell the whole story in either case. Nevertheless, the analytic Ivantsov solution is useful as a limiting case and for examining overall trends regarding different parameters and growth behaviors. As with the other analytic solutions presented earlier, the Ivantsov solution helps build one's intuition and understanding about which physical processes are important and which can be safely neglected in different circumstances. Reproducing actual snow crystal structures and growth measurements with any real fidelity, however, will require computational modeling.

SOLVABILITY THEORY

For roughly a decade around the 1980s, a concerted effort was made to create a full analytical model of free-dendrite growth, and the result became known as *solvability theory* [1988Kes, 1988Sai, 1989Lan, 1991Bre]. The primary goal of this endeavor was to derive v_{tip} and

FIGURE 3.19. A series of photographs showing the tips of free dendrites growing during the solidification of liquid succinonitrile (a clear, waxy material that melts at 57°C). As the supercooling ΔT of the liquid increases, R_{tip} decreases while v_{tip} increases, while their constant product $R_{tip}v_{tip}$ satisfies the Ivantsov relation for thermal diffusion. The overall growth behavior and dendrite tip morphology remain essentially independent of ΔT. Image adapted from [1981Hua].

R_{tip} directly from basic physical principles and intrinsic material properties, reproducing measurements from a broad range of materials. This research effort was stimulated in part by a series of beautiful, quantitative observations of dendritic solidification from the melt by Martin Glicksman and others, with one experimental example shown in Figure 3.19.

It was realized early on that the Ivantsov relation provides the appropriate solution to the diffusion equation, but it provided only a relation between v_{tip} and R_{tip} without specifying either, as seen in Equation 3.43 for the snow crystal case. This physical indeterminacy became known as the *selection problem*. If diffusion alone does not specify v_{tip} and R_{tip} uniquely, what does?

The Selection Problem

Resolving the selection problem requires some additional physics beyond diffusion alone, and the only two viable possibilities (in most realistic cases) are surface energy and attachment kinetics. In solidification from the melt, surface energy turns out to be the dominant effect, and many theoretical treatments in the literature ignore attachment kinetics for that reason. Solidification from the vapor phase has been much less studied, but here it appears that attachment kinetics are more important than surface energy, as we will see shortly in the snow crystal case.

My goal in this chapter is not to provide an in-depth review of all aspects of solvability theory, but rather to outline its basic results as applied to snow crystal formation. To this end, I ignore all heating effects, as particle diffusion is more important than heat diffusion, and the former by itself is enough to develop a crude version of vapor-growth solvability theory. Using this theory, I then show that attachment kinetics are likely more important than surface energy in the theory, which is opposite to the melt-growth case.

I begin with a perturbation expansion of the spherical solution, Equation 3.34, neglecting heating

($\alpha_{diff,heat} = \alpha_{diff}$) and assuming that the growth is mainly diffusion limited ($\alpha_{diff} \ll \alpha$), which gives

$$v_n \approx \frac{X_0}{R} v_{kin} \left(\sigma_\infty - \frac{2d_{sv}}{R} - \frac{\sigma_\infty}{\alpha} \frac{X_0}{R} \right). \qquad 3.44$$

For a typical fernlike dendrite tip in air (taking $X_0 \approx 145$ nm, $R = R_{tip} \approx 1$ μm, $\sigma_\infty \approx 1$, and $\alpha \approx 1$), we find that the second and third terms in this equation are roughly 0.002 and 0.15, respectively, so both are indeed small compared to σ_∞, justifying the perturbation expansion. To take the next step, assume that the near-hemispherical tip of a parabolic ice dendrite behaves much like spherical growth, so doing an analogous perturbation expansion of the Ivantsov solution, Equation 3.43, yields

$$v_{tip} \approx \frac{2X_0}{BR_{tip}} v_{kin} \left(\sigma_{far} - \frac{R_{GT}}{R_{tip}} - \frac{\sigma_{far}}{\alpha} \frac{R_{kin}}{R_{tip}} \right), \qquad (3.45)$$

where $R_{GT} = 2d_{sv} \approx 2$ nm, and $R_{kin} = 2X_0/B \approx 35$ nm. (Choosing $B \approx 8$ is a reasonable approximation for typical snow crystal dendrites.)

From this expansion, we can begin to see the essential physics underlying the dendrite selection problem. Referring to Figure 3.7, we see that the Mullins-Sekerka instability generally promotes the growth of bumps on top of broad, flat surfaces. Zooming in on the end of a dendrite tip, it stands to reason that the Mullins-Sekerka instability would also promote the growth of a smaller bump on top of a broad dendrite tip. Taking this reasoning to its logical conclusion, we see that the Mullins-Sekerka instability would, if no other forces intervened, sharpen a dendrite tip indefinitely, driving $R_{tip} \to 0$. The available intervening forces are those found in Equation 3.45, specifically in the second and third terms of this expression. As $R_{tip} \to 0$, these terms both become so large that they are no longer small compared to σ_{far}. At some value of R_{tip}, therefore, these forces halt any further tip sharpening.

The nature of these two stabilizing effects can also be reasonably well understood from the underlying physics. The Gibbs-Thomson effect states that the equilibrium vapor pressure increases as $1/R$ on a spherical surface, and this effectively reduces the driving supersaturation at the tip. Following the math through gives the R_{GT} term in Equation 3.45. The negative sign means that this is a stabilizing force that prevents runaway tip sharpening. Put another way, there is a Gibbs-Thomson "penalty" for fast growth (which requires a small R_{tip}), which serves to prevent $R_{tip} \to 0$.

The kinetics term R_{kin} arises because a finite surface supersaturation $\sigma_{surf} > 0$ is needed to drive crystal growth, as the growth must become identically zero if $\sigma_{surf} = 0$. Moreover, the necessary σ_{surf} increases with v_{tip}, which is proportional to $1/R_{tip}$ to first order in this perturbation analysis. Here again, there is a supersaturation penalty for fast growth, which also serves to prevent $R_{tip} \to 0$. The fact that $R_{GT} \ll R_{kin}$ suggests that the kinetics term in Equation 3.45 is more important than the surface energy term for selecting the final dendrite tip radius in typical snow crystals.

Although the attachment kinetics usually dominate, one can imagine a snow crystal scenario in which surface energy is more important in the tip selection problem. From Equation 3.45, this would happen if σ_{far} were low and α remained high. While this scenario is conceivable, it is not particularly realistic. As we will see in Chapter 4, α depends strongly on σ_{surf}, going to zero rapidly at low supersaturations. As a result, faceting becomes a dominant factor at low σ_{surf}, making the Ivantsov solution inappropriate in that regime. Computational models (see Chapter 5) generally support this conclusion. Snow crystal dendrites are typically found only at high σ_{far}, where the surface energy term is small compared to the attachment kinetics term. When considering free dendrites and solvability theory, therefore, it is reasonable to neglect the R_{GT} term while keeping the R_{kin} term in Equation 3.45.

Snow Crystal Dendrites

Extending this qualitative discussion into a rigorous theory is not a simple task, which is why it took a significant effort to develop solvability theory. Although my comprehension of this highly mathematical theory is not thorough, it appears that the final result can be expressed in a fairly simple form [1988Kes, 1988Sai, 1989Lan, 1991Bre, 2002Lib]. The answer differs for melt growth and snow crystal growth, however, because of the relative importance of the surface energy and attachment kinetics terms in Equation 3.45. For the snow crystal case, we neglect surface energy (on the grounds that $R_{GT} \ll R_{kin}$), and solvability theory then yields the relationship

$$v_{tip} R_{tip}^2 \approx \frac{4\sigma_{far} v_{kin} X_0^2}{s_0 B\alpha}, \tag{3.46}$$

where s_0 is a dimensionless constant called the *solvability parameter*. This second mathematical relationship, in addition to the Ivantsov solution, allows one to uniquely determine both R_{tip} and v_{tip} as a function of intrinsic material properties and external growth conditions. Combining Equations 3.45 and 3.46 yields

$$R_{tip} \approx \frac{2X_0}{s_0\alpha}; \; v_{tip} \approx \frac{s_0}{B}\alpha v_{kin}\sigma_{far}, \tag{3.47}$$

and at this point, it is beneficial to compare the theory with experimental observations. Figure 3.20 show measurements of v_{tip} as a function of σ_{far} for fernlike free dendrites growing near −15°C. The data support a linear dependence $v_{tip} \sim \sigma_{far}$, and the low-resolution tip images are at least consistent with R_{tip} being independent of σ_{far}, so both observed trends agree with Equation 3.47. A fit to the data assuming $B \approx 8$ yields $R_{tip} \approx 1$ μm and $\alpha s_0 \approx 0.25$ [2002Lib]. Similar data for fishbone free dendrites yields $R_{tip} \approx 1.5$ μm and $\alpha s_0 \approx 0.2$ [2002Lib].

These values of R_{tip} were measured in air, and we see from Equation 3.47 that theory indicates $R_{tip} \sim X_0$, implying that R_{tip} should be roughly inversely proportional

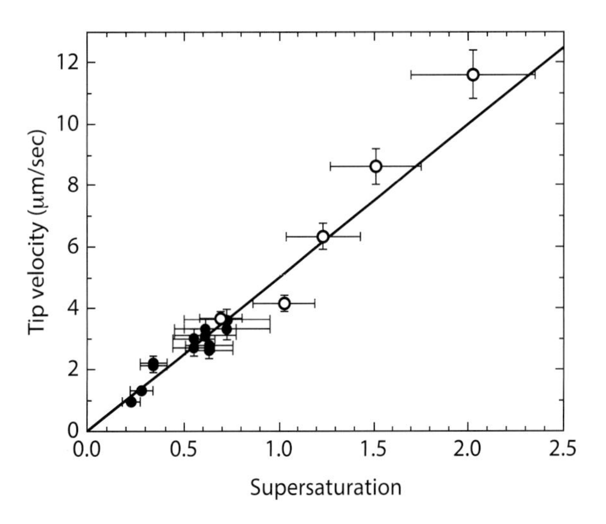

FIGURE 3.20. Measurements of the tip velocity of fernlike free dendrites growing near −15°C as a function of the faraway supersaturation. The data indicate a linear relationship between these variables, and the line shows $v_{tip} = 5\sigma_{far}$ μm/sec [2002Lib].

to the background air pressure, while v_{tip} should be roughly independent of pressure. It would be straightforward to confirm this prediction, but to my knowledge, it has not yet been done. However, experiments have revealed finer structural details in snow crystals grown at higher pressures, supporting the result from solvability theory [1976Gon].

This analysis of snow crystal free dendrite growth comes with some caveats, however. Solvability theory indicates that the value of s_0 depends on the detailed properties of the most important stabilization term, namely, the attachment kinetics in this case. This is problematic, because the attachment kinetics are not well known from independent measurements, and they may depend on growth conditions at the tip surface, specifically the near-surface supersaturation. Thus, the theory is somewhat underconstrained due to a poor knowledge of material properties, so we should perhaps not read too much into the linear trend seen in Figure 3.20. This issue is a manifestation of a more general problem with solvability theory in all experimental systems: there is no easy way to calculate s_0, so an analytic theory including just a few basic parameters might not be sufficient to describe a complex phenomenon like free dendrite growth.

Note that had we ignored the attachment-kinetic perturbation and instead kept the surface energy perturbation in solvability theory, the result would have included the scaling $R_{tip} \sim \sigma_{far}$ and $v_{tip} \sim \sigma^2_{far}$. The above caveats notwithstanding, Figure 3.20 does not agree with such a quadratic dependence, supporting the notion that attachment kinetics provide the more important stabilizing mechanism, in agreement with expectations.

Anisotropy and Tip Splitting

Another important discovery from solvability theory is that s_0 depends on the anisotropy of the surface physics that stabilizes the dendrite tip radius. For perfectly isotropic systems, even the initial premise of a stable, Ivantsov-like parabolic tip structure turns out to be incorrect. With perfect isotropy, the Mullins-Sekerka instability brings about not only sidebranches, but also tip splitting. This phenomenon is best seen in computer simulations of dendritic growth, and Figure 3.21 shows a growing dendritic system for which the anisotropy was varied in different runs. With no anisotropy, the dendritic branches exhibited frequent tip splitting that resulted in a complex "seaweed-like" structure. Above some threshold anisotropy, dendrites with stable tip structures appeared.

That stable free-dendrite growth requires anisotropy appears to be a general property of diffusion-limited dendrite formation, present over a broad range of different physical systems. When the surface stabilization forces are sufficiently anisotropic, Ivantsov-like dendrites appear with stable tip structures, as seen in snow crystal dendrites. As the anisotropy is turned down, tip splitting begins to occur only occasionally, increasing with lower anisotropy. Eventually the growth transitions to completely random seaweed-like structures as the anisotropy decreases to zero.

Tip splitting is largely absent in snow crystal dendrites, owing to the exceptionally large underlying anisotropy in the attachment kinetics. Nevertheless, Figure 3.22 shows an example of tip splitting in a rapidly growing fernlike stellar dendrite, indicating $\alpha_{prism} \approx 1$ when the supersaturation is sufficiently high. This general behavior fits the model for nucleation-limited attachment kinetics presented in Chapter 4. I have also witnessed some dendritic tip splitting at temperatures near 0°C when the supersaturation is high and chemical contaminants are present, again indicating $\alpha_{prism} \approx 1$ under those conditions. The basal anisotropy is relatively high under essentially all growth conditions, owing to a finite basal step energy at 0°C (see Chapter 4). Thus, one expects a complete absence of basal tip splitting, and this expectation is consistent with observations.

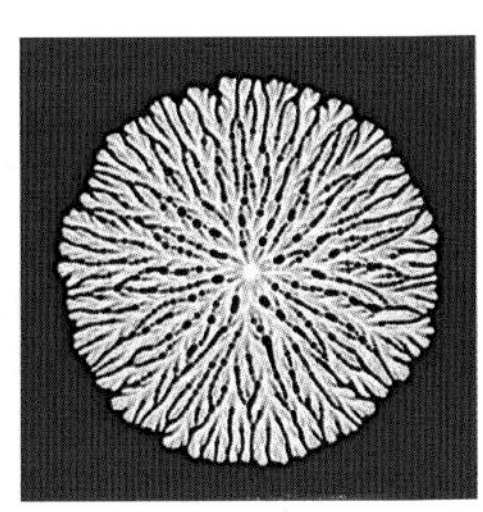

FIGURE 3.21. In crystal growth, some degree of surface anisotropy in the attachment kinetics or surface energy is necessary to prevent tip splitting and create stable free-dendrite growth (far left). For perfectly isotropic systems, seaweed-like structures emerge (far right). This numerical simulation illustrates a morphological transition between these two states as the underlying anisotropy is reduced, in this case for a system with sixfold symmetry. Image adapted from [2006Gra1].

FIGURE 3.22. An example of tip splitting in rapidly growing fernlike dendrites at $\sigma_{far} \approx 1.3$ near −15°C. In these conditions, α_{prism} is close to unity, and the anisotropy in the attachment coefficient becomes quite low. Notably, the tip splitting in this example occurred early in its growth, when the surface supersaturation was higher than at later times. Chemical vapor additives that increase α_{prism} can also result in increased tip splitting.

SNOW CRYSTAL AERODYNAMICS

We next turn our attention to how aerodynamics can affect snow crystal growth and morphologies [1982Kel, 1997Pru, 1999Fuk, 2002Wan, 2009Lib3]. In normal air, the motion of falling crystals can align their orientation relative to the horizon, change their growth rates, and even alter their growth morphologies, although typically all these effects are rather small perturbations compared to normal growth behaviors. Our main goal in this section is to outline the basic physical processes by using analytic models and estimating the importance of the various effects over a range of growth conditions. Throughout this discussion, it is important to remember that wind speed relative to the ground is not the relevant parameter in the problem, but rather wind speed relative to the crystal in question. A small snow crystal may be carried by the wind for long distances, but it mostly travels along with the moving air around it. Thus, while wind blowing over a stationary snow crystal in the lab may strongly perturb its growth [1982Kel], we cannot apply these results until we understand the velocity of air flow around a freely falling snow crystal.

Drag and Terminal Velocity

Gravity creates a net velocity between crystals and air, and a falling snow crystal quickly reaches its terminal velocity in still air. The viscous drag on a snow crystal is well described by Stokes drag at low velocities, given by

$$F_{Stokes} = 6\pi\mu R_H u, \qquad (3.48)$$

where F_{Stokes} is the drag force, R_H is the hydrodynamic radius of the object, μ is the dynamical viscosity of air, and u is the flow velocity. For a spherical particle, R_H equals the radius R of the sphere.

At the velocity increases, the flow becomes turbulent, adding a component to the drag that is proportional to u^2. Assuming a thin disk morphology with radius R and thickness T (a satisfactory model for a platelike snow crystal), the drag force becomes

$$F_{drag} \approx 6\pi\mu R u + \frac{\pi}{2}\rho_{air} R^2 u^2 \qquad (3.49)$$

to a reasonable approximation, where ρ_{air} is the density of air [2009Lib2]. The two terms in this expression are equal when the Reynolds number R_e is about 24, where I take

$$R_e = \frac{2\rho_{air} uR}{\mu} = \frac{2uR}{v_{kinematic}}, \quad (3.50)$$

and $v_{kinematic} = \mu/\rho_{air}$ is the kinematic viscosity.

The falling thin-disk crystal reaches its terminal velocity u_{term} when $F_{drag} = mg$, where $m = \pi R^2 T \rho_{ice}$ is the mass of the crystal, giving

$$\begin{aligned} u_{term} &\approx \frac{1}{6}\frac{\rho_{ice} g}{\mu} RT \quad (\text{low } R_e) \\ &\approx 8\left(\frac{R}{100\ \mu\text{m}}\right)\left(\frac{T}{10\ \mu\text{m}}\right) \text{cm/sec} \end{aligned} \quad (3.51)$$

for the case of small crystals falling at low Reynolds number and

$$\begin{aligned} u_{term} &\approx \left(\frac{2T\rho_{ice} g}{\rho_{air}}\right)^{1/2} \quad (\text{high } R_e) \\ &\approx 40\left(\frac{T}{10\ \mu\text{m}}\right)^{1/2} \text{cm/sec} \end{aligned} \quad (3.52)$$

for larger crystals moving at high Reynolds number. For these thin-disk crystals, the transition from low to high Reynolds number terminal velocity occurs when the crystal radius exceeds

$$R_{transition} \approx 450\left(\frac{10\ \mu\text{m}}{T}\right)^{1/2} \mu\text{m}. \quad (3.53)$$

Figure 3.23 shows an example of the terminal velocity of a 2-μm thick disk as a function of its radius.

Comparing terminal velocity calculations with observations is not especially fruitful, unfortunately. The theory is well understood for small crystals with simple shapes, while most measurements have been obtained using larger crystals with complex, rather poorly characterized morphologies and sizes. Nevertheless, the extensive measurements of fall velocities in a vertical flow chamber made by Fukuta and Takahashi [1999Fuk] seem to be consistent with the above theory, given the substantial uncertainties involved.

Horizontal Alignment

Over a range of snow crystal sizes and morphologies, drag forces can align falling crystals relative to the horizon. The resulting alignment is well known in natural snow crystals, as it is essential for explaining many distinctive features in atmospheric halos [1980Gre, 1990Tap, 2006Tap]. For example, thin disks often align with a vertical c-axis, while slender columns align with a horizontal c-axis. In some instances, columns may align further with two prism facets in a horizontal orientation, known as the Parry orientation [2006Tap]. In some rare halo observations, models suggest widespread crystal alignments as precise as a few degrees relative to the horizon.

Focusing on plates, theory suggests that the smallest crystals will not align unless their terminal velocities are larger than surrounding turbulent air flows that perturb their fall and orientation. Moreover, large plates are unstable to various fluttering and tumbling instabilities when the Reynolds number exceeds $R_e \approx 100$. The latter regime applies to crystals with sizes of about 1 mm or more, while Figure 3.23 shown two models for turbulent air velocities. Unless the air is exceptionally still, snow crystals are likely to exhibit good alignment only in roughly the 0.1–1 mm size range.

The Ventilation Effect

When supersaturated air flows around a snow crystal, the latter's growth rate increases as the flow essentially enhances the diffusion of water vapor molecules to its surface, and this phenomenon is called the *ventilation effect* [1982Kel, 1997Pru]. The magnitude of the growth change can be estimated by comparing the diffusion time and the flow time. The diffusion timescale for water molecules diffusing a distance L through the air is

$$\tau_{diffusion} \approx \frac{L^2}{D}, \quad (3.54)$$

and a growing crystal significantly reduces the supersaturation in its vicinity only out to a distance comparable

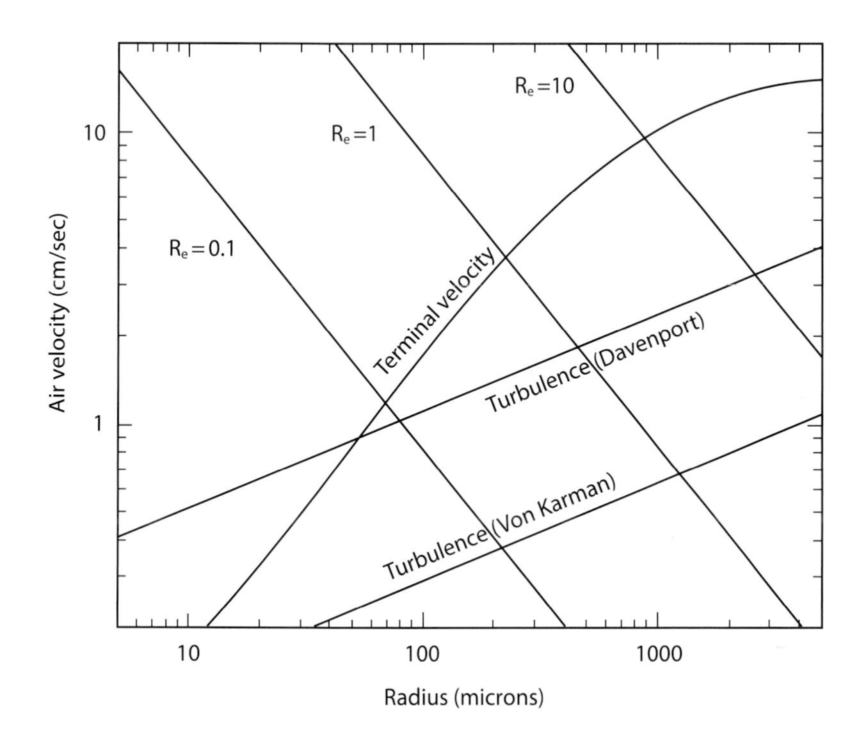

FIGURE 3.23. The terminal-velocity curve in this plot shows the fall velocity of a 2-μm-thick snow crystal disk as a function of disk radius, while approximate scaling with disk thickness is given by Equations 3.51 and 3.52. The turbulence curves show two models of root-mean-squared air velocities when the average air speed in 1 m/sec, and three lines of constant Reynolds number R_e are also shown. When the disk radius R is large enough that the terminal velocity curve is above the turbulence curves, then gravity can align the crystal horizontally [2009Lib2].

to its size, so we take L to be the approximate size of the crystal. Meanwhile, the time it takes for air to flow the same distance L is

$$\tau_{flow} \approx \frac{L}{u}. \tag{3.55}$$

If the flow velocity is low and $\tau_{diffusion} \ll \tau_{flow}$, then diffusion creates a depleted region around the crystal before the air flows by it. In this case, the ventilation effect becomes negligible, as must be the case when $u \to 0$. We expect, therefore, that air flow significantly affects the crystal growth only when $\tau_{flow} < \tau_{diffusion}$, which is equivalent to the regime $R_e > 1$. In a somewhat more in-depth analysis incorporating studies of liquid droplet growth in the literature, I found that that the growth rate of a spherical snow crystal is enhanced by a factor

$$\begin{aligned} f_v &\approx 1 + 0.1 R_e \quad (R_e < 1) \\ f_v &\approx 0.8 + 0.3 R_e^{\frac{1}{2}} \quad (R_e > 1), \end{aligned} \tag{3.56}$$

to a reasonable approximation [2009Lib2].

Applying these results to a specific example, consider the fernlike stellar dendrite shown in Figure 3.9. This is a common snow crystal morphology, and an examination of the calibrated photo reveals that the initial branching instability occurred when the crystal radius was no larger than $R \approx 30\ \mu m$. From Figure 3.23, the Reynolds number of the air flow around this nascent crystal was likely about $R_e \approx 0.1$, giving an enhancement factor of $f_v \approx 1.01$, meaning that the ventilation effect was likely negligible when the first branching event occurred. Strong turbulence might have increased this enhancement, but well-formed crystals like the one in Figure 3.9 rarely survive long in windy, turbulent conditions.

As this crystal grew larger, the Reynolds number of the flow around it increased, and the crystal morphology became dominated by the six fernlike dendritic branches. Then the air flow likely aligned the crystal so that its basal faces were nearly horizontal, and the flow past each tip was roughly perpendicular to the growth direction. The ventilation effect is more difficult to analyze in this case, but the sharp-tipped geometry leads to a substantially higher ventilation effect compared to the spherical case. At terminal velocity for this crystal, the ventilation effect would produce roughly a 25 percent increase in tip growth velocity [2009Lib2].

Through a combination of alignment and ventilation effects, it is possible that aerodynamics plays a role in promoting the high symmetry of some snow crystal structures. This is likely a small effect, a supposition that is supported by the fact that most snow crystals do not exhibit a high degree of sixfold symmetry. Nevertheless, aerodynamic alignment can lead to tumbling instabilities that would tend to enhance symmetrical growth of several crystal morphologies. Some possibilities along these lines have been discussed in the literature [1999Fuk, 2009Lib3]. It has also been suggested that aerodynamic effects may promote the growth of triangular snow crystals through a combination of alignment and ventilation effects [2009Lib3]. However, as discussed in the next section, the origin of triangular plate snow crystals is still a bit of a mystery, and it is not yet clear if aerodynamics plays a major role in their development.

The bottom line in this discussion is that aerodynamics can play a role in snow crystal growth dynamics, but it is usually a rather minor one. Small crystals are the least susceptible to aerodynamic effects, although remarkably precise crystal alignments are possible in especially calm conditions.

ORDER AND CHAOS

Snow crystal morphologies are determined mainly by the interplay of two physical processes: attachment kinetics and particle diffusion. Attachment kinetics bring about ordered, faceted surfaces with sharp edges and corners, defined precisely by the crystal lattice structure. Diffusion brings about instability, yielding complex structures and the chaotic sidebranching seen in dendritic growth. These are the competing forces of order and chaos that drive the formation of snow crystals. In some circumstances, additional physical effects from heat diffusion, surface energy, aerodynamics, and other factors may be significant as well. But attachment kinetics and particle diffusion are usually the main players.

In this section, I examine a selection of snow crystal morphological features in some detail and attempt to describe how each originates. This undertaking would be best accomplished with the help of corresponding numerical simulations, but here the state of the art is somewhat unreliable. Although computational models are improving rapidly (see Chapter 5), they cannot yet reproduce real snow crystal structures with good fidelity. Thus, one motivation in this section is to provide qualitative descriptions of growth behaviors that might be explained more quantitatively in future numerical investigations. Another motivation is to develop an overarching physical intuition regarding the underlying causes of snow crystal formation, as this is helpful for making additional progress in the field. And last, but not least, it is simply pleasing to have an essential understanding of some of the puzzling characteristics often found in natural snow crystals.

Aspect Ratios and Anisotropy

As a general rule, the large-scale aspect ratio of a snow crystal—here defined as the ratio of the overall size of a crystal along the c-axis to that along an a-axis—reflects the anisotropy in the underlying attachment kinetics. For example, the formation of thin plates invariably requires $\alpha_{basal} \ll \alpha_{prism}$, while the formation of slender columns requires $\alpha_{basal} \gg \alpha_{prism}$. Although qualitative in nature, this aspect-ratio rule applies throughout the menagerie of different snow crystal types.

One reason this rule exists is because diffusion-limited growth alone cannot yield structures with extreme aspect ratios, like thin plates or slender columns. Numerical models reveal that while the Mullins-Sekerka instability often drives complex dendritic branching, the overall aspect ratios of the resulting crystals are still mainly determined by anisotropies in the underlying attachment kinetics. A second reason the rule exists is that extreme aspect ratios in crystal growth do not usually arise from surface energy anisotropy. Highly anisotropic surface energies exist in the realm of exotic materials, but simple solids (such as ice or metals) generally exhibit modest anisotropies that are too small to produce large aspect ratios during solidification. In particular, the extreme aspect ratios seen in snow crystals and platelike pond crystals are the result of highly anisotropic attachment kinetics and not of highly anisotropic surface energies.

Figure 3.24 shows an illustration of what low and high anisotropy in the attachment kinetics can look like around a faceted snow crystal. In the case of a nearly isometric faceted prism (left panel in the figure), α_{basal} and α_{prism} are roughly equal in magnitude, and σ_{surf} is highest at the corners of the prism (the Berg effect). The presence of basal and prism faceting on this crystal indicates anisotropy in the sense that both α_{basal} and α_{prism} are smaller than α_{rough}. But the overall aspect ratio of the crystal is near unity, because $\alpha_{basal}/\alpha_{prism} \approx 1$. The figure also shows a model with $\alpha_{basal} \ll \alpha_{prism}$, which resulted in the formation of a thin platelike crystal, again supporting the anisotropy rule (right panel). In this case, we see that σ_{surf} is substantially higher on the basal facet than on the prism facet, contrary to the usual expectation from the Berg effect. Even though the prism edge sticks out farther into the supersaturated air, σ_{surf} is lowest there. Note also that the supersaturation gradient is highest near the fast-growing edge of the plate. This makes sense, because the fast growth needs a high flux of water vapor molecules, which can only happen in a steep supersaturation gradient. In contrast, the particle flux and supersaturation gradient are low near the center of the basal facets, reflecting their lower growth rates.

Additional models like these reveal that the strong correlation between aspect ratio and anisotropy in the attachment kinetics applies over a broad range of growth conditions. If, for example, one begins with a thin plate crystal and then changes parameters so that $\alpha_{basal} \approx \alpha_{prism}$, then the subsequent growth will not maintain the thin-plate structure. Instead the edges of the plate thicken over time, and the overall aspect ratio will tend toward unity as the crystal continues to grow. Diffusion-limited growth generally pushes morphologies toward small overall aspect ratios, and this trend is usually countered only by a strong anisotropy in the attachment kinetics.

Morphological Complexity in the Nakaya Diagram

Another general rule in snow crystal growth is that faceting requires a high anisotropy in the attachment kinetics, specifically, $\alpha_{basal} \ll 1$ for basal faceting or $\alpha_{prism} \ll 1$ for prism faceting. Turning this around, the higher α_{facet} becomes, the more likely it will be that a faceted surface

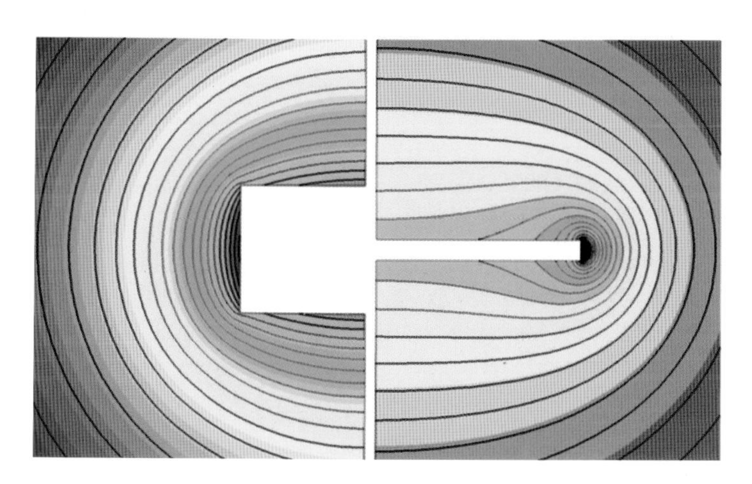

FIGURE 3.24. Calculated contour plots of supersaturation levels around two growing platelike ice crystals, shown here in (r, z) coordinates. Around a nearly isometric crystal (left), σ_{surf} is highest near the corners of the faceted prism, a phenomenon called the "Berg effect." Around a thin-plate crystal, the contour lines are tightly bunched at the fast-growing plate edge, while σ_{surf} is highest at the centers of the basal facets. The model on the left assumes $\alpha_{basal} \approx \alpha_{prism} < 1$, while the model on the right assumes $\alpha_{basal} \ll \alpha_{prism}$.

will be susceptible to some form of the branching instability. When $\alpha_{facet} \approx 1$, faceting is no longer possible at all, yielding rounded (unfaceted) surfaces and highly branched structures.

This fundamental feature of the Mullins-Sekerka instability provides a straightforward explanation for the increasing morphological complexity with increasing σ_∞, which is one of the principal characteristics of the Nakaya diagram. To see this, note that σ_{surf} generally increases with σ_∞, and further note that α_{facet} typically increases strongly and monotonically with σ_{surf} (Chapter 4). Together these statements imply that α_{facet} must increase with increasing σ_∞. Because branching invariably increases as α_{facet} increases, the unavoidable conclusion is that branching must increase at higher σ_∞. Of course, this basic reasoning glosses over many details, and computational modeling will be required to fully comprehend how morphology depends on supersaturation. But the overarching conclusion is that snow crystal morphologies will be more complex at higher supersaturations, as seen in the Nakaya diagram.

Interestingly, the tip radius and sidebranching spacing in snow crystal dendrites does not change substantially with supersaturation once the dendritic structure is well established. This follows from Equation 3.47, which shows that R_{tip} is independent of σ_∞, a result that generally agrees with observations. At sufficiently low σ_∞, however, α drops and R_{tip} increases, eventually yielding faceted morphologies. Adding an additional pressure axis to the Nakaya diagram would reveal that R_{tip} decreases with increasing pressure.

Stellar Dendrites Near –15°C

Although often overlooked, the primary morphological feature of a stellar dendrite snow crystal is the fact that it is thin and flat. Aspect ratios can be as low as 0.01 for thin plates, and I like to say that this extreme aspect ratio is what puts the "flake" in "snowflake." Following the discussion above, the platelike aspect ratio immediately demands that $\alpha_{basal} \ll \alpha_{prism}$. Moreover, rounding on the branch tips indicates that $\alpha_{prism} \approx 1$ there, because rounding indicates that rough and faceted surfaces are growing at nearly the same rate. If α_{prism} were substantially below unity, prism faceting would be more prevalent. Thus, just looking at a large stellar dendrite with rounded branch tips reveals that $\alpha_{prism} \approx 1$ at the tips and $\alpha_{basal} \approx 0.01$. In the words of Yogi Berra, you can observe a lot just by watching.

Stellar dendrites are also a good illustration of the complex interplay between branching and faceting. Because $\alpha_{basal} \ll 1$, basal faceting dominates the c-axis dimension of the crystal structure. At the opposite extreme, the fact that $\alpha_{prism} \approx 1$ means that prism faceting is quite weak and susceptible to branching and sidebranching. Thus, both the aspect ratio and the degree of sidebranching are determined by the attachment coefficients on the two primary facet surfaces.

If σ_∞ around a growing stellar dendrite is high, then σ_{surf} becomes relatively high as well, sending $\alpha_{prism} \rightarrow 1$ via the Edge-Sharpening Instability (Chapter 4), stimulating copious sidebranching. But if σ_∞ and σ_{surf} are lower, then α_{prism} is lower and the branches exhibit greater prism faceting. Thus, the ESI mechanism contributes to why higher σ_∞ yields more complex branched structures. Similarly, σ_{surf} is typically highest near the branch tips, so these are often rounded, while σ_{surf} is lower near the crystal center, yielding more prism faceting in the central region. Indeed, photographs of stellar dendrite crystals often reveal greater prism faceting in the inner parts of the crystals.

We see that many morphological characteristics of stellar dendrites can be explained from the detailed behavior of α_{basal} and α_{prism} as functions of σ_{surf}, along with subtle effects like the ESI that can greatly affect the morphological development of a crystal. The big challenge in making computational models is to reproduce the full range of observed growth behaviors. Doing this requires a comprehensive model of the attachment kinetics (Chapter 4) together with numerical algorithms that can accurately model the growth process (Chapter 5). And confirming that the models work correctly necessitates making quantitative comparisons with laboratory measurements of complex morphologies (Chapter 8). Putting all these pieces together is a task that has not even begun to an appreciable degree.

Hollow Columns and Needles Near −5°C

Once again, the primary morphological feature of snow crystal columns and needles is their large aspect ratio, which can be 20 or more for an especially slender needle. The anisotropy rule applies here as well, so an overall columnar shape indicates $\alpha_{prism} \ll 1$ for moderate supersaturations near −5°C. One can create a simple diffusion model for a faceted column, analogous to the models in Figure 3.24, and the results are similar to what was discussed above. If the anisotropy in the attachment kinetics is sufficiently high, then the supersaturation around a slender column is lowest near the basal surfaces, accompanied by steep supersaturation gradients needed to achieve the faster basal growth.

Fully faceted, platelike prisms are the norm when σ_∞ is sufficiently low at −5°C (Chapter 7), but hollow columns form when the supersaturation increases to intermediate values. The basic hollowing mechanism is a form of the Mullins-Sekerka instability illustrated in Figure 3.25. Diffusion-limited growth causes σ_{surf} to be higher at the edges of a basal facet compared to the facet center, and soon the facet edges grow upward and leave the center behind, resulting in conical hollow regions on both ends of the column. This behavior is analogous to the formation of branches on platelike crystals, except now the edges of the basal surface remain faceted, or nearly so, as the hollows develop. As described in Chapter 4, the Edge-Sharpening Instability augments the normal Mullins-Sekerka instability in this case, adding an additional layer of subtly to the overall problem. Here again, computational modeling will be needed to fully comprehend and reproduce laboratory measurements of hollow-column growth but the basic interplay of diffusion-limited growth and anisotropic attachment kinetics is clearly a repeating theme in the symphony of snow crystal growth dynamics.

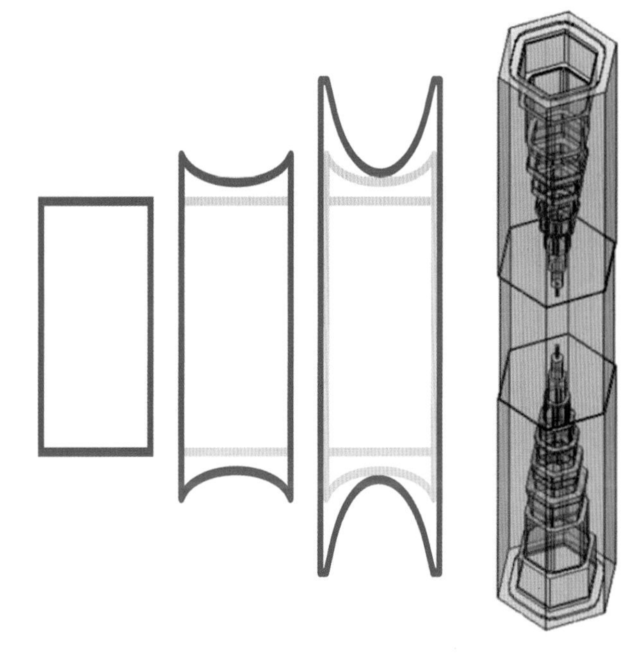

FIGURE 3.25. A schematic diagram illustrating the transition from solid columnar growth (first sketch, showing a side view of a solid column) to the formation of a hollow columnar snow crystal (third sketch) via the Mullins-Sekerka instability. The image on the right shows a 3D numerical simulation of hollow-column growth, adapted from [2009Gra].

Hollow columns appear when three conditions are met: 1) $\alpha_{prism} \ll \alpha_{basal}$, 2) $\alpha_{basal} \approx 1$, and 3) the supersaturation is not too low (which would yield solid prisms) and not to high (which would yield needle-like crystals). These conditions are often found at temperatures around −5°C, so this is why hollow columns are most prevalent at this region of the snow crystal morphology diagram. The conditions can also be met at much lower temperatures, around −40°C, and these low-temperature hollow columns can also be found in the atmosphere.

Figure 3.26(a) shows an example of a −5°C hollow columnar snow crystal that exhibits long conical voids. In this crystal, the hollow regions changed their growth behavior slightly as the external conditions changed,

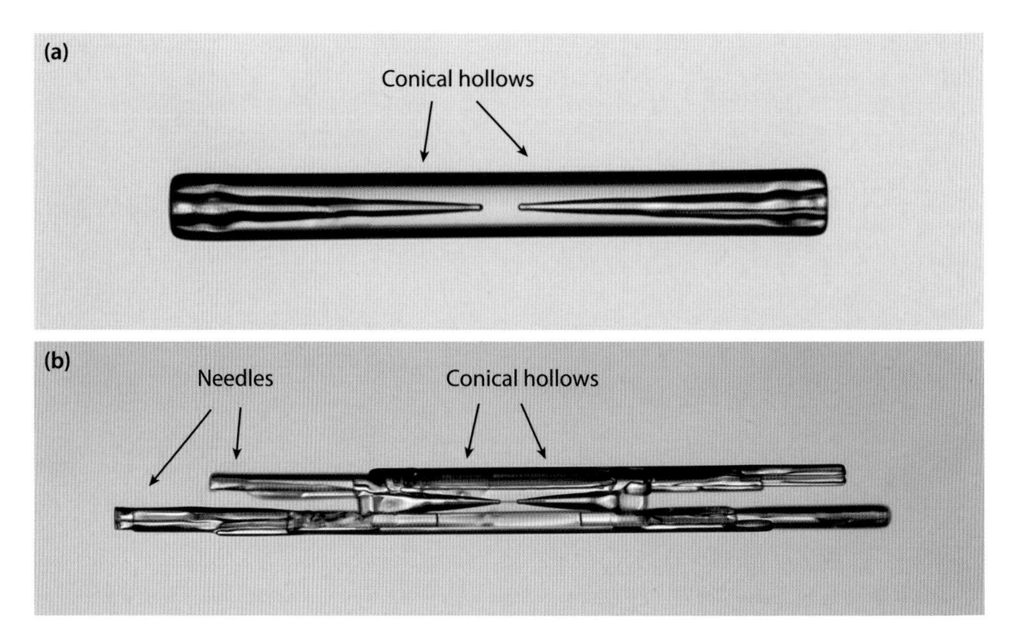

FIGURE 3.26. (a) This hollow column snow crystal shows a characteristic matched pair of conical hollow voids in the ice. (b) This crystal began as a solid prism when it was small but soon transformed into a hollow column, leaving behind central conical voids in the ice as it grew. Later, the corners of the basal edges sprouted branches that developed into a set of slender ice needles.

yielding a wavy structure in the shape of the hollows. Because both ends of the column experienced the same growth conditions as a function of time, the shape variations on the two ends of the column are nearly symmetrical.

As the growth of a hollow column continues, often the basal edges are no longer able to maintain their faceted shape, as they too succumb to the branching instability. When this happens, the basal edges can split into slender needles, as shown in Figure 3.26(b). Note how the initial conical voids are still present near the center of this crystal, illustrating the transition from a solid column at the earliest stage of growth to a hollow column, and finally to a set of needle-like branches sprouting from the basal corners. Note also that the very center of a hollow column can never itself be hollow, as there would be no mechanism that would yield such a structure from a small seed crystal.

The successful numerical simulation of faceted hollow columns was an excellent early achievement for the 3D cellular-automata method [2009Gra], which I describe in Chapter 5. Moreover, hollow columnar behavior can be reproduced and studied in the lab using electric needle crystals, as described in Chapter 8. Making quantitative comparisons between laboratory observations and numerical models is thus quite feasible, but work along those lines is only just beginning.

Bubbles in Columns

Under appropriate circumstances, the conical hollows in a hollow-column snow crystal can develop into enclosed bubbles, as illustrated in Figure 3.27. I have also created enclosed columnar bubbles in the lab using electric needles, and an example is shown near the end of Chapter 8. To form such bubbles, the first step is to create a hollow column, as described previously, followed by a period of growth at lower supersaturation that seals off the hollow ends.

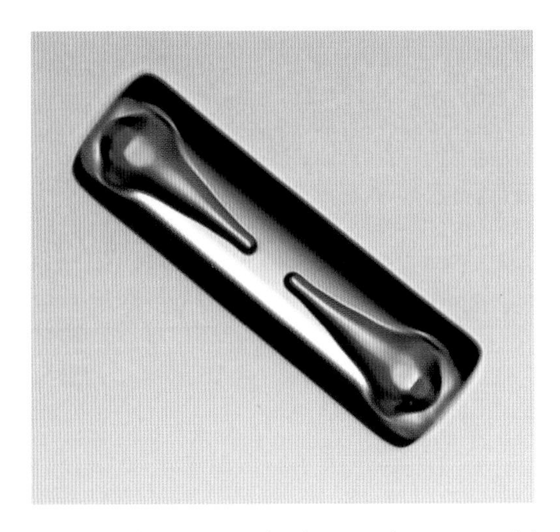

FIGURE 3.27. This photograph of a natural snow crystal shows enclosed bubbles in an ice column.

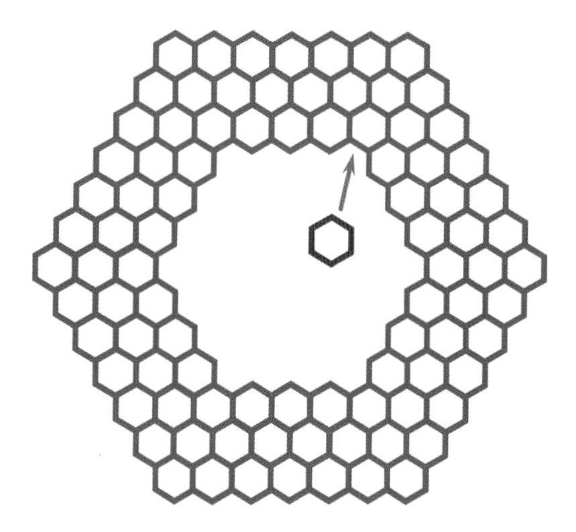

FIGURE 3.28. Diagram of the end of an idealized nanoscopic hollow columnar snow crystal, with hexagons representing molecular cells in the ice lattice. While a nucleation barrier prevents growth of the outer faceted surfaces, there is no nucleation barrier on the inner surfaces, because molecules can always attach at interior corners (red hexagon).

Figure 3.28 illustrates how different nucleation dynamics on convex and concave surfaces facilitates the sealing-off process. The outer surfaces of a growing column soon become faceted because of the usual nucleation barrier that makes $\alpha_{prism} \ll \alpha_{rough} \approx 1$, yielding a hexagonal column. Inside the hollow region, however, there are always interior corners at which there is no nucleation barrier. In the figure, for example, the red hexagon (representing an idealized molecular cell) can readily attach at the corner shown, as this position is essentially the same as the edge of a terrace step on a faceted surface. Because of this mechanism, the growth of concave surfaces is never limited (in a global sense) by a nucleation barrier. When a hollow column is exposed to a relatively low supersaturation, therefore, the strong nucleation barrier on the outer faceted surfaces slows additional growth. But the inner surfaces lack this nucleation barrier, so they grow readily under the same conditions. At the same time, diffusion brings more water vapor molecules to the columnar ends than to regions deep inside the hollows, so growth at the ends is preferred. Putting all this together, the inner surfaces near the columnar ends grow fastest, soon sealing off the conical hollow regions to form enclosed bubbles.

Hollows and Bubbles in Plates

Hollow plates are essentially like hollow columns, except with the basal and prism roles reversed. The underlying physics is essentially the same as with hollow columns, and Figures 3.29 and 3.30 show two examples of hollow plates, the latter also exhibiting some enclosed bubbles. The formation of "corner pockets" [2019Nel] shows some similar characteristics. If a hollow region evolves into an enclosed bubble, the void becomes essentially a closed system unaffected by the supersaturation field surrounding the crystal. In this isolated state (neglecting any temperature gradients in the crystal), the bubble would naturally evolve toward its equilibrium shape, which is nearly spherical (Chapter 2). However, relaxation toward equilibrium is significantly hindered by a nucleation barrier on the interior faceted surfaces, as shown in Figure 3.31. Here we see that while the growth of interior concave surfaces is never limited by a nucleation barrier, evaporation from the interior basal surfaces is strongly limited by a hole-nucleation barrier. For this

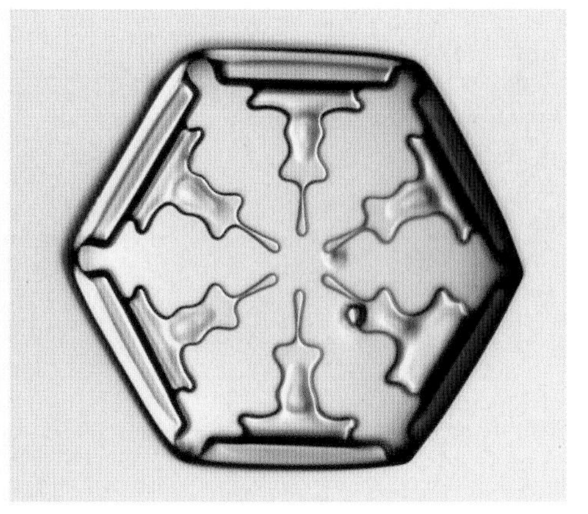

FIGURE 3.29. The essential geometry of a hollow-plate snow crystal (left) and a photograph of a natural snow crystal with deep hollows (right). Note that the six hollow regions are separated by solid ice at the hexagonal corners. The photo exhibits oddly shaped hollow regions that reflect the changing conditions the crystal experienced during its growth.

FIGURE 3.30. A natural snow crystal exhibiting deep hollow regions in each of the prism surfaces. Near the center of the crystal, some of the hollows have closed off to form thin bubbles in the ice. The colors arise from optical interference between reflections off the top and bottom surfaces of the hollows/bubbles, which are separated by about one wavelength of light (see Chapter 11). Photo courtesy of Don Komarechka [2013Kom].

reason, even an exceedingly thin bubble in a platelike snow crystal may retain its nonequilibrium shape for long periods of time.

Ridges and Sectored Plates

Ridge structures are commonly found in both natural and synthetic snow crystals, and Figure 3.32 shows a particularly simple example. Here the six ridges are thick linear features on an otherwise thin plate. The ridges divide the hexagonal plate into six equal sectors, like slices of a hexagonal pie, so snow crystals displaying these distinctive features are called sectored plates. Sectored-plate snow crystals typically grow near −15°C at intermediate supersaturations when the growth conditions are nearly constant in time. Laboratory observations reveal that ridges are associated with slightly convex basal surfaces, such as those shown in the sketch in Figure 3.32.

Figure 3.33 shows a diagram of the growth process that leads to ridge formation. The convex basal surface includes a series of regularly spaced molecular terrace steps, and the step spacing defines the slope of the surface, like a set of contour lines on a topographic map. As

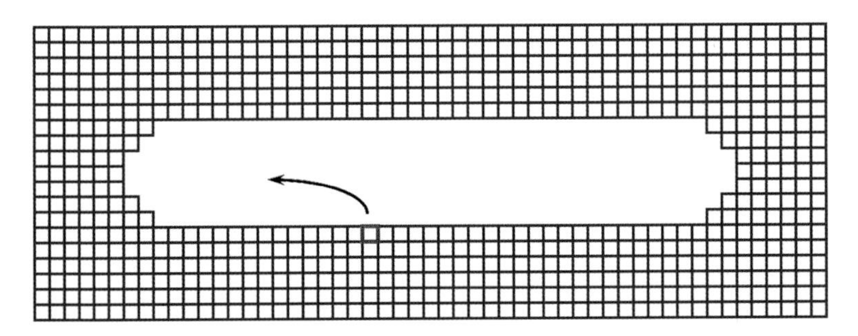

FIGURE 3.31. Even if the initial shape of an enclosed bubble is highly nonspherical, a faceted bubble may evolve exceedingly slowly toward its nearly spherical equilibrium shape. It is difficult to remove molecules from a fully faceted surface (red), and this presents a strong hole-nucleation barrier that can greatly slow equilibration. For this reason, even very thin bubbles in platelike crystals can retain their nonequilibrium shapes for long periods of time.

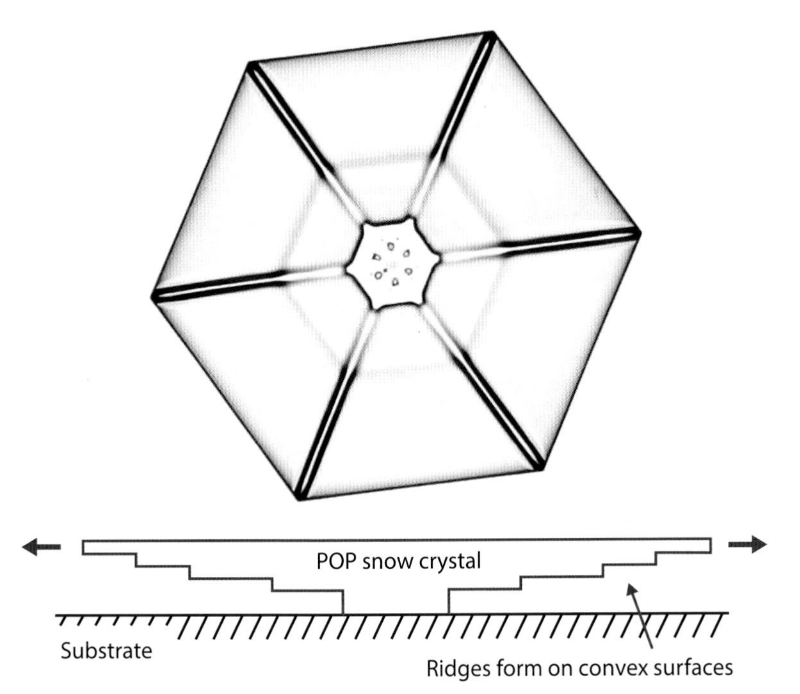

FIGURE 3.32. This small PoP snow crystal (top) exhibits simple ridges that originate at the six faceted corners as the plate grows outward. As shown in the accompanying sketch (bottom), the top basal surface of this crystal is essentially flat and featureless, while the ridges and other visible structural features exist on the convex lower surface of the crystal.

the faceted prism edges grow outward (arrows in Figure 3.33) the lower terrace edges grow outward also, although the step velocity need not be the same as the edge velocity. A 2D manifestation of the Mullins-Sekerka instability comes into play on the molecular steps, enhancing the corner growth, as diffusion brings a greater supply of water vapor to the step corners. Each terrace corner thus sprouts a one-molecule-high "branch," as shown in the figure, and these linear branches combine to form a macroscopic ridge. Note that the closely spaced contours around the ridge indicate its steep vertical sides, like a ridge on a topographic map.

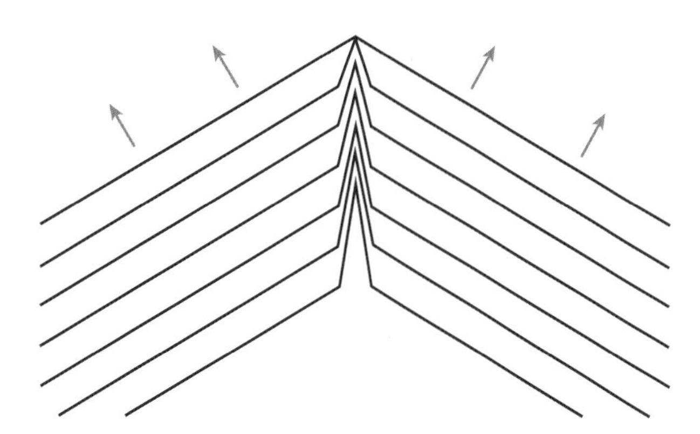

FIGURE 3.33. The development of a snow crystal ridge on a convex basal surface. Lines represent molecular steps defining individual terraces. As the prism facet edges grow outward (arrows), diffusion enhances the growth of the terrace corners, leading to ridge formation.

There are two ways to establish and maintain the convex basal shape needed to form ridges: 1) nucleation of new basal terraces near the center of the convex basal facet, or 2) nucleation of new basal terraces near the edges of the opposing basal surface. The second mechanism dominates in Figure 3.32, and the necessary conditions for this to happen are imposed by the laboratory environment. The first mechanism is likely responsible for formation of natural sectored plates, although their growth may also be affected by the aerodynamic effects described above. The overall structure of ridges on natural snow crystals has not been well studied [1954Nak] but it appears to fit this basic convex-basal-surface picture.

FIGURE 3.34. This numerical model of a growing stellar crystal exhibits clear ridging on the primary branches [2014Kel]. The micron-scale steps in the model are orders of magnitude larger than molecular steps, but the underlying diffusion-driven ridge growth is essentially the same. Image courtesy of James Kelly.

The relatively simple nature of ridge formation means that ridges are readily found in 3D numerical simulations of snow crystal growth using cellular automata, as shown in Figures 3.34 and 3.35. It is not necessary to have completely accurate attachment kinetics in these models to produce ridgelike structures, as ridging requires only that diffusion limits the growth of basal terraces. These examples of ridge formation suggest that many new insights await when we move beyond the demonstration phase in computational modeling and are finally able to make direct quantitative

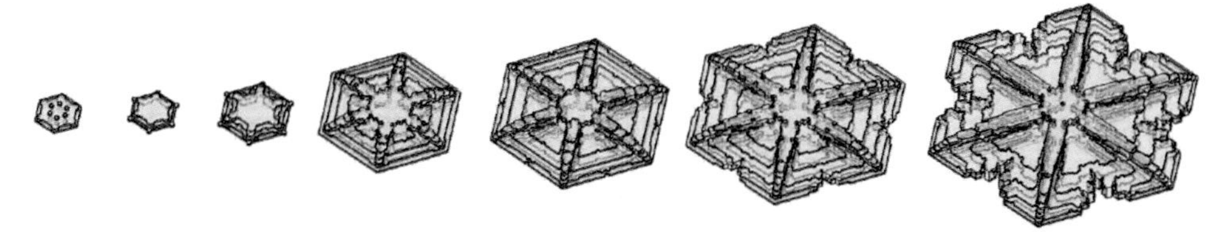

FIGURE 3.35. This 3D numerical model also shows ridge structures on slightly convex basal surfaces [2009Gra]. It appears that ridge formation is readily seen in both natural snow crystals and numerical modeling, at least when using the cellular automata method. The robustness of ridging seems to reflect the basic diffusion physics underlying the phenomenon, which is insensitive to other material parameters.

comparisons between snow crystal models and laboratory experiments.

Ridges with Grooves

In some instances, snow crystal ridges are flanked by linear "grooves" that are long, shallow depressions in the ice on either side of a ridge, as shown in Figure 3.36. The formation of these grooves appears to be yet another example of the Mullins-Sekerka instability relating to step growth, an additional feature on top of basic ridge formation described above. Once a ridge begins to form, as shown in Figure 3.37, it sticks up above the basal surface surrounding it, and α on the sides of the ridges is close to unity. The ridge growth thus attracts a great deal of water vapor, depleting the air nearby. As diagrammed in the figure, the presence of the high-α ridge means that the growth velocity of a step far from the ridge is larger than the velocity of the same step adjacent to the ridge. This rather subtle dance of step advancements, choreographed by particle diffusion around the growing crystal, results in ridges flanked by parallel grooves.

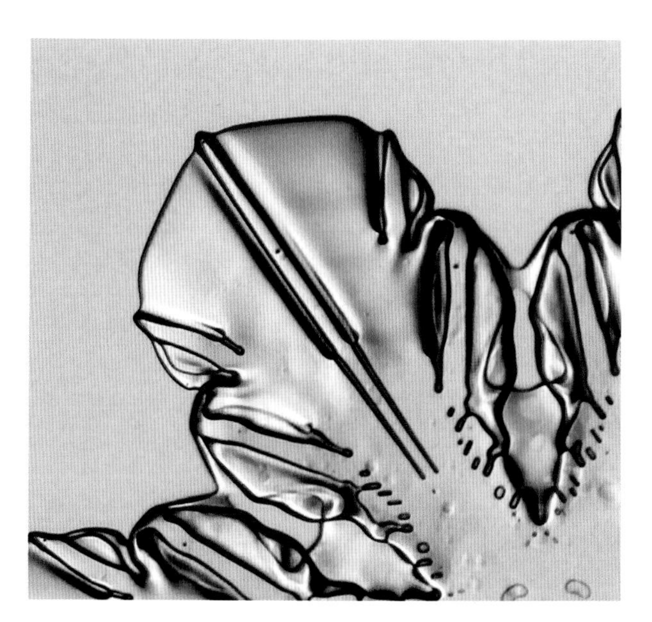

FIGURE 3.36. This partially sublimated sectored-plate snow crystal exhibits a pair of grooves flanking the central ridge. Similar features can be found in many snow crystals, although usually they are not as distinctive as in this example.

Chemical impurities in the air may further aid in the formation of these grooves, along with perhaps some of the small pits seen in Figure 3.36 and other snow crystals. Impurities are not readily incorporated into the ice lattice, and an advancing molecular step will tend to push impurity molecules ahead of it as it grows. Thus, although the average density of impurity molecules on the ice surface may be quite low, step motion will redistribute and concentrate those impurities that are present. Looking at ridge and groove formation, as thousands of steps march along during the process, their collective motion will tend to push impurities into the grooves and pits, where they will remain, stuck on the ice surface. The concentrated region of chemical crud could then substantially impede further ice growth, and the grooves would remain unfilled in the ice. Additional laboratory experiments would be needed to investigate whether chemical impurities really have such effects on snow crystal surface features.

Ridges on Cones and Cups

The detailed structure of ridges in snow crystals depends a great deal on the "cone angle" of the plate on which they grow. The previous discussion assumed a small cone angle, by which I mean a nearly flat basal surface that is slightly convex in overall shape. (Of course, this is not a true cone in the strict geometrical sense, but a roughly conelike shape made from six slightly tapered flat surfaces.) This morphology includes the trains of propagating steps shown in Figures 3.33 and 3.37 that are necessary to produce ridge structures.

One can extend the discussion further to include steeper cone angles, progressing from nearly flat plates to cuplike structures like those shown in Figure 3.38. The outer surfaces of a cup also include trains of molecular steps, but with much higher step densities. The same ridge-formation instability applies, but now the ridges

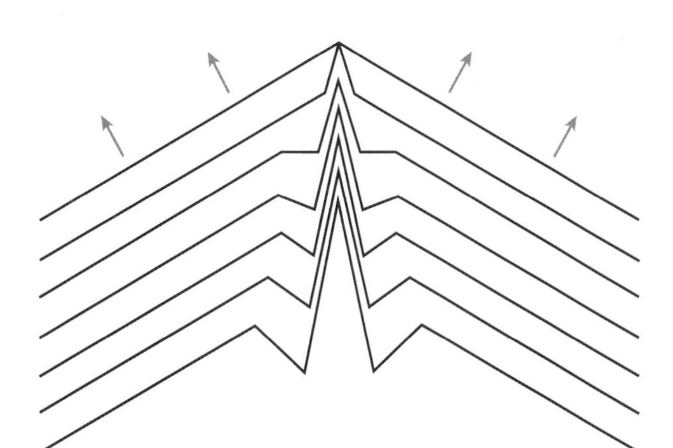

FIGURE 3.37. The development of a snow crystal ridge flanked by two grooves on a convex basal surface. As the ridge structure develops, it depletes the water vapor supply in its proximity. This suppresses the advancement of steps near the ridge, thereby creating grooves.

develop into the pronounced "fins" shown in the figure. Note that because the supersaturation is relatively low below the fast-growing cup edge, the fins develop nearly faceted prism surfaces. Figure 3.39 shows another ridge morphology that readily occurs on e-needles over a certain range of conditions when the cone angle is intermediate between plates and cups. Here the ridges grow out to form what are essentially stubby fins, but then platelike extensions grow out from the base of the fins, yielding what I call an "I-beam" structure. This feature can be found in natural snow crystals as well, but it is especially clear on e-needles because the growth conditions can be kept constant for long periods of time. The lower plates on the I-beams are another example of how readily thin plates emerge in snow crystal growth.

All the ridge structures described above are easily created in the laboratory under constant growing conditions, especially on e-needles. Moreover, the ridge morphology is quite robust, with different variations appearing over a broad range of temperatures and supersaturations. It is difficult to explain their structure simply, owing to the complex balance of faceting, branching, and step motions that must be happening. For this reason, however, ridges should prove to be a good test of future 3D numerical modeling techniques. Once mod-

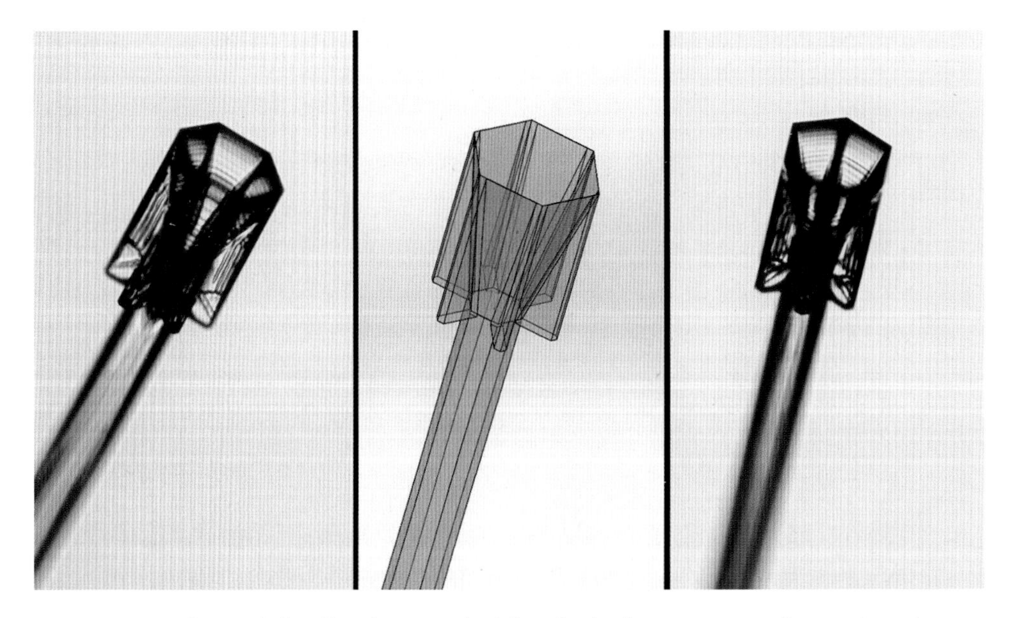

FIGURE 3.38. Cups with fins. The photos on the left and right show two views of a cup-shaped snow crystal growing on the end of a slender ice column. The 3D drawing (center) illustrates the main structural features, including six platelike fins that are related to snow crystal ridges. The laboratory crystals were grown on an e-needle (see Chapter 8) near −7°C. SolidWorks drawings by Ryan Potter.

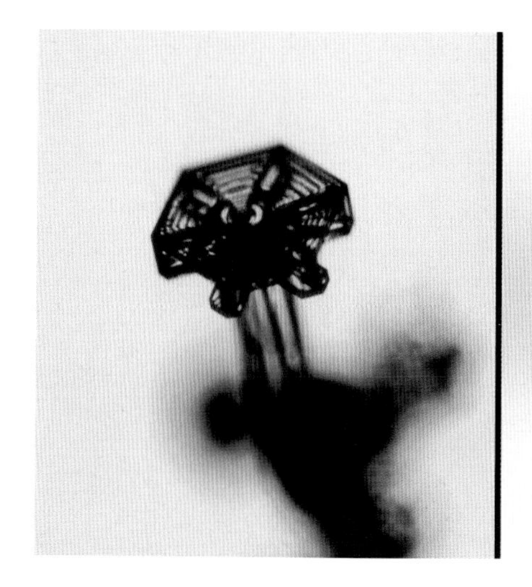

FIGURE 3.39. I-beams. The photo on the left shows a snow crystal growing on an e-needle near $(T, \sigma) = (-9°C, 16 \text{ percent})$, and its overall structure is illustrated in the drawing on the right. Here the plate has an intermediate cone angle, yielding short fin-ridges that subsequently developed platelike extensions, resulting in an overall I-beam ridge structure. This morphology is remarkably robust on e-needles, occurring over a considerable range of growth conditions (see Chapter 8). SolidWorks drawings by Ryan Potter.

els are able to reproduce these kinds of complex structures, especially with growth rates that match observations, we will finally be able to say that we have made serious progress toward providing realistic simulations of snow crystal structure formation.

Ribs on Plates

While ridges readily appear under constant environmental conditions, other common snow crystal structural features require changing conditions for their formation. One prominent example is the creation of hexagonal "ribs" like those shown in Figure 3.40. In the first of these crystals, the ribs form a set of hexagonal rings where the ice is a bit thicker than elsewhere in the plate. In the second example, the ribs are restricted to the crystal's outer platelike extensions, where they exhibit the same overall hexagonal structure. In both cases, the ribs are accompanied by ridges that divide the plates into sectors. Both ribs and ridges are frequently found in natural snow crystals as well, as described in Chapter 10. As with other growth phenomena in this section, ribs are especially nicely demonstrated using laboratory-grown Plate-on-Pedestal (PoP) snow crystals, where the growth conditions can be well controlled and quickly modified at will (see Chapter 9).

Figure 3.41 illustrates how a temporary drop in supersaturation surrounding a snow crystal can lead to the formation of a rib on a growing plate. The top sketch shows the outer edge of a PoP crystal growing at a relatively high supersaturation near −15°C. These conditions result in the formation of thin plates, and the sketch shows a plate that is flat on the upper surface and slightly convex on the lower surface, which is typical of PoP crystals. This thin-plate morphology continues as long as the supersaturation remains high. Upon lowering the supersaturation (second sketch in Figure 3.41), the edge-sharpening instability (ESI) is diminished, yielding subsequent growth as a thicker plate. Note that the faceted basal and prism surfaces grow slowly at low σ because of the usual nucleation barriers on those surfaces. The underside of the plate begins as a vicinal surface, however, on which there is no nucleation barrier. Thus, the underside grows relatively quickly, especially near the edge of the plate, as illustrated in the figure. Soon a thick "rim" of ice emerges on the edge of the plate. Increasing the supersaturation to its previous high level (third sketch), the ESI again kicks in and a thin plate grows out from the upper edge of the thicker rim. As this thin plate grows outward, it shields further growth below it, leaving a thick rib structure behind.

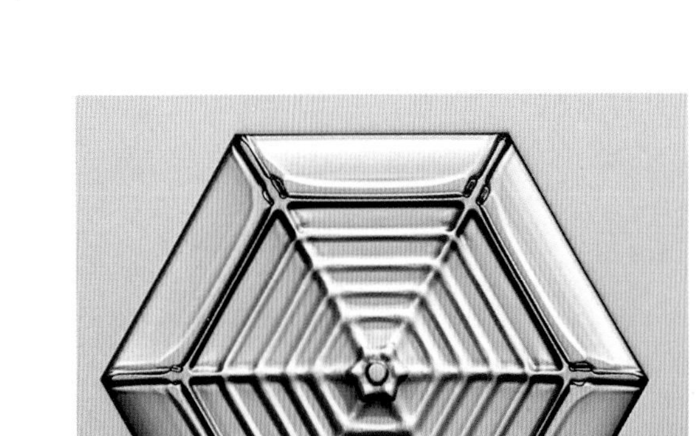

FIGURE 3.40. While growing these PoP snow crystals, I periodically reduced and then increased the supersaturation, yielding a spider-web structure of ribs and ridges. Hexagonal rib patterns like these are typically associated with changes in external growth conditions.

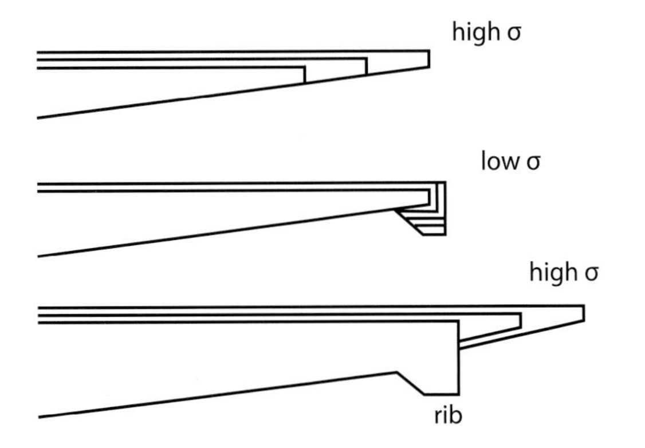

FIGURE 3.41. This series of sketches chronicles the formation of a snow crystal rib on the underside of a PoP snow crystal. Starting with a thin plate (top), lowering the supersaturation yields a thicker edge (middle). Restoring the high supersaturation yields a thin plate again, leaving a rib structure behind on the underside of the plate (bottom).

The qualitative explanation of rib formation in Figure 3.41 was easily confirmed by observing the growth of PoP crystals in real time while adjusting the supersaturation in the process. Both ribs and ridges were easily created, and it was straightforward to confirm that both these features were confined to the lower convex basal surfaces. The multitude of features seen in natural snow crystals are generally much more difficult to interpret, as natural crystals often exhibit a cacophony of ridges and riblike features that reflect the ever-changing and unknown conditions in which they grew.

Inwardly Propagating Rings

Just as ridgelike structures are common on slightly convex basal plates growing under constant environmental conditions, inward-propagating rings readily form on slightly concave basal plates, and two examples are shown in Figure 3.42. Similar rings appear quite frequently on PoP snow crystals (see Chapter 9), owing to their unique geometry of thin plates that are slightly conical in overall shape, as illustrated in the accompanying sketch in the figure. Inwardly propagating rings can also be found on natural snow crystals, but they are somewhat rare because quite specific growth conditions are required.

Under uniform growth conditions, one might naively expect that a steady creation of new terraces on the top basal surface might yield a simple vicinal surface with roughly uniform spacing between steps. In fact, while such a surface is a valid solution to the diffusion equation, it is not a stable solution. The ubiquitous Mullins-Sekerka instability, along with possible additional effects from molecular surface diffusion, results in a phenomenon called *step bunching*. As the name implies, isolated steps soon bunch together to form *macrosteps* that are large enough that they can be seen using optical microscopy, as illustrated in Figure 3.42.

Step bunching can be the result of several different physical effects, so disentangling these for the case of ice growth is not a trivial task. Bulk diffusion almost certainly plays a role via the Mullins-Sekerka instability, but surface diffusion effects might be important also. Because there is no clear model of step bunching in ice, it is not yet possible to calculate the average macrostep height or (equivalently) the spacing between macrosteps for a given vicinal angle. As with so many features in snow crystal growth, macrostep phenomena are easily observable, but not so easily understood in detail.

Note that diffusion effects on inwardly propagating rings causes them to evolve toward a generally circular shape. Because water vapor diffuses in from the supersaturated air surrounding the crystal, the step growth is faster for steps nearer the outer edges. Thus, any deviation from a circular shape is corrected by the growth dynamics. For inward-propagating steps (on slightly con-

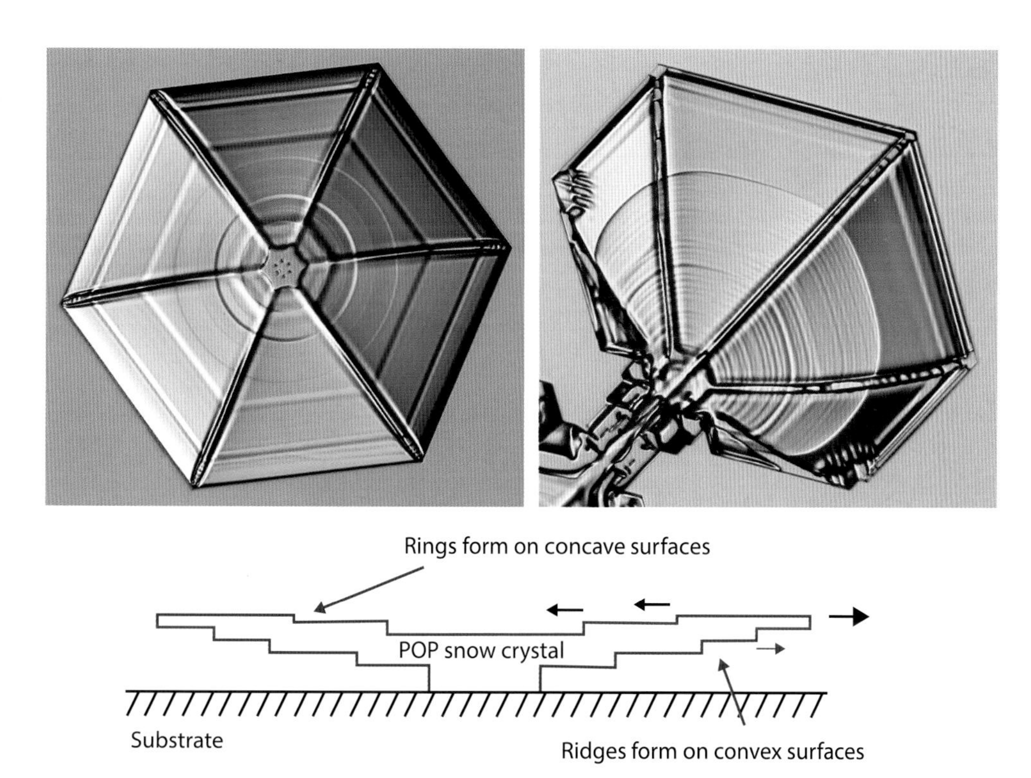

FIGURE 3.42. (Top left) This small hexagonal PoP crystal exhibits ridges, ribs, and a pair of inwardly propagating rings. (Top right) A series of inwardly propagating rings appears on this sectored plate branch, while the top basal surface is flat beyond the outermost ring. (Bottom) The sketch illustrates the growth of inwardly propagating rings on the upper concave surface of a PoP crystal along with ridges and ribs on the lower convex surface. This interpretation of these features is made somewhat easier by the clear distinction between the concave upper surface and the convex lower surface in PoP crystals. Patterns on natural snow crystals can be substantially more complex and difficult to decipher.

cave basal surfaces), therefore, diffusion-limited growth brings about a stabilizing effect that maintains a simple circular shape. In contrast, Figures 3.33 and 3.37 shows how diffusion-limited growth on outward-propagating steps (on a slightly convex basal surfaces) yields a form of the branching instability. The water vapor supersaturation is highest near the outer corners of the crystal, so terrace step branches soon form at each corner, and repeated branching on multiple steps leads to the formation of macroscopic ridges, as described previously. A small change in the basal surface geometry, from slightly concave to slightly convex, thus yields a large change in overall growth behavior.

The intrinsically asymmetrical PoP construction often produces outward-propagating terrace steps on the lower basal surface and inward-propagating steps on the upper basal surface. Ridges then develop on the lower convex surface, as described above, while rings appear on the upper concave surface. There is essentially no interaction between the ridges and rings, and indeed these structures appear to be independent of one another when observed as a function of time.

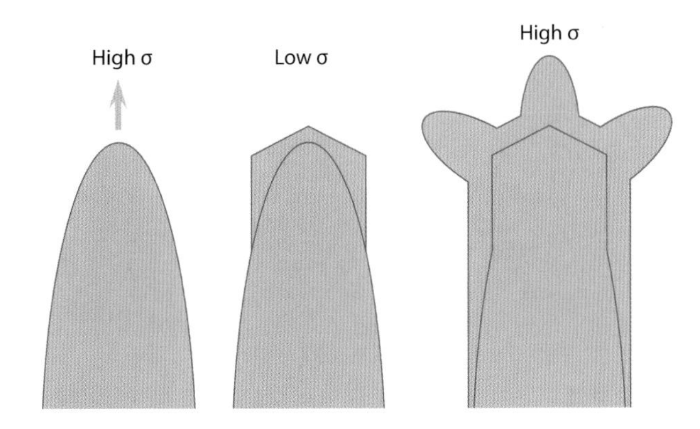

FIGURE 3.43. Induced sidebranching. When the supersaturation is sufficiently high near −15°C (left), α_{prism} will be near unity at the tip of a growing dendrite branch, yielding a rounded shape with little prism faceting. Upon lowering the supersaturation (middle), α_{prism} is reduced, and the tip becomes faceted. Increasing the supersaturation once again (right), branches sprout from the three exposed corners of the faceted tip. One branch continues in the primary direction, while the other two become sidebranches. Looking at the whole crystal, this mechanism creates a coordinated set of sidebranches on all six primary branches. Induced sidebranching is thus responsible for much of the complex symmetry seen in stellar dendrite snow crystals.

Induced Sidebranching and Complex Symmetry

Changes in environmental conditions can profoundly affect the growth of many snow crystal forms, and one particularly important phenomenon is *induced sidebranching*, illustrated in Figure 3.43. During this process, a short reduction in the supersaturation yields prism facets on a growing branch tip, and a subsequent increase in the supersaturation will stimulate the growth of sidebranches from the prism corners. If this series of events happens with a growing stellar dendrite, the result can be the coordinated appearance of sidebranches on all the main branches. In other words, induced sidebranching is one method for producing complex dendritic structure that includes an overall sixfold symmetry. Figure 3.44 illustrates this same process on a laboratory-grown PoP crystal.

One can contrast the induced-sidebranching process with the growth of a fernlike stellar dendrite like the one shown earlier in Figure 3.9. This crystal grew under nearly constant growth conditions with a high supersaturation, so there were no induced sidebranching events. Instead, the sidebranches were stimulated by random perturbations that occurred separately on each of the six main branches. Even the sidebranches on either side of a single main branch were not synchronized.

The induced-sidebranching process shows in detail that no direct internal communication between the different branches is needed to induce simultaneous sidebranching and large-scale snow crystal symmetry. Changes in the externally applied environmental conditions are sufficient to bring this about. It is straightforward to introduce many abrupt changes as a PoP snow crystal is growing, and one result is shown in Figure 3.45. Essentially all the large-scale symmetry seen in this ex-

FIGURE 3.44. This composite photograph shows the phenomenon of induced sidebranching on a laboratory-grown PoP snow crystal. The left image shows the crystal after long branch tips were first grown out at high supersaturation. Then the branch tips became faceted after a short period of low supersaturation (center image). Increasing the supersaturation again then caused central branches and sidebranches to sprout simultaneously on the tips of all the primary branches (right image).

ample was created by induced-sidebranching events or other longer-time changes in the growth conditions. This kind of large-scale complex symmetry generally does not arise if the growth conditions are held constant in time.

Having made numerous movies of growing PoP snow crystals, they have an almost magical final appearance because the viewer cannot discern the temperature or humidity from the images alone. Watching the video, sidebranches appear simultaneously on all the primary branches from no apparent cause. Making the movie is a different experience, however, as I consciously change the growth conditions to produce different effects at different times, with predictable outcomes that can be seen in real time. This experience makes it abundantly clear that the choreography and symmetry of a complex snow crystal is almost entirely determined by time-varying externally applied growth conditions.

The natural snow crystal shown in Figure 3.1 exhibits a great deal of chaotic dendrite growth that produced the helter-skelter sidebranching typical of fernlike stellar dendrites. But the largest set of sidebranches, occurring at the same location on all six primary branches, were almost certainly the product of an induced-sidebranching event.

Tridents and Triangular Snow Crystals

While sixfold symmetry is a snow crystal hallmark, small plates occasionally exhibit a threefold symmetry like that shown in Figure 3.46. Note that the angles of the prism facets are the same as with a normal hexagonal prism, but now there are alternating long and short facets, giving the overall appearance of a truncated equilateral triangle. Triangular plates like these typically appear together with hexagonal plates in natural snowfalls, although the latter are always much more common [1973Yam].

We did a brief study looking at the statistics of triangular plates by growing small platelike snow crystals in a free-fall growth chamber in air near −10°C with $\sigma_\infty \approx 1.4$ percent [2009Lib3]. Small hexagonal plates are the normal morphology under these conditions [2008Lib1, 2009Lib], but about five percent of the crystals exhibited a truncated triangular morphology. Figure 3.47

FIGURE 3.45. The high degree of complex symmetry seen in this PoP snow crystal did not emerge spontaneously; I imposed it using a series of induced-sidebranching events. Induced sidebranching is the primary mechanism that coordinates the growth of sidebranching on stellar snow crystals, both in the lab and in nature.

shows some examples of these and other non-hexagonal morphologies observed. Crystals with nearly perfect equilateral-triangle morphologies were also readily found in this sample.

We first measured an unbiased sample of all simple platelike crystals and defined a "hexagonality" parameter $H = L_1/L_6$ as the ratio of the length of the shortest side to that of the longest side. While $H = 1$ for a perfect hexagonal prism, we found that any crystal with $H > 0.75$ had a generally hexagonal appearance by eye. While many crystals in our sample exhibited a roughly hexagonal appearance, these data indicated that near-perfect hexagons ($H \approx 1$) were somewhat rare. We then examined a larger sample from which we rejected crystals with $H > 0.33$, and in this non-hexagonal sample we defined a "triangularity" parameter $T = L_3/L_4$ as the ratio of the lengths of the third and fourth longest sides. A truncated triangular morphology would have a small value of T, while $T \rightarrow 0$ for a near-perfect equilateral triangle.

The T distribution we measured from our data exhibited a sizable peak at low values, quantifying our visual impression that truncated triangular morphologies were the most common among the nonhexagonal shapes. We then devised a Monte Carlo model in which we generated crystals where the perpendicular growth velocity of each facet was chosen from the same random distribution. From these crystals, we selected ones with $H<0.33$ and calculated the T parameter for each. Comparing the model and data, we concluded that crystals with a triangular morphology (small T) are much more common than one would expect from random growth perturbations of normal hexagonal crystals.

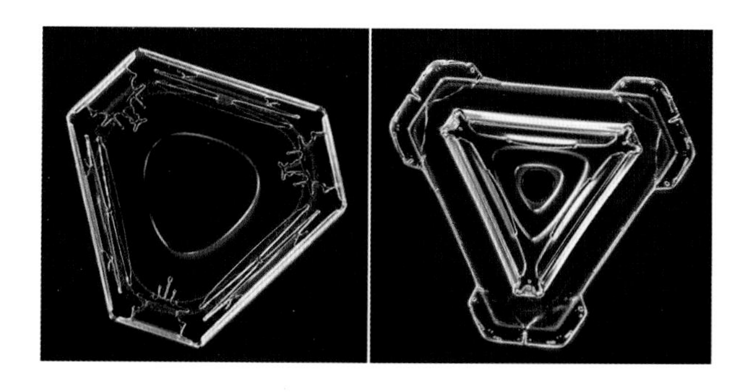

FIGURE 3.46. These natural snow crystals are shaped like small truncated triangular plates. Although not common, they can sometimes be found together with normal hexagonal plates.

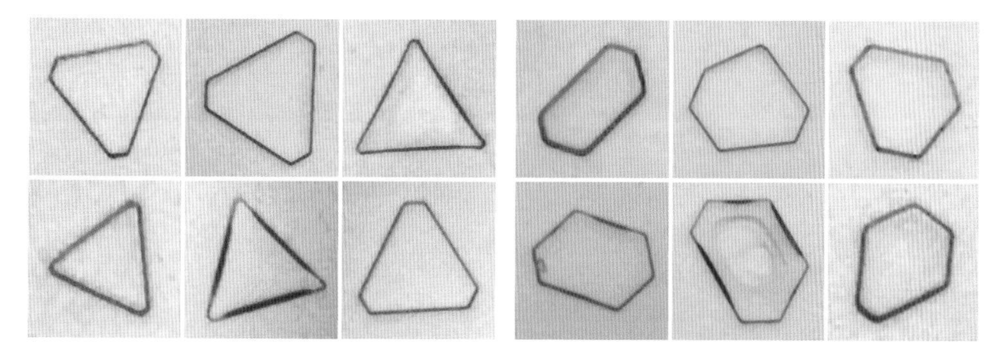

FIGURE 3.47. A selection of nonhexagonal plates observed in a free-fall growth chamber at −10°C with $\sigma_\infty \approx 1.4$ percent [2009Lib3]. The six images on the left show crystals with an overall triangular symmetry, while the six images on the right show "scalene" crystals having sides with somewhat random widths. While hexagonal crystals were most common in this sample, triangular crystals far outnumbered scalene crystals.

The point of this exercise is to show, quantitatively, that there really is something special about the triangular morphology. Of all the other possible nonhexagonal shapes (some of which are shown in Figure 3.47), those with overall threefold symmetry are by far the most numerous. Given that the underlying ice crystal symmetry is unchanged, so all six facets are essentially identical at the molecular level, the question then becomes what forces guide the development of triangular plates?

The answer appears to be another diffusion-limited snow crystal growth instability that favors triangular crystals over hexagonal ones. For example, if a small perturbation produced a slight triangularity, then particle diffusion would amplify it via the usual Mullins-Sekerka instability. This would be an unusual perturbation, so it is unlikely to happen during the short growth span of a hexagonal plate. But clearly the hexagonal form is unstable to a triangular perturbation.

A similar triangular instability is more readily found in the formation of dendritic crystals forming from hexagonal columns near −5°C, as illustrated in Figure 3.48. Here the high-σ crystal developed into a six-pronged "witch's broom" shape as fishbone dendrites sprouted from each of the six corners of the initial hexagonal column. In contrast, only three branches developed in the low-σ crystal, giving it a three-pronged "trident" shape. Here again the latter crystal exhibits a threefold symmetry, and more than half of all crystals grown under these conditions exhibited the same trident morphology.

FIGURE 3.48. (Left) A "trident" snow crystal forms on the end of an ice e-needle (see Chapter 8) in air with $(T, \sigma) = (-5°C, 32$ percent). Starting from a hexagonal column, only three branches grew to a discernable length. (Right) A similar crystal grows with $(T, \sigma) = (-5°C, 64$ percent), but the opening angle between the branches is larger and all six grew out from the initial columnar crystal.

The formation of tridents can be explained from the diffusion-driven competition between the different branches, as illustrated in Figure 3.49. Beginning with six identical branches, assume that one grows out a bit faster than the others, just by random chance. This branch then sticks out farther into the supersaturated air and shields the growth of its nearest neighbors slightly. The larger branch thus grows faster while its nearest neighbors are soon left behind, this process being yet another manifestation of the Mullins-Sekerka instability.

Of the remaining three branches, the outer two receive slightly more water vapor because of their two stunted neighbors, so they too grow out faster, leaving their middle neighbor behind. Assuming this diffusion dance plays out quickly, a trident crystal emerges. It is left as an exercise for the reader to show that a trident also results if initially one branch grew slightly slower than the others. This mechanism also explains why the low-σ crystal in Figure 3.48 developed into a trident while the high-σ crystal retained all six branches. In the low-σ case, the opening angle of the branches was small, so the competition between branches was strong and persistent. In the high-σ case, the opening angle was greater, and the growth rate was faster, so the branches quickly grew apart and the competition between them was weaker.

A key feature in this discussion is that threefold symmetry is generally more stable than sixfold symmetry, at least regarding diffusion-limited growth. If a hexagonal plate is perturbed slightly toward a trigonal symmetry, then the Mullins-Sekerka instability will reinforce this perturbation, growing it to larger scales. But this process does not work in reverse; perturbing a triangular crystal slightly cannot produce a hexagonal crystal via this mechanism. In the same vein, tridents are more stable than six-branched witch's brooms. Given this one-way stability feature, all that is needed to turn a hexagonal crystal into a trigonal one is the initial perturbation. In the case of tridents, the likely

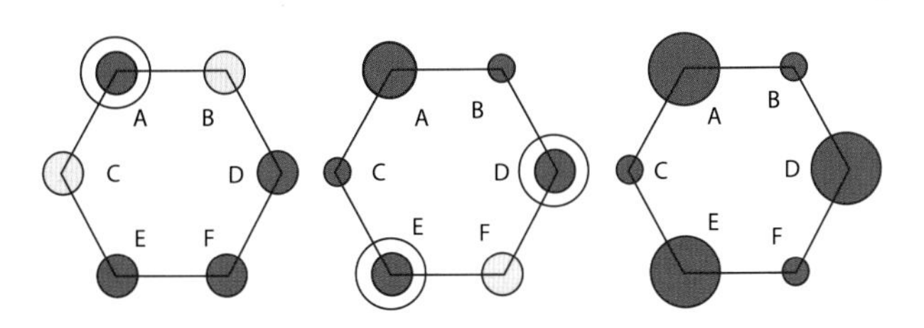

FIGURE 3.49. Trident formation. (Left) If one of six primary branches (branch A) extending from a columnar crystal becomes a bit taller than the others, then the Mullins-Sekerka instability will enhance its growth, while its immediate neighbors (branches B and C) will be shielded. (Center) As branch A grows taller and branches B and C are left behind, D and E will be more exposed to the supersaturated air and will thus grow faster, shielding branch F. (Right) In time, branches A, D, and E will dominate, while B, C and F are shielded, yielding a trident-shaped crystal.

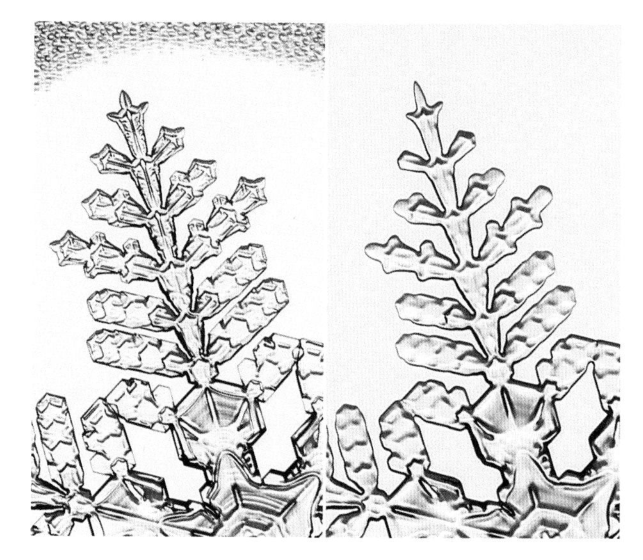

FIGURE 3.50. (Left) A photograph of a single branch of a growing PoP snow crystal. (Right) The same PoP crystal a few minutes later, after reducing the humidity to subsaturation levels, at which point the ice stopped growing and began sublimating away.

perturbation mechanism is illustrated in Figure 3.49. The case for triangular plates is not so clear, but we described a possible aerodynamic mechanism in [2009Lib3]. Regardless of the specific details, it appears that threefold symmetry in snow crystal formation generally arises from diffusion-limited growth.

Sublimation

If the water vapor pressure in air surrounding a snow crystal is lower than the equilibrium vapor pressure of ice, then $\sigma_\infty < 0$ and sublimation will begin removing molecules from the ice surface, as illustrated in Figure 3.50. In sub-saturation conditions in air, the excess vapor near the surface must be carried away by diffusion, yielding diffusion-limited sublimation.

One substantial difference between sublimation and deposition is that there are no nucleation barriers for the sublimation of convex surfaces, including most of the surfaces seen in Figure 3.50. Thus, while faceting is a major player in snow crystal growth, it is largely absent in snow crystal sublimation and most surfaces exhibit $\alpha \approx 1$. For this reason, the sublimating crystal in the figure exhibits mostly rounded surfaces, while the growing crystal has a generally sharper appearance. In diffusion-limited sublimation, crystal features that stick out farthest into the sub-saturated air sublimate fastest, so sharp corners and edges quickly become rounded.

Occasionally people will capture multiple photographs of a natural snow crystal as it sublimates away under the camera lens, and then show the time series running in reverse and call it a growing snowflake. Of course, taking a set of photographs of a sublimating crystal is much easier than growing a snow crystal in the lab, but a trained eye can quickly identify the telltale signs of sublimation.

Photographs of natural snow crystals often show rounded edges because they begin sublimating once they leave the supersaturated clouds for their final descent through sub-saturated air. When snow clouds are quite high in the sky, falling crystals often have a "travel-worn" appearance for this reason. Snow crystal photography can be especially rewarding when the clouds are close to ground level, revealing sharply faceted features. Laboratory-grown PoP crystals exhibit generally sharper, more vibrant structural features because they are photographed as they are growing.

A Panoply of Growth Instabilities

Nature abounds with dynamical instabilities in nonequilibrium systems, even though they are not much discussed in early science teaching. For example, when sunlight heats the ground, the air warms and becomes unstable to convection. This convective instability drives the wind, the clouds, and much of our weather. When the resulting wind blows over a still lake, the surface of the lake becomes unstable to the formation of ripples and waves. When waves reach the shore, they become unstable and break. Whenever you see any kind

FIGURE 3.51. This PoP snow crystal exhibits broad, platelike extensions growing on the ends of narrow branches, decorated with ridges and ribs. The two prominent sets of ribs in these plates did not appear spontaneously but were induced by twice lowering the supersaturation briefly.

of complex structure in nature, it is a good bet that some dynamical instabilities were involved in its formation. We tend to skip over this topic in science courses because it is so complicated and difficult to understand, and it seems that each case has its own unique characteristics. Nevertheless, instabilities should be appreciated more, because they are everywhere the natural world.

The Mullins-Sekerka instability (MSI) is clearly a big player in snow crystal formation, as described throughout this chapter, manifesting itself in remarkable variety of pattern-forming phenomena:

- Dendrite formation might be called the standard form the MSI, as diffusion-limited branching is almost synonymous with this instability. Fernlike and fishbone dendrites result when branching is additionally influenced by attachment kinetics.

- Sidebranch competition is another MSI variant, resulting in a gradual thinning of the sidebranch forest as some individuals dominate over the rest.
- The triangular instability leading to the development of tridents and triangular crystals is a more orderly version of sidebranch competition involving just six players.
- Hollow columns appear when the MSI is combined with strong prism faceting. Hollow plates likewise arise when strong basal faceting is present.
- Ridge formation involves a 2D version of the MSI that guides the growth of terrace steps, resulting in the spontaneous development of these common surface features. Related structures include ridges on cups and I-beam crystals.
- The grooves flanking ridges are essentially a higher-order effect in the same terrace instability responsible for ridges.
- Inwardly propagating rings require step bunching to be observable, again resulting from the Mullins-Sekerka instability.
- When strong electric fields are added to the mix, "electric" needle crystals arise, as described in Chapter 8.
- When structure-dependent attachment kinetics play a role, the Edge-Sharpening Instability arises (Chapter 4), playing a large role in the formation of thin plates and hollow columns.

One likes to think of snow crystal growth as a relatively simple phenomenon, being little more than ice condensing from water vapor. But, as illustrated in Figure 3.51, when particle diffusion limits ice growth, a rich exhibition of pattern-forming processes can result.

FIGURE 4.1. These photographs show a series of laboratory-grown PoP snow crystals (see Chapter 9). Their strong faceted features and platelike forms are largely defined by molecular attachment kinetics.

FOUR

Attachment Kinetics

Like a great poet, Nature knows how to produce the greatest effects with the most limited means.

—HEINRICH HEINE, *PICTURES OF TRAVEL*, 1871

In snow crystal growth, the *attachment kinetics* describe how water vapor molecules striking an ice surface become incorporated into the crystal lattice. The attachment process is ultimately determined by the complex molecular interactions that jostle incident water vapor molecules into position, so they can bind to the existing lattice structure. Because this many-body molecular dance is both intricate and unseen, there is much about it that we do not understand, even at a basic qualitative level. Nevertheless, the attachment kinetics represent one of the most important parts of the snow crystal story, driving the formation of faceted ice surfaces and other large-scale structural features. It has been well known for many decades that particle diffusion and surface attachment kinetics are the two primary physical processes governing snow crystal formation [1982Kur, 1984Kur1, 1990Yok, 2017Lib].

Whether a snow crystal develops into a thin stellar plate or a slender columnar form is determined by how rapidly water vapor molecules bind to different ice surfaces. With a platelike crystal, water molecules attach readily to the edges of the plate but attach only slowly to the basal faces. The opposite is true for long columns. The appearance of plates and columns at different temperatures in the Nakaya diagram (see Chapter 1) is thus an especially intriguing aspect of snow crystal formation that derives mainly from the anisotropy in the molecular attachment kinetics.

My principal goal in this chapter is to define and quantify a comprehensive physical model of the molecular attachment kinetics at the ice/vapor interface. I outline a suitable theoretical framework, examine what has been learned from experiments, and develop at least a qualitative picture of the detailed molecular processes involved. Many-body molecular dynamics is a complex subject, however, and previous attempts to understand the attachment kinetics have not fared particularly well (as discussed in this chapter). But a good model provides a springboard for future discussion and makes quantitative predictions that can be tested in future experiments. In the science of complex systems, progress often begins with a model, even an imperfect one. For the lack of a better name, I refer to the specific model advanced in this chapter as the *comprehensive attachment kinetics* (CAK) model.

Because this book is all about the growth and formation of snow crystals, our model of the attachment kinetics should provide a foundation for explaining the Nakaya diagram as well as quantitative ice growth experiments (see Chapters 7 and 8), and it should provide a suitable parameterization for creating computational models of growing snow crystals (Chapter 5). As we will quickly discover, however, achieving these goals is a tall order. The underlying molecular physics is often unclear, and a complex model will be needed to encompass a variety of different growth regimes. Like it or not, understanding snow crystal formation is not a trivial undertaking.

I believe that the model presented below represents significant progress toward understanding the many nuances of the Nakaya diagram and other aspects of snow crystal growth. However, some of its core ideas are still relatively new and not thoroughly tested. The model has some important merits (in my opinion), but it is far from being widely accepted canon. Indeed, some aspects of the model are speculative and may not withstand additional experimental and theoretical scrutiny. Nevertheless, this is the best solution I have found after considerable study of this thorny problem, so I present it here in its current form. Developing a comprehensive model of the ice/vapor attachment kinetics remains very much a work in progress, as one can say about many scientific endeavors.

ICE KINETICS

We begin our discussion with the Hertz-Knudsen relation [1882Her, 1915Knu, 1990Yok, 1996Sai], which can be written as

$$v_n = \alpha v_{kin} \sigma_{surf}, \tag{4.1}$$

where v_n is the crystal growth velocity perpendicular to the growing surface, α is a dimensionless *attachment coefficient*, $\sigma_{surf} = (c_{surf} - c_{sat})/c_{sat}$ is the water vapor supersaturation at the surface, c_{surf} is the water vapor number density just above the surface, $c_{sat} = c_{sat}(T)$ is the saturated number density of an ice/vapor surface in equilibrium at temperature T, and

$$v_{kin} = \frac{c_{sat}}{c_{ice}} \sqrt{\frac{kT}{2\pi m_{mol}}} \tag{4.2}$$

is the *kinetic velocity*, in which m_{mol} is the mass of a water molecule, $c_{ice} = \rho_{ice}/m_{mol}$ is the number density of ice, and ρ_{ice} is the mass density of ice. Values of several of these quantities as a function of temperature are given in Chapter 2 and in the Appendix. Throughout the discussion, I assume that any background gases surrounding a growing snow crystal, such as air and water vapor, are well described by the ideal gas laws in statistical mechanics. Given this assumption, which is highly accurate in most situations, one can work in terms of the water vapor molecular number density or the water vapor partial pressure, as the two are proportional (at constant temperature). I prefer the former, so c_{sat}, c_{surf}, and c_{ice} appear throughout this book.

Equations 4.1 and 4.2 derive from the basic tenets of statistical mechanics [1965Rei; 1996Sai], and I assume that the reader is generally familiar with this area of fundamental physics. In a nutshell, these equations come from considering two water vapor fluxes: the flux of molecules incident on a surface, equal to $c_{surf} v_{mol}$, where v_{mol} is an average molecular velocity, and the flux leaving the surface from sublimation, equal to $c_{sat} v_{mol}$. The difference between these two fluxes defines the growth velocity, and the statistical physics of ideal gases gives the appropriately weighted average velocity v_{mol} that appears in v_{kin}. Note that Equation 4.1 includes the trivial case of a vapor/solid interface in equilibrium: if the supersaturation σ_{surf} is zero, then the growth rate must also be zero.

Much of the interesting molecular physics involved in snow crystal growth is wrapped up in the attachment coefficient α, whose value lies in the range $0 \le \alpha \le 1$. One can think of α as a sticking probability, equal to the probability that a water vapor molecule striking the ice surface becomes assimilated into the crystal lattice. The value of α may depend on σ_{surf}, T, surface orientation relative to the crystal axes, and perhaps other factors. Ice surfaces that are "rough" at the molecular level typically exhibit $\alpha_{rough} \approx 1$, as water vapor molecules striking a rough surface usually become immediately indistinguishable from those in the existing ice lattice. Meanwhile, it is common to find $\alpha_{facet} \ll 1$ on "smooth" faceted surfaces, as these expose fewer open molecular binding sites, reducing the average sticking probability. With this definition of the attachment coefficient, it is generally true that thin platelike crystals (including stellar plates and stellar dendrites) form only when $\alpha_{basal} \ll \alpha_{prism}$, while slender columnar forms require $\alpha_{prism} \ll \alpha_{basal}$.

Molecular Processes

Theoretical models of the attachment kinetics typically begin with an atomistic (or, in our case, molecular) picture of the crystal surface structure and dynamics, as illustrated in Figure 4.2. This sketch depicts several molecular processes that can occur on a growing ice surface:

1) Deposition (a.k.a. adsorption)—when a water vapor molecule strikes the surface and sticks. Molecules that are loosely attached to faceted surfaces are called *admolecules*, while tightly bound molecules are simply considered part of the underlying solid.
2) Sublimation (a.k.a. desorption)—when thermal fluctuations cause molecules to leave the surface and join the vapor phase. Isolated admolecules are especially likely to sublimate, often doing so before ever becoming tightly bound to the crystal lattice.
3) Surface diffusion—random motions of admolecules along a crystal surface. Diffusion along faceted surfaces can be especially substantial, as lateral motion is only weakly inhibited by molecular binding.
4) Attachment—when a diffusing admolecule encounters a terrace step (typically from the lower terrace) and becomes incorporated into the ice lattice. Detachment from a terrace step (typically onto the lower terrace) yields an isolated admolecule, and this process is often a precursor to sublimation.
5) Ehrlich-Schwoebel barrier—a potential barrier that inhibits admolecule motion between terraces. For an admolecule to leave an upper terrace and attach to a lower terrace step (see Figure 4.2), it would first have to detach from the upper terrace. The attachment to the lower edge is energetically favorable, but the initial detachment from the upper terrace is not. The resulting Ehrlich-Schwoebel barrier tends to suppress these over-step transitions. Thus, surface diffusion between separate terraces is suppressed compared to diffusion confined to the surface of a single terrace.

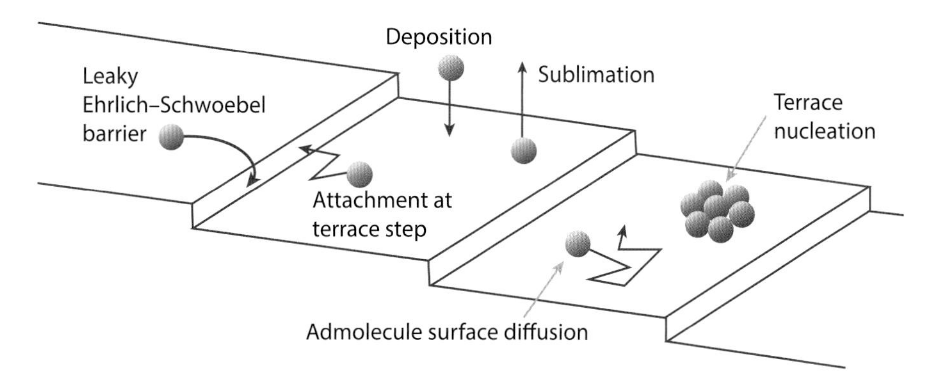

FIGURE 4.2. The attachment kinetics are governed by a variety of molecular processes occurring on the surface of a growing crystal, including those illustrated here.

6) Terrace nucleation—when several admolecules on a faceted surface come together to form a new molecular layer, or terrace. This process is required to form new terraces on the topmost terrace of a facet surface.

This cartoon molecular picture of a crystal surface is too simplistic to provide a full kinetic description of ice, even at a qualitative level. Ice has a high vapor pressure, is not made from simple spherical molecules, and the ice surface experiences surface premelting at high temperatures (see Chapter 2). As a result, the above list of molecular processes likely leaves out much important many-body physics. Molecular dynamics simulations can help develop our understanding of real ice surfaces, but we are still far from fully understanding the structure and dynamics of the ice/vapor interface. Nevertheless, experiments suggest that this cartoon picture can aid our intuition when describing many important aspects of snow crystal growth, so we adopt it as a reasonable starting point.

Surface Characteristics

Three types of surfaces play especially important roles in the discussion of snow crystal attachment kinetics: *faceted*, *rough*, and *vicinal* surfaces:

Faceted Surfaces. A faceted crystalline surface is defined by its low Miller indices, as described in Chapter 2. Low-index facets tend to have well-defined molecular terraces, and these surfaces best resemble the sketch shown in Figure 4.2. A perfect faceted surface can be thought of as being molecularly "flat" in that it contains no terrace steps or dislocations. The detailed molecular structures of the principal basal and prism facets are described in Chapter 2, as are the bilayer terrace steps on those surfaces.

An important feature of any faceted surface is that its molecular structure includes fewer dangling molecular bonds than does a nonfaceted surface. One result of this tighter molecular structure is a lower surface energy, but this fact plays a relatively minor role in snow crystal growth dynamics. The overall surface energy anisotropy is just too weak to significantly affect snow crystal growth or facet development. However, the anisotropy in the attachment kinetics can be enormous, as faceted surfaces often exhibit α_{facet} values that are orders of magnitude lower than on nonfaceted surfaces. In contrast to the surface energy, the highly anisotropic attachment kinetics typically play a major role in determining the development of snow crystal structures.

Rough Surfaces. A rough surface contains a high density of terrace steps, giving it a high density of dangling molecular bonds and a high attachment coefficient. In the case of ice surfaces, the experimental evidence suggests that water vapor molecules striking a rough surface are immediately indistinguishable from molecules in the ice lattice, which is another way of saying that incident molecules are immediately incorporated into the ice lattice. And this, by definition, means $\alpha_{rough} \approx 1$. This commonly encountered rough-surface limit is sometimes called *fast kinetics*. The attachment kinetics at a water/vapor interface is also near unity.

Vicinal Surfaces. A vicinal surface is essentially a flat surface cut at a slight angle relative to a faceted surface. Figure 4.2 illustrates a vicinal surface, which includes a series of terrace steps with an average spacing $\ell_{vicinal}$, where $\ell_{vicinal}$ depends on the vicinal angle. If $\ell_{vicinal}$ is less than the mean diffusion length x_{diff} (the typical distance admolecules on a faceted surface will diffuse before sublimating), then most admolecules will encounter a terrace step and attach. Therefore, $\alpha_{vicinal} \approx 1$ when $\ell_{vicinal} < x_{diff}$, and $\alpha_{vicinal} \rightarrow \alpha_{facet}$ when $\ell_{vicinal} \rightarrow \infty$. Note that the process of surface diffusion is usually implicitly incorporated into the attachment coefficient. It is also generally assumed, as part of the local form of the attachment kinetics defined by Equation 4.1, that x_{diff} can be considered small compared to most macroscopic snow crystal structures.

Facet-Dominated Growth

In ice growth from water vapor, the attachment kinetics are often highly anisotropic with deep cusps in $\alpha(\theta, \phi)$ at the principal facet angles, where (θ, ϕ) is the angular orientation of the surface normal relative to the crystal lattice axes. It is not uncommon to have $\alpha_{facet} < 0.01$, while $\alpha_{vicinal} \approx 1$ at a vicinal angle of just a degree or two. As we will see in Chapter 5, such a deep, cusplike anisotropy can lead to challenging numerical stability problems in computational snow crystal modeling. But this strong anisotropy also helps bring about the rich variety of observed snow crystal patterns.

As we saw in Chapter 1, this basic model of anisotropic molecular attachment kinetics immediately explains the formation of snow crystal facets. Because $\alpha_{facet} \ll \alpha_{rough}$, the rough surfaces quickly accumulate material and fill in, while the faceted surfaces accumulate material at a much slower rate. The appearance of macroscopic facets in most natural crystalline materials,

FIGURE 4.3. Facet-dominated growth. The attachment kinetics on any snow crystal can be divided into three classes of surfaces: top basal terraces described by α_{basal}, top prism terraces described by α_{prism}, and all other surfaces described by $\alpha \approx \alpha_{rough} \approx 1$. The overall platelike structure of this crystal, as well as much of its dendritic branching, is determined by the first two surface classes. Thus, the attachment kinetics on the topmost facet terraces strongly influence the morphological development of the entire structure.

including mineral crystals, typically results from highly anisotropic attachment kinetics via this mechanism. (Although on most commercial gemstones, facets are fabricated with a grinder.)

Beyond facet formation, the attachment kinetics on the basal and prism surfaces also play a large role in guiding the formation of even complex, large-scale snow crystal structures (Figure 4.3). Essentially all nonfaceted ice/vapor surfaces can be described with $\alpha \approx \alpha_{rough} \approx 1$ to a reasonable degree of accuracy. Thus, the only parameters left to define the overall morphology of a snow crystal are α_{basal} and α_{prism}. On dendritic structures exhibiting complex curved surfaces, it is often the attachment kinetics on the topmost, sometimes tiny, basal and prism terraces that guide the overall growth behavior. If not for the high anisotropy in the attachment kinetics, expressed mainly in α_{basal} and α_{prism}, the menagerie of snow crystal forms would be absent its marvelous diversity. I call this situation *facet-dominated growth*, because the slowest growing, faceted surfaces tend to define the overall growth morphology, even when the growth is also strongly diffusion limited. In this circumstance, understanding snow crystal morphologies largely comes down to creating a suitable comprehensive model of the attachment kinetics on the basal and prism facets.

LARGE-FACET ATTACHMENT KINETICS

Because the basal and prism facets play such an important role in defining the growth behavior and structural development of many snow crystals, our first task is to examine the attachment kinetics on these surfaces. We begin with the ideal case of a large-faceted surface of effectively infinite lateral extent, so we can ignore any nonlocal, edge-related effects. We further assume a perfect crystalline structure, free from dislocations and other lattice defects. This is a good place to begin the discussion, as large-faceted surfaces are common in many snow crystals. Moreover, both natural and laboratory-grown snow crystals frequently (although not always) appear to be essentially free from significant lattice imperfections.

As illustrated in Figure 4.2, isolated admolecules on a faceted ice surface are not yet fully incorporated into the crystalline lattice, owing to their relatively weak binding. In the absence of nearby terrace edges to bind to, a typical admolecule will reside on the surface for only a short time before thermal fluctuations send it back into the vapor phase. On large-faceted surfaces, sustained crystal growth requires the *nucleation* of new molecular terraces. In this circumstance, α_{facet} is mainly determined by *nucleation-limited attachment kinetics*.

The facet surface is typically a maelstrom of molecular activity, as admolecules are continually coming and going, diffusing along the surface, and interacting with one another. Small terrace islands are constantly forming and disintegrating, growing as admolecules attach to their edges and shrinking as molecules thermally detach and diffuse away. Small terrace islands are the least stable and frequently break up via thermal fluctuations, but these are also the most likely to form via chance encounters. Larger islands are less likely to form but generally survive longer before breaking up. The nucleation of a new, permanent terrace occurs when an island appears with a radius larger than some critical size R_{crit} that depends on the local supersaturation. Once such a stable terrace forms, it will usually continue to grow indefinitely as more admolecules diffuse to its edges and attach. The facet attachment coefficient depends on the value of R_{crit}, the rate at which stable terraces nucleate, and how rapidly stable terraces accumulate additional admolecules.

If the terrace nucleation rate is exceptionally low, then a single terrace may nucleate and grow until it covers the entire facet surface before the next new terrace appears. This is called *layer-by-layer* growth, and it is generally not so important when considering large facets (and certainly not for our ideal facets of infinite extent). For most growing snow crystals, a large-faceted surface will contain many stable terraces of various sizes at all times, and this situation is called a *multinucleation model.*

Terrace Nucleation Theory

On a faceted ice surface, the equilibrium vapor pressure of a small island of admolecules is higher than the normal saturated vapor pressure. Using an argument like that used to derive the Gibbs-Thomson effect in Chapter 2, the equilibrium vapor pressure of a circular island terrace of radius R is

$$c_{eq} \approx c_{sat}\left(1+\frac{a^2\beta}{RkT}\right), \tag{4.3}$$

where a is the molecular size, k is the Boltzmann factor, T is the surface temperature, and β is the step energy of the terrace edge. Thus, for a terrace island to be stable against sublimation, the supersaturation near the surface must be at least $\sigma_{surf} = a^2\beta/RkT$. Turning this around, an island terrace will achieve long-term stability only if its radius is greater than $R_{crit} = a^2\beta/\sigma_{surf}kT$. Putting in some typical numbers, $\beta \approx 10^{-12}$ J/m and $\sigma_{surf} \approx 1$ percent, a marginally stable island terrace might have a radius of about $R_{crit} \approx 10a$ and contain roughly 300 water molecules.

The growth rate v_n of a faceted surface, and thus the attachment coefficient α_{facet}, is tied directly to the rate at which new terraces appear and the rate at which existing terraces grow via admolecule attachment. The statistical mechanics describing these processes has been much studied over many decades, yielding a well-established *classical nucleation theory* that is described in detail in essentially all textbooks on crystal growth [e.g., 1994Ven, 1996Sai, 1999Pim, 2002Mut]. In three dimensions, classical nucleation theory describes the homogeneous nucleation of liquid droplets, while in two dimensions, the same theory applies to the nucleation of island terraces on faceted crystal surfaces. The derivation of nucleation theory is quite involved, and I cannot improve on the existing textbook treatments. In this book, therefore, I simply quote several salient features of the theory and apply it to the case of snow crystal growth.

Jumping straight to the main result in a multinucleation model, the creation and growth of new terraces yields an attachment coefficient that can be written, to a reasonable approximation, as [1996Sai]

$$\alpha(\sigma_{surf}) = Ae^{-\sigma_0/\sigma_{surf}} \tag{4.4}$$

for the growth of a faceted surface, where A and σ_0 are dimensionless parameters, with

$$\sigma_0(T) = \frac{S\beta^2a^2}{k^2T^2}. \tag{4.5}$$

Here I have included a dimensionless geometrical factor $S \approx 1$ to absorb several small theoretical factors (for example, the difference between a and the actual terrace thickness). Given the substantial uncertainties in our current knowledge of β, the exact value of S is not of great concern at this time, so I assume $S = 1$.

Nucleation theory generally indicates that A will depend weakly on σ_{surf}, but I will mostly neglect any such dependence, as it is dwarfed by the strongly varying $\exp(-\sigma_0/\sigma_{surf})$ factor. Therefore, I take the parameters $A(T)$, $\sigma_0(T)$, and $\beta(T)$ to all be independent of σ_{surf}, and the resulting function in Equation 4.4 seems to provide a good fit to existing measurements of faceted ice growth (see Chapter 7).

As a note of caution, I point out that theoretical models of terrace nucleation invariably include a variety of implicit, simplifying assumptions regarding molecular surface structure and dynamics, and some of these assumptions may not be justified for ice. For example, the theory usually begins with the basic surface molecular picture illustrated in Figure 4.2, which does not include surface premelting. While it is well known that premelting is an important structural characteristic of ice crystal surfaces near 0°C, we do not know how this phenomenon modifies the dynamics of terrace nucleation. Nucleation theory was developed mainly for low-vapor-pressure solids like metals and semiconductors, and it is,

I believe, not thoroughly tested experimentally outside this realm.

These caveats notwithstanding, some aspects of nucleation theory appear to be quite robust in the sense that they are largely insensitive to many surface characteristics. The exponential factor $\exp(-\sigma_0/\sigma_{surf})$ is an especially robust feature in nucleation theory, along with the relationship between σ_0 and β given in Equation 4.5. These aspects of the theory are essentially independent of details pertaining to how admolecules diffuse along a faceted surface, the admolecule residence time, how terraces grow, and the number of stable terraces that are present on the surface at any given time. Importantly, over a broad range of surface characteristics, the terrace step energy β is the only parameter that has a substantial effect on σ_0.

Because of this robust feature in nucleation theory, observing the functional form $\alpha \sim \exp(-\sigma_0/\sigma_{surf})$ in ice growth experiments is a strong indication that the growth rate is limited primarily by the nucleation of new terraces. In this case, one can use measurements of $\alpha(\sigma_{surf}, T)$ to extract the step energy function $\beta(T)$, which is otherwise difficult to measure. Moreover, β is a fundamental material property of any faceted surface, like the surface energy. In principle, the step energy could be determined from purely equilibrium measurements (i.e., independent of crystal growth dynamics) or even by detailed calculations that determine crystal structure and energetics from known molecular interactions. To date, however, growth-rate experiments have yielded the best measurements of β on the ice facets (see Chapter 7).

One of the marvelous aspects of terrace nucleation theory is that it reduces a complex surface dynamical process to essentially a single equilibrium quantity β. All the specific molecular dynamics details regarding admolecule deposition, sublimation, surface diffusion, and bonding at terrace steps become largely irrelevant. The distinctive functional form $\alpha \sim \exp(-\sigma_0/\sigma_{surf})$ depends only on the terrace step energy. Terrace nucleation theory thus provides a simple, accurate, and quantitative physical model that plays an important role in the ice/vapor attachment kinetics.

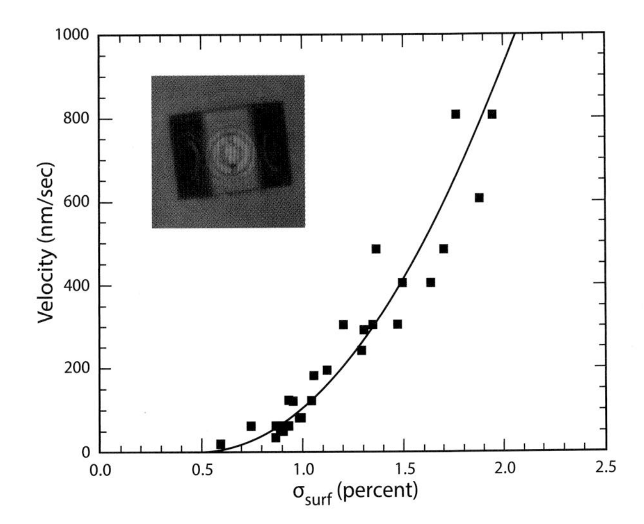

FIGURE 4.4. Ice growth measurements (data points) are well described by a terrace nucleation model (curve) [2013Lib]. Chapter 7 describes these data in detail.

Model Parameters from Measurements

How well the terrace nucleation model applies to real ice growth can only be determined by detailed comparisons with experimental observations. Early measurements clearly exhibited some of the characteristic traits of nucleation-limited growth [1972Lam, 1983Bec, 1989Sei, 1998Nel], and the evidence became ever stronger as improved experimental techniques yielded greater measurement precision [2013Lib, 2019Lib3]. For example, Figure 4.4 illustrates the exponential rise in growth velocity with σ_{surf} that is the hallmark of a nucleation-limited model. Here the data points show measurements of the perpendicular growth velocity of a prism facet surface as a function of the near-surface supersaturation σ_{surf}. These data were taken at −15°C with a background air pressure of 20 mbar using the vacuum ice growth apparatus described in Chapter 7. The curve shows $v_n = \alpha v_{kin} \sigma_{surf}$ with $\alpha(\sigma_{surf}) = \exp(-\sigma_0/\sigma_{surf})$ and $\sigma_0 = 3$ percent. The inset image shows the test crystal immediately after the growth measurements were completed [2013Lib]. Many additional experimental results like this suggest that the terrace nucleation model accurately describes the growth of large, defect-free basal and prism faceted surfaces over essentially all conditions relevant to snow crystal formation. This is a strong statement, and it represents my experienced, albeit perhaps not entirely unbiased, interpretation of the available published data.

As a first step in developing the comprehensive attachment kinetics (CAK) model in this chapter, I combine results from the best available ice growth experiments to produce the best-estimate extracted model parameters $A(T)$, $\sigma_0(T)$, and $\beta(T)$ shown in Figures 4.5 and 4.6. For the sake of pedagogy, I display these largely empirical results in the form of smooth curves, as this approach allows a less cluttered discussion of the underlying physical processes, thus facilitating our continued development of the CAK model below. This somewhat oversimplified presentation glosses over experimental uncertainties and the inevitable inconsistencies between published data sets, and I examine some of these issues in Chapter 7. For now, suffice it to say that these curves are probably reasonably accurate in an absolute sense but may require some modification in the light of future improved experiments.

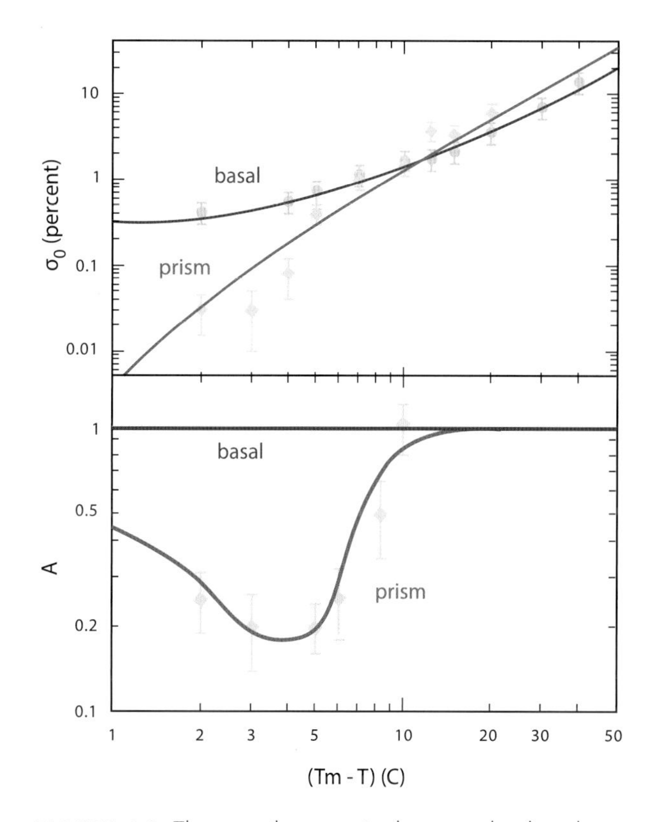

FIGURE 4.5. The smooth curves in these graphs show best-estimate CAK model functions $\sigma_0(T)$ and $A(T)$ on large basal and prism facets as a function of temperature, where $T_m = 0°C$ is the ice melting point. The respective attachment coefficients are then given by $\alpha(\sigma_{surf}) = A \exp(-\sigma_0/\sigma_{surf})$. The underlying experimental data and measurement techniques used to obtain these curves are described in Chapter 7. Because these curves parameterize the large-facet attachment kinetics over a broad range of growth conditions, they provide the foundation for the CAK model of snow crystal attachment kinetics.

The set of curves in Figures 4.5 and 4.6 parameterize the attachment kinetics describing large, defect-free faceted surfaces in the CAK model, and they are little more than parameterized fits to experimental data. Measurements of $v_n(\sigma_{surf}, T)$ like those shown in Figure 4.4 are first converted to $\alpha(\sigma_{surf}, T)$ using Equation 4.1, and then fits to Equation 4.4 are used to extract the parameters $A(T)$ and $\sigma_0(T)$. The step energies $\beta(T)$ are further obtained from $\sigma_0(T)$ using Equation 4.5. The assumed functional forms come from terrace nucleation theory, and they seem to provide quite a good fit to ice growth data over a broad range of (σ_{surf}, T). The CAK model includes an implicit assumption that the attachment kinetics do not depend on background gas pressure or other chemical effects. In the spirit of an ideal large-facet kinetics model, I assume that water is the only chemical species in the problem. How well this pure model describes real snow crystal growth is an important experimental question, and I discuss this issue later in this chapter and in Chapter 7.

The measured terrace step energies are a foundational element in the CAK model, and these represent equilibrium material properties of the ice surface, as discussed in Chapter 2. At low temperatures, surface premelting is essentially absent on both the basal and prism facets, so the step energies should tend toward the rigid-

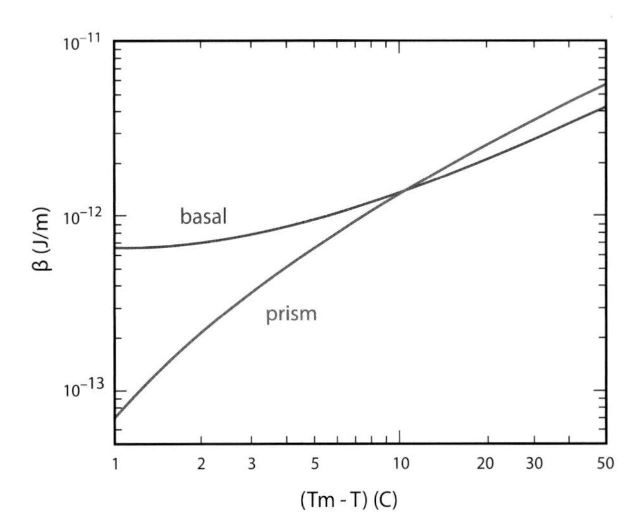

FIGURE 4.6. These curves show CAK model estimates for the step energies $\beta(T)$ for large basal and prism facets as a function of temperature, extracted from $\sigma_0(T)$ using Equation 4.5. While derived from dynamical growth measurements, these step energies represent fundamental equilibrium properties of the ice crystal surface.

lattice value of $\beta_0 = a\gamma_{sv} \approx 3 \times 10^{-11}$ J/m, and this statement is reasonably consistent with the data. At high temperatures, the basal step energy should tend toward the ice/water step energy $\beta_{sl,basal} \approx 5.6 \pm 0.7 \times 10^{-13}$ J/m (see Chapter 2), while the prism step energy should tend toward $\beta_{sl,prism} \ll \beta_{sl,basal}$. Again, these statements appear to fit the data. Put another way, although we do not have a precise model of the terrace step energies for either the ice/vapor or ice/water interfaces, the measurements seem to paint a reasonably consistent picture of the underlying physics. Thus, while the step energies in Figure 4.6 reflect empirical measurements, the results seem sensible overall.

Frustrated QLL Kinetics

The measured $A(T)$ behavior is not as important as the terrace step energies, but it is a significant feature in the measurements, and it does not readily suggest a physical explanation. Going back to the ice growth data that yielded Figure 4.5, the values of both A_{basal} and A_{prism} generally reflect the growth behaviors at high σ_{surf}, specifically when $\sigma_{surf} \gg \sigma_0$ and the $\exp(-\sigma_0/\sigma_{surf})$ factor is near unity. In this fast growth regime, terrace nucleation is rapid, so the surface contains a relatively high density of terrace steps, even though it remains faceted on large scales. The abundance of terrace steps means that the surface begins to resemble a molecularly rough surface. In this situation, surface diffusion will quickly transport admolecules to nearby terrace steps, where they will be incorporated into the ice lattice. Thus, on both the prism and basal facets, it is natural to expect $\alpha \rightarrow \alpha_{rough} \approx 1$ when $\sigma_{surf} \gg \sigma_0$, and this expectation implies $A \approx 1$. And, indeed, we see $A_{basal} \approx 1$ at all temperatures in Figure 4.5 (see Chapter 7), along with $A_{prism} \approx 1$ at temperatures below $-10°$C. Consequently, it appears that the high-temperature region with $A_{prism} < 1$ is something of an anomaly.

To explain this anomalous behavior, the short answer is that this is an empirical observation with no obvious physical explanation. Nevertheless, the temperature dependence suggests that surface premelting may be involved, because the transition to $A_{prism} < 1$ happens right about where surface premelting becomes important. Therefore, undeterred by mere ignorance, let us follow this thread a bit further and look at the attachment kinetics in the presence of surface premelting when the nucleation barrier is low. In the extreme case of a thick QLL, assume that the surface has an overall ice/QLL/vapor structure (as inferred from MD simulations), and further assume that attachment is fast at the QLL/vapor interface. If these reasonable assumptions are true, then the relevant physics giving us $A_{prism} < 1$ must be happening at the ice/QLL interface. And this suggests that we take a quick foray into the ice/water attachment kinetics.

While the fundamental theory of solid/liquid attachment kinetics is not well developed, the Wilson-Frenkel model of melt growth provides an overview of the underlying physics [1900Wil, 1932Fre, 1996Sai]. In a nutshell, the solid/liquid kinetics is limited by the mobility that allows a near-surface liquid molecule to move into position so it can be incorporated into the solid lattice. In the Wilson-Frenkel model, this mobility translates into an attachment coefficient that is inversely proportional to the liquid viscosity, which is proportional the liquid diffusion constant [1996Sai].

In ice growth from liquid water, prism growth rates are extremely fast, which suggests that the ice/QLL kinetics would be fast as well, again implying $A_{prism} \approx 1$. Because this is not observed to be the case above −10°C, it suggests that the molecular diffusion constant in the QLL is substantially slower than in bulk water. This is not a shocking statement to make, as the QLL is a nanoscale thin layer, plus its structure is somewhat constrained by its proximity to the ordered ice lattice [2018Lou]. However, neither theory nor experiments tell us much that is definitive about QLL diffusion, so we cannot say how this might give us the observed $A_{prism} < 1$ behavior seen in Figure 4.5, or why there is no similar behavior for A_{basal}. Nevertheless, a crude "frustrated QLL kinetics" model (the word "model" being a bit generous here) may provide at least a qualitative physical picture of the underlying molecular processes.

Figure 4.7 shows a graphical representation of this frustrated-kinetics model, comparing the ice/water case with that for an ice/QLL/vapor interface. Although this sketchy line of reasoning provides us no method for calculating A_{prism}, it does agree with at least one recent MD simulation [2018Lou] showing that the near-surface QLL viscosity is up to 1,000 times higher than bulk water, and is especially high on the prism facet. Although simulating ice surfaces is not an exact science at present, this result suggests that it may be possible to investigate the frustrated-QLL model further using MD simulations. Moreover, the model immediately makes a significant experimental prediction. Because the mobility must increase with QLL thickness in this model, it predicts that $A_{prism} \to 1$ as $T \to 0°C$. The upward trend in A_{prism} at the highest temperatures in Figure 4.5 largely reflects this CAK model bias, although there is only a weak indication that this trend is present in the data. If the upward trend is confirmed in experimental investigations at higher temperatures, this would support the model, crude as it is. Moreover, the frustrated-kinetics model becomes useful once again for interpreting the growth of ice dendrites at high temperatures, which I discuss further in the section on structure-dependent attachment kinetics later in this chapter.

An alternative explanation for $A_{prism} < 1$ might come from classical terrace nucleation theory [1996Sai], which indicates that A_{prism} is proportional to the surface diffu-

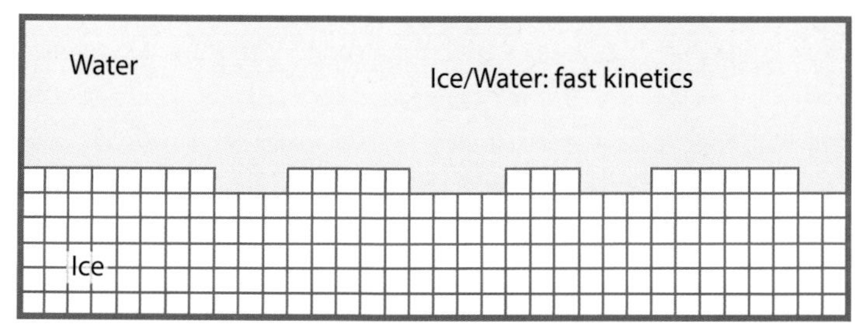

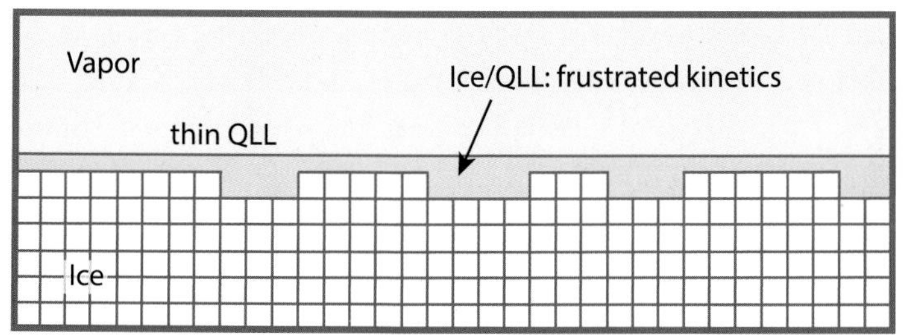

FIGURE 4.7. An illustration of how the inhibited mobility of QLL molecules might explain the observed $A_{prism} < 1$ behavior seen in Figure 4.5. In the absence of a nucleation barrier (when $\sigma_{surf} \gg \sigma_0$), ice growth is limited by how quickly liquid (or quasiliquid) molecules can diffuse into position to be incorporated into the ice lattice. The data suggest that perhaps there is a high mobility in the ice/water system (top) but a substantially lower mobility near the ice/QLL interface (bottom).

sion length x_{diff} on prism facets. If x_{diff} is substantially lowered by the onset of surface premelting, this could also explain the observed $A_{prism} < 1$ behavior. However, a reduction in x_{diff} is not altogether dissimilar to a reduction in QLL mobility, so perhaps the overall picture is roughly the same in both explanations. And again, it remains a mystery why A_{prism} is substantially different from A_{basal} at high temperatures. But this behavior can certainly be investigated further with additional targeted experiments and MD simulations [2018Lou, 2020Llo].

We can add an additional telling observation to this story, as prism faceting is readily observed at −0.5°C in air, as illustrated in Figure 4.8. This faceting could not result from a high nucleation barrier, because the growth data indicate that σ_0 is too low to cause faceting at the estimated σ_{surf}. Instead, this high-temperature prism faceting appears to arise because $A_{prism} < 1$ at high temperatures, providing a direct, independent confirmation of this feature in the data [1991Elb].

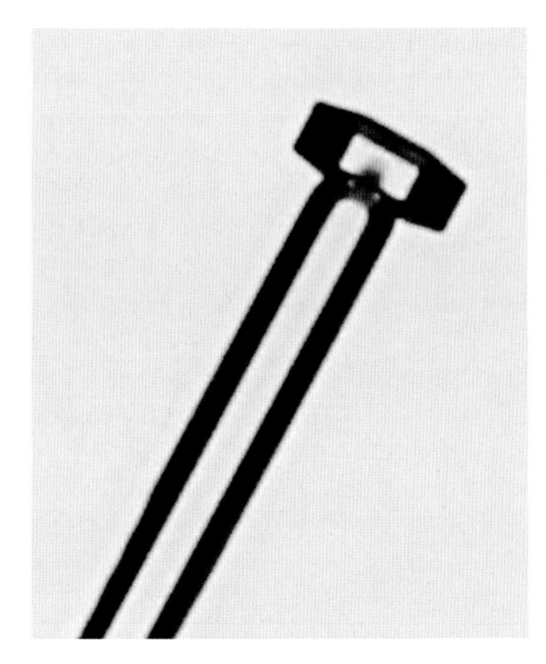

FIGURE 4.8. A blocky plate crystal grows on the end of a c-axis electric ice needle (see Chapter 8) in air at −0.5°C, with $\sigma_{surf} \approx 0.1$ percent. Because $\sigma_{surf} \gg \sigma_0$ under these conditions, the strong prism faceting did not result from a nucleation barrier but instead arose because $\alpha_{prism} \approx A_{prism} < 1$ at this temperature, as shown in Figure 4.5. The CAK model proposes a "frustrated QLL kinetics" phenomenon to explain this low A_{prism} behavior, although the underlying molecular physics in this case is not known with certainty.

Because Figures 4.5 and 4.6 describe the attachment kinetics on perfect faceted surfaces having infinite lateral extent, the CAK model curves are relatively clean and well defined. There must exist, for example, definite values for $\beta_{basal}(T)$ and $\beta_{prism}(T)$, just as there are definite values for the facet surface energies $\gamma_{basal}(T)$ and $\gamma_{prism}(T)$. These are all fundamental properties of the ice crystal, determined by the lattice structure and water molecular interactions. The actual values of $\beta_{basal}(T)$ and $\beta_{prism}(T)$ shown in Figure 4.6 are imperfect to some degree, owing to experimental uncertainties, and the step energies may vary slightly with the orientation angles of the steps, step curvature, and other minor factors. But well-defined step energies must exist in the defect-free, large-facet limit.

The parameters $A(T)$ and $\sigma_0(T)$ are somewhat model dependent, as they assume that a terrace nucleation model is generally correct for the growth of faceted ice surfaces. But this assumption is quite strongly supported by a fair amount of experimental evidence at this point (see Chapter 7), and defect-free crystals seem to be the norm over this temperature range. The somewhat odd behavior of $A_{prism}(T)$ is a complication, but this is a relatively minor issue that mainly affects fast-growing prism surfaces at high temperatures. Even including that regime, ice growth measurements generally tell us that the terrace nucleation model parameterized in Figures 4.5 and 4.6 provides quite a good approximation for the growth of the principal basal and prism faceted surfaces in the large-facet limit.

The large-facet limit, however, is only one aspect of the attachment kinetics. A great variety of snow crystal morphologies—including thin plates, hollow columns, and essentially all dendritic structures—exhibit sharply curved features that contain tiny basal and prism facets with last-terrace dimensions that are often as small as 50 nanometers. But even these minute facets can be important drivers of the overall growth morphology, as essentially all nonfaceted surfaces exhibit $\alpha \approx \alpha_{rough} \approx 1$. One might imagine that the addition of edge effects or other factors on such small facets could yield attachment kinetics that are quite different compared to large facets. In our quest to understand the full range of snow crystal morphologies, therefore, we must address the attachment kinetics on these diminutive faceted surfaces.

STRUCTURE-DEPENDENT ATTACHMENT KINETICS

To begin our examination of the growth of small faceted surfaces, consider the edge of a thin hexagonal plate, as illustrated in Figure 4.9. The edge terminates on a narrow prism facet, and, from basic geometry, the width of the last prism terrace on a curved edge is $w \approx \sqrt{8Ra}$, where R is the radius of curvature of the edge, and a is the size of a water molecule. Experimental measurements, together with input from solvability theory (see Chapter 3) suggest edge and tip radii that are often in the 1–2 μm range for a broad range of snow crystal morphologies growing in normal air. Slower growth generally yields larger R, as does growth at lower air pressures. But fast-growing structures in air do not generally develop structures with R values much lower than 1–2 μm. Thus w can be as low as about 50 nm on many snow crystal structures, which is less than 200 molecules across. (Although electric ice needles have substantially sharper tips, as described in Chapter 8.)

This lower limit puts us squarely in the mesoscopic regime, as a 200-molecule-wide facet would likely have roughly the same molecular surface structure as a large facet, even in the presence of significant surface premelting. However, lateral molecular transport from the facet edges might affect the overall growth behavior in comparison to that of a large-faceted surface. Theoretical considerations alone cannot tell us whether edge effects will significantly alter the attachment kinetics, but rather basic observations of snow crystal morphologies indicate that the large-facet attachment kinetics described above cannot be the whole story.

Thin-plate crystals like that shown in Figure 4.9 are common near −15°C, and their very existence requires $\alpha_{prism} \gg \alpha_{basal}$. Basic numerical modeling further indicates that the large aspect ratio of a thin-plate crystal cannot be a simple diffusion effect but must be caused by highly anisotropic attachment kinetics (see Chapter 3). The large-facet curves in Figure 4.5, however, clearly indicate that $\alpha_{prism} \approx \alpha_{basal}$ at −15°C, thereby precluding the formation of thin plates. This discrepancy is immediately problematic, because our large-facet attach-

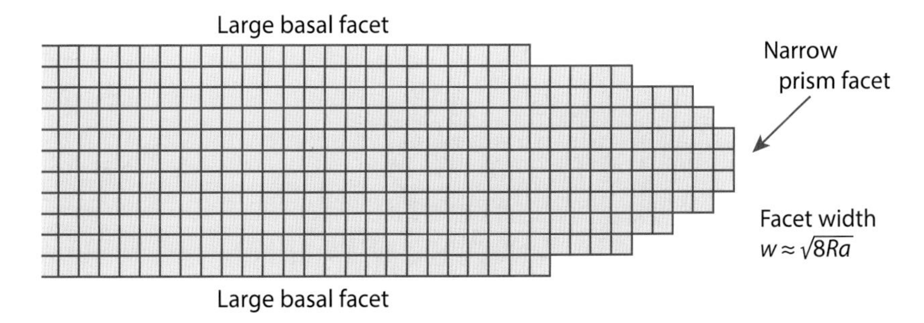

FIGURE 4.9. The shape of a thin hexagonal platelike crystal is largely defined by the growth of the basal and prism facets. While the basal facets are quite large, the six prism facets have widths $w \approx \sqrt{8Ra}$, where R is the radius of curvature of the edge and a is the molecular size. In many common circumstances, the attachment kinetics on these extremely narrow facets are markedly different from the attachment kinetics on large-faceted surfaces.

ment kinetics model is largely empirical, supported by quite solid experimental measurements. Either something is very wrong with the large-facet measurements, or small facets have a different story to tell. Delving into this discrepancy over some years, I reached the conclusion that the facet attachment kinetics must depend strongly on the local mesoscopic crystal structure, specifically the facet width w, a phenomenon I call *structure-dependent attachment kinetics* (SDAK) [2003Libl].

Supposing that this SDAK hypothesis is correct, our next challenge is to understand the physical origin of a facet-width-dependent attachment kinetics. Ice surface physics is already challenging in the large-facet limit, and it does not become simpler when any number of edge-related physical effects may come into play. While this might be a good point to throw up our hands and declare that no clear solution is possible, such a declaration provides little insight into possible paths forward. Instead I will plunge ahead and describe two possible mechanisms I have developed for SDAK effects. Both have merits in different growth regimes, but neither is solidly rooted in established molecular physics. These models should therefore be considered as hypotheses to be tested with additional experimental and theoretical investigations. I have been going down this path of hypothesis testing recently using numerous precision ice growth measurements, and so far, the results are quite promising. But a long road lies ahead to fully develop and verify these SDAK models.

SDAK-1: Enhanced Terrace Nucleation

As a first venture into developing an SDAK model, let us consider more carefully the structure and growth of a prism edge on a thin platelike crystal. If we establish the thin-plate morphology and then turn down the supersaturation, so the crystal sits in near-equilibrium conditions, then Figure 4.10a illustrates the resulting rounded edge structure. The value of R_{edge} is set by the initial conditions, and the discussion in the previous several paragraphs suggests that 1–2 μm is a typical value for an especially thin edge. If we set $\sigma_\infty \approx \sigma_{surf} \approx 2d_{sv}/R_{edge}$ around this crystal, then the Gibbs-Thomson effect provides that the edge of the plate will neither grow nor shrink with time. If we further assume a large nucleation barrier having $\sigma_{0,basal} \gg \sigma_{surf}$, then the basal facets will remain static as well. The resulting lack of any significant growth is consistent with our assumption of $\sigma_\infty \approx \sigma_{surf}$ in this near-equilibrium situation.

The shape of the edge will be rounded, as shown in Figure 4.10a, as this minimizes the local surface energy. And, again from Chapter 2, the time needed for the edge to reach this rounded state is just a few seconds in air. Eventually, the overall platelike shape would evolve toward the (roughly spherical) equilibrium crystal shape, but the time required for this relaxation is extremely long. Thus, Figure 4.10a represents a quasi-static, near-equilibrium condition.

Next let us increase the supersaturation, as illustrated in Figure 4.10b. Assume that the nucleation barriers are still quite large on both facets, so they experience little growth. The only significant change is that the Gibbs-Thomson effect now yields a new quasi-equilibrium state in which the corner radii are smaller than R_{edge}. This means that the increased supersaturation now prevents the plate edge from assuming its rounded, near-equilibrium shape. Thus, the high applied supersaturation constitutes a nonequilibrium condition that then supports a static, nonequilibrium edge shape, in this case one with sharpened corners.

For our final step, illustrated in Figure 4.10c, we now introduce the possibility of surface diffusion onto the prism facet. The nonequilibrium edge shape means that a transfer of molecules from the sharpened corners to the prism facet (arrows) will reduce the total surface energy of the system. Thus, if significant surface diffusion is allowed, there will be an energetically driven flow of water molecules onto the prism surface. These additional admolecules will not immediately attach to the prism surface, as a nucleation barrier still must be overcome. But they will increase the overall admolecule surface density, and this disturbs the normal deposition/sublimation balance on the surface. With a greater admolecule surface density, terrace nucleation is statistically more likely,

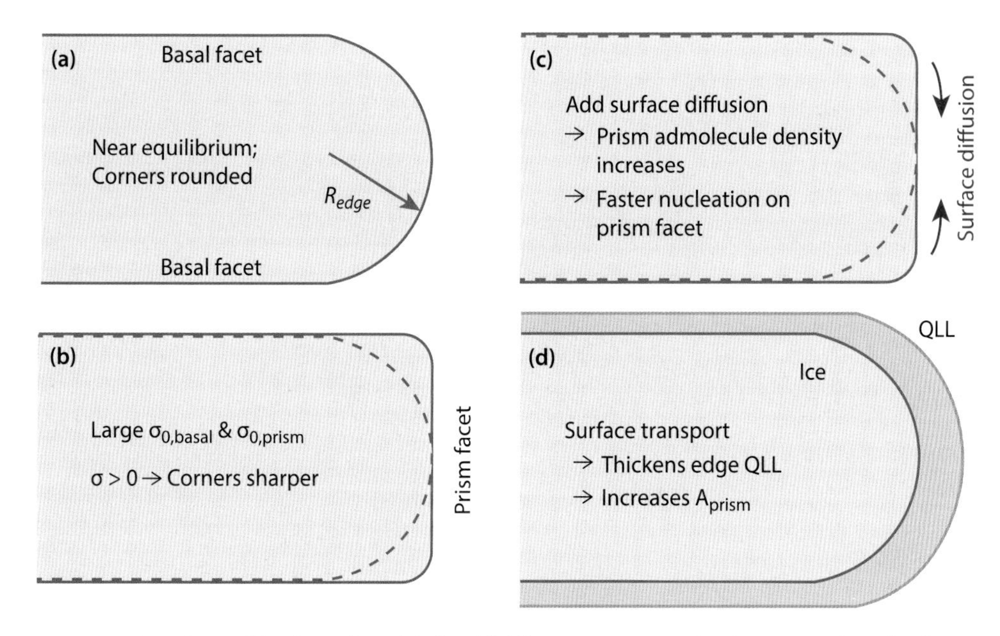

FIGURE 4.10. A possible physical model of an SDAK phenomenon on the thin prism edge of a plate-like snow crystal [2019Lib]. As described in detail in the text, a surface diffusion driven by surface energy can increase the admolecule density on the narrow prism facet, enhancing the normal rate of terrace nucleation. As a result, the attachment kinetics on the narrow prism facet is different from that on a large prism facet, which is the essence of the SDAK phenomenon.

which results in a lower effective nucleation barrier, and thus a higher α_{prism}. With some additional reasonable assumptions [2019Lib], the growth increase behaves essentially as if β_{prism} were lower. Note that the real β_{prism} is unchanged, as the lattice structure of the facet is essentially equal to that of a large facet. But the nucleation rate goes up nevertheless, behaving much like a surface with a lower β_{prism}. As a result, because of this facet-edge effect, the effective $\sigma_{0,prism}$ goes down even when β_{prism} remains unchanged. This phenomenon cannot occur on large facets, because large facets, by definition, are not influenced by edge effects.

This picture also suggests that the increase in admolecule surface density will be more pronounced as the facet width w decreases, because the additional admolecules will quickly disperse over the available top-terrace surface. The narrower the facet surface is, the higher the resulting admolecule density will be. Once again, this cartoon picture does not allow us to calculate the SDAK effect in any detail, as this is a complex surface dynamical process. But it does suggest at least a plausible molecular mechanism for an SDAK phenomenon.

Notably, this SDAK mechanism cannot yield a value of α_{prism} above unity. As the nucleation barrier drops and $\alpha_{prism} \rightarrow 1$, the Gibbs-Thomson effect will no longer support the nonequilibrium edge shape with sharpened corners. This only happens when the prism nucleation barrier is significant enough to suppress prism growth. Once $\alpha_{prism} \approx 1$, the edge reverts to an overall rounded shape that no longer provides an energetically driven surface flow. In a nutshell, this SDAK mechanism effectively lowers $\sigma_{0,prism}$ until $\alpha_{prism} \rightarrow 1$, but it cannot yield $\alpha_{prism} > 1$.

Surface Premelting and the Ehrlich-Schwoebel Barrier

Continuing with this line of reasoning, our next question is when to expect a significant amount of corner-to-facet surface diffusion. With a traditional rigid-lattice surface structure, this type of edge-crossing surface diffusion is suppressed by the Ehrlich-Schwoebel barrier (Figure 4.11), a phenomenon that is commonly discussed in crystal growth textbooks [1996Sai, 1999Pim, 2002Mut]. From Chapter 2, the ice surface tends to resemble this rigid structure at especially low temperatures, when surface premelting is absent. We expect, therefore, that surface diffusion like that shown in Figure 4.10c would be effectively suppressed by the Ehrlich-Schwoebel barrier at low temperatures. In this case, terrace nucleation on narrow facets would be no different than on large facets.

As the temperature increases, however, surface premelting eventually begins to develop, increasing the amount of disorder in the top molecular layers of the crystal surface. When the temperature reaches some critical value T_{onset} and surface disorder becomes sufficiently large, it stands to reason that the Ehrlich-Schwoebel barrier will become quite leaky. At this temperature, I have proposed that the above SDAK mechanism becomes important, effectively lowering the nucleation barrier and thereby increasing α_{prism} [2019Lib].

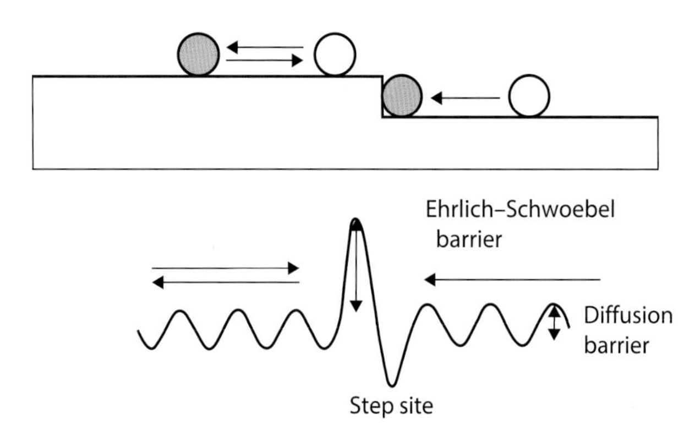

FIGURE 4.11. On a rigid lattice, admolecule diffusion over abrupt terrace steps is suppressed by the Ehrlich-Schwoebel barrier. An admolecule crossing over the step edge will (temporarily) be in a position nearly removed from the surface, which is energetically quite unfavorable. This creates a potential barrier that impedes such crossings. The step structure will become less rigid in the presence of surface premelting, however, which should substantially reduce the Ehrlich-Schwoebel barrier. Image adapted from [2002He].

At still higher temperatures, surface diffusion becomes even more prevalent, and a thick QLL develops atop the crystalline lattice, as illustrated in Figure 4.10d. In this regime, the ice/vapor terrace nucleation model is replaced with an ice/QLL model that starts to resemble terrace nucleation at an ice/water interface. The concept of an admolecule surface density begins to lose its meaning at the ice/QLL interface, as the molecular dynamics is more influenced by QLL bulk diffusion, as in the Wilson-Frenkel picture discussed earlier in the chapter. Although there should be plenty of surface diffusion at temperatures well above T_{onset}, when the Ehrlich-Schwoebel barrier is essentially absent, this additional transport will have a relatively minor effect on the nucleation barrier. Thus, at temperatures substantially above T_{onset}, we again expect that terrace nucleation on small facets will be little changed from the large-facet model.

Putting these ideas together, this (admittedly inexact) model predicts the existence of an "SDAK dip" on each of the large-facet σ_0 curves, as illustrated in Figure 4.12. Here I have chosen different values of T_{onset} for the basal and prism facets to impose agreement with the known ice growth behaviors from the Nakaya diagram and other growth measurements. Moreover, the values of the SDAK curves must depend on w and perhaps other factors, as the very nature of the SDAK hypothesis is that the attachment kinetics depend on the mesoscopic crystal structure. The specific curves shown in Figure 4.12 represent rough estimates for R_{edge} in the 1–2 μm range, as these are typical for snow crystal growth in air, as discussed above. These portions of the curves are drawn as dotted lines to signify their substantial uncertainties.

Although the SDAK model is certainly wanting in that it does not make precise calculable predictions for α_{basal} and α_{prism}, it nevertheless makes some surprisingly concrete predictions. For example, the CAK model with the SDAK phenomenon indicates that thick plates should be the primary habit at −5°C when the supersaturation is low, because $\alpha_{prism} > \alpha_{basal}$

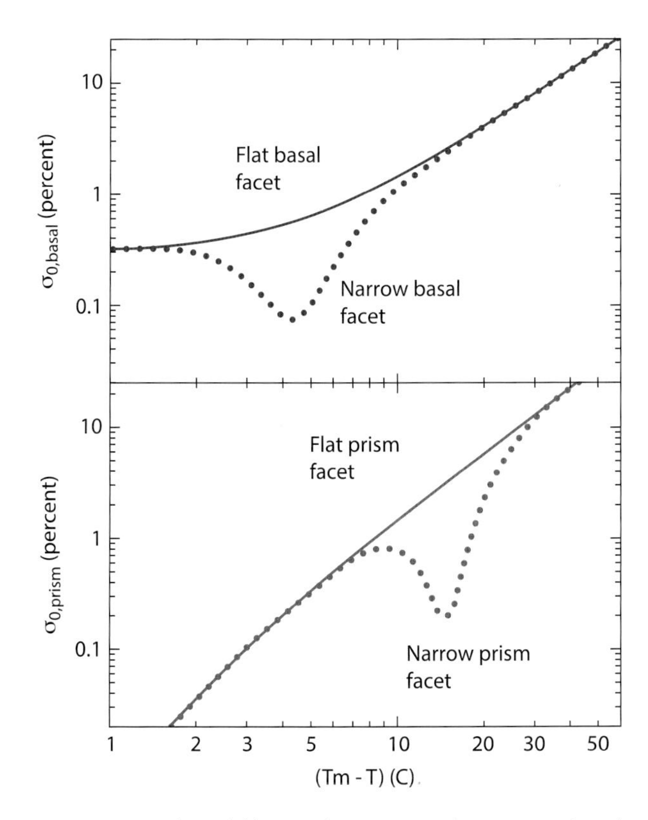

FIGURE 4.12. The solid lines in these two graphs are reproduced from Figure 4.5, again illustrating the CAK terrace-nucleation model for the attachment kinetics on large basal and prism facets. The dotted lines show the "SDAK dips" described in the text. Near −5°C, the narrow basal facets on hollow columns exhibit a much reduced effective $\sigma_{0,basal}$, explaining the rapid growth of these features. Near −15°C, narrow prism facets exhibit a similarly low effective $\sigma_{0,prism}$, explaining the growth of thin plates at that temperature.

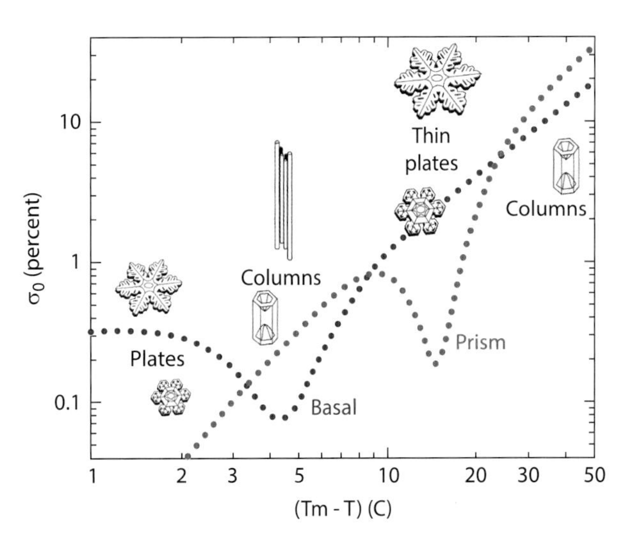

FIGURE 4.13. Plotting the SDAK curves from Figure 4.12 together, we can compare the basal and prism nucleation barriers as a function of temperature. A lower nucleation barrier (that is, a lower σ_0) means faster attachment kinetics on the corresponding facet surface. Thus the CAK model exhibits the same transitions between platelike and columnar growth that are seen in the Nakaya diagram (see Chapter 1). Beyond this qualitative agreement with morphologies, the CAK model also provides good quantitative agreement with many precision ice growth measurements [2019Lib1, 2020Lib].

on large facets at that temperature. But hollow columns can also grow at −5°C if the basal edges are thin, owing to the SDAK effect. Indeed, as described below, a close look at a broad range of ice growth data indicates that thick plates do readily form at −5°C, as do hollow columns, and the two forms can even grow concurrently under certain conditions [2012Kni, 2019Lib2].

Plotting the two narrow-facet curves together yields the result shown in Figure 4.13, revealing the full impact of this SDAK modeling exercise. Using suitable choices for T_{onset} on the basal and prism facets results in a set of model curves that reproduces the overall temperature dependence seen in the Nakaya diagram. According to this CAK model, the various transitions between platelike and columnar growth are caused largely by changes in the effective nucleation barriers on *narrow* basal and prism facets. If correct, the SDAK phenomenon thus plays a crucial role in defining the long-mysterious temperature transitions observed in snow crystal morphology.

Of course, the critical reader may balk at the speculative nature of this model, along with its complexity and ill-defined curves. Fair enough. What the reader may not immediately see, however, is that the model is largely driven by empirical data from a variety of ice growth measurements. The large-facet curves are well grounded in experiments, and the SDAK dips, or something similarly complex, are required to explain a substantial body of additional data. Creating a comprehensive model of the attachment kinetics that can explain the Nakaya diagram, along with a variety of additional ice growth measurements, is a challenging task that requires a complex solution. As I discuss later in this chapter, there are essentially no other viable alternatives to the CAK model at present, at least not if one requires a reasonable consistency with recent experimental data. Although "something else" is always a viable alternative hypothesis, the CAK model, including the SDAK effect, provides at least a plausible working model of the attachment kinetics over a broad range of conditions. Moreover, the CAK model makes abundant quantifiable predictions regarding growth rates and morphological behaviors that can be tested with straightforward experimental investigations, as I describe below.

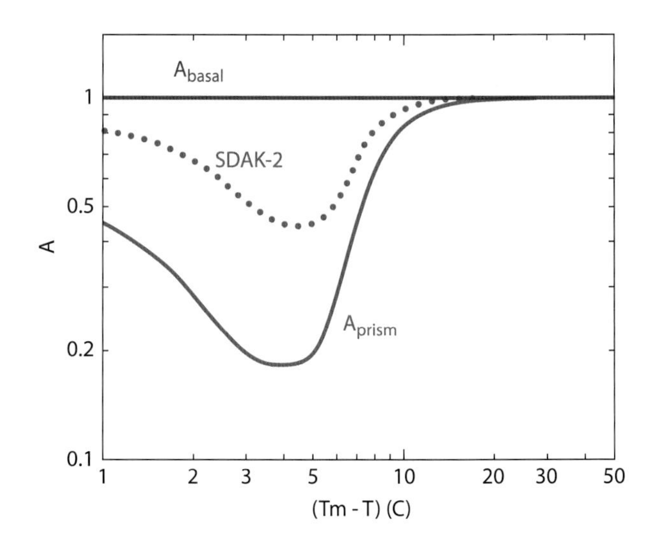

FIGURE 4.14. The SDAK-2 mechanism increases the value of A_{prism} on small prism facets at temperatures above −10°C, as approximated here by the dotted curve. This curve is expected to depend on supersaturation, such that $A_{prism} \rightarrow 1$ at especially high σ_{surf}. The solid curves are reproduced from Figure 4.5. This change in A_{prism} mainly affects the growth of sharp-tipped dendritic structures, which are often observed in high-supersaturation experiments.

SDAK-2: Enhanced QLL Kinetics

The SDAK mechanism illustrated in Figure 4.10 suggests an additional consequence of enhanced surface diffusion that might change the attachment kinetics at high temperatures. When the QLL is well developed, the transport of water molecules from the facet corners to narrow prism facets could become great enough to increase the nominal thickness of the QLL on the prism surfaces. If so, then the increased QLL thickness should tend to alleviate the frustrated QLL kinetics described in the large-facet model above. This additional mechanism—call it SDAK-2—will thus increase A_{prism} on small prism facets, as illustrated in Figure 4.14. Here again, the SDAK-2 curve is drawn as a dotted line, signifying that its exact location is not well defined, and it will change with w and σ_{surf}. The model mainly makes the qualitative prediction that $A_{prism} \rightarrow 1$ on especially small, fast-growing prism facets on the ends of sharply tipped dendritic structures. The SDAK-2 phenomenon thus reconciles $A_{prism} < 1$ seen in large-facet growth with additional experimental evidence pointing to $A_{prism} \rightarrow 1$ on narrow facets [2019Lib1]. Perhaps other physical mechanisms could produce a similar growth behavior, but this model provides a ready explanation for A_{prism} being especially large on the smallest prism facets.

Note that the SDAK-2 effect only becomes important on small prism facets at high growth rates, and it has little effect on the overall morphology diagram shown in Figure 4.13. As a result, SDAK-2 is overall less important than SDAK-1, as the former is mostly needed to reproduce the observed growth of sharply tipped dendritic structures at temperatures above −10°C. This is an easily studied region, however, as such crystals grow reliably in air at high supersaturations.

The Edge-Sharpening Instability

A particularly fascinating consequence of the SDAK hypothesis is a phenomenon I call the *edge-sharpening instability* (ESI), which is illustrated in Figure 4.15. As the name implies, the ESI tends to sharpen basal and prism

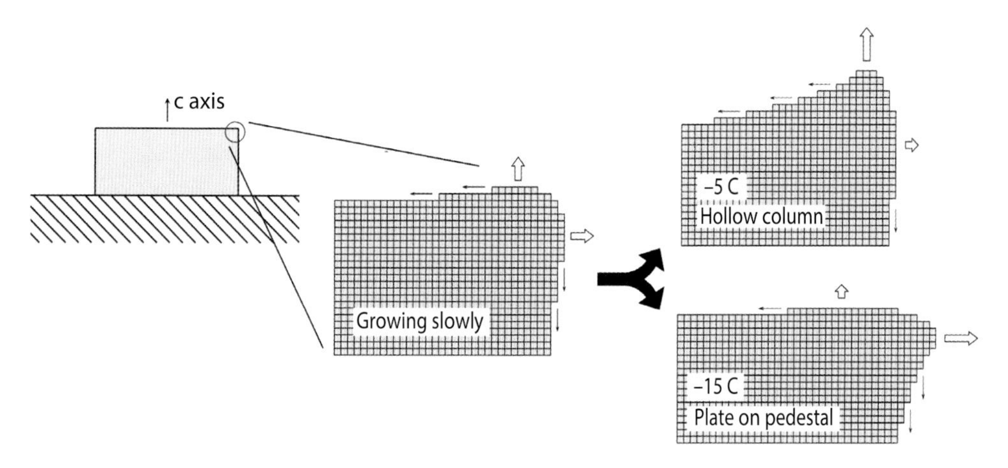

FIGURE 4.15. The edge-sharpening instability (ESI). This sketch illustrates how the SDAK phenomenon can bring about an edge-sharpening growth instability on either the basal or prism facets. As described in detail in the text, as the uppermost terrace becomes narrower, the attachment coefficient on that terrace can increase via the SDAK mechanism, thus increasing the growth and further narrowing the upper terrace. This brings about a positive feedback effect, and thus a growth instability, that promotes the formation of sharp edges, either as hollow columns (upper right sketch) or thin plates (lower right sketch). In the comprehensive attachment kinetics (CAK) model presented in this chapter, the ESI mechanism in air is largely responsible for the formation of thin plates at −15°C and hollow columns at −5°C.

edges via a positive feedback effect that takes place during diffusion-limited growth. I believe that this growth instability is largely responsible for the remarkably robust appearance of thin plates near −15°C and the similarly robust formation of hollow columns near −5°C. If this addition to the CAK model is correct, then the ESI is one of the most important physical processes shaping the growth of atmospheric snow crystals.

Looking at thin-plate growth near −15°C in Figure 4.15 (shown as a plate-on-pedestal morphology), the essential starting point for the ESI is the hypothesis that α_{prism} on an edgelike prism facet depends strongly on the width of the top prism terrace (the SDAK hypothesis). For a broad facet, α_{prism} is well described by the nucleation-limited model with the parameters given in Figure 4.5. On a narrow prism edge, $\alpha_{prism} \to 1$ as $w \to 0$, where w is the edge width. The SDAK mechanism described above would bring about this behavior, but the precise physical mechanism is not important as long as it operates over a narrow range of temperatures near −15°C.

If the SDAK hypothesis is correct, then diffusion-limited growth naturally brings about the positive-feedback effect illustrated in Figure 4.15. For example, when the applied supersaturation σ_∞ is low and the facet growth rates are corresponding low, then stable facets form, as shown in the middle sketch in the figure. On both the basal and prism facets, new terraces mostly nucleate near the exposed corner, where the supersaturation is highest owing to diffusion effects. This yields trains of basal and prism terrace steps propagating from the corners to the facet centers. The facet surfaces both become slightly concave in the process, and this overall growth morphology describes stable, faceted crystal growth (see Chapter 3).

As σ_∞ is increased, the crystal grows faster as terraces nucleate more readily at the corners, and the facets become more concave. The faster growth means that the terrace steps become more closely spaced, and the width of the uppermost basal and prism terraces becomes smaller. Because $\sigma_{0,prism}$ decreases as the top prism terrace width decreases (our hypothesized SDAK behavior), the nucleation rate increases as well, so more prism terraces appear, and thus the width of the top prism terrace decreases further still. The result is a growth instability brought about by a positive feedback effect—the prism edge sharpens, σ_0 decreases, the edge sharpens more, σ_0 decreases more, and so forth. Above some threshold value of σ_∞, the process runs away, and a thin plate forms atop the initial blocky crystal. This is the ESI, which I exploit to create the PoP laboratory snow crystals described in Chapter 9.

Note that the ESI is *not* equivalent to the normal Mullins-Sekerka instability (MSI) that is a part of diffusion-limited growth (see Chapter 3). Instead, the ESI can be thought of as an extension of the MSI that comes about when the SDAK phenomenon is incorporated into the overall growth process. The MSI by itself (without SDAK) will certainly cause branching that results in the formation of complex dendritic structures. But diffusion-limited growth and the MSI alone will not produce morphologies with extreme aspect ratios, like thin plates. The formation of thin plates requires highly anisotropic attachment kinetics, specifically, $\alpha_{prism}/\alpha_{basal} \gg 1$, which results from the SDAK phenomenon. The underlying physics is both complicated and somewhat subtle, because several physical processes are happening simultaneously, but I believe that the ESI is the best explanation for the robust formation of thin plates and hollow columns in atmospheric snow crystals.

Being a growth instability, the ESI helps explain the somewhat extreme morphological features one often observes in natural snow crystals, as illustrated in Figure 4.16. As the left image illustrates, it is not uncommon to find capped columns with surprisingly thin endplates, suggesting an abrupt transition from columnar to platelike growth. These crystals likely experience gradually changing environmental conditions in the clouds, so one might expect more gradual morphological transitions. But the presence of a bone fide growth instability can handily explain these abrupt transitions, as the ESI naturally brings about rapid changes in $\alpha_{prism}/\alpha_{basal}$, which the MSI cannot accomplish by itself. Moreover, once the transition from columnar to platelike growth has occurred, the SDAK mechanism keeps α_{prism} high on the plate edges, reinforcing the development of thin plates on the capped columnar crystal.

In the right image in the figure, the ESI also provides a ready explanation for the formation of thin-walled hollow columns near −5°C. These crystals appear over a narrow range of supersaturations, as σ_∞ must be high enough to initiate the ESI mechanism but not so high that the sheathlike edges split to yield a complex needle cluster. Note also that natural snow crystals can be a bit tricky to interpret, as the final morphology of a given crystal depends on its entire growth history. In the case of hollow columns, sheathlike crystal edges may develop a thicker character if the supersaturation later decreases, thus complicating the final interpretation. Therefore, laboratory observations are generally better suited to examining the dynamics of the ESI under controlled conditions.

As a growth instability, the ESI naturally yields quite thin basal and prism edges in air, as are often observed. Once the instability kicks in, it often does not stop until the edge thickness reduces to a width of 1–2 microns. As a result, this edge thickness appears over a broad range of growth conditions, including thin-plate crystals, thin-walled hollow columns, and fast-growing dendritic structures. Solvability theory (see Chapter 3) provides some insights as to why this particular size scale is so prevalent, but computational modeling in tandem with experimental observations under controlled conditions will be needed to investigate this phenomenon in detail.

The nature of the ESI mechanism also suggests that interesting effects should be found when growing complex snow crystal structures as a function of background gas pressure. Because the ESI couples the attachment kinetics to diffusion-limited growth, this opens a new dimension in the exploration of structures that are both faceted and branched. Takehiko Gonda found that both structural complexity and overall aspect ratios were pressure dependent [1976Gon], but it appears that the supersaturation may have also varied with applied pressure in that work, making it difficult to obtain a quantitative interpretation of the observations. Simultaneously controlling the tempera-

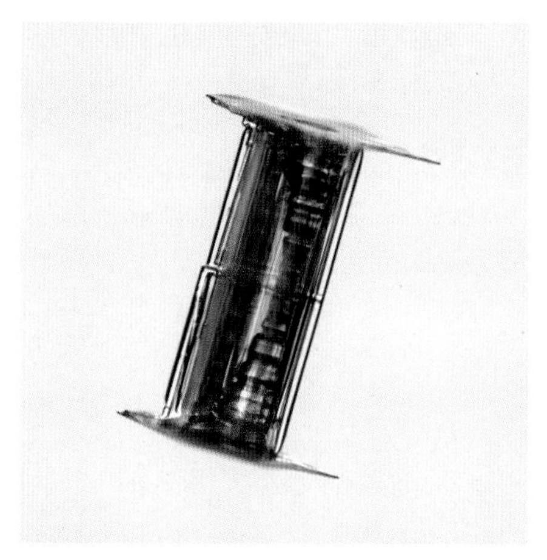
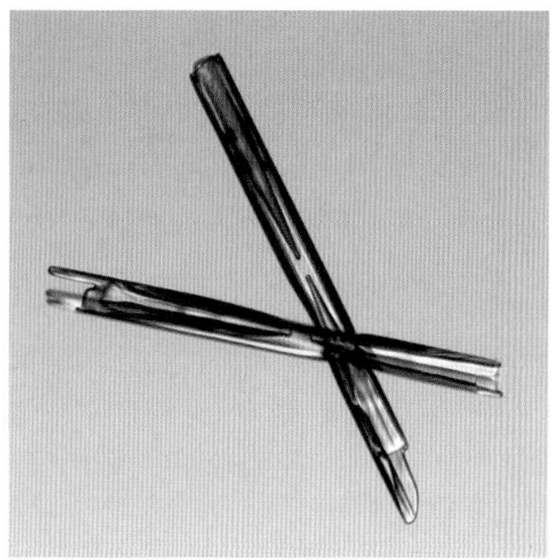

FIGURE 4.16. The ESI can explain many distinctive morphological features in natural snow crystals. For example, the often abrupt transition from columnar to platelike growth seen in many capped columns (left) suggests the ESI operating on the prism surfaces near −15°C. Similarly, the remarkably thin, sheathlike edges on many hollow columns (right) can arise when the ESI operates on basal surfaces near −5°C. It is difficult to explain these commonly observed snow crystal features without invoking the ESI mechanism.

ture, supersaturation, and background gas pressure is challenging in the laboratory, but exploring the ESI over a broad range of conditions could yield many interesting insights regarding our general understanding and modeling of snow crystal growth.

Parameterized Kinetics

While the large-facet attachment kinetics can be examined by measuring the growth of simple ice prisms (see Chapter 7), observations of complex snow crystal structures will be needed to investigate the SDAK and ESI phenomena. These can be readily grown in air on the tips of electric ice needles (see Chapter 8), but analyzing these complex structures will be challenging. Simple platelike or hollow-column structures can be approximated by cylindrically symmetrical forms, but complex dendritic crystals will require fully 3D computational modeling (see Chapter 5). In either case, parameterized expressions describing the attachment kinetics will be needed as model inputs.

As described earlier in the chapter, the large-facet attachment kinetics can be parameterized by Equation 4.4, as this functional form provides good agreement with experimental measurements. Including SDAK effects necessitates parameterizing the attachment coefficient as $\alpha(\sigma_{surf}, T, w)$ or $\alpha(\sigma_{surf}, T, \kappa)$, where w is the width of the top terrace, and κ is the surface curvature. Note that w and κ are related, so the choice of independent variable is largely a matter of computational convenience. However, unlike the case for terrace nucleation, theory does not provide any clear functional form for the SDAK effect.

Because the ESI tends to yield structures with tip radii in the 1–2 μm range in air, the precise functional form for the SDAK effect may not be essential for producing realistic computational models. As long as the chosen $\alpha(\sigma_{surf}, T, w)$ results in a suitable growth instability, the details may not matter as much as the attachment coefficients on the narrow 1–2 μm tips and edges. For example, in [2015Lib2] we used the functional form $\sigma_0 = \sigma_{0,\infty}[1 - \exp(-w/w_0)]$, where $\sigma_{0,\infty}$ is the broad-facet value and w_0 is an adjustable model parameter. This allowed us to reproduce the ESI transition to platelike growth at −15°C reasonably well, as the behavior that mattered most is having $\sigma_0 \rightarrow 0$ as $w \rightarrow 0$.

At present, 3D computational models are not sufficiently advanced to reproduce the full menagerie of snow crystal structures with good fidelity, but that situation is improving rapidly (see Chapter 5). On a parallel track, our ability to create complex snow crystals in controlled environments with reproducible initial conditions is also improving rapidly (see Chapter 8). Soon it should be possible to combine these technologies and fully explore the SDAK and ESI phenomena, no doubt generating many new insights into how these peculiar aspects of the attachment kinetics affect snow crystal structure formation.

EXPLAINING THE NAKAYA DIAGRAM

At this point, it is beneficial to take a step back and examine just how difficult it has been to understand the enigmatic Nakaya diagram, even at a basic qualitative level. Ever since its discovery (see Chapter 1), researchers have struggled to create a comprehensive physical model that describes its overall features. The increase in morphological complexity with greater supersaturation generally arises from diffusion-limited growth (Chapter 3), so this aspect is reasonably well explained. But the various transitions between platelike and columnar growth with temperature are brought about by changes in the attachment kinetics, and this aspect of the Nakaya diagram presented a challenge right from the outset. The overall empirical behaviors of the morphology transitions were quickly reproduced and extended by several researchers, so the observational side was quickly well established. But the quest for a comprehensive physical model continues to this day.

Previous Attempts

Basil Mason and collaborators made an early attempt at developing a suitable model of the attachment kinetics in the 1960s by reporting measurements of admolecule diffusion lengths $x_{diff}(T)$ on both the basal and prism facets as a function of temperature [1958Hal, 1963Mas]. These observations suggested a temperature dependence in the relative growth rates of the primary facets that roughly explained the several transitions between platelike and columnar growth seen in the Nakaya diagram. The Mason et al. model enjoyed some early popularity, but it has generally not withstood the test of time for several reasons. On the experimental side, the underlying measurements relied on observations of macrostep growth velocities in air to extract x_{diff} values, and this technique is now known to be prone to systematic errors. Macrostep growth rates are strongly affected by bulk diffusion and other subtle effects, and even today, it is exceedingly difficult to measure admolecule surface diffusion lengths with good absolute accuracy [2014Asa, 2015Lib]. Moreover, recent measurements have achieved a reasonable degree of consensus that terrace nucleation, not surface diffusion, is the primary factor limiting facet growth in most circumstances [2013Lib, 2017Lib, 2019Har].

On the theory side, measurements of $x_{diff}(T)$ alone would not constitute a comprehensive model of the attachment kinetics unless the model included at least some sketch of the underlying molecular physics that explained the observations. In this same vein, while empirical measurements of α_{basal} and α_{prism} as a function of growth conditions are an important step forward, such measurements alone do not constitute a comprehensive physical model.

In the 1980s, Toshio Kuroda and R. Lacmann (KL) created a new comprehensive model incorporating a molecular picture of how attachment kinetics might vary with temperature-dependent changes in surface premelting [1982Kur, 1984Kur]. The authors postulated several speculative, but physically plausible, transitions in the ice surface structure as a function of temperature and further postulated that these transitions occurred at different temperatures on the basal and prism facets. By adjusting several model parameters, rough agreement with the observed plate/column transitions in the Nakaya diagram was obtained.

In many respects, the KL model was prescient in several of its features. Kuroda and Lacmann clearly recognized the separation between large-scale diffusion effects and the localized attachment kinetics, with the latter depending mainly on the near-surface supersaturation σ_{surf}. The authors further realized the importance of determining σ_{surf} in ice growth experiments via careful diffusion modeling. The KL model was also the first to incorporate surface premelting as a major physical component and to postulate a facet-dependent premelting behavior as an integral part of the model. My selective positioning of the SDAK dips, postulated from facet-dependent premelting, was an idea adopted from the KL model.

Beyond these clear successes, the KL model does not provide an adequate comprehensive description of the attachment kinetics. With its focus on surface roughening phenomena at the ice/QLL/vapor interface, the KL model is generally inconsistent with subsequent observations identifying terrace nucleation as the primary mechanism limiting the growth of large facets. Moreover, it now appears that a more complex set of physical mechanisms, including substantial differences between large and small facets, is necessary to explain all the different snow crystal growth behaviors that have been revealed by newer experiments.

The CAK model presented in this chapter builds on these past efforts, responding to much additional input from recent precision ice growth experiments. The model thus does a better job reproducing a broader spectrum of measurements, but at a cost of considerably increased physical complexity. Although the CAK model may not be correct in every detail, I believe that the following features will likely withstand the test of time:

1) Terrace nucleation on large facets. The experimental evidence has become quite strong that ice growth data are well represented by a terrace nucleation model over a broad range of growth conditions. The specific curves in Figures 4.5 and 4.6 may require some tweaking as experimental measurements improve, but the importance of terrace nucleation seems to be a solid conclusion.
2) On most nonfaceted surfaces, $\alpha_{rough} \approx 1$. Experiments generally support this blanket statement, and there are no definitive measurements (to my knowledge) indicating α_{rough} values significantly below unity. Here again, this may change with improved measurements, and the transition from rough to faceted (i.e., the attachment kinetics on vicinal surfaces) has not been well explored.
3) Structure-dependent attachment kinetics (SDAK). The realization that the attachment kinetics on small facets cannot be entirely the same as on large facets constituted something of a breakthrough in my thinking on this subject. Once I began examining large and small facets separately, I encountered many "ah, that makes sense now" moments. The specific SDAK models presented above may be overthrown or discarded by future research, but I have come to believe that the SDAK concept is essential for explaining snow crystal growth.
4) A rich SDAK phenomenology. If the preceding points are correct, then it follows that the full SDAK picture will be complicated and difficult to understand in detail. The edge-sharpening instability (ESI) is one example. Developing an accurate model of the attachment kinetics in this regime will be challenging, likely requiring a substantial research effort that combines carefully controlled laboratory observations with fully 3D computational modeling of complex snow crystal structures.
5) Facet-dependent surface premelting. This seems to be the only reasonable way to explain the observed transitions between platelike and columnar growth as a function of temperature. The concept was postulated in the KL model, and I incorporated it into the CAK model as well. Surface premelting likely exhibits some differences on the basal and prism facets, so this assumption is plausible. MD simulations generally exhibit similar premelting behaviors on the two facets, but the calculations may not be accurate enough

to see small facet-dependent effects. Moreover, no one has suggested a viable alternative that would yield a physically sensible, yet sufficiently complex, model of the attachment kinetics.

6) Data-driven features. Beyond terrace nucleation, there is little solid theoretical guidance for modeling the attachment kinetics, and cartoon sketches can only take one so far. Progress must be driven largely by empirical reasoning, especially using precision measurements of ice growth rates under well-controlled conditions. I have begun a critical examination of the CAK model using a variety of ice growth measurements [2019Lib1, 2020Lib], and the results so far are quite promising. But much additional work needs to be done along these lines.

My objective with the CAK model presented in this chapter is not to declare a full solution to the attachment kinetics problem but rather to move the ball forward while stimulating additional progress in this area. A quantitative working model, even an imperfect one, is far better than no model at all, as it can serve to suggest specific, targeted experimental and theoretical investigations.

As described in Chapter 1, a primary motivator for this book is a desire to grow computational snow crystals that accurately reproduce what one finds in nature. Developing a physical model of the attachment kinetics is an important prerequisite for this task, as computational models require physical inputs. Because tracking every molecule is clearly impossible, even the best model can only provide approximate parameterizations of the functions α_{basal} and α_{prism} as a function of σ_{surf}, T, w, and perhaps other variables. While no such parameterizations can attain absolute accuracy, one can hope to capture the essential physics and create plausibly realistic computational snow crystals. At the very least, a good model of the attachment kinetics should agree with ice growth experiments and provide a reasonable explanation of the Nakaya diagram. The CAK model presented in this chapter strives to attain these objectives.

Beyond the Nakaya Diagram

One fact that becomes glaringly obvious when thinking about a comprehensive model of the attachment kinetics is that the traditional Nakaya diagram cannot capture the full richness of snow crystal formation. Describing typically observed growth morphologies in air on a temperature/supersaturation plane simply cannot do justice to such a complex physical phenomenon. Background pressure is another interesting environmental factor, and initial conditions play a role via the SDAK effect. Overall crystal size is important as well, as dendritic structures are generally less prevalent on smaller crystals. Therefore, it is important to move beyond morphologies and develop new ways to visualize the physics and phenomenology describing snow crystal growth.

I have found that plotting α_{basal} and α_{prism} as a function of σ_{surf} at constant temperature provides one useful complement to the Nakaya diagram, and some illustrative examples are shown in Figure 4.17. These curves were derived from the CAK model and include several separate "branches" that describe different growth regimes. Solid lines show the large-facet kinetics, while dotted lines indicate the two SDAK effects. Drawn this way, one begins to see some of the morphological complexity inherent in the CAK model.

At −5°C, for example, the appearance of four separate branches leads to several distinct morphological behaviors. Both platelike and columnar forms can develop at σ_{surf} = 0.15 percent, the former being simple plates with large facets, the latter being hollow columns or needles with narrow basal facets. And both forms are readily observed, offering persuasive support for the CAK model [2012Kni, 2019Lib2]. The ESI tends to stimulate hollow-column growth at relatively high σ_{∞} in normal air, while platelike forms are more prevalent at lower σ_{∞} and lower air pressures. At the highest σ_{surf}, the two SDAK curves at −5°C yield "fishbone" dendritic structures (Chapter 3) with a tip growth axis that varies with growth rate (see Chapter 8). The full range of observed growth behaviors at −5°C is reasonably consistent with the CAK model but is difficult to explain otherwise.

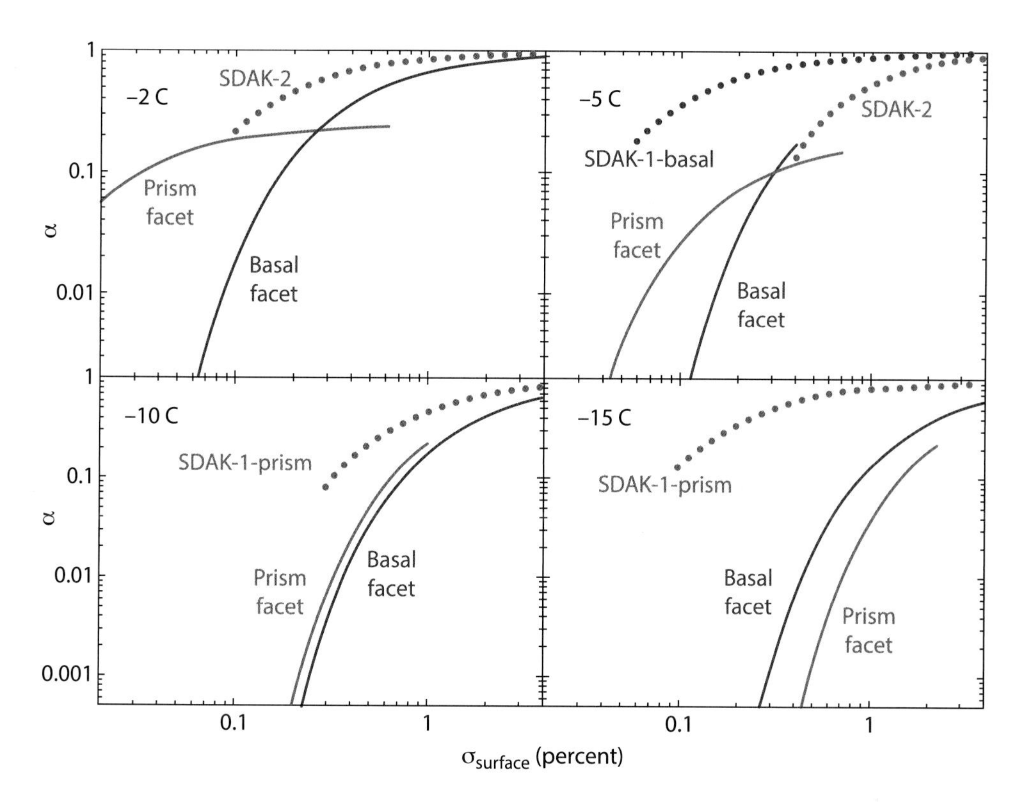

FIGURE 4.17. The inherent richness of the CAK model, and the attachment kinetics more generally, becomes apparent when displayed as constant-temperature $\alpha(\sigma_{surf})$ plots. The different "branches" represent different growth regimes identified by their corresponding physical effects. The graph shown for −5°C is particularly noteworthy, as it indicates that both platelike and columnar forms can appear at σ_{surf}= 0.15 percent, a feature that is supported by observations. Plots like these are especially useful for comparing with all different types of ice growth experiments.

At −15°C, Figure 4.17 indicates that simple prisms with large facets should be nearly isometric in form, and such crystals are readily observed in a vacuum. But the ESI quickly yields rapidly growing prism edges in air, resulting in the formation of thin plates. Dendritic forms remain nearly platelike even at high growth rates, owing to the strong SDAK effect on prism facets together with the large nucleation barrier on basal facets. Morphologies at −2°C are generally platelike at all supersaturations.

While morphology studies are a good starting point, there is much more to be learned by exploring snow crystal growth as a function of background gas pressure and generally covering a broader range of environments and initial conditions. And, of course, quantitative experiments are a must for making detailed comparisons with models of the attachment kinetics. Graphical representations of growth phenomenology may soon become even more sophisticated, once we begin making fully 3D structural comparisons between computational models and advanced laboratory observations of complex snow crystal structures.

THE MORPHOLOGICAL NEXUS AT −5°C

Because snow crystal growth at −5°C involves a confluence of all the varied growth behaviors identified in the CAK model, this temperature provides an especially thorough test of many of the most important aspects of the attachment kinetics, including large-facet growth and both SDAK effects. In this section, I examine data from a variety of ice growth experiments, including large and small facets, all focusing on the morphological nexus that appears at −5°C.

Large-Facet Growth

Beginning with large facets, both the vacuum ice growth (VIG) and variable pressure growth (VPG) experiments (see Chapter 7) provide measurements of α_{basal} and α_{prism} as a function of σ_{surf}, and these results are shown together in Figure 4.18. Several immediate conclusions can be reached from these data:

1) The VIG and VPG experiments show excellent agreement at −5°C, even though they use significantly different measurement strategies and completely different hardware. Moreover, the basal and prism facets were measured separately in the VIG experiment, using different crystals in different runs, while both facets were measured simultaneously on smaller ice prisms in the VPG experiment. Obtaining clear agreement from such different experiments is itself a noteworthy achievement, given the considerable measurement uncertainties and discrepancies found in earlier ice growth experiments.
2) There is no obvious air-pressure dependence in either α_{basal} or α_{prism} as a function of σ_{surf} in these data, at least over the limited range that was measured (see Chapter 7). This supports the assumption in the CAK model that air at a pressure of 1 bar has little effect on the attachment kinetics.
3) The data all strongly support the terrace nucleation mechanism for the large-facet attachment kinetics in the CAK model. Note also that the data support $A_{basal} \approx 1$ and $A_{prism} < 1$ at this temperature, as described above. Constraining the model to have $A_{prism} = 1$ is not excluded completely by the measurements, but both data sets show quite similar trends that prefer $A_{prism} < 1$.

As shown in these examples at −5°C, precision measurements of simple prisms provide the foundation for the CAK model, in which large-facet growth is limited by terrace nucleation over a broad range of environmental conditions. The functional form of the model curves comes from terrace nucleation theory, but the model parameters $A(T)$ and $\sigma_0(T)$ are obtained entirely from empirical data.

Bimodal Behavior

The VPG experiment at −5°C allows another look at how the SDAK phenomenon can yield remarkably different growth morphologies at a single temperature. In a near-vacuum environment at low supersaturations, platelike growth is the norm, and Figure 4.19 illustrates that even quite extreme aspect ratios can be observed. According to the CAK model, even thinner plates should appear at lower supersaturations, but residual substrate interactions may be preventing these from forming in the VPG apparatus [2019Lib1]. Thin plates at low supersaturations require an exceedingly low basal nucleation rate (with growth rates below 1 nm/sec), and substrate-induced basal nucleation could set a practical limit on making thin plates on any substrate. A levitation experiment at −5°C might yield thinner plates than those shown in the figure, but small amounts of chemically induced basal nucleation may also be problematic at the lowest growth rates. In any real experimental environment, there may be some small rate of heterogeneous terrace nucleation that is difficult to eliminate.

Growing crystals in air at higher supersaturations yields the hollow columnar crystals shown in Figure 4.19, their overall shape being somewhat distorted because they are resting on the substrate. The sharp basal edges observed on these crystals suggests that the ESI played a major role in determining their growth morphology. As discussed earlier in this chapter, the development of the CAK model was driven largely by the empirical necessity to explain the diversity of observations like these.

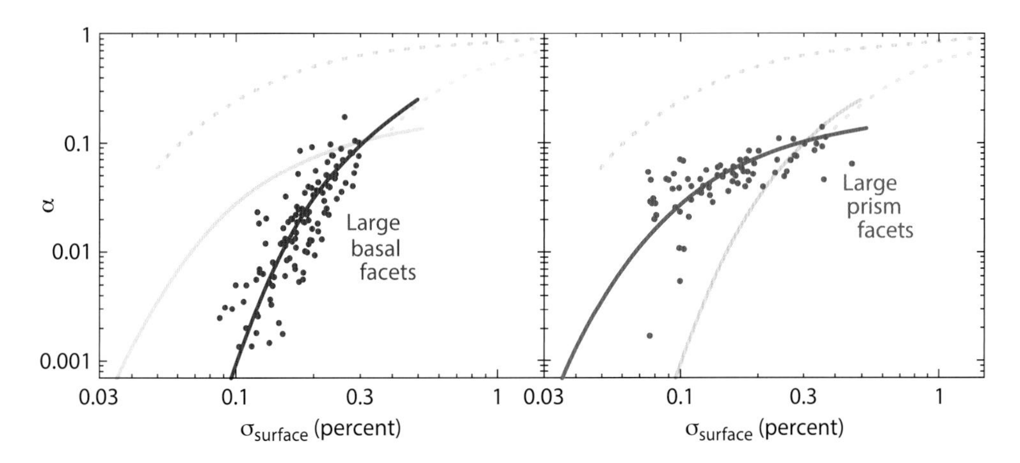

FIGURE 4.18. Data points in these two graphs show measurements of $\alpha_{basal}(\sigma_{surf})$ and $\alpha_{prism}(\sigma_{surf})$ at −5°C in a low-pressure environment. Lines in both graphs are from the CAK model for −5°C (less relevant curves have been faded in the different panels). These observations reveal that platelike behaviors are the norm for simple ice prisms at −5°C, with thinner plates forming at lower supersaturations. These plots combine measurements from the VIG and VPG experiments described in Chapter 7.

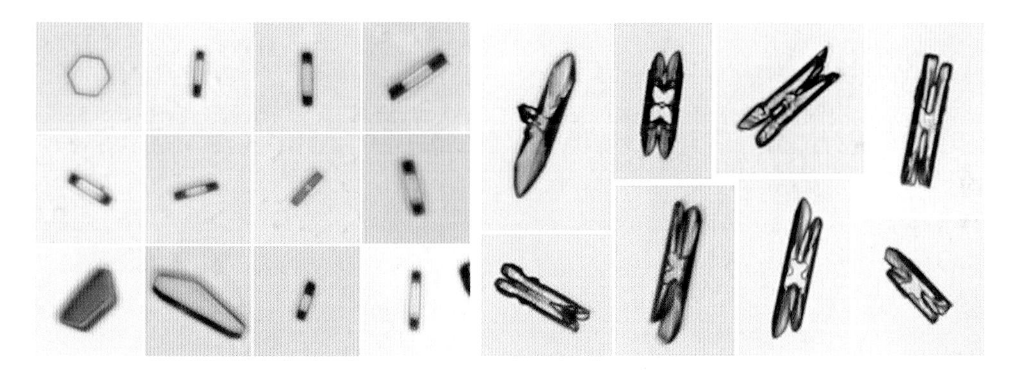

FIGURE 4.19. (Left) A selection of thin platelike crystals growing at −5°C at low pressure in the VPG apparatus with $\sigma_{surf} \approx 0.1$ percent. These examples illustrate that quite thin plates can be grown at −5°C. (Right) Hollow columnar crystals grown in air at −5°C using the VPG apparatus. Although the overall morphologies are distorted by their intersection with the substrate, it appears that the ESI is largely responsible for the formation of thin basal edges on these crystals.

Fishbones, Tridents, Needles, and Columns

The −5°C story becomes especially intriguing when one also considers the small prism and basal facets appearing on complex crystal structures growing in air. Figure 4.20 illustrates several examples, including hollow columns (4 percent and 8 percent supersaturation), needles (16 percent), and fishbone dendrites (32 percent, 64 percent, and 128 percent). Although we cannot yet model the growth of these elaborate morphologies in detail, it is still possible to glean some useful information from growth ratios. For the hollow columnar and needle crystals, for example, it is straightforward to measure v_{basal} and v_{prism} at the end of each column from a time series of images. Because the sides of the columns contain large-faceted prism surfaces, the CAK model suggests that their growth should be determined by terrace

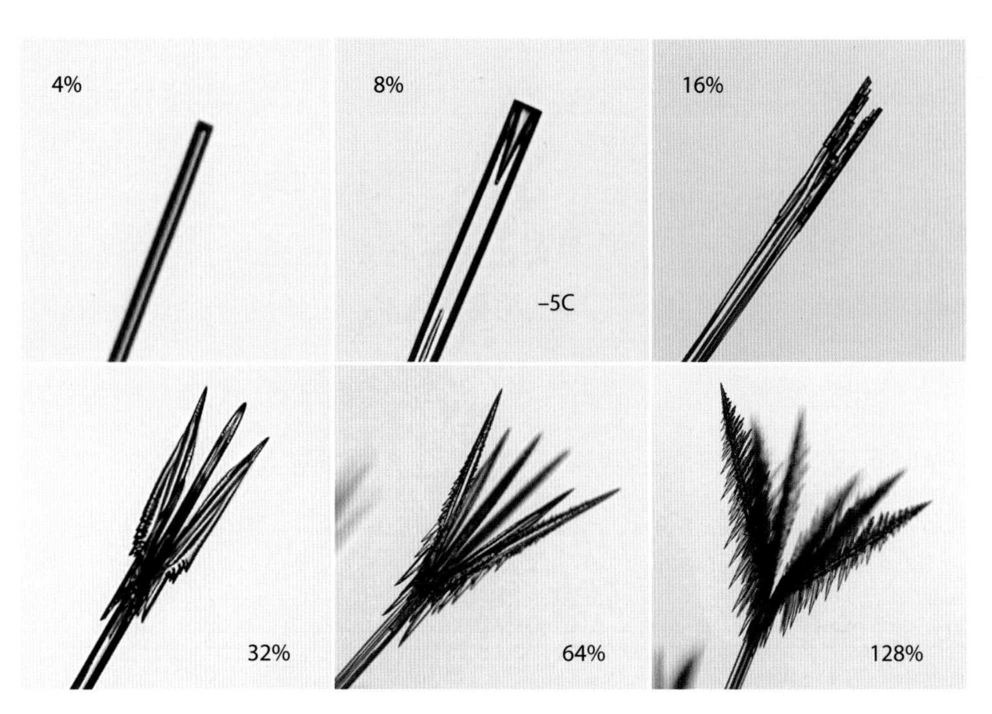

FIGURE 4.20. Several examples of crystals growing at −5°C in air on the ends of "electric" ice needles (see Chapter 8). Each image is labeled with the supersaturation σ_∞ far from the needle tip. At 4 percent, the initial needle thickens to form a solid column with slight basal hollowing, while hollow-column growth develops quickly at 8 percent. At 16 percent, a hollow column has split to form needle clusters. At 32 percent, trident structures are common (see Chapter 3). At 64 percent and 128 percent, fishbone dendrites (Chapter 3) exhibit rapid growth.

nucleation, so the known $\alpha_{prism}(\sigma_{surf})$ in that case means we can ascertain σ_{surf} from the measured v_{prism}. Of course, this is a model-dependent statement, and it would be better if a full diffusion model determined σ_{surf} directly, as we did with the VIG and VPG experiments. But the diffusion corrections are too large for this to be practical at present, so the indirect, model-dependent determination will have to suffice. I refer to this as a "witness-surface" analysis, as the (presumably) known large-facet prism growth behavior provides a witness that allows a determination of σ_{surf}.

The next step in the witness-surface analysis is to assume that σ_{surf} on the prism facet near the top corner of the columnar crystal is approximately equal to σ_{surf} on the basal surface at the same corner. This is reasonable when the two surfaces are small and proximate, although the equality might be somewhat off if large supersaturation gradients are present. With this approximation for σ_{surf}, it then becomes possible to determine $\alpha_{basal}(\sigma_{surf})$ from the measured v_{basal}, thus yielding the first three data points in Figure 4.21. Note that $v_{basal}/v_{prism} \gg 1$ for columnar and needle growth, fixing $\alpha_{basal}/\alpha_{prism}$ in this figure, and our main model–dependent statement was estimating σ_{surf} by putting the α_{prism} points on the large-facet terrace nucleation model curve. Note also that we could not have applied the witness-surface analysis in the opposite direction, as the basal facet is small on a hollow column, so its growth is not well known from the CAK model.

Perhaps the most interesting part of this analysis is that it quantifies the bimodal behavior at −5°C, showing that both platelike and columnar crystals can form under quite similar conditions. Platelike crystals grow readily in near-vacuum and in air, while columnar crystals only appear when the basal facets are small, which only happens in air because of the ESI. With the CAK model including the ESI, it all makes a reasonable amount of sense. But without the ESI, explaining the full range of columnar and platelike growth behaviors would be quite challenging.

The analysis is a bit trickier for the three fishbone dendrites shown in Figure 4.20, but again we can extract some useful information from growth ratios. Each of these crystals exhibits a measurable tip velocity and growth direction, where the latter is measured relative to the c-axis defined by the initial electric-needle crystal. Thus, from a time series of images, it is straightforward to measure v_{basal} and v_{prism} at the fishbone tips. However, the witness-surface analysis runs into problems, because now both the basal and prism surfaces are tiny on these sharply pointed crystals, so neither would be well described by large-facet growth. Nevertheless, starting with the 128 percent fishbone, the fast growth points to α_{basal} being near unity, which is consistent with the SDAK-1 basal curve in Figure 4.21. Assuming $\alpha_{basal} \approx 1$, the witness surface analysis then yields $\alpha_{prism} \approx 1$ as well, simply reflecting the large opening angle of this set of fishbone dendrites (or, equivalently, $v_{basal} \approx v_{prism}$). This allows us to bootstrap our way along, as the α_{basal} points can be interpolated onto a reasonable SDAK-1 curve, in turn giving a set of α_{prism} points from an approximate witness-surface analysis. Note that the fishbone angle changes with σ_{surf}, which means that the SDAK-2 points diverge from the SDAK-1 points at lower supersaturations. With the trident morphology at 32 percent, for example, we see growth that is midway between columnar and fishbone behaviors.

Also note that the σ_{surf} values in Figure 4.21, as determined from the witness-surface analysis, drop by about a factor of two between data points, as one would expect from the corresponding drop in σ_{∞} between the different crystals. The Ivantsov solutions for parabolic growth (Chapter 3) predicts $\sigma_{surf}/\sigma_{\infty} \approx 2X_0/BR_{tip}$, and

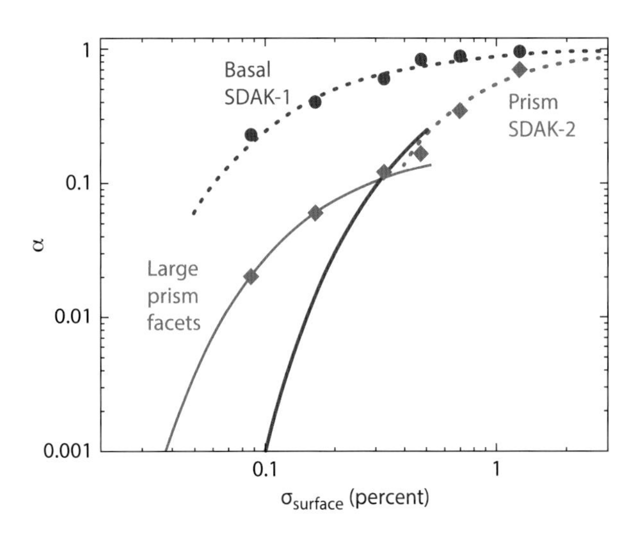

FIGURE 4.21. Analysis of the snow crystals growing at −5°C in air on the ends of electric ice needles shown in Figure 4.20. These data all show fast basal kinetics compared to large basal facets, resulting from the basal SDAK-1 mechanism. Prism growth also increases at high supersaturations from the SDAK-2 phenomenon. The lines are again from the CAK model at −5°C.

using $B \approx 10$ and $R_{tip} \approx 1.5$ μm gives $\sigma_{surf}/\sigma_{\infty} \approx 1/50$, which is not too far from what comes out of the witness-surface analysis. Thus, the numbers are reasonably consistent in an absolute sense, even if it is not yet possible to solve the diffusion equation with sufficient accuracy to make an accurate measurement. Although the witness-surface analysis is model dependent and only uses growth ratios, the results describe a self-consistent pattern of growth behaviors that support the CAK model overall. Figure 4.22 shows all the different CAK model growth behaviors present at −5°C.

Figure 4.23(a) shows another data set, this time for columnar crystals growing in ordinary air in a free-fall growth chamber (see Chapter 6). Although the experiment produced estimates of σ_{∞}, the diffusion corrections are so high that it is not possible to determine σ_{surf} with good accuracy. For this reason, the graph shows the result of another witness-surface analysis, once more assuming that the prism growth of the columnar crystals is given by the terrace nucleation curve in the CAK model. And again, we see that the basal growth lies near the SDAK-1 curve, reflecting the fact that the hollow columnar crystals exhibit sharp basal edges. The free-fall measurements again support the CAK model to a reasonable degree, considering the various data and analysis uncertainties.

Figure 4.23(b) shows a crude analysis of some observations at −5°C reported by Knight [2012Kni], in which he describes the simultaneous growth of platelike and needle crystals in air. The nature of the observations makes it impossible to estimate σ_{surf} with any real accuracy, so the data points shown in the figure should be taken as rough estimates only. Nevertheless, with some commonsense reckoning, it is possible to place these observations in the CAK model, even if this cannot be done with great accuracy. The exercise mainly shows that these observations are at least consistent with the CAK model, and the simultaneous observation of platelike and needle forms certainly confirms the bimodal growth behavior at −5°C.

Summarizing all these results at −5°C, we find:

1) When large prism or basal facets are present, their growth fits the large-facet behavior provided by the CAK model, as illustrated in Figure 4.18.
2) On hollow columns, the narrow basal facets grow rapidly in a manner that is consistent with the SDAK-1 mechanism, while the large prism facets are still described by large-facet growth (Figures 4.21 and 4.23).
3) On the tips of fishbone dendrites, the basal and prism growth behaviors are consistent with the SDAK-1 and SDAK-2 phenomena (Figure 4.21).

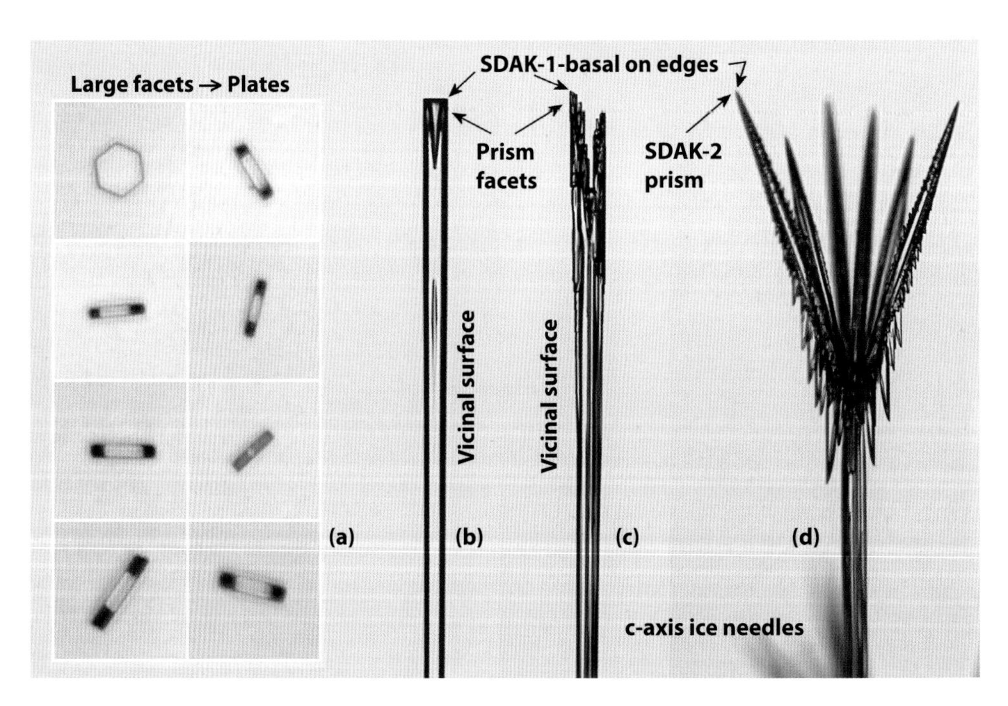

FIGURE 4.22. This photo collage illustrates several snow crystal growth behaviors at −5°C. (a) With simple prisms grown in vacuum, thick plates are the norm. (b) Starting with a c-axis needle in air (see Chapter 8), the basal ESI from the SDAK-1 phenomenon brings about sharp basal edges and a hollow column. (c) At a higher supersaturation, the hollow column splits into a needle cluster. (d) At still higher growth rates, the SDAK-2 mechanism increases the prism kinetics, yielding dendritic structures where the opening angle increases with supersaturation. All four CAK branches in Figure 4.21 are needed to reproduce these different growth behaviors.

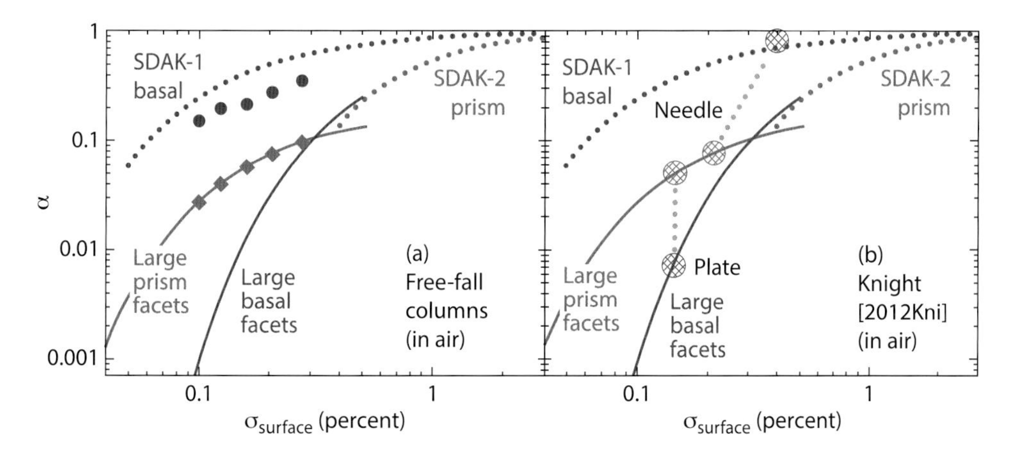

FIGURE 4.23. Additional snow crystal growth data obtained at a fixed temperature of −5°C [2019Lib1]. (a) Data from a free-fall growth chamber, again analyzed using the witness-surface method. (b) A rough analysis of observations [2012Kni] that demonstrated the simultaneous growth of platelike and needle structures at −5°C. The data can also be explained using the CAK model, albeit with substantial experimental uncertainties. Together with Figures 4.18 and 4.21, these data exhibit a remarkably diverse range of growth behaviors, and all can be reasonably well explained with the CAK model.

4) The CAK model explains the simultaneous formation of platelike and needle crystals under certain conditions (Figure 4.23).
5) By treating broad and narrow facets differently, the CAK model reasonably explains a convoluted tangle of morphological observations and quantitative ice growth data at −5°C.

SNOW CRYSTAL CARTOGRAPHY

The full spectrum of measurements at −5°C illustrates how several quite different growth behaviors can all be explained with the same comprehensive model of the attachment kinetics. These results at −5°C are especially intriguing, as so many physical effects come into play at this temperature, thus providing an excellent overall test the CAK model. I have begun to call this analysis "snow crystal cartography," as one is able to map out the different growth regimes in remarkable detail using a variety of different experimental and data-analysis strategies. Remarkably, even with just a simple witness-surface analysis for the fast growth data, requiring no complex 3D diffusion modeling, we can explore the different SDAK branches in the CAK model. Although much remains to be done, a compelling picture of snow crystal growth phenomenology is beginning to emerge from these studies.

Figure 4.24 provides another example using data compiled at −2°C [2020Lib]. The CAK model is generally simpler at this temperature, as the SDAK-1 effect is no longer significant, so only platelike and dendritic morphologies appear under all measured growth conditions. In keeping with expectations from the Nakaya diagram, no columnar forms are present at −2°C. Once again, displaying the measurements on a log-log plot showing α as a function of σ_{surf} provides a useful tool for exploring the different growth regimes, and again the data generally support the CAK model. At the time of this writing, my colleagues and I are examining additional temperatures following this same methodology, a process that will likely take some years to complete.

DISLOCATION-MEDIATED GROWTH

Up to this point, we have ignored lattice dislocations and other lattice imperfections in this chapter, focusing first on the ideal case of the attachment kinetics on a flawless underlying crystal. But real crystals inevitably have some

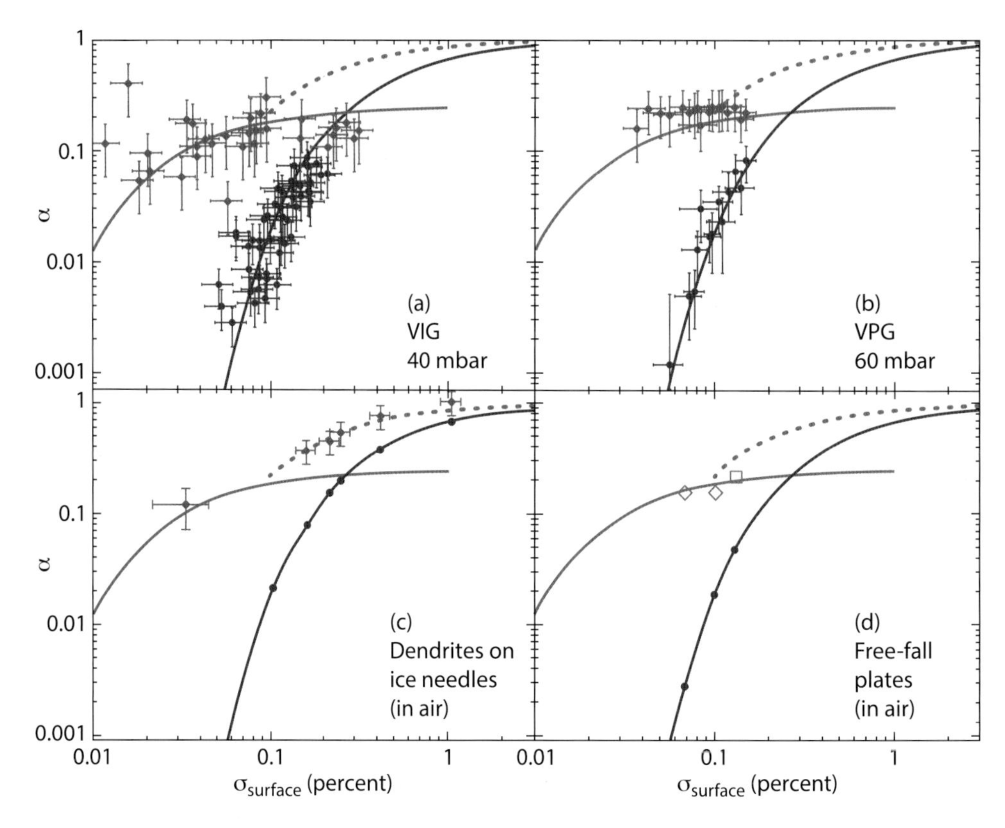

FIGURE 4.24. A collection of snow crystal growth data obtained at a fixed temperature of −2°C [2020Lib]. (a) Data from the VIG experiment at low background pressure, showing good agreement with the CAK model. (b) Similar data from the VPG experiment, confirming the earlier VIG results and the CAK model. (c) Observations of ice crystals growing on the ends of electric ice needles, analyzed using the witness-surface method on the basal facets. These data show rapid prism growth at high supersaturations, in agreement with the SDAK-2 phenomenon in the CAK model. (d) Additional data from a free-fall growth chamber, again analyzed using the witness-surface method. Although the overall phenomenology at −2°C is not as rich as at −5°C, the data show quite good agreement with the CAK model on all fronts.

flaws, so let us now examine to what extent these flaws can affect the attachment kinetics. If lattice dislocations are present on a faceted surface, then the terrace nucleation mechanism described above may be augmented by a faster mechanism involving screw dislocations. The left sketch in Figure 4.25 shows the lattice structure of a screw dislocation, which begins as a defect in the ideal lattice configuration. As admolecules attach to the exposed terrace edge, it develops into the characteristic spiral pattern shown in the right sketch in the figure. This spiral pattern can persist indefinitely, as the dislocation will not "heal" as the crystal grows. By providing a continuous source of molecular steps, a screw dislocation can thus yield substantial growth rates even when σ_{surf} is far below that required for normal terrace nucleation. The theory behind this mechanism is described in most crystal growth textbooks [1996Sai, 1999Pim, 2002Mut], yielding $\alpha \sim \sigma_{surf}$ and a perpendicular growth velocity $v_n \sim \sigma_{surf}^2$.

F. C. Frank suggested that the simple observation of symmetrical hexagonal prisms likely indicates the absence of dislocation-mediated growth on the six prism facets

[1982Fra]. Because the facet surfaces all grow at equal rates, either there are no dislocations present on any of the facets or there must be at least one dislocation on each of them. If screw dislocations were present on some of the facet surfaces but not all, then the growth rates would vary, and the overall prism morphology would not show hexagonal symmetry. Although asymmetrical prisms can be found [1979Kik], most prisms are nearly symmetrical. As a high dislocation density seems improbable on small crystals with simple morphologies, the logical conclusion is that most prism facet surfaces are free of dislocations, or at least those that greatly alter the facet growth rates.

Beyond this simple observation, there is also considerable experimental evidence indicating that many laboratory-grown snow crystals, particularly simple prisms and other forms exhibiting especially clean morphologies, are generally free from dislocations that affect their growth. The example in Figure 4.4, which is representative of many similar crystal specimens, is quite well described by a nucleation-limited model, while a dislocation-mediated growth model gives a substantially poorer fit to the data. Some simple prisms do exhibit anomalous growth behaviors, but up to 90 percent of well-formed samples are consistent with nucleation-limited growth [2013Lib, 2019Lib2], at least at temperatures above −20°C. As with polycrystalline forms, dislocations appear to be more common as the temperature decreases [2004Bai, 2009Bai].

While many fascinating crystal growth phenomena are mediated by dislocations [2013Men], this subject is somewhat beyond the scope of this book. Most of this chapter has examined the attachment kinetics on essentially perfect single crystals, and the resulting CAK model is already quite complex. For this reason, I leave a fuller discussion of the effects of dislocations on snow crystal growth for another day. Instead, throughout most of this book, I focus on the growth of defect-free ice crystals.

CHEMICAL VAPOR EFFECTS

In the preceding sections of this chapter, I made the implicit assumption that the ice/vapor attachment kinetics are completely unaffected by any physical or chemical effects associated with the surrounding medium. In our discussion of the ESI, for example, particle diffusion influenced the nucleation of new terraces by modifying σ_{surf} around the crystal, but the presence of air did not change the attachment kinetics directly. In other words, I assumed that α_{basal} and α_{prism} had no intrinsic dependence on the background gas pressure or composition. Let us now examine this assumption more closely.

Air Effects

A first question is whether ordinary clean air affects the attachment kinetics in any significant way. This matters not only for natural snow crystals but also for the many laboratory ice growth experiments that have been performed in laboratory air, sometimes at reduced pres-

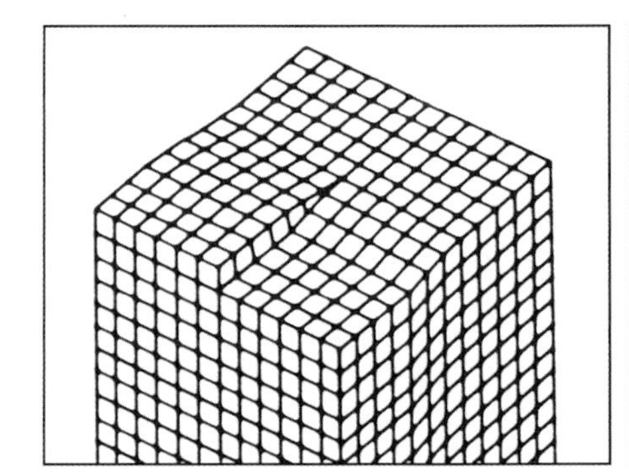

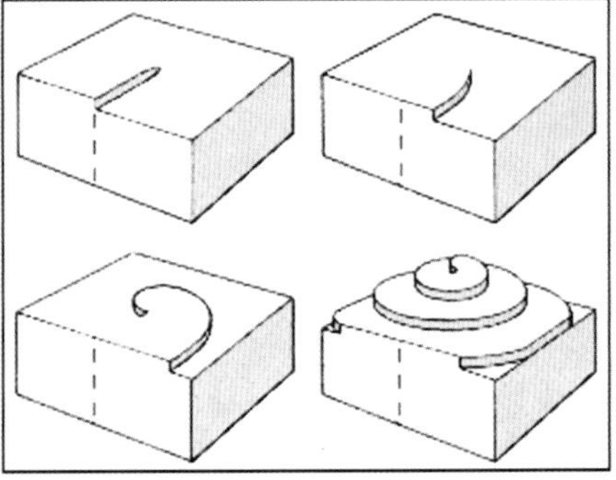

FIGURE 4.25. (Left) The lattice structure of a screw dislocation. (Right) As the dislocation edge grows from admolecule attachment, it creates a spiral pattern that can propagate indefinitely, yielding growth in the absence of additional terrace nucleation.

sure. Beginning with nitrogen and oxygen, Henry's Law data show that both these gases have rather low solubilities in water, and adsorption data reveal that they do not stick well to ice surfaces either. Moreover, recent molecular dynamics simulations indicate that nitrogen gas at 1 bar appears to have little direct effect on the structure and dynamics of surface premelting [2019Llo]. These considerations, although somewhat indirect, all suggest that normal air probably has a negligible influence on the attachment kinetics. Carbon dioxide has a higher solubility in water, but it has a low concentration in air, and snow crystals grown in CO_2 at 1 bar do not show any obvious abnormalities (see below).

Nevertheless, some experiments have indicated that ice growth rates are influenced by the presence of air, to an extent beyond that caused by normal particle diffusion-limited growth. For example, Beckmann et al. [1983Bec] found that the slower growth rates of ice crystals in air were due in part to a reduction in the attachment kinetics, and similar conclusions were reached by Kuroda and Gonda [1984Kur1]. Recently I suggested that air-dependent effects were especially strong on prism facets at −5°C, which could explain the growth of columnar crystals at that temperature [2016Lib1]. Looking more closely at these experiments, however, I now believe that all three suffered from faulty diffusion analyses. The diffusion corrections are quite large in air, and relatively small systematic errors can easily be mistaken for a pressure-dependent attachment kinetics. The best way around this problem is to examine especially small crystals, and then only in a low-α regime, where diffusion effects are relatively minor. In that restricted regime, newer measurements show no significant dependence of the attachment kinetics on air pressure at −5°C [2019Lib1]. Given our current state of knowledge, several pieces of evidence appear to be reliable:

1) Basic chemical physics reasoning suggests that nitrogen, oxygen, and other molecular species in air are generally inert with respect to ice adsorption. MD simulations appear to support this conclusion as well.
2) When diffusion effects are especially small (low α, small crystals), recent experiments suggest negligible changes in α with gas pressure (see Chapter 7). In this regime, the attachment kinetics are dominated by terrace nucleation on large facets.
3) When $\alpha_{diff} \ll \alpha$ in ice growth experiments, it becomes exceedingly difficult to separate diffusion-limited growth from kinetics-limited growth, so results can be influenced by quite small systematic errors.
4) The CAK model can explain the morphology diagram and other ice growth measurements without requiring that the attachment kinetics change with gas pressure. Although this model may not be entirely correct, it suggests that gas-dependent kinetics are not obviously required to explain the full range of observed snow crystal growth behaviors.

There is a great need for additional investigation in this area, as large swaths of parameter space have not yet been adequately explored by experiments. At present, however, it appears that the molecular dynamics determining the ice/vapor attachment kinetics is little affected by a background gas of ordinary air.

Chemical-Dependent Attachment Kinetics

In contrast to inert gases, including clean air, the presence of chemically active vapor additives can dramatically change ice growth rates and morphologies. There is much experimental evidence supporting this general conclusion, and little understanding of any of it. This is a fascinating area for continued investigation and a constant concern that unwanted chemical impurities can lead to erroneous experimental conclusions.

This area of research has a long history, as Vonnegut, Hallett, and Mason [1948Von, 1958Hal, 1971Mas] found that the addition of just 10 ppm of butyl alcohol in air yielded columnar growth at −20°C instead of the usual platelike growth. Schaefer [1949Sch] further observed that vapors of ketones, fatty acids, silicones, aldehydes, and alcohols could all change ice growth morphologies to varying degrees. Nakaya, Hanajima, and Mugurama [1958Nak] observed that even trace silicone vapor in air caused columnar crystals to grow at −15°C instead of thin plates. Hallett and Mason [1958Hal] found that the addition of camphor vapor in air could yield columnar ice crystals at all temperatures in the range $-40°C < T < 0°C$. These authors also observed that isobutyl alcohol in air changed ice growth near −15°C from plates to columns and then back to plates again as the concentration was increased. Anderson, Sutkoff, and Hallett [1969And] found that methyl 2-cyanoacrylate in air could change the morphology from platelike dendrites to needles at −15°C. Libbrecht, Crosby, and Swanson [2002Lib] found that acetic acid and other vapors promoted the c-axis growth of "electric" needle crystals in air near −5°C, thus yielding a useful tool for studying snow crystal growth (see Chapter 8). Knepp, Renkens, and Shepson [2009Kne] observed various morphological changes caused by acetic acid vapor in air, even in concentrations as low as 1 ppm. Libbrecht and Bell [2011Lib] examined snow crystal morphologies as a function of temperature for a range of chemical additives as a function of concentration. From all these reports, we can summarize some of the principal findings:

1) In nitrogen gas at 1 bar, most chemical additives at concentrations below 10 ppm produce no clearly observable changes in ice crystal growth morphologies [2011Lib]. As a rough rule of thumb, if the air has no discernable odor, it likely has little effect on snow crystal formation at 1 bar.
2) Ice growth in air, nitrogen, helium, argon, hydrogen, carbon dioxide, and methane gases at a pressure of 1 bar yield roughly identical crystal morphologies as a function of temperature [1959Heu, 2008Lib1], suggesting that these gases are essentially chemically inert.
3) Growth in ultraclean nitrogen gas was not significantly different from growth in ordinary laboratory air [2011Lib]. These first three points suggest that trace impurities in ordinary air do not play a large role in snow crystal growth.
4) Chemical additives generally tend to promote the growth of columnar crystals over platelike crystals.
5) Nitric acid vapor or nitrous oxide tends to promote the growth of triangular crystals near −15°C [1949Sch].
6) The most effective chemicals for producing growth modification are those having strong polar properties [1949Sch].
7) Chemical additives generally have an especially large, detrimental effect on thin plates and platelike dendritic growth at −15°C [2011Lib]. In general, thin plates at −15°C seem to grow best in inert gases at pressures of 1 bar or higher.
8) Chemical effects are generally more pronounced at lower temperatures, and ice growth experiments performed at $T < -20°C$ are especially prone to unwanted influences from trace chemical contaminants.

Given our poor understanding of ice growth without the complicating effects of chemical additives, it should come as no surprise that there is essentially no theoretical understanding, even at a qualitative level, of how chemical additives alter growth rates and change growth morphologies. It appears likely that these impurities adsorb onto the ice surface, where they modify the attachment kinetics in a variety of ways, but there is little theoretical guidance relating to how. While the adsorption of chemical vapors on ice has been much studied from the standpoint of atmospheric chemistry, relatively little attention has been given to chemical modification of the attachment kinetics. This remains a fascinating research direction, although somewhat hampered by a lack of theoretical guidance. Antifreeze proteins strongly affect the attachment kinetics at the ice/water interface, but that fascinating topic is only peripherally related to the ice/vapor interface.

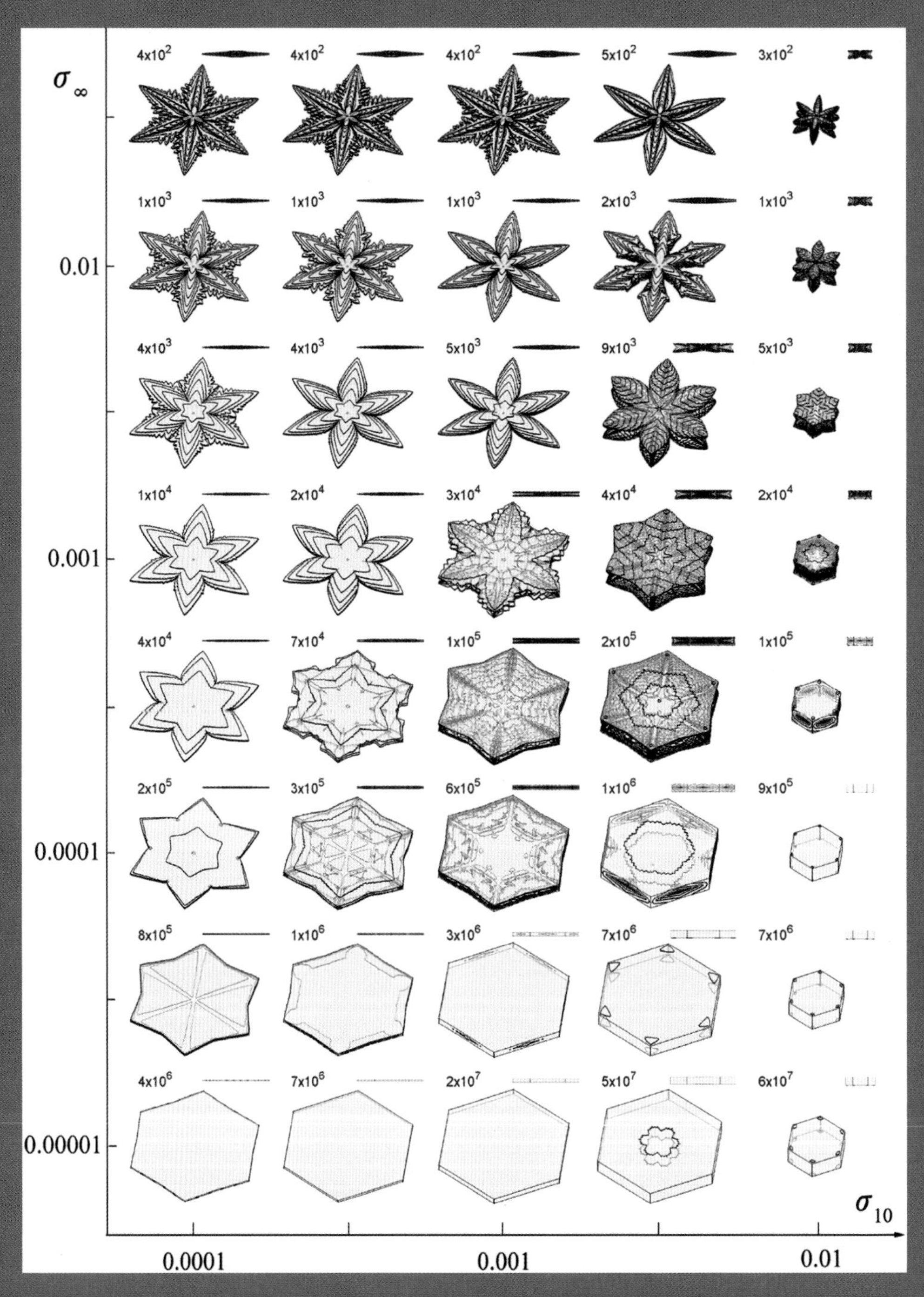

FIGURE 5.1. An array of snow crystal models generated using the cellular-automaton method [2014Kel]. Different values of the supersaturation σ_∞ (vertical axis) and σ_{10} (horizontal axis) were used, the latter being a nucleation parameter in $\alpha_{prism}(\sigma_{surf})$. The basal attachment coefficient $\alpha_{basal}(\sigma_{surf})$ was the same for all models shown. Numbers give physical growth times in seconds. Image courtesy of James Kelly.

FIVE

Computational Snow Crystals

Nature is an endless combination and repetition of a very few laws. She hums the old well-known air through innumerable variations.

—RALPH WALDO EMERSON, ESSAYS, LECTURES AND ORATIONS, 1851

Computational modeling has become an important tool in contemporary science, and once again we find that the snow crystal presents a fascinating microcosm of modern scientific investigation. Being an intrinsically complex phenomenon, snow crystal formation cannot be described by a fundamental "theory," at least not in the simplest sense of the word. Instead it is necessary to first break the problem down into its constituent parts to better understand the variety of physical processes acting over different scales. One then reassembles those parts into a computational model to replicate the full range of complex behaviors.

The scientific method remains intact in this view, but now the computational model becomes the hypothesis to be tested, as it predicts specific snow crystal structures for given environmental and physical inputs. If the hypothesis agrees with experimental measurements over a broad range of conditions, then we can rightly say that we generally understand the phenomenon. In the preceding chapters, we focused on the individual pieces of the puzzle, examining the ice crystal structure, diffusion physics, and the attachment kinetics in isolation. Using physical insights gained from these studies, we now examine numerical techniques that allow us to grow computational snow crystals.

In principle, building a computer model of a growing snow crystal is straightforward enough. Starting with a small digital ice crystal, first numerically solve the diffusion equation around it, assuming all the proper boundary conditions. From this solution, extract the growth rate at all points on the surface, and then use this information to "grow" the crystal a small amount to yield a slightly larger crystal. Repeat. After many iterations, the crystal develops into a complex morphology that hopefully resembles a laboratory snow crystal produced under the same physical conditions. Alas, although the task is relatively straightforward in principle, developing appropriate numerical algorithms that can accomplish this objective is remarkably difficult. Current computational models are just beginning to look promising, and there have been almost no quantitative comparisons between models and laboratory observations.

When setting out to create a snowflake simulator, one soon encounters a host of thorny technical issues to deal with, such as numerical instabilities, highly anisotropic boundary conditions, nonlocal effects, and numerous other problems that must be addressed. Moreover, a variety of shortcuts and approximations are required if one is to produce a realistic code with finite spatial resolution and a reasonable running time. Moreover, beyond these computational issues, accurately reproducing snow crystal growth requires a thorough understanding of the molecular attachment kinetics, which is a remarkably challenging problem all by itself (see Chapter 4). As is often the case in science, the devil is in the details, and producing realistic computational snow crystals involves a lot of details. Several different classes of computational strategies have been developed over the years for simulating a range of solidification problems, and each technique comes with its own strengths and weaknesses. Several of these methods have been applied to the specific problem of snow crystal growth, but with limited success to date. Developing a robust numerical method that reproduces crystal growth that is both branched and faceted remains very much a work in progress.

Because this book is about the science of snow crystal formation, the present chapter will focus on numerical modeling techniques that strive to produce physically realistic simulations, not just pretty pictures that vaguely resemble snowflakes. Our objective is to create computational models that can be compared with experimental observations in a quantitative fashion, reproducing both growth rates and morphologies over a broad range of conditions. Moreover, the model underpinnings should derive from sound molecular and statistical physics to the greatest possible degree, rather than ad hoc parameterizations.

Importantly, a successful computational model should reproduce the full menagerie of snow crystal structures as a function of external growth conditions, including temperature, supersaturation, background gas pressure, and any other factors we care to include. Generating digital structures that resemble stellar snowflakes is a fine start, but this alone is not a bone fide scientific objective. The ultimate goal in this chapter, and indeed this book, is to develop a comprehensive physical model of snow crystal formation, which will require physically realistic computational modeling.

A PROGRESSION OF SNOW CRYSTAL MODELS

The overarching topic of structure formation during solidification has received much attention in the scientific literature, and numerous reviews are available [2002Boe, 2016Kar, 2017Jaa, 2018Che]. The various algorithms and computational techniques can be mathematically quite sophisticated, and I am by no means an expert in this broad and technical field. Thus, to limit the scope of this chapter, I mostly restrict the discussion to research efforts that have examined the specific problem of snow crystal growth.

Packard Snowflakes

In 1986, Norman Packard described one of the first attempts to model structure formation during solidification using cellular automata (CA) methods [1986Pac]. Although Packard's CA rules were not physically derived, they revealed a rich variety of morphological structures that developed during growth, including the *Packard snowflake* shown in Figure 5.2. Packard's iterative CA could be considered something of an extension of the ideas behind the Koch snowflake [1904Koc] (see Chapter 3), enabled by the widespread availability of personal computers in the 1980s. These early models were intriguing for their ease in generating complex structures from simple governing rules, but they contained only a superficial relation to the actual physical process of solidification.

FIGURE 5.2. A Packard snowflake (left), generated using simple nearest-neighbor rules in a cellular automaton, is compared to a photograph of a natural snowflake (right). Although there are obvious structural similarities between the two images, the model has little basis in the physical processes underlying snow crystal growth. Image adapted from [2008Gra].

Diffusion-Limited Aggregation

In a landmark early paper on modeling diffusive transport, Thomas Witten and Leonard Sander [1981Wit] examined the formation of metal-particle aggregates via a random-walk process that they called diffusion-limited aggregation (DLA). In their model, individual particles traverse a fixed grid in random small steps until they encounter a solid surface and stick to it, thus simulating a crude form of diffusion-limited solidification. Being especially simple to implement on small computers, the DLA method was quickly adapted and applied across many fields to a wide range of physical phenomenon.

Rong-Fu Xiao, J. Iwan Alexander, and Franz Rosenberger carried the DLA method a step further by incorporating CA rules that attempted to simulate anisotropic attachment kinetics and molecular surface diffusion [1988Xia]. With a suitable adjustment of their model parameters, the authors demonstrated a clear transition from faceted to dendritic growth morphologies, as seen in Figure 5.3. Moreover, this transition resulted from the competing processes of particle diffusion and attachment kinetics, which is essentially the current paradigm of snow crystal formation. When applied to a fixed triangular grid, the Xiao et al. DLA model was the first to convincingly demonstrate this central snow crystal morphological transition using rational (albeit not entirely accurate) physical foundations.

Improving Physical Inputs

In 1990, Etsuro Yokoyama and Toshio Kuroda presented the first significant attempt to create a comprehensive physical model of snow crystal growth dynamics [1990Yok]. By combining a novel molecular model of temperature-dependent attachment kinetics [1982Kur] with a numerical method for solving the diffusion equation, the authors sought to recreate the growth behavior of actual snow crystals under realistic environmental conditions.

While previous investigators had simulated general growth behaviors using ad hoc parameterizations, Yokoyama and Kuroda modeled the specific phenomenon of snow crystal growth from water vapor, including a careful examination of all the physical processes involved. Notably, the authors incorporated the known physical properties of ice and water vapor, allowing a direct quantitative comparison between simulated snow crystals and laboratory experiments. As the authors stated in their abstract [1990Yok, p. 2038]: "We propose a model of pattern formation in the growth of snow crystals that takes into account the actual elemental pro-

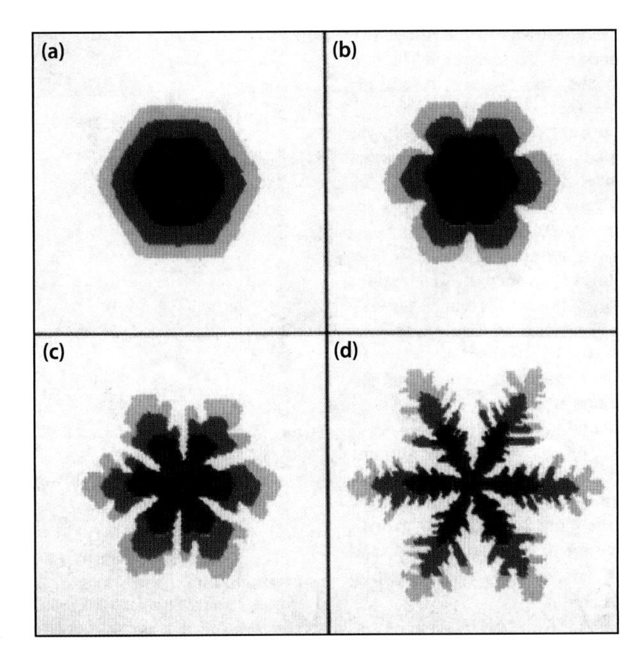

FIGURE 5.3. A progression from faceted prism growth (a) to dendritic growth (d) in a 2D DLA model. This work was the first to demonstrate a morphological transition of this nature resulting from the competing processes of diffusion-limited growth and surface attachment kinetics. Image adapted from [1988Xia].

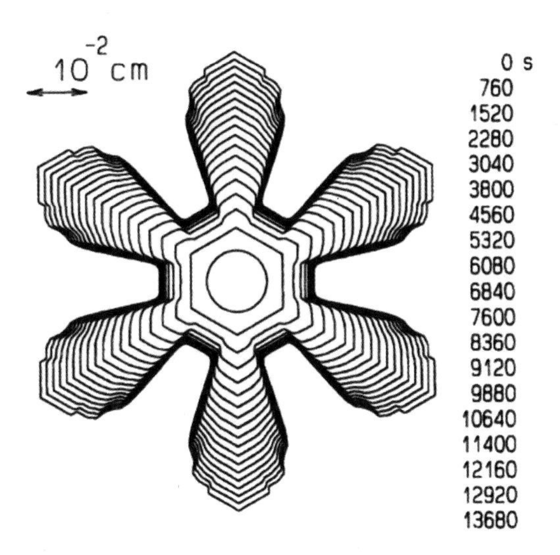

FIGURE 5.4. This numerical model by Yokoyama and Kuroda exhibits an initial transition from a round seed crystal to a faceted plate, followed by the formation of six primary branches. Unlike earlier investigations, these authors created the first detailed physical model of the specific phenomenon of snow crystal growth from water vapor. Image adapted from [1990Yok].

cesses relevant to the growth of crystals, i.e., a surface kinetic process for incorporating molecules into a crystal lattice and a diffusion process."

In terms of numerical techniques, the authors began with the differential equations describing the diffusion of water vapor molecules in air along with a reasonable estimate for the boundary conditions at the crystal surface, including an attachment coefficient with deep cusps at the facet angles. The diffusion equation was solved using a Green's function method that generated the supersaturation field around the crystal along with the growth velocity at each point on the surface. The solidification front was then propagated in small steps to grow a 2D snow crystal, as illustrated in Figure 5.4. This simulation exhibits the initial growth of a faceted prism followed by the development of six primary branches, both well-known phenomena in snow crystal formation. Note also that the model depicts actual physical sizes at real physical times in seconds, as is needed for comparison with experimental measurements.

The Yokoyama and Kuroda paper was pioneering in recognizing the central importance of a detailed molecular model of the attachment kinetics for understanding snow crystal formation, along with an early realization that latent heating and surface energy effects were relatively minor compared to the dominant processes of particle diffusion and surface attachment kinetics, as I discussed in detail in Chapter 3. This was only a 2D simulation, using a largely incorrect model of the attachment kinetics (see Chapter 4), so it was not yet suitable for direct comparisons with experiments. Moreover, the Green's-function method used to solve the diffusion equation was inefficient compared to modern numerical techniques. Nevertheless, the authors' careful examination of the relevant physical processes was a substantial step toward developing a physically accurate model of snow crystal growth.

Front Tracking

Soon after these early modeling efforts, the field expanded rapidly, as several innovative mathematical techniques were developed and applied to investigations of a variety of solidification phenomena. At first these studies focused mainly on freezing from the melt, which has metallurgical applications and involves relatively small anisotropies in surface physics (see Chapter 3). Quite recently, however, the field has begun developing models of crystal systems that exhibit both faceting and branching, including snow crystal growth. With sharp cusps in surface anisotropies at the facet angles, this solidification problem introduces additional computational challenges that are not present in the unfaceted case.

In 2012, Barrett, Garcke, and Nürnberg presented 3D numerical simulations of growing snow crystals using a finite-element method, in which the ice surface was approximated using an adaptive polygonal mesh [2012Bar]; Figure 5.5 shows an example from this paper. This technique demonstrates one variant of a *front-tracking* strategy, as it defines a sharp solidification front between solid ice and the water vapor field surrounding it [1996Sch, 2010Bar]. As with the Yokoyama and Kuroda model, Figure 5.5 again exhibits the initial formation of a faceted prism followed by a transition from faceted to branched growth. Figure 5.6 shows another example modeling the growth of a hollow column.

FIGURE 5.5. A 3D front-tracking model of a platelike snow crystal, showing a transition from faceted to branched growth [2012Bar]. Image courtesy of Harald Garcke.

The authors concluded in this study that a substantial surface energy anisotropy was necessary to produce faceted growth in their models, while anisotropy in the attachment coefficient was not enough to produce faceting. I believe that this conclusion is likely not correct, as the simulations in Barret et al. [2012Bar] examined only a rather weak anisotropy in the attachment kinetics, far weaker than what is indicated in recent experiments (see Chapter 4). It appears to be generally the case in materials science that highly anisotropic attachment kinetics are usually the dominant cause of faceting in crystal growth, and this appears to be the case for snow crystal faceting as well. These issues aside, the work presented in Barrett et al. [2012Bar] demonstrated that a modern front-tracking numerical model can generate 3D structures that are both faceted and branched, which is a substantial step forward.

Phase-Field Snow Crystals

In 2017, Demange, Zapolsky, Patte, and Brunel demonstrated a novel phase-field technique for simulating snow crystal growth [2017Dem, 2017Dem1]. In contrast to front-tracking approaches, this method defines an artificial *phase-field* parameter that equals −1 for the water vapor phase and +1 for the ice phase, and this parameter varies smoothly between these values across a spatially diffuse interfacial region (spanned by at least several pix-

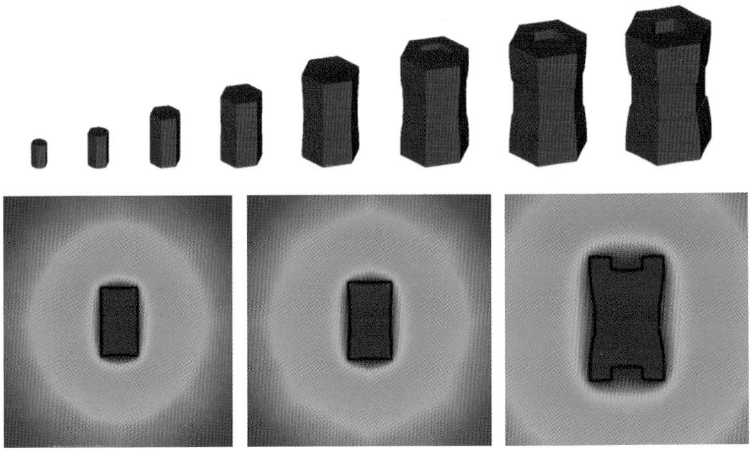

FIGURE 5.6. A 3D front-tracking model of a hollow columnar snow crystal, showing a transition from faceted to hollow growth [2012Bar]. The top images show renderings of the full 3D model as it developed, while the bottom images show cross-sectional views that also depict the supersaturation field around the crystal. Image courtesy of Harald Garcke.

els in the model). By eliminating the sharp solidification boundary in this way, phase-field models can employ generally simpler numerical propagation algorithms [1996Kar, 1998Kar, 2002Boe, 2017Jaa].

In the phase-field technique, the diffusion equation and its accompanying boundary conditions are replaced with a set of nonconservative phase-field equations. These equations represent a phenomenological description of the underlying microscopic interfacial physics that reduces to the correct physical description of the growth problem in the sharp-interface limit [1998Kar]. Once the proper phase-field equations have been determined, they are used to evolve the entire phase field in a uniform fashion, so no explicit front tracking is required.

Figure 5.7 shows an example of a 3D simulation of a growing snow crystal from Demange et al. [2017Dem] that again illustrates several features representative of stellar snow crystals, including the transition from faceted to branched growth, well-developed sidebranching, and rib-like surface markings. The authors were also able to reproduce several other commonly observed snow crystal structures, as illustrated in Figure 5.8 [2017Dem1]. As with Barrett et al. [2012Bar], however, the underlying physical parameters used in Demange et al. [2017Dem] were not entirely realistic. For example, the attachment kinetics function was only weakly anisotropic and did not include the known basal nucleation barrier described in Chapter 4. The high degree of surface anisotropy needed to produce faceting instead came from the surface energy, which is likely not an accurate physical model for snow crystal dynamics. The Peclet number was also orders of magnitude higher in the model than in real snow crystal growth. These significant technical points notwithstanding, the authors clearly demonstrated the potential of the phase-field method for modeling growth that is both faceted and branched, a necessary condition for creating accurate simulations of snow crystal growth.

Cellular Automata

Of the various computational strategies that have been applied to solidification problems so far, the CA method has demonstrated the most promise (in my opinion) for providing a powerful research tool for investigating the physical dynamics of snow crystal growth. Much like the early Packard snowflakes [1986Pac], a CA model begins by defining a fixed grid having the same hexagonal symmetry as the ice crystal lattice. Individual cells (also known as pixels) on the grid are labeled as either ice or vapor, with vapor pixels having a value proportional to the water vapor supersaturation. A set of CA "rules" evolves the supersaturation field

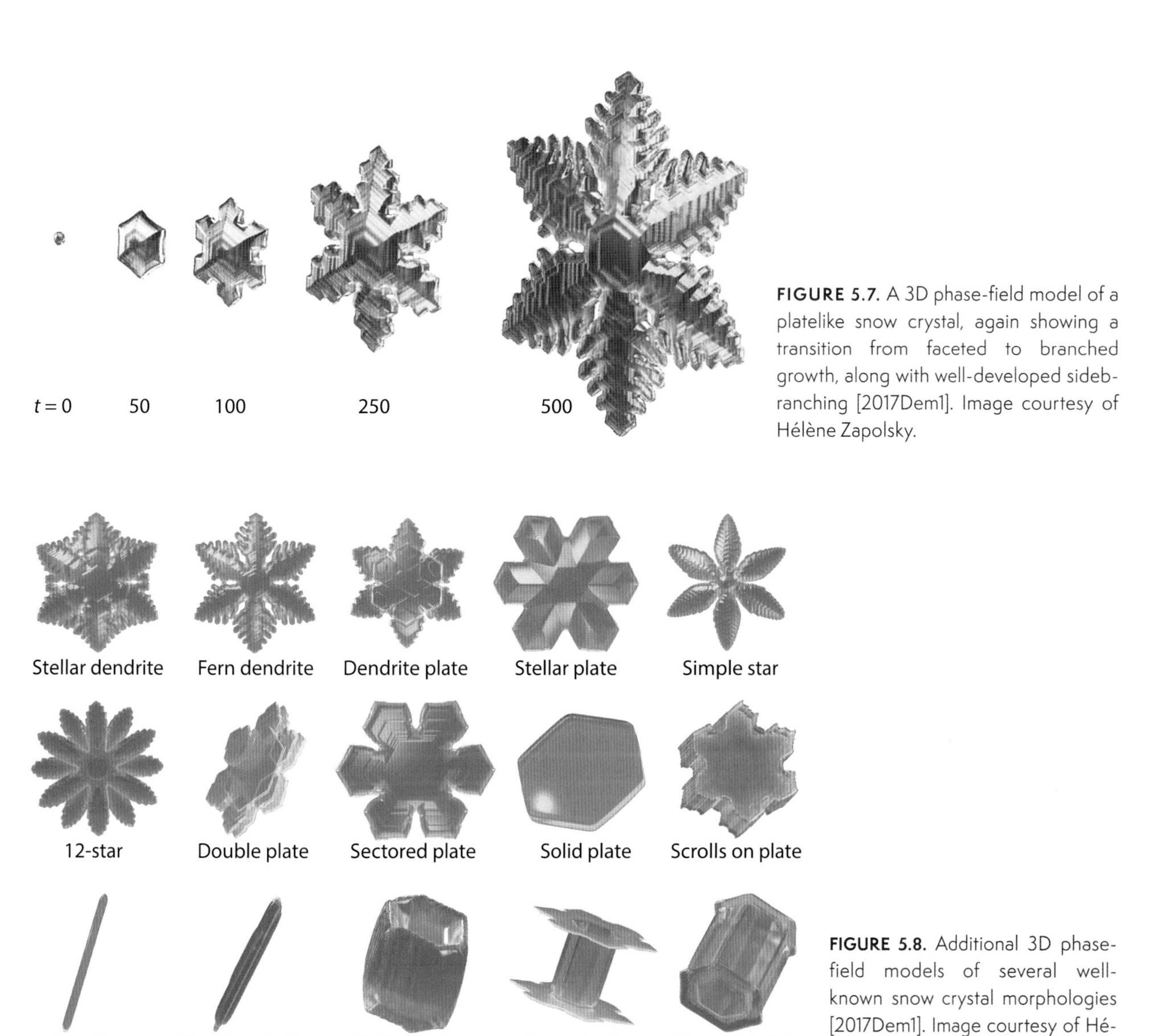

FIGURE 5.7. A 3D phase-field model of a platelike snow crystal, again showing a transition from faceted to branched growth, along with well-developed sidebranching [2017Dem1]. Image courtesy of Hélène Zapolsky.

FIGURE 5.8. Additional 3D phase-field models of several well-known snow crystal morphologies [2017Dem1]. Image courtesy of Hélène Zapolsky.

over time and determines when pixels change their state from vapor to ice.

A CA model can describe physically realistic snow crystal growth if the rules are carefully chosen to simulate the actual physical processes involved. Both the mathematical structure and the numerical implementation of CA techniques are generally simpler than other simulation strategies, plus the results to date suggest that the CA method is rather well suited for handing the highly anisotropic attachment kinetics present in snow crystal growth.

Clifford Reiter first demonstrated the potential for creating realistic CA simulations of snow crystal growth when he presented a simple 2D model that yielded several snowflake-like structures, including those shown in Figure 5.9 [2005Rei]. Ning and Reiter described additional 3D models [2007Nin]. Reiter's algorithms implemented nearest-neighbor rules that solved Laplace's

FIGURE 5.9. Several 2D CA models of diffusion-limited growth on a sixfold symmetrical lattice, exhibiting faceted and branched structures [2005Rei]. Image courtesy of Cliff Reiter.

equation in the region surrounding the snow crystal, thus accurately modeling the diffusion of water vapor toward the growing crystal. The rules governing the conversion of vapor to ice had little basis in solidification physics, however, so the Reiter model did not describe the formation of actual snow crystals in a meaningful way. But it was a first step in what has turned out to be a fruitful direction.

Janko Gravner and David Griffeath greatly expanded these ideas in a series of influential papers [2006Gra, 2008Gra, 2009Gra], the latest of which demonstrated a fully 3D snow crystal simulator that generated a remarkable diversity of realistic morphologies, including details that had hitherto not been seen in any numerical simulations. The appearance of robust ridge-like structures on several stellar-plate morphologies is especially noteworthy, as these are also robust features in real snow crystals (see Chapter 3). A few representative examples are shown in Figure 5.10. Rendering a 3D model to produce a 2D image is also a nontrivial challenge, and Figures 5.11 and 5.12 show some particularly artistic renderings of Gravner-Griffeath snow crystals done by Antoine Clappier.

The Gravner-Griffeath work was a significant breakthrough in modeling snow crystal growth, as it so clearly demonstrated the great potential of the cellular-automata method, especially for fully 3D simulations. In

FIGURE 5.10. Several 3D CA models by Gravner and Griffeath [2009Gra] demonstrating structures that are both faceted and branched, reproducing many morphological features found in real snow crystals. Image courtesy of Janko Gravner.

additional to modeling several common snow crystal types, many surface structural details matched those seen on natural crystals to a remarkable degree. All previous numerical models of solidification had shown little or no adeptness for generating structures that are simultaneously faceted and branched, and this problem is still present to some degree in several of the computational strategies described above. In contrast, the CA method appears to be almost ideally suited for handling faceted + branched structures produced by diffusion-limited growth with highly anisotropic attachment kinetics.

However, a substantial remaining problem with the Gravner-Griffeath model [2009Gra] was that it used a parameterized set of CA rules that were largely ad hoc and not entirely physically realistic. For example, the model imposed an artificial boundary condition setting $\sigma_{surf} = 0$ on all ice surfaces. This is technically true only in equilibrium and thus is not correct for any growing snow crystal. Setting $\sigma_{surf} = 0$ may be a reasonable approximation for diffusion-limited growth in some circumstances (Chapter 3), but it is not a suitable general assumption to make. Kelly and Boyer [2014Kel] further pointed out that the Gravner-Griffeath model does not always obey mass conservation in the attachment step, in that crystal growth is not accompanied by a corresponding removal of water vapor from the air. Moreover, no clear relationship exists between the parameterized CA rules and the known physical properties of ice crystal attachment kinetics. Thus, while the parameters in the Gravner-Griffeath algorithm could be adjusted to yield remarkably realistic snow crystal structures, the surface boundary conditions were not appropriate for a physically accurate model.

FIGURE 5.12. An abrupt change in model parameters can yield the formation of capped columns with axial hollows, as illustrated in this Clappier-rendered Gravner-Griffeath 3D model. Image courtesy of Janko Gravner and Antoine Clappier.

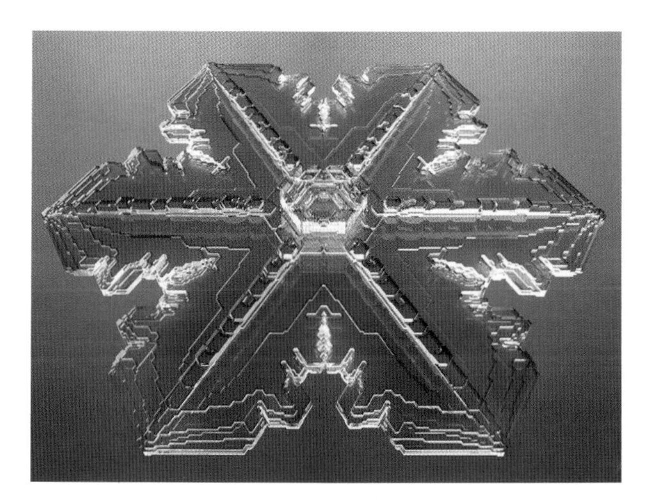

FIGURE 5.11. A Gravner-Griffeath 3D snow crystal model rendered by Antoine Clappier. Image courtesy of Janko Gravner and Antoine Clappier.

Physically Realistic Cellular Automata

The problem of creating a CA snow crystal model with physically derived rules was soon addressed by Libbrecht [2008Lib, 2013Lib1], who further investigated the incorporation of surface energy effects, surface diffusion [2015Lib1], and the edge-sharpening instability (ESI) [2015Lib2] in a CA model with anisotropic attachment kinetics. Using a 2D model of cylindrically symmetrical 3D growth, this allowed some of the first direct, quanti-

FIGURE 5.13. A few representative 3D snow crystal models created by Kelly and Boyer using CA rules derived from physically realistic calculations [2014Kel], exhibiting several morphological features found in real snow crystals. Image courtesy of James Kelly.

tative comparisons of simulated snow crystal growth with laboratory measurements, as I describe below.

James Kelly and Everett Boyer made substantial additional progress by developing a fully 3D CA model with sound physical foundations, thus beginning a systematic study of 3D snow crystal growth as a function of parameterized attachment kinetics [2013Kel, 2014Kel]. Some results from this work are shown in Figures 5.1 and 5.13. With these recent model studies, it is now becoming clear that perhaps the biggest impediment to creating accurate computational snow crystals is simply incorporating an accurate and comprehensive model of the surface attachment kinetics. Given recent progress in this area (see Chapter 4), the path forward appears to be quite open and accessible. As further physical insights and model improvements are realized [2016Li], I expect that the CA technique will become the method of choice for modeling snow crystal growth, at least in the near term, and I discuss the specific algorithms and physical underpinnings in more detail below.

Comparing Computational Methods

Although CA models have produced the most impressive snow crystal results to date, other techniques show great promise as well. Numerical modeling of structure formation during solidification is a rapidly evolving field, so it makes sense at this point to briefly compare the different computational approaches.

The first thing to note is that all the existing computational techniques can solve the particle diffusion equation in free space with ease, especially as the Laplace approximation simplifies the problem considerably (see Chapter 3). The phase-field, front-tracking, and CA models all take different mathematical approaches to solving the diffusion equation, but the results are all basically the same, and all are highly accurate. The main differences between techniques lie not in solving the free-space diffusion equation but rather in how the surface boundary conditions are handled and how surface growth is propagated.

In many respects, polygonal front-tracking methods seem the most natural when dealing with a continuum phenomenon like crystal growth. A distinct solidification front makes perfect sense for snow crystal growth, as there is an extremely sharp transition between the vapor and solid phases at the ice surface, just a few molecular layers in thickness. Because the molecular size is so small compared to even the smallest morphological features being modeled, a continuum model with a sharp interface is an excellent approximation. Moreover, a polygonal surface is a reasonable computational model for almost any morphological situation, and the underlying surface physics is well defined on such a surface. If the grid is small enough, a front-tracking algorithm should be capable of modeling all manner of solidification problems, including snow crystal growth.

One disadvantage with front tracking, however, is the algorithmic complexity involved with deriving and continually adapting the polygonal solidification surface and the polygonal mesh that surrounds it. I have not worked in this area myself, but my impression is that it took many years to develop the computational tools needed to manage the diffusion equation with its nontrivial surface boundary conditions on an ever-adapting polygonal mesh. However, now that the required algorithms have been established, perhaps it is straightforward to apply them to a new physical system like snow crystal growth. But it does appear to be a nontrivial undertaking.

Perhaps the biggest uncertainty in creating a suitable front-tracking code for snow crystal growth involves dealing with highly anisotropic attachment kinetics and faceting. The facet planes are unusual in that α_{basal} and α_{prism} can be much smaller than $\alpha_{vicinal}$, even when the vicinal angle is extremely low. Put another way, the attachment coefficient $\alpha(\theta_{surf})$ as a function of surface angle may have extremely sharp and deep cusps at the facet angles. This likely requires some special treatment of the facet surfaces, as Yokoyama and Kuroda noted even in their early examination of snow crystal modeling using front tracking [1990Yok]. It certainly does not seem unfeasible that one could incorporate highly anisotropic attachment kinetics into a front-tracking model. However, it has not been done to date, so we cannot say for sure how difficult such a task might be.

In contrast to front-tracking models, phase-field techniques are typically applied on a fixed coordinate grid, thus avoiding the use of complex polygonal meshes. This may also facilitate the preferred treatment of faceted surfaces, as the grid coordinates can easily be defined to be along facet planes. As with front-tracking methods, however, the issue of faceted growth remains a substantial uncertainty regarding our desire to model snow crystal growth, as highly anisotropic attachment kinetics have not yet been adequately explored in phase-field models.

One disadvantage with phase-field models is the difficulty inherent in defining the phase-field equations. Once again, I have not worked in this area, but my impression is that deriving an appropriate set of phase-field equations from a specific parameterization of the surface boundary conditions is not a trivial assignment. It is unclear (to me) that finding phase-field equations that properly incorporate highly anisotropic attachment kinetics will be entirely straightforward. And when addition physical features are added, like surface energy effects and surface diffusion, the problem becomes even more challenging. Once again, this is not necessarily a showstopper for phase-field models, but it is an uncertainty, given that the problem has not yet been adequately studied to date.

One excellent advantage of the phase-field method is that it nicely handles the merging of solidification fronts. In metallurgical applications, for example, one often requires that a melt will eventually fully solidify into a solid block, including the merging of numerous independent solidifying structures into a final matrix of solid domains. Moreover, the material properties of the solidified metal may depend strongly on the size, structure, and arrangement of the domains. This kind of domain merging happens naturally in a phase-field model but is something of a computational nightmare in a front-tracking model.

In snow crystal growth, the merging of separate solidification fronts can happen, but it is not usually an important consideration. Sidebranches are especially prone to colliding trajectories, as illustrated in Figure 5.14. But merging events like these are not a central feature in snow crystal growth compared to more basic morphological features. As Barrett et al. succinctly described [2014Bar1, p. 550], "The main advantage of phase field methods over direct front-tracking methods is that they intrinsically allow for topological changes. However, for the problem

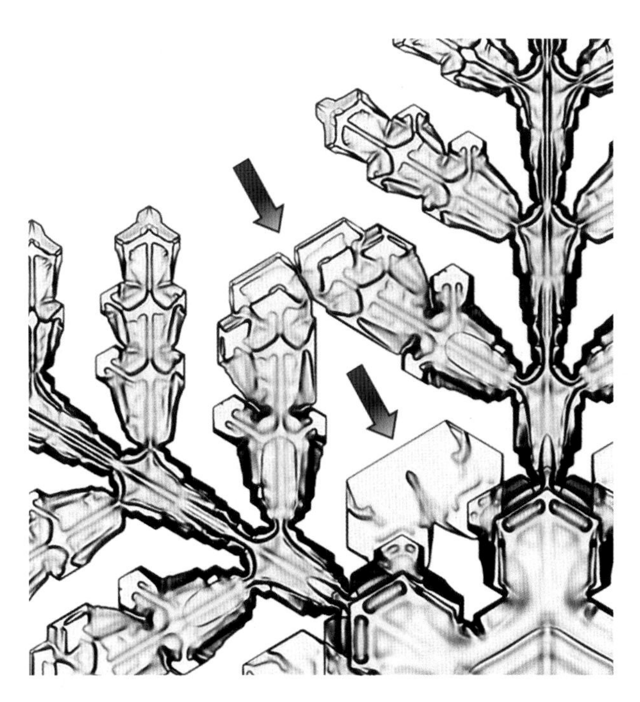

FIGURE 5.14. A laboratory-grown PoP snow crystal with two instances of sidebranch mergers (arrows). The upper merger occurred relatively recently before the photo was taken, so the separate branch edges are still clearly seen. The lower merger is older, and the individual sidebranches have grown together into a single flat plate.

of solidification and dendritic growth as considered in this paper, topological changes are rare."

CA models are not especially popular in metallurgical solidification modeling, and yet they have demonstrated a remarkable aptitude for modeling snow crystal structures. A big reason for this success is that CA models can easily incorporate highly anisotropic attachment kinetics. By defining a grid with the same symmetry as the underlying ice crystal lattice, it is straightforward to give special treatment to faceted surfaces, as this is practically built into the model structure. However, one important downside of this rigid grid structure is that it is nearly impossible to create a CA model that does *not* include some level of intrinsic numerical anisotropy in the surface boundary conditions, as I describe below.

Another outstanding feature of CA models is that they are remarkably simple to define and build, plus the run times are relatively short. Moreover, the CA rules can be derived fairly easily from physical foundations, allowing straightforward parameterizations of the attachment kinetics and other physical effects. This will become apparent as I focus the remainder of this chapter mainly on developing CA techniques specifically for modeling snow crystal growth.

Facet-Dominated Growth

Much of the scientific literature on solidification modeling focuses on metallurgical systems, where the material anisotropies (mostly in the surface energy) are quite small, perhaps a few percent. However, as we learned from solvability theory (Chapter 3), it is important that these small anisotropies be treated correctly, as they are critical in determining dendritic growth morphologies. Because CA techniques tend to exhibit numerical anisotropies that are nearly impossible to eliminate, they are a poor choice for modeling metallurgical solidification. In the opposite extreme, however, when the intrinsic material anisotropies are high, modest levels of numerical anisotropy are not huge problem. Thus, different classes of materials may call for different computational techniques.

Snow crystal growth is somewhat unique in the field of solidification modeling because of the importance of highly anisotropic attachment kinetics. To my knowledge, snow crystal growth is the only highly anisotropic system that has received much attention, either theoretical or experimental, from the standpoint of understanding the basic physics of solidification and structure formation. Beginning with the careful studies of dendritic growth by Glicksman and others in the 1980s (see Chapter 3), nearly all substantial scientific efforts aimed at numerical solidification modeling were focused on weakly anisotropic metallurgical systems. In these systems, the Peclet number is high, growth is largely limited by thermal diffusion, weakly anisotropic surface energy dominates the surface boundary conditions, and attachment kinetics are either weakly anisotropic or ignored altogether. Dendritic structures typically exhibit no faceting in these materials, so it is imperative that computational models include low intrinsic numerical anisotropies.

Snow crystal growth is, in many ways, a completely different problem. The Peclet number is extremely small, particle diffusion is more important than heat diffusion, surface energy effects are almost negligible, and anisotropic attachment kinetics play a central role in bringing about highly faceted dendritic structures. In snow crystal growth modeling, both α_{prism} and α_{basal} are often small and highly dependent on σ_{surf}, while one can reasonably assume $\alpha \approx 1$ on nearly all nonfaceted surfaces. In this "facet-dominated" growth regime, the overall growth rates and morphologies are largely defined by the growth of the faceted surfaces.

Modeling facet-dominated growth requires an especially accurate treatment of the facet dynamics, which means a careful handling of anisotropic attachment kinetics. In contrast, a somewhat sloppy treatment of nonfaceted surfaces may be tolerable. Thus, although it is not clear that one can build a perfect CA model even in principle, it may nevertheless be possible to build a CA model that reproduces most snow crystal morphologies with reasonable fidelity. The early results look quite promising, but the only real way to know for sure is to start building physically accurate models that allow quantitative comparison with careful experimental observations.

SPHERICAL CELLULAR AUTOMATA

I focus the remainder of this chapter on CA models, as they are the leading contender for creating realistic computational snow crystals, at least in the short term. Opinions may differ on this, as other numerical strategies are promising as well, and we may uncover serious inherent limitations in the CA technique with additional study. Nevertheless, given how little modeling effort has focused on strongly faceted solidification, and how rapidly the field is evolving, I suspect the best strategy at this point is simply to dive in and see how far the CA method can take us.

The discussion below derives mainly from work I have done in developing CA models for snow crystal growth using physically derived CA rules [2013Lib1, 2015Lib2], combined with significant model improvements from Kelly and Boyer [2013Kel, 2014Kel], as well as a few additional (unpublished) tweaks I have been investigating recently. My focus with these models has been not just on morphologies but also on developing quantitative CA rules that accurately reflect the underlying physical processes that govern snow crystal growth.

This field is evolving rapidly at present, especially because our understanding of the attachment kinetics is only beginning to converge on a workable model (Chapter 4). As targeted experiments continue to develop (see Chapter 7), my hope is that better computational models will soon allow detailed comparisons with complex morphologies grown under well-known conditions, especially on electric ice needles (see Chapter 8). As these experimental and computation efforts move forward in parallel, they will eventually combine to yield a truly comprehensive model of snow crystal formation.

The Diffusion Equation

For pedagogical reasons, I like to begin the discussion by creating a numerical model of the simplest possible physically interesting one-dimensional (1D) problem—the growth of a spherical crystal governed solely by particle diffusion and attachment kinetics. There is hardly any need for numerical modeling of this system, of course, as an exact analytic solution exists and is described in Chapter 3. But fully understanding spherical growth is always a good beginning before modeling more complex systems.

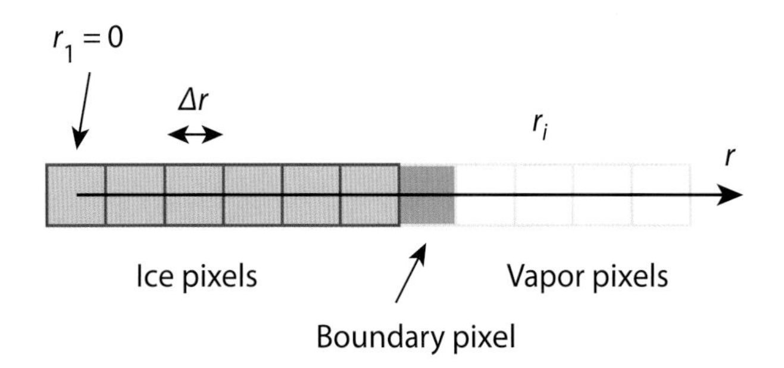

FIGURE 5.15. The radial layout of cells (a.k.a. pixels) for a 1D spherical CA model. At any given time, the model consists mainly of ice pixels (blue) and vapor pixels (white). The red "boundary" pixel is a vapor pixel that borders an ice pixel.

The first step in any CA model is to define the cells, which I usually call "pixels," as shown in Figure 5.15. For our spherical system, assume a set of radial pixels for which the pixel center is located at $r_i = (i-1)\Delta r$ for integer i with $1 \le i \le N$. (Of course, other numbering conventions could be used to equal effect.) We further assume that pixels with $r_i < r_b$ are filled with ice, while pixels with $r_i \ge r_b$ are filled with vapor. We call the $i = b$ pixel a "boundary" pixel, because it is filled with vapor but is adjacent to an ice pixel. The vapor pixels typically include a background gas of air, and each is labeled with the water vapor supersaturation $\sigma_i = \sigma(r_i)$ at that location. In the vapor surrounding the crystal, we write the particle diffusion equation in spherical coordinates

$$\frac{\partial \sigma}{\partial t} = D\nabla^2 \sigma = D\left(\frac{2}{r}\frac{\partial \sigma}{\partial r} + \frac{\partial^2 \sigma}{\partial r^2}\right), \tag{5.1}$$

and on our radial grid, this becomes

$$\sigma(r_i, \tau + \Delta\tau) = (1 - 2\Delta\tau)\sigma(r_i) + \Delta\tau\left[\left(1 + \frac{\Delta r}{r_i}\right)\sigma(r_{i+1}) + \left(1 - \frac{\Delta r}{r_i}\right)\sigma(r_{i-1})\right], \tag{5.2}$$

where $\tau = Dt/(\Delta r)^2$. Note that the $(1 \pm \Delta r/r_i)$ terms arise from the spherical coordinate system, reflecting the fact that the volume in a Δr shell increases with r_i. These terms introduce a potential problem when dividing by $r_i = 0$, but we will ignore this issue, because the central pixel will always be part of the seed crystal in our model. A 1D Cartesian model would avoid the $(1 \pm \Delta r/r_i)$ terms, but I prefer to work with a model that describes a real physical system, in this case the growth of a spherical ice crystal.

At this point, we recognize that snow crystal growth is described by a very low Peclet number, as described in Chapter 3. Thus the supersaturation field around a crystal relaxes very rapidly compared to the crystal growth time, so we can solve the particle diffusion equation while assuming a nonmoving crystal surface. In our CA model, this means we can iterate Equation 5.2 with fixed boundaries until $\sigma(r_i)$ converges to a stationary solution of Laplace's equation. We do this without yet worrying about the actual growth of the crystal, because, as far as particle diffusion is concerned, the crystal is growing so slowly that it is essentially stationary.

For computational efficiency, we would like to relax $\sigma(r)$ using the smallest possible number of iterations of Equation 5.2, so we want to choose $\Delta\tau$ to be as large as possible. Taking $\Delta\tau = D\Delta t/(\Delta r)^2 = 1/2$ seems to be about optimal, as larger values can lead to numerical instabilities. As a bonus, this choice sets one term in Equation 5.2 equal to zero, so the optimal propagation equation becomes

$$\sigma(r_i, k+1) = \frac{1}{2}\left(1 + \frac{\Delta r}{r_i}\right)\sigma(r_{i+1}, k) + \frac{1}{2}\left(1 - \frac{\Delta r}{r_i}\right)\sigma(r_{i-1}, k), \tag{5.3}$$

where here we have replaced τ with a simple integer indexing variable k. At each instant in time, we simply iterate Equation 5.3 to determine the correct supersaturation field $\sigma(r)$ surrounding the crystal at that time.

From a computational perspective, note that Equation 5.3 can be performed using highly efficient vector processing. The vectors $(1 \pm \Delta r/r_i)$ are constant and need only be calculated once at the beginning of a modeling run. The vectors $\sigma(r_{i\pm1})$ are rapidly computed using a simple permutation of $\sigma(r_i)$, and vector operators can perform the arithmetic in Equation 5.3 using optimized parallel-processing algorithms built into the compiler. While optimizing efficiency is of little concern for a 1D spherical calculation, it becomes quite important in 3D codes with high spatial resolution.

Boundary Conditions

For the outer boundary condition far from our growing spherical crystal, we assume a constant value $\sigma(r_N) = \sigma_{far}$, where $r_N = r_{far}$ is the position of the outer boundary. This is easily implemented in our CA algorithm by simply applying Equation 5.3 to all $\sigma(r_i)$ out to $\sigma(r_{N-1})$. As described in Chapter 3, we have a mixed boundary condition at the crystal surface:

$$X_0\left(\frac{\partial\sigma}{\partial n}\right)_{surf} = \alpha\sigma_{surf}, \tag{5.4}$$

where

$$X_0 = \frac{c_{sat}}{c_{ice}}\frac{D}{v_{kin}} \tag{5.5}$$

is a fundamental length scale in the diffusion problem. On our CA model grid, Equation 5.4 becomes

$$\sigma_b = \sigma_{b+1}\left(1+\alpha(\sigma_b)\frac{\Delta r}{X_0}\right)^{-1} \tag{5.6}$$

to first order in Δr, where σ_b is the supersaturation in the boundary pixel.

Note that determining the value of $\alpha(\sigma_b)$ can be included in the convergence process by generalizing Equation 5.6 to give the propagation equation

$$\sigma(r_b,k+1) = \sigma(r_{b+1},k)\left(1+\alpha(\sigma(r_b,k))\frac{\Delta r}{X_0}\right)^{-1}. \tag{5.7}$$

Doing this allows one to assume any desired functional form for $\alpha(\sigma_{surf})$ without having to solve Equation 5.6 analytically.

In summary, calculating the supersaturation field means iterating Equation 5.3 for $b<i<N$ using $\sigma(r_N) = \sigma_{far}$ while simultaneously iterating Equation 5.7 for $i=b$. For any physically realistic scenario, this should converge to give the full solution $\sigma(r_i)$ that satisfies Laplace's equation with the proper boundary conditions.

Convergence Criterion

A next question is how long to continue the iterative propagation of Equations 5.3 and 5.7. The supersaturation field $\sigma(r)$ typically converges exponentially with time, which unfortunately means that it never actually reaches the exact solution. An informative worst-case test is easily done using a 1D Cartesian system and setting $\sigma(r_i, k=1)=1$ for all i while setting both boundaries equal to zero. In this test case, iterating the propagation equation yields a center-pixel value of $\sigma(r_{N/2}, k) \approx A\exp(-k/k_0)$ for large k, where $A=\exp(0.25)$, and $k_0=0.2N^2$. Thus, a 10 percent convergence is obtained after about $N^2/2$ steps, while a 1 percent convergence takes N^2 steps. This sets a good benchmark for the initial relaxation of the supersaturation field around the seed crystal, but fewer steps may be sufficient after a boundary pixel converts to ice (see below), as a one-pixel change may not substantially alter the supersaturation at large distances. Determining an optimal convergence criterion is often best accomplished by comparing numerical simulations with known analytical results in simple test cases. Of course, there is always a trade-off between accuracy and code running time.

Growth Steps

Once we have calculated the supersaturation field $\sigma(r)$ at some instant in time, the next step is to use this solution to grow the crystal out a small amount. In the spherical CA model, this means turning a boundary pixel into an ice pixel using the known surface growth rate (Chapter 4)

$$v_n = \alpha v_{kin}\,\sigma_{surf} = \alpha(\sigma(r_b))\, v_{kin}\,\sigma(r_b), \tag{5.8}$$

which indicates that ice growth would "fill" the boundary pixel in a time

$$\delta t = \frac{\Delta r}{v_n}, \tag{5.9}$$

and this time interval is easily calculated from the known supersaturation field $\sigma(r_i)$.

Putting everything together, growing a spherical snow crystal using this 1D CA model involves the following steps:

1) Set up the physical parameters and initial conditions, including the initial seed crystal.
2) Iterate Equations 5.3 and 5.7 until reaching convergence, yielding the supersaturation field $\sigma(r_i)$ around the crystal.
3) Promote the boundary pixel (there is only one in this 1D model) into an ice pixel while advancing the real time by δt in Equation 5.9. Promote the next vapor pixel to a boundary pixel.
4) Either stop the model at this point or go back to step 2.

The result of this process is a series of time steps giving $R(t)$, the size of the crystal as a function of time, along with $\sigma(r_i, t)$. Note that although the growth steps have a uniform size Δr, the time steps are not uniform in duration. Note also that this CA model is completely deterministic, including no random processes of any kind. It also excludes evaporation, so once a vapor pixel becomes an ice pixel, it cannot go back again. Finally, the model only includes bulk diffusion in air together with surface attachment kinetics, ignoring all heating and surface energy effects. These model attributes are put in place mainly to simplify the calculations at this point, and other choices are possible. In particular, I discuss surface energy, surface diffusion, and other physical effects in more detail below. However, I neglect thermal effects entirely in this discussion, as they play only a relatively minor role in snow crystal growth (see Chapter 3). One could make a dual-diffusion CA model that would incorporate both particle and thermal diffusion, but that requires a significant increase in complexity that is best left for another day.

Adaptive Boundary Matching

One problem with any computational diffusion model is that the outer boundary is only a finite distance from the growing crystal, while often the outer boundary condition is specified at infinity, $\sigma(r \to \infty) = \sigma_\infty$. If we use the outer boundary condition $\sigma(r_{far}) = \sigma_{far}$ as described above, then the best results will be obtained by making r_{far} as large as possible. However, for computational efficiency, one would like to keep r_{far} small, as that decreases the total volume of modeled space.

One way to address this problem is with an adaptive grid, increasing the pixel size with distance from the crystal. This works because $\sigma(r)$ changes rapidly only near the crystal surface, so a coarser grid can be used far from the surface. With an adaptive grid, a small number of pixels can be used to model a large volume of space efficiently. But an adaptive grid introduces additional computational complexity and overhead, and it may interfere with one's ability to make full use of highly efficient parallel processing algorithms. Simply changing Δr with r_i is a fine approach with a 1D model, but extending this idea to higher dimensions becomes problematic.

Another relatively easy approach to the far-away boundary problem is to keep Δr constant with r_{far} reasonably small, and then adjust σ_{far} appropriately as the crystal grows. To see how this works, start with the analytic solution for spherical growth presented in Chapter 3. Because the full supersaturation field $\sigma(r)$ is known in the spherical model, it is straightforward to show that

$$\sigma(r_{far}) = \sigma_\infty - \frac{dV/dt}{4\pi r_{far} X_0 v_{kin}}, \tag{5.10}$$

where

$$\frac{dV}{dt} = 4\pi R^2 v_n \tag{5.11}$$

is the volume change per unit time for a spherical crystal with radius R. This expression is dictated by conservation of mass, which requires that the flux of water vapor diffusing toward the crystal must equal the rate at which vapor turns into ice.

From this knowledge of the exact spherical solution, we can write a propagation equation for $\sigma(r_{far})$

$$\sigma(r_{far}, k+1) = \sigma_\infty - \frac{dV/dt(k)}{4\pi r_{far} X_0 v_{kin}}, \tag{5.12}$$

and this operation would be performed between steps 3 and 4 listed above. As the model crystal develops, $\sigma(r_{far})$ adapts to the changing crystal size and growth behavior.

Note that this is an iterative procedure; once the supersaturation field $\sigma(r_i)$ around the crystal is known, this allows a calculation of dV/dt at that point in time. Performing Equation 5.12 then sets up $\sigma(r_{far})$ for the next time step in the series. If the crystal grows slowly, the process will converge to yield a reasonable approximation of $\sigma(r_{far})$ at each time step. Once again, the overall accuracy of this adaptive outer boundary can be examined by comparing model results with the analytic solution of the diffusion equation.

For typical conditions, we expect (from the analytic solution for a spherical crystal) that $dV/dt \sim R$ for diffusion-limited growth, and $dV/dt \sim R^2$ for kinetics-limited growth (Chapter 3). In both cases, dV/dt is small at early times, so Equation 5.12 yields $\sigma(r_{far}) \approx \sigma_\infty$. This makes sense, as presumably r_{far} is much larger than the initial seed crystal. Then $\sigma(r_{far})$ decreases as the crystal grows larger, as one would expect. This adaptive outer boundary method essentially "matches" the CA solution to the known analytic solution beyond r_{far}. While this procedure is somewhat trivial for the 1D spherical model (because the analytic solution is already known for all r), it becomes useful when working in higher dimensions, as we will see in the next section.

CYLINDRICALLY SYMMETRIC CELLULAR AUTOMATA

Having set the stage by exploring the simplest 1D model, the next obvious step is to move up to a 2D model. A 2D model introduces additional complexity and new physical effects compared to the 1D model, and it introduces additional model developments along several fronts. While it is tempting to jump straightaway to a full 3D model, we will soon find that the 2D case provides a valuable test system for addressing many nontrivial issues. Also, from the standpoint of practical pedagogics, a 2D model can be easily described using 2D sketches, which display well on a printed page and are relatively easy to comprehend. In contrast, communicating ideas relating to full 3D structures can be something of a visual challenge.

Focusing, therefore, on 2D models, I have found that a cylindrically symmetrical system is the best choice for exploring the physics of snow crystal growth. A planar model is another 2D option that is often explored, but such a model is not as useful for examining realistic solidification physics. To see this, consider 2D models of stellar crystals like those shown in Figure 5.9. These models to not describe real snow crystals, but they are adequate models of infinitely long bars with snowflake-shaped cross-sections. Solving the infinite-bar problem in three dimensions is identical to solving just the 2D cross-section. While this is a fine exercise, there are no real snow crystals that have anything like this kind of complex extruded morphology. Thus, this kind of 2D flat-plate model is of little actual use when examining the physics of real snow crystals.

In contrast, 2D cylindrically symmetric models can include simple disks, simple columns, hollow columns, disks on columns, and other morphologies that serve as reasonable proxies for real snow crystals. While cylindrical symmetry has its limitations, I have found that it works quite well for describing simple snow crystal morphologies. Thus, unlike the flat-plate 2D model, a cylindrically symmetrical 2D model connects much better to the real physics of snow crystal growth.

In a 2D cylindrically symmetrical model, a simple hexagonal plate is approximated by a thin disk. The six prism facets on the hexagonal plate are thus replaced by a single cylindrical "facet," while the basal facets are essentially unchanged. Particle diffusion around a thin disk is about the same as that around a hexagonal plate, and there is a good correspondence between the basal and prism attachment coefficients in the two cases [2015Lib2]. In particular, the same α_{prism} describing growth on the six prism surfaces of a hexagonal plate can be used for the single edge of the circular disk.

Transforming a hexagonal plate to a thin disk involves a small geometrical correction, but otherwise, the cylindrically symmetric disk is a tolerably good physical representation of a hexagonal plate. The same is true for snow crystal columns, hollow columns, and capped columns. For all these simple morphologies, cylindrically symmetrical models are quite well suited for investigating growth dynamics and attachment kinetics. Snow crystals grown on electric needles are also well suited for study using cylindrically symmetric models (see Chapter 8). Of course, dendritic structures and other complex morphologies will require full 3D modeling, but I have always found that difficult physics problems are best solved one step at a time.

Figure 5.16 shows a typical pixel geometry for a 2D cylindrically symmetrical CA model. The position of the center of each pixel is (r_i, z_j), where $r_i = (i-1)\Delta r$ and $z_j = (j-1)\Delta z$ for all (i, j) ranging from (1, 1) to (N_r, N_z). (Once again, different coordinate conventions are also possible.) I usually choose $\Delta r = \Delta z = \Delta x$, thus defining Δx, as this simplifies the mathematics and is also a reasonable choice for a realistic snow crystal model. The physical size of Δx is somewhat arbitrary, but we will see below that Δx should not be much greater than X_0 if one wishes to accurately reproduce small-scale snow crystal structures. It appears that there is little to be gained, however, in choosing $\Delta x < X_0$.

As with the 1D spherical case, most pixels in a CA model are vapor pixels, and each of these is assigned a supersaturation $\sigma_{i,j} = \sigma(r_i, z_j)$. The supersaturation field is determined by particle diffusion together with the appropriate boundary conditions, and the diffusion equation in cylindrical coordinates is

$$\frac{\partial \sigma}{\partial t} = D\nabla^2 \sigma = D\left(\frac{1}{r}\frac{\partial \sigma}{\partial r} + \frac{\partial^2 \sigma}{\partial r^2} + \frac{\partial^2 \sigma}{\partial z^2}\right) \quad (5.13)$$

for $\sigma(r, z, t)$. Projecting this equation onto the 2D CA grid yields

$$\sigma_{i,j}(\tau+\Delta\tau) = (1-4\Delta\tau)\sigma_{i,j} + \Delta\tau\left[\left(1+\frac{\Delta r}{2r_i}\right)\sigma_{i+1,j} + \left(1-\frac{\Delta r}{2r_i}\right)\sigma_{i-1,j}\right] + \Delta\tau[\sigma_{i,j+1} + \sigma_{i,j-1}], \quad (5.14)$$

and choosing $\Delta\tau = 1/4$ simplifies this expression by eliminating the first term, giving the propagation equation

$$\sigma_{i,j}(k+1) = \frac{1}{4}\left[\left(1+\frac{\Delta r}{2r_i}\right)\sigma_{i+1,j}(k) + \left(1-\frac{\Delta r}{2r_i}\right)\sigma_{i-1,j}(k)\right] + \frac{1}{4}[\sigma_{i,j+1}(k) + \sigma_{i,j-1}(k)]. \quad (5.15)$$

In the limit of low Peclet number, iterating this equation to convergence will yield the static supersaturation field $\sigma(r_i, z_j)$ that satisfies Laplace's equation. As we saw with the 1D case, the choice of $\Delta\tau = 1/4$ is a good one in that it produces rapid convergence without introducing numerical instabilities that can be problematic with higher values of $\Delta\tau$.

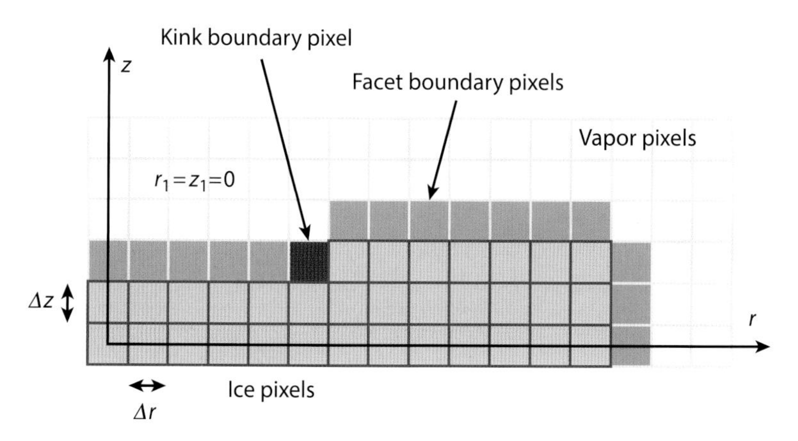

FIGURE 5.16. A pixel geometry for a 2D cylindrically symmetric CA model. Ice pixels are shown as blue, while vapor pixels are white. The "facet" boundary pixels (adjacent to faceted ice surfaces) are pink, while this particular model includes a single red "kink" boundary pixel that touches two ice pixels. For convenience, I often take $\Delta r = \Delta z = \Delta x$, which defines the generic pixel size Δx.

A Facet-Kink Model

A central feature of any finite-element computational model is that one must define a mathematical system that operates at finite resolution while providing a good physical representation of what is essentially a continuum system. Thus for crystal growth, the mathematics must somehow deal with both nanometer physics at scales much smaller than X_0 (e.g., the molecular dynamics that governs attachment kinetics) and mesoscale physics at scales at and above X_0 (including particle diffusion around the crystal and other processes).

With a CA model, the specific CA rules need to derive from nanoscale physics but run accurately on a mesoscale grid. In particular, the attachment kinetics rules must be parameterized so that they can be applied at the much larger scale of the CA, and how one does this is not always immediately obvious. Dealing with this broad range of physical scales is one of the most difficult aspects of modeling snow crystal growth.

These issues mostly play out at the crystal boundary, and in two-dimensions, there is no obvious, simple choice for accurately specifying the boundary conditions or growth rules. One of the simpler ways to tackle this problem using CA is with what I call a *facet-kink* model. In the 2D cylindrically symmetric case, this means defining the two classes of boundary pixels shown in Figure 5.16. Each *facet boundary pixel* is a vapor pixel bordered by exactly one nearest-neighbor ice pixel (out of four nearest-neighbor positions, neglecting all farther positions). Meanwhile *kink boundary pixels* are vapor pixels bordered by exactly two ice pixels. Boundary pixels with three or even four neighboring ice pixels are also possible, but only the facet and kink boundary pixels play important roles in simple growth morphologies. How one treats three- and four-neighbor boundary pixels is not so important; even just turning them immediately into ice as soon as they appear does not greatly affect the overall model dynamics, except perhaps in somewhat convoluted morphologies.

The facet-kink CA model is especially simple in that the character of each boundary pixel is defined solely from its nearest neighbors. This local definition makes for easy bookkeeping, but we will soon find that this simplicity requires some compromises in physical accuracy.

Boundary Conditions

The outer boundary is typically defined by a constant far-away supersaturation, so we set $\sigma_{i,j} = \sigma_{far}$ when $i = N_r$ or $j = N_z$, and this is easily implemented in the model by applying Equation 5.15 only out to $i = N_r - 1$ and $j = N_z - 1$. There are some numerical issues that must be dealt with along the $(0, z)$ and $(r, 0)$ axes, but these are minor bookkeeping details that are discussed elsewhere [2013Lib1]. It is typical to use reflection boundary conditions at $z=0$, so the physically modeled space then includes $-z_{max} \le z \le z_{max}$ and $0 \le r \le r_{max}$.

The surface boundary conditions for faceted boundary pixels are similar to the 1D case described above, and Equation 5.6 becomes

$$\sigma_{i,j} = \sigma_{i+1,j}\left(1+\alpha_{prism}(\sigma_{i,j})\frac{\Delta x}{X_0}\right)^{-1}$$
$$\sigma_{i,j} = \sigma_{i,j+1}\left(1+\alpha_{basal}(\sigma_{i,j})\frac{\Delta x}{X_0}\right)^{-1}, \qquad (5.16)$$

where α_{prism} and α_{basal} are the attachment coefficients for the two principal facets. For simplicity, the index notation here is for facets that face in the $+r$ and $+z$ directions, and we have assumed $\Delta r = \Delta z = \Delta x$. These then become propagation equations that are similar in form to Equation 5.7.

For a kink boundary pixel, the optimum boundary condition can be estimated by examining the growth of the 45-degree surface orientation shown in Figure 5.17. Because only kink boundary pixels are present on this surface, the continuum boundary condition Equation 5.4 can be expressed in two essentially equivalent forms

$$\sigma_{i,j} = \sigma_{i+1,j+1}\left(1+\alpha_{kink}(\sigma_{i,j})\frac{\sqrt{2}\Delta x}{X_0}\right)^{-1}$$

or

$$\sigma_{i,j} = \sigma_{opp}\left(1+\alpha_{kink}(\sigma_{i,j})\frac{\Delta x}{\sqrt{2}X_0}\right)^{-1}, \qquad (5.17)$$

where $\sigma_{opp} = (\sigma_{i+1,j} + \sigma_{i,j+1})/2$, and α_{kink} is the appropriate attachment coefficient. Both expressions can be derived from Equation 5.4 using the geometry illustrated in Figure 5.17, and the two equations are equal to first order in Δx. If the 45-degree surface is essentially flat (the small Δx limit), these expressions provide an accurate model of the surface boundary condition. Because a surface made from kink sites is molecularly rough, we expect $\alpha_{kink} \approx 1$. Note that the additional factor of $\sqrt{2}$ compared to Equation 5.16 is simply a geometrical factor coming from the tilted geometry of the 45-degree surface on the CA grid. As with the 1D model, the 2D propagation equations can be iterated until some suitable convergence criterion is satisfied. This process solves Laplace's equation in the space surrounding the growing crystal, thus yielding the supersaturation field $\sigma(r_i, z_j)$ in all vapor and boundary pixels at a fixed time.

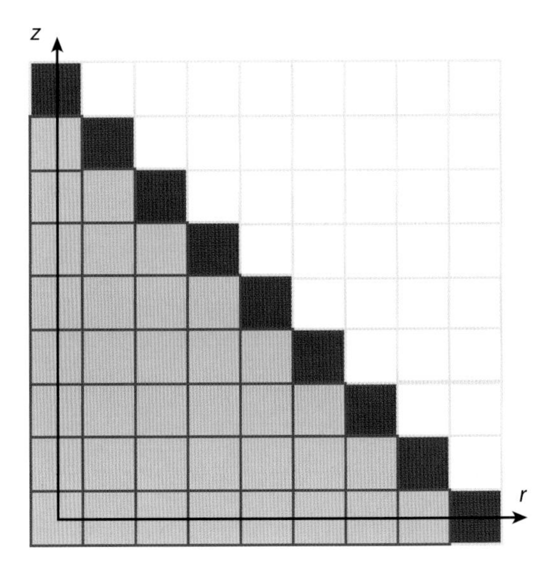

FIGURE 5.17. The 45-degree surface depicted here includes only kink boundary pixels. This can be taken as an essentially flat surface in the limit of a large crystal, giving a good approximation for determining the CA rules governing kink boundary pixels.

Growth Steps

The next step in the model is to use the known supersaturation field to calculate the crystal growth rates at each point and define appropriate CA rules for turning boundary pixels into ice pixels. If all goes well, these the CA propagation equations and growth rules will generate physically accurate computational snow crystals.

As with the 1D model, we define a growth step as occurring when a single boundary pixel transforms into an ice pixel [2014Kel], and the newly defined surface then requires a new calculation of the supersaturation field. The main difference between the 1D and 2D models is that now there are many boundary pixels to consider simultaneously.

To keep an ongoing account of the crystal growth at each point on the ice surface, we assign a numerical "filling factor" f_b to each boundary pixel, where we assign an integer index b to label the boundary pixels. Whenever a vapor pixel becomes a new boundary pixel, f_b for that pixel is set to zero. As the model develops, each f_b increases with time at a rate that derives from the crystal growth rate at its position. When a filling factor increases to unity, then that boundary pixel turns to ice.

After relaxing the supersaturation field to produce $\sigma(r_i, z_j)$ throughout the space above the crystal, we can again use Equation 5.8 to calculate the growth velocity along the surface normal. For a facet boundary pixel, the time required to "fill" the remainder of each boundary pixel becomes

$$\delta t_b = \frac{\Delta x}{v_n}(1 - f_b), \tag{5.18}$$

while for kink boundary pixels, we again examine the 45-degree surface to obtain

$$\delta t_b = \frac{\Delta x}{\sqrt{2} v_n}(1 - f_b), \tag{5.19}$$

and the additional $\sqrt{2}$ is again a geometrical factor associated with the 45-degree surface. From the entire set of time intervals δt_b, we choose the smallest one, $\delta t_{b,min}$, and then fill each boundary pixel for this amount of time, giving

$$f_b \rightarrow f_b + \frac{v_n}{\Delta x}\delta t_{b,min} \tag{5.20}$$

for all facet boundary pixels, and

$$f_b \rightarrow f_b + \frac{\sqrt{2} v_n}{\Delta x}\delta t_{b,min} \tag{5.21}$$

for all kink boundary pixels. In doing this, one filling factor will reach $f_b = 1$, while all the others will increase but remain below unity. After updating the filling factors and turning one boundary pixel to ice, we then locate the new boundary pixels (assigning to them a filling factor of zero) and proceed with calculating the next supersaturation field.

One pleasant feature of the CA method is that it is remarkably easy to write down physically realistic (albeit not entirely accurate) rules and transcribe them into relatively simple iterative algorithms. In general, other front-tracking and phase-field techniques require a substantially greater mathematical sophistication and a commensurate increase in programming effort. Unfortunately, the relative simplicity of a CA model brings with it some deficiencies in terms of accuracy, which we examine next. How serious all these problems are, and how well they can be addressed by developing more advanced CA rules, remains a topic for additional research.

Numerical Anisotropy

If cylindrically symmetrical snow crystal growth could be modeled accurately using only faceted surfaces together with the 45-degree surface shown in Figure 5.17, then the facet-kink model described above would be adequate to solve this 2D problem. As $\Delta x \rightarrow 0$, the facet-kink CA rules satisfy the surface boundary conditions and growth rates to high accuracy on these surfaces. Problems arise, however, when one considers other surfaces.

Consider, for example, the 2:1 surface shown in Figure 5.18. In the small-Δx limit, this is a simple vicinal surface, so a solution of the diffusion equation (assuming an infinite surface and ignoring the Mullins-Sekerka instability) would yield uniform planar growth with a growth velocity $v_n = \alpha v_{kin} \sigma_{surf}$. However, the facet-kink model cannot reproduce this simple result, even if α is constant on all surfaces. The model growth rates are typically off by about 10 percent, depending on the model

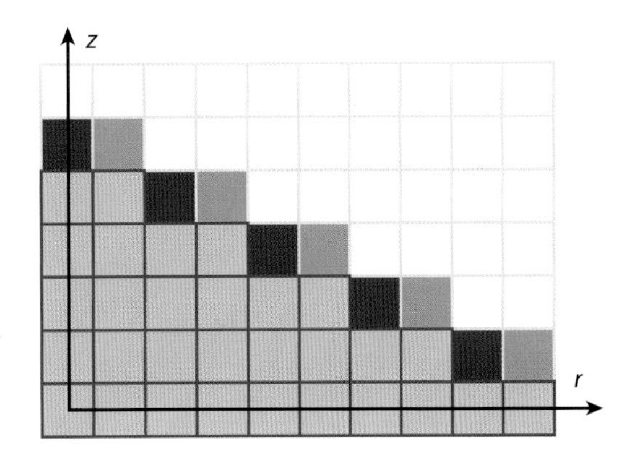

FIGURE 5.18. This 2:1 vicinal surface includes equal numbers of facet and kink boundary pixels. A facet-kink CA model with a constant α yields growth-rate errors of about 10 percent, depending on model parameters. These errors present an intrinsic anisotropy in the facet-kink model that cannot be corrected by increasing the model resolution.

details. With the facet-kink model, the growth rates are essentially exact for the facet surfaces and the 45-degree surface, but systematic errors are unavoidable on other vicinal surfaces. If one removes all the $\sqrt{2}$ factors in the above discussion, the maximum error can be as high as 40 percent.

In an absolute sense, a 10 percent growth-rate error may not be terrible, as experiments are typically not able to determine σ_{far} to this level of accuracy. The problem arises because this is an anisotropic error. If snow crystal growth is sufficiently facet-dominated, as described above, then perhaps a small intrinsic anisotropy in the model will have little importance in its overall morphological development. Even a $\sqrt{2}$ anisotropic error may not have much of a detrimental effect [2014Kel]. Some features in snow crystal growth, however, may simply be impossible to reproduce with this level of intrinsic anisotropy. One example might be tip splitting (see Chapter 3), as this phenomenon arises when the attachment kinetics are especially isotropic. Subtle features in ridge formation and other common snow crystal morphological features may also be adversely affected by built-in anisotropies in the facet-kink model. The only way to answer these questions will be to make detailed comparisons between computational and laboratory snow crystals over a broad range of growth conditions.

A Facet-Vicinal Model

One way to reduce the intrinsic anisotropies is to devise an improved set of CA rules. The facet-kink model uses only nearest-neighbor interactions to determine the boundary conditions, and we can do better by incorporating nonlocal effects, at the expense of increased algorithmic complexity. One possibility in two dimensions is what I call a *facet-vicinal model.*

The basic idea in the facet-vicinal model is to define a new parameter L, equal to the width of the terrace "ledge" associated with each boundary pixel, in integer pixel units. Figure 5.19 shows one example of a vicinal surface with $L=3$ for each boundary pixel. In this example, all the terrace ledges are basal surfaces facing the $+z$ direction. On a more complex surface, one would simply count how many adjacent boundary pixels make up a single ledge, and that value of L would be assigned

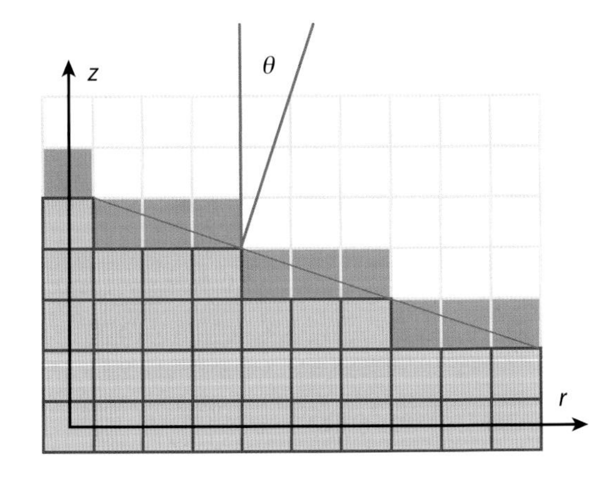

FIGURE 5.19. A vicinal surface in which all the terrace ledges have a ledge width $L=3$, and the vicinal angle θ is given by $\tan(\theta)=1/L$. For any individual terrace facing the $+z$ direction (as shown here), the ledge width L is defined as the number of adjacent boundary pixels in its row, and that value of L is assigned to all boundary pixels in that row.

to all the boundary pixels making up that ledge. Although more complicated than the facet-kink model, this degree of nonlocal bookkeeping is not an onerous computational task.

From the ledge width L, the vicinal angle θ is given by $\tan(\theta) = 1/L$, and the attachment coefficient is specified as $\alpha_{vicinal}(\sigma_{i,j}, \theta)$, being a function of both the surface supersaturation and the vicinal angle. For simplicity, I now assume $0 < \theta < 45$ degrees, as generalization to angles outside this range is straightforward. For any vicinal surface, the boundary conditions are again derived from the continuum boundary conditions

$$X_0 \left(\frac{\partial \sigma}{\partial n} \right)_{surf} = \alpha \sigma_{surf}, \tag{5.22}$$

with

$$\left(\frac{\partial \sigma}{\partial n} \right)_{surf} = \hat{n} \cdot \nabla \sigma = A_c \frac{\sigma_{i,j+1} - \sigma_{i,j}}{\Delta x} + A_s \frac{\sigma_{i+1,j} - \sigma_{i,j}}{\Delta x}, \tag{5.23}$$

and again we are assuming $\Delta r = \Delta z = \Delta x$ and $\sigma_{i,j} = \sigma(r_i, z_j)$, and we define $A_c = \cos(\theta)$ and $A_s = \sin(\theta)$. With these assumptions, the propagation equation for the supersaturation in any boundary pixel becomes

$$\sigma_{i,j} = \frac{A_c \sigma_{i,j+1} + A_s \sigma_{i+1,j}}{A_c + A_s + \alpha_{vicinal}(\sigma_{i,j}, \theta) \frac{\Delta x}{X_0}}. \tag{5.24}$$

This expression is a generalized form of the boundary conditions described in Equations 5.16 and 5.17 (and agreement can be seen by noting that $\sigma_{i,j+1} + \sigma_{i+1,j} = \sigma_{i,j} + \sigma_{i+1,j+1}$ to first order in Δx). However, while Equations 5.16 and 5.17 applied with high accuracy to only three surfaces, Equation 5.24 is accurate for all flat vicinal surfaces with any vicinal angle. Thus, with some increase in bookkeeping, we have a new boundary condition with substantially reduced intrinsic anisotropy.

The growth algorithm is like that described above, but with Equations 5.18 and 5.19 replaced by

$$\delta t_b = A_c \frac{\Delta x}{v_n} (1 - f_b) \tag{5.25}$$

for the growth of +z boundary pixels with $0 < \theta < 45$ degrees, like those shown in Figure 5.19.

For the special case of a flat vicinal surface, we see that all the boundary pixels shown have identical properties. Because σ_{surf} will be nearly constant along this flat surface, all the boundary pixels will turn to ice pixels at essentially the same time, and this will preserve the vicinal character of the surface. In the limit of small Δx, I expect that the facet-vicinal model will provide an improved model behavior compared to a facet-kink model. The facet-vicinal CA model has not yet been tried, as only facet-kink models have so far been demonstrated for snow crystal growth. I suspect that facet-vicinal, or some improved version of this model, may someday displace the facet-kink model for snow crystal growth. But the additional level of complexity is probably not desirable until facet-kink models have been better explored.

Monopole Matching

Extending the outer boundary to infinity can again be accomplished, to a reasonable approximation, using the known analytical solution for spherical growth. The essential idea is the same as was described above, but in place of Equation 5.11, we use

$$\frac{dV}{dt} = \sum 2\pi r_b \frac{\Delta x^2}{\delta t_b} \tag{5.26}$$

where the sum is over all boundary pixels, and δt_b is evaluated with $f_b = 0$. This yields the adaptive outer boundary

$$\sigma_{i,j}(\rho_{far}) \to \sigma_\infty - \frac{dV/dt}{4\pi \rho_{far} X_0 v_{kin}}, \tag{5.27}$$

where $\sigma_{i,j}$ refers to an outer boundary pixel, and $\rho_{far} = \sqrt{r^2 + z^2}$ is the distance to the outer boundary point.

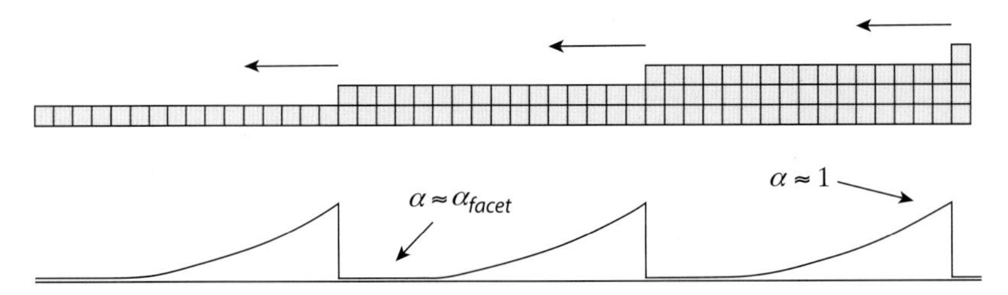

FIGURE 5.20. On a low-angle vicinal surface, molecules can diffuse along a faceted surface to reach kink sites where they readily attach. Thus $\alpha \approx \alpha_{facet}$ far from a kink site, while $\alpha \approx 1$ within one surface diffusion length from a kink site. A high Ehrlich-Schwoebel barrier that prevents diffusion over the tops of terrace steps is assumed here.

This iterative outer boundary assignment matches the outer boundary to an optimal spherical solution, so it would be quite accurate for the case of nearly isometric crystals. For the general case, it can be considered a monopole approximation of the correct outer boundary. One can imagine extending this to higher-order multipole matching, but I will not elaborate further on that possibility here. For a sufficiently distant model boundary, the monopole approximation is probably good enough for most purposes, allowing a reasonable first estimate for extending the model to an infinite outer boundary.

Surface Diffusion and the Fast Surface Diffusion Approximation

In its most basic form, the facet-kink CA model described above does a rather poor job describing the growth of low-angle vicinal surfaces. As illustrated in Figure 5.20, surface diffusion on faceted surfaces transports admolecules to kink sites, and this process can greatly increase the attachment coefficient near terrace steps (assuming $\alpha_{kink} \gg \alpha_{facet}$). This bit of physics is absent in the facet-kink model, where all facet boundary pixels are described by α_{facet}, even if they are right next to kink sites [2015Libl]. When incorporating this physical effect into a CA model, the increase in the attachment coefficient extends over a distance of approximately (x_{surf}/a) pixels, where x_{surf} is the surface diffusion length, and a is the size of a water molecule. Although x_{surf} is not well known on faceted ice surfaces, one expects $(x_{surf}/a) > 30$, so this factor presents a sizable increase in the attachment coefficient over large vicinal surfaces in a CA model. Here we see that the relevant surface diffusion parameter in a CA model is not x_{surf}, but x_{surf}/a.

Given the finite resolution in a CA model, it is not unreasonable to speak of a fast surface diffusion (FSD) approximation that assumes $x_{surf}/a \rightarrow \infty$. In this approximation, any terrace that includes a kink site would have $\alpha \approx 1$ over the entire terrace surface, while only small faceted "island" terraces (which I also call "upper terraces") would be described by α_{basal} or α_{prism}. I suspect that the FSD approximation may be a reasonable representation of the actual attachment kinetics, but the use of this approximation has not yet been explored in CA models. In contrast, the bare facet-kink CA model could be called a low surface diffusion approximation, as it neglects surface diffusion entirely. This is almost certainly a poor approximation of the actual ice surface physics, although, once again, the actual surface diffusion lengths have not been well determined.

Concave Growth

The growth of shallow concave plates is another interesting testing ground for exploring the accuracy of CA models. As shown in Figure 5.21, the issue of insufficient spatial resolution becomes especially acute with this

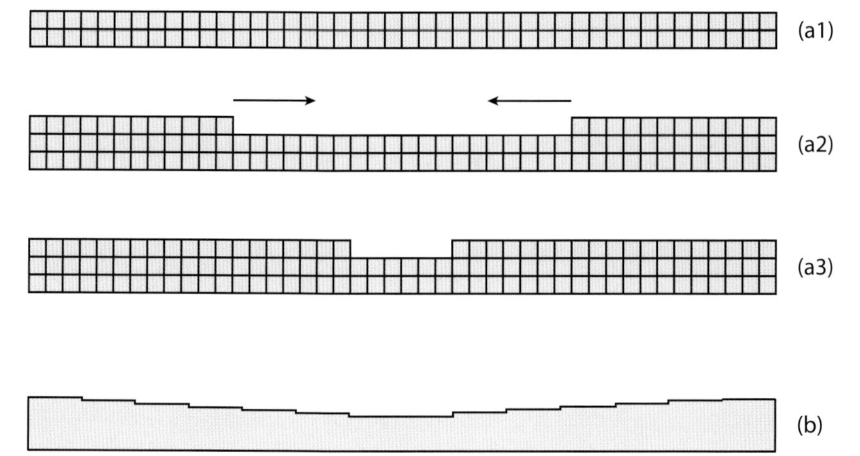

FIGURE 5.21. The growth of a shallow concave basal surface is difficult to reproduce with good accuracy in a CA model. In the model (a1, a2, a3), a new terrace nucleates at the edges of the plate and then grows inward. In real life (b), a continuous series of terraces nucleate at the plate edge and propagate inward. The finite resolution of the model precludes a fully accurate representation of such surfaces.

geometry. Watching a CA model evolve in real time, one sees a peculiar time dependence in the model that does not happen in real life. With a perfectly faceted plate (a1 in the figure), α_{basal} is low across the entire plate, and the slow growth of this surface yields a high σ_{surf} above it (see Chapter 4). As soon as a kink site appears, however (a2), α increases substantially (especially with an FSD model), and thus σ_{surf} drops significantly. Moreover, this happens essentially instantaneously, as that is how fast the supersaturation field responds in the Laplace approximation. Once the terrace fills in and is replaced by a fully faceted surface, α and σ_{surf} again change instantaneously. This is a curious sight to watch, as clearly it does not accurately model what must happen around an actual crystal (sketch b in Figure 5.21).

With full molecular resolution, even a shallow concave surface contains hundreds of terrace steps all marching inward. The attachment coefficient is thus α_{basal} near the plate edge and $\alpha \approx 1$ elsewhere, while σ_{surf} over the surface changes little with time as the steps progress. This issue would go away with sufficient resolution, and it would not be a problem if the concave depression is deep enough, but it does identify a deficiency in the CA technique. In contrast, a front-tracking model would likely handle this scenario much better. Even at low spatial resolution, the front-tracking model would look more like (b) in the figure, with a faceted region at the outer edge and a segmented concave region within. The freedom to build a surface out of short line segments, rather than small blocks, gives a substantial advantage to the front-tracking model in this case.

The Gibbs-Thomson Effect

So far in our CA modeling discussion, we have ignored surface energy effects, focusing mainly on vapor diffusion and attachment kinetics, the latter depending on surface diffusion. The resulting models are likely reasonable approximations in many situations, but our examination of solvability theory in Chapter 3 suggests that surface energy effects become important when α is large and σ is low. This is borne out in model investigations like that shown in Figure 5.22.

In this modeling exercise, $\alpha_{basal} \ll \alpha_{prism} = 1$, and σ_{∞} is quite low, yielding the growth of a one-pixel-thick plate from the edge of a columnar crystal. A single pixel measured 0.15 μm in this model, and the high curvature of the plate edge would create a large Gibbs-Thomson effect. Clearly the emergence of this ultrathin plate would have been suppressed by surface energy effects, so this result is not physically plausible.

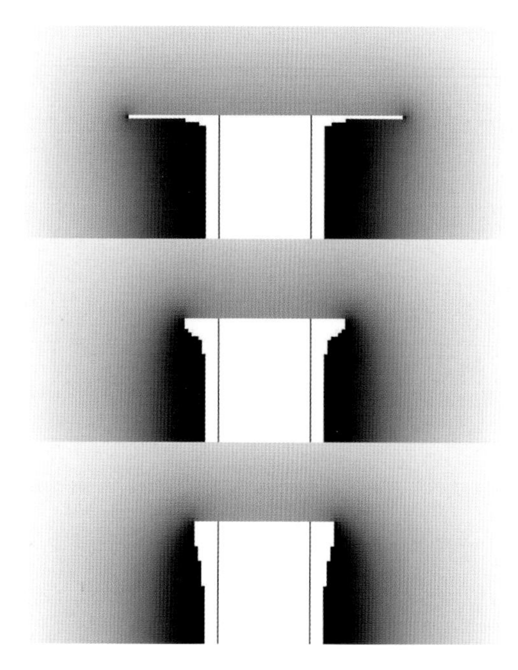

FIGURE 5.22. An illustration of the Gibbs-Thomson effect in a cylindrically symmetric CA model of a plate growing from the edge of a column [2013Lib1]. With zero surface energy (top), a one-pixel-thick (0.15 μm) plate grows from the edge of the column, which is not a physically plausible solution. A Gibbs-Thomson length of d_{sv}=0.3 nm (middle) or d_{sv}=1 nm (bottom) suppresses the thin-plate growth. Vertical lines show the original seed crystal, and the supersaturation around the crystal is proportional to the image brightness.

There are several ways to avoid this problem in a CA model. One is simply to avoid regions of parameter space where low σ and high α can occur simultaneously. This is not especially difficult to arrange, and the model crystal in Figure 5.22 was something of a physically unrealistic case. Another approach is to increase the CA pixel size Δx to the point that the Gibbs-Thomson effect is negligible even with one-pixel-thick structures. And again, this is not especially difficult to arrange, but it is not a very satisfying approach to the problem. Of course, a better solution is to add the correct surface energy physics to the CA model, thereby obviating the need to avoid certain areas of parameter space.

Because the Gibbs-Thomson effect is quite small in snow crystal growth, it is sufficient to approximate it rather crudely, as this is enough to eliminate one-pixel-wide plates and other nonphysical model manifestations. One way to accomplish this quite easily in the CA model is to use the widths of the outermost terraces, which, owing to their extreme positions, do not include any kink pixels. The values of L_{rmax} and L_{zmax} in Figure 5.23 can be used as proxies to estimate the edge curvatures, and the precise algorithm used is not very important.

The Gibbs-Thomson effect can be ignored in calculating the supersaturation field, as its effect is negligibly small. It need only be included in the calculation of the pixel growth, specifically replacing the usual $v_n = \alpha v_{kin}\sigma_b$ with $v_n = \alpha v_{kin}(\sigma_b - d_{sv}k)$, using the known Gibbs-Thomson parameter d_{sv} and a roughly estimated curvature κ. Although the outer terrace widths are not extremely accurate curvature indicators, this method is sufficient to suppress the formation of structures with especially high surface curvature.

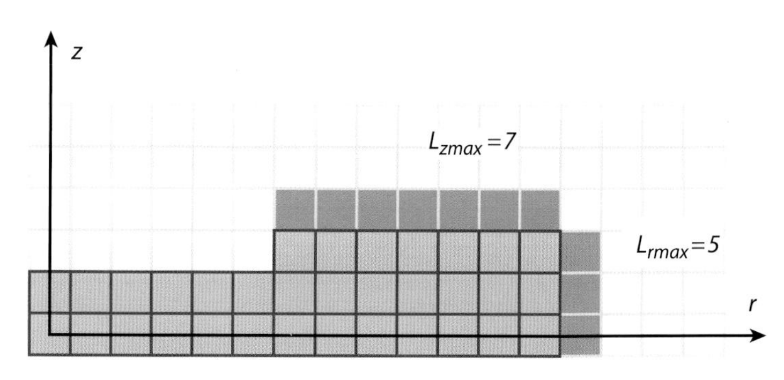

FIGURE 5.23. A CA model in which the outer basal and prism boundary pixels are shown in pink. These pixels represent upper terraces with no adjoining kinks sites, and the terrace widths can be used to roughly estimate the edge curvature for including the Gibbs-Thomson effect.

Structure-Dependent Attachment Kinetics

The outer facet widths defined in Figure 5.23, with the surface curvatures derived from them, can also be used to incorporate the SDAK phenomenon (see Chapter 4). The basic idea here is to make the attachment coefficient depend on curvature κ, just as the effective supersaturation depends on κ via the Gibbs-Thomson effect. Although these physical effects are quite different, both can be included in the CA model using the outer facet widths. This idea was explored somewhat [2015Lib2] by comparing CA models with experimental measurements of thin plates forming on electric needles near −15°C. Some additional details are presented in Chapter 8, with the results generally supporting the SDAK model and its CA models. Additional work is certainly needed in this area, but this result suggests that much could be learned from continued careful comparisons between CA models and experimental measurements.

Minimum Feature Sizes

While the CA grid size Δx is somewhat arbitrary, it should not be made too large if one expects to reproduce realistic snow crystal structures. As we saw in our discussion of solvability theory (Chapter 3), the characteristic radius of curvature of a growing dendrite tip is roughly

$$R_{tip} \approx \frac{2X_0}{s_0\alpha}, \tag{5.28}$$

and measurements of ice dendrites in air have yielded $R_{tip} \approx 1$ μm and $\alpha s_0 \approx 0.25$ for fernlike dendrites near −15°C, and $R_{tip} \approx 1.5$ μm and $\alpha s_0 \approx 0.2$ for fishbone dendrites growing near −5°C (see Chapter 3). In both cases, the tip structure was quite rounded, suggesting $\alpha \approx 1$.

In addition to dendrite tips, interferometric measurements of thin plates growing in a free-fall chamber showed thicknesses down to 1 μm at temperatures near −2°C and again near −12°C [2008Lib1]. Column diameters as low as a few microns were also observed near −5°C [2009Lib]. These observations all suggest that a grid size of a few times X_0 should be sufficient to reproduce essentially all snow crystal structures, where $X_0 \approx 0.15$ μm in normal air. Finer structures are likely suppressed by the Gibbs-Thomson effect and perhaps by additional surface diffusion effects.

A Scaling Relation

If we switch variables from physical dimensions (r, z) to scaled dimensions $(\xi_r, \xi_z) = X_0^{-1}(r,z)$, the surface boundary condition, Equation 5.6, is converted to dimensionless form. Laplace's equation is essentially unaltered by this variable change, so our CA models will proceed equally well in dimensionless coordinates, while physical time intervals change from $\delta t = \Delta x / \alpha v_{kin}\,\sigma$ to $\delta t = X_0\,\Delta\xi / \alpha v_{kin}\,\sigma$. What this all means is that we have a scaling relation for growth as a function of the diffusion constant D, or equivalently, a scaling relation with air pressure P [2013Lib1]. Because $X_0 \sim D \sim P^{-1}$, we see that increasing the air pressure by a factor of two will result in a crystal that grows half as large in a time that is twice as long compared to growth at the original pressure. This assumes that all other aspects of the model (for example, the attachment coefficients) are unchanged as a function of pressure and crystal size.

This scaling relation directly explains some prominent characteristics of snow crystal growth as a function of pressure. At low pressures, for example, crystals grow rapidly into faceted, prismatic shapes, even when the crystals are quite large. However, in normal air, initially faceted prisms quickly branch into dendritic morphologies, and dendritic shapes at higher pressures have been observed to show generally finer structural features [1976Gon]. At least at a qualitative level, the observed pressure dependence is nicely explained by this simple scaling relation. Besides just morphologies, however, the scaling relation also makes clear predictions regarding growth rates as a function of pressure, although these have not yet been experimentally confirmed.

But this scaling relation comes with numerous caveats, as it assumes that all other factors (other than particle diffusion) are independent of pressure and physical scale. For example, the scaling relation requires that α be independent of pressure, which contradicts the fact that α depends strongly on σ_{surf}. In addition, as particle diffusion becomes rapid at lower pressures, heat diffusion begins to dominate as a factor that limits growth, and this complicating factor negates the simple scaling relation. Other physical effects, including surface energy, surface diffusion, and the ESI, may also affect pressure scaling in various regions of parameter space. Thus, although expressing the problem in dimensionless coordinates brings some mathematical appeal, I find it tends to obscure the physics as well. And there is no getting around the fact that snow crystal growth is not just a mathematical problem, as it involves a variety of physical processes acting over many length scales.

Comparison with Experiments

Although CA models have some inherent shortcomings, they have already been shown to produce realistic snow crystal morphologies and growth rates. This first became abundantly apparent when 3D models yielded morphological structures that resembled real snow crystals to a much higher degree than did other models, including ridging and other features. In this section, I describe how CA models have fared quite well in comparisons with experimental observations as well, at least to the limited degree to which the models have been tested.

To date, no 3D models of snow crystal growth, of any kind, have been subjected to detailed comparisons with experimental observations. That day is coming, but so far only 2D cylindrically symmetrical models have been examined in conjunction with quantitative

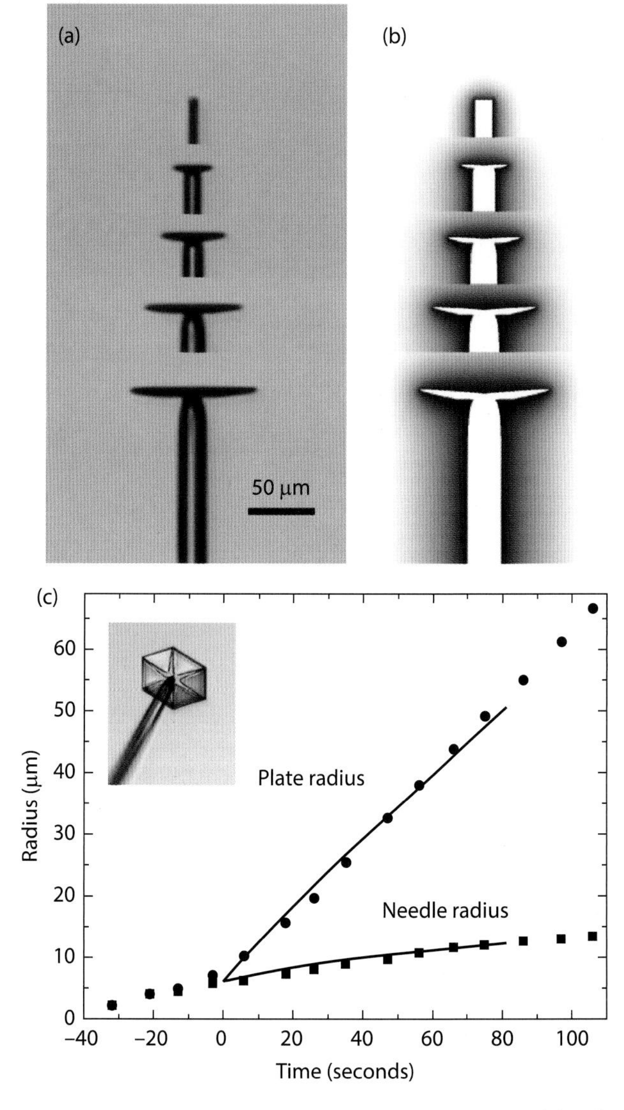

FIGURE 5.24. A quantitative comparison a 2D CA model with experimental data [2013Lib]. (a) A composite image made from five photographs shows the growth of a platelike snow crystal on the end of an electric ice needle, viewed from the side. (b) A cylindrically symmetrical 2D CA model reproduces the observations, also showing the water vapor diffusion field around the crystal. (c) A quantitative comparison of experimental data (points) and the computational model (lines) shows good agreement. The inset photo shows the crystal in (a) from a different angle.

snow crystal growth measurements. Figure 5.24 shows one of my favorite early examples of a CA model matching the formation of a thin-plate snow crystal growing on the end of an electric ice needle (see Chapter 8). The cylindrically symmetric CA model cannot reproduce the hexagonal faceting or ridge features, as these would require fully 3D modeling. But it does reproduce the slightly concave plate growth and the shielding of the columnar growth just below the plate. In terms of overall morphological features, the model seems to get the details right.

Moreover, both the morphology and growth measurements were adequately reproduced using one set of model parameters. After some tweaking of the outer-boundary supersaturation and the attachment coefficients, the model could be made to fit the growth measurements quite well, as can be seen in the figure. A clear result from this exercise was that $\alpha_{prism} \approx 1$ was essential to fit the data and morphology, matching our expectations based on the observation that the plate sprouts branches if the supersaturation is raised only slightly higher than was used in this experiment. Libbrecht et al. [2015Lib2] performed a series of measurements like this, investigating the formation of thin plates on electric needles as a function of the faraway supersaturation level. This experiment nicely illustrates the potential for using electric needles in quantitative studies of snow crystal growth.

This experiment is also a good example of how cylindrically symmetrical CA models can be used to analyze precise growth measurements to reach substantial, quantitative physical conclusions. As described elsewhere [2015Lib2], the data support the ESI described in Chapter 4, indicating the need for structure-dependent attachment kinetics. The ESI model is still a hypothesis in need of additional testing, but it is abundantly clear that comparing CA models with experimental observations has much potential for yielding interesting scientific progress regarding the physics of snow crystal growth.

The Two-Dimensional Future

Although clearly 3D modeling will be preferred in the long run, I believe that 2D cylindrically symmetrical models have some substantial advantages in the short term for investigating the physics underlying snow crystal growth dynamics, including:

1) With one fewer dimension, the run times for a 2D code are much faster than a 3D code. This allows one to run dozens or hundreds of models quickly, which is highly beneficial when making detailed comparisons between models and experiments.
2) With a simpler geometry and fewer special cases to deal with, a 2D code is easier to write and modify than a 3D code. This makes it generally easier to incorporate additional physics, such as the Gibbs-Thomson effect, the ESI, and surface diffusion. Thus, a 2D model can be more practical for investigating the overall importance of these effects in conjunction with experimental observations.
3) A 2D cylindrically symmetric model can provide a reasonably accurate approximation for simple snow crystal morphologies, including simple plates and columns, hollow columns, capped columns, and simple forms growing on electric needles. Here again, the 2D model is well suited for examining basic morphological changes in growth behavior with temperature, supersaturation, and background gas pressure. Low-pressure growth, exhibiting overall simpler structures than at higher pressures, is especially amenable to 2D modeling.
4) Most of what we have learned to date about snow crystal attachment kinetics has been from measurements of small crystals with relatively simple morphological structures. Once more, a 2D model should be sufficient for further investigations along these lines.
5) Working with a 2D CA model is a good prelude to building a full 3D CA model, as the 2D model al-

ready exhibits interesting behaviors and puzzling quirks, so learning about these will benefit future efforts with full 3D modeling.

THREE-DIMENSIONAL CELLULAR AUTOMATA

As described in the previous sections, 1D and 2D models provide an instructive perspective on many of the good and bad aspects of modeling snow crystal growth using cellular automata. CA models are generally simple to construct and fast to run, but it is difficult to remove the mathematical anisotropies and other quirks that are essentially hardwired into the fixed grid and CA rules. Several of these issues are relatively easy to see and understand in 2D cylindrically symmetric models, as I attempted to describe above. Three-dimensional models have not yet been abundantly explored, and the extra dimension will likely introduce even more hidden foibles that have not yet been discovered. Nevertheless, 3D modeling is the ultimate goal, so I examine this next step now. Much of this section is based on work done by Gravner and Griffeath [2009Gra] and by Kelly and Boyer [2014Kel], together with some additional embellishments derived from my own research.

A 3D Hexagonal Grid

Figure 5.25 illustrates a hexagonal grid of cells appropriate for a 3D CA snow crystal model. In this grid, each pixel has eight nearest neighbors: two in the vertical direction and six in the horizontal direction, where here we use "horizontal" as a somewhat generic term referring to all directions perpendicular to the c-axis. As with the previous 1D and 2D models, the cells are labeled as ice, vapor, or boundary pixels.

Figure 5.26 shows a convenient mapping that takes a honeycomb structure in the horizontal plane to a simple rectilinear grid, which can be useful for bookkeeping purposes in the various CA algorithms. Note the definition of the spacing Δx between prism-facet terraces shown in the figure. This is different from the definition used by Kelly and Boyer [2014Kel], for reasons that will become apparent when we discuss boundary conditions below. We also define Δz to be the spacing between basal terraces, and we usually assume $\Delta z = \Delta x$. Like the previous CA models in this chapter, our 3D model will be completely deterministic, including no random walks or random probabilities of any kind, and evaporation will not be included. Running the model twice with the same initial conditions will produce the identical results. Such a deterministic model must always

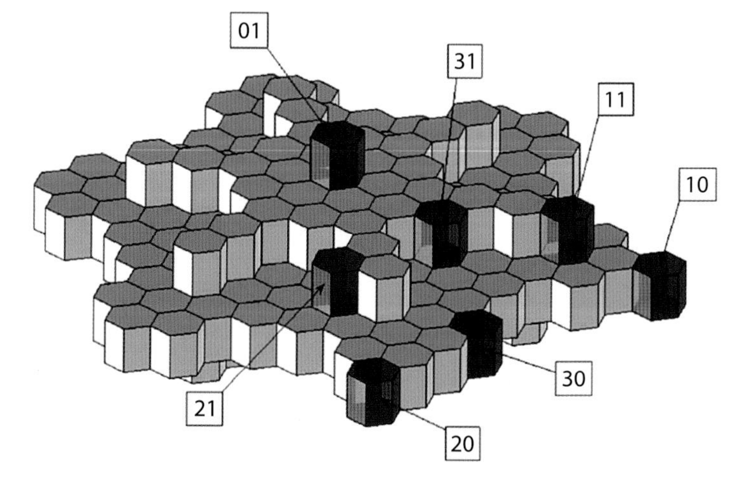

FIGURE 5.25. A 3D hexagonal grid of cells for a CA snow crystal model. Here the blue-green pixels represent ice, and the red pixels show a few representative boundary pixels. Vapor pixels are not shown. The boundary pixels are labeled with [HV] nearest-neighbor data, where H is the number of adjacent horizontal ice pixels, and V is the number of adjacent vertical ice pixels. Image adapted from [2009Gra].

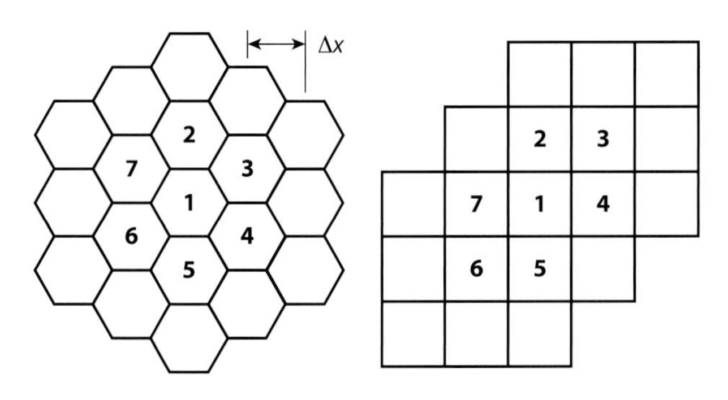

FIGURE 5.26. A coordinate mapping that connects a 2D hexagonal grid to a 2D Cartesian grid, with numbers showing corresponding pixels. Note the definition of the horizontal coordinate scale Δx, equal to the spacing between prism facet terraces. We typically assume $\Delta x = \Delta z$, the latter being the distance between basal facet terraces.

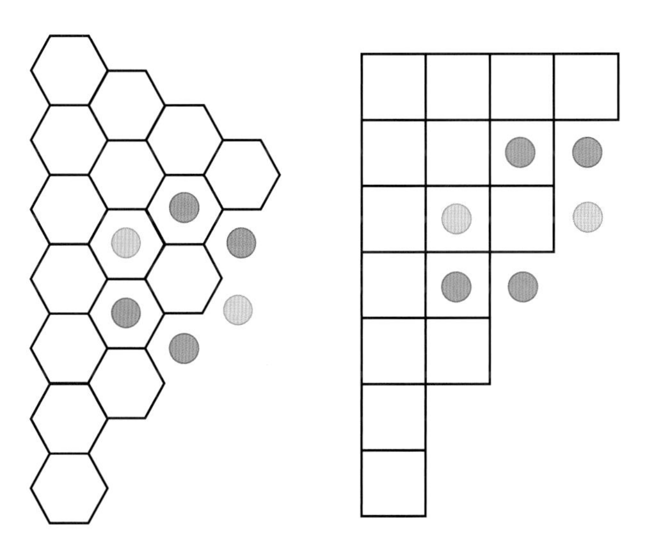

FIGURE 5.27. The 1/12 slice of horizontal space needed to model a symmetrical snow crystal. Reflection boundary conditions apply on the two long edges, and corresponding dots illustrate some reflected pixels on the stepped edge. The upper edge is the faraway boundary of the model.

exhibit perfect sixfold bilateral snow crystal symmetry, simply because the input physics and the external boundary conditions are defined to have this same symmetry.

Because of this intrinsic symmetry, the CA model need only include 1/24 of the total physical space being modeled, with reflection boundary conditions recreating the full space from the 1/24 slice. A reflection boundary condition is applied about the $z = 0$ plane, as it was with the 2D model discussed earlier in the chapter, and similar boundary reflections occur at the edges of the 30-degree wedge shown in Figure 5.27. As one might expect, there is considerably more bookkeeping involved in a 3D model than in a 2D model, which is simply the price one has to pay for the added complexity.

Boundary Pixel Attributes

Another complicating issue with a 3D CA model is the plethora of different boundary pixel types, as illustrated in Figure 5.25. Moreover, to encompass all the varied physical processes governing snow crystal growth, it is necessary to consider both nonlocal and nearest-neighbor interactions, as I briefly discussed for the 2D model. For this reason, it is necessary to describe the different types of boundary pixels and their various attributes with some care. For example, a [01] boundary pixel indicates a position on a basal facet, and we also want to label this pixel with information relating to its surroundings beyond its nearest neighbors. One approach to accomplishing this is to count the number of boundary pixels in each of the six directions out from the pixel in question, staying in the same basal plane, as illustrated in Figure 5.28. Doing so yields six integer pixel lengths $\pm L_i$, where the value is positive if the line of boundary pixels ends with a ledge (a terrace step approached from the top) and the value is negative if the line ends in a kink (a terrace step approached from below).

From these six lengths, one can extract quite a lot of useful information about the crystal structure near that boundary point:

1) If the L_i are all positive, then the boundary pixel in question lies on an upper terrace, which is identified as having no ice terraces on top of it. On the faceted upper-terrace surface, α_{basal} is the appropriate attachment coefficient. As described in Chapter 4, the best functional form for this term is

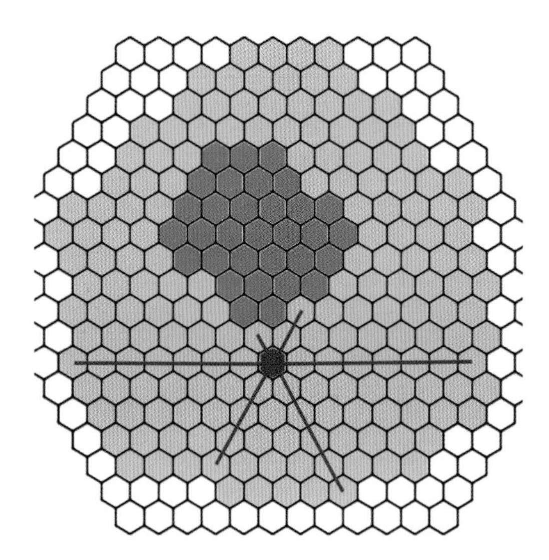

FIGURE 5.28. This diagram of several basal terraces shows a topmost "upper" basal terrace in dark blue, the next lowest terrace in light blue, and the terrace below that in white. The red cell illustrates a representative boundary pixel that lies atop the light-blue terrace. Counting the number of same-terrace boundary pixels in the six directions shown yields the L_i vector [−1, −2, +7, +5, +4, +7].

$\alpha_{basal} = A\exp(-\sigma_0/\sigma_{surf})$, where A and σ_0 are physical parameters included in the model.

2) If at least one of the L_i is negative and small, then surface diffusion can carry admolecules to kink sites, where they are readily adsorbed. In this case, the boundary pixel is best described by $\alpha \approx 1$.
3) If the L_i are all positive and two opposing L_i are both small, then the boundary pixel may lie on a thin basal edge, for example, on the edge of a hollow column, which may change the value of α_{basal} through the SDAK phenomenon (see Chapter 4). Additionally, a thin edge suggests a high surface curvature, which lowers the effective supersaturation via the Gibbs-Thomson effect.
4) The six $\pm L_i$ can also be used to infer something about the vicinal angle of the surface near the boundary pixel in question. As described in the 2D model above, such knowledge is useful for turning a facet-kink model into a facet-vicinal model, and the latter has substantially reduced intrinsic model anisotropy.

All these possibilities exist for every [01] boundary pixel, which is just one type of boundary pixel, and perhaps the simplest type at that. One must face an unfortunate reality in snow crystal growth that the underlying growth physics is complicated, so many different effects must be considered before a computational model will reproduce realistic crystals. It is not obvious at this point which physical effects must be included to high precision, which can be ignored altogether, and which are necessary but only to a rough approximation. Labeling each [01] boundary pixel with the six $\pm L_i$ is one way to incorporate a fair amount of flexibility into the model, which can then be used to explore different physical effects.

Moving on, the [20] boundary pixels describe prism facets, so, like the [10] boundary pixels, it is important to characterize their surroundings carefully. Again, we can define four lengths $\pm L_i$ by measuring the distances to the nearest ledges or kinks in each of the four vertical and horizontal directions. The discussion is then essentially identical to that for the [10] boundary pixels, except that we know that modeling of the prism-facet edges of thin plates is even more likely to involve some unusual physics, such as the SDAK effect or the Gibbs-Thomson effect. Suppressing the growth of one-pixel-thick plates via the Gibbs-Thomson effect could be incorporated into the model via these parameters.

Although the facet surfaces must be described carefully in any realistic snow crystal model, we can be a bit more cavalier regarding many of the remaining boundary pixels. For example, simply setting $\alpha = 0$ for all [10] pixels should be fine, as the molecular attachment at isolated [10] tips will be weak. At the same time, one can likely assume $\alpha = 1$ for all "kink-dominated" boundary pixels (such as [30], [40], and [21]), as these are all tight-binding sites. Snow crystal growth is largely facet dominated, so these model simplifications are likely acceptable over a broad range of growth conditions.

Note that the boundary pixel attributes must be recalculated after each growth step in the model. Every time a single boundary pixel turns to an ice pixel, the boundary geometry changes along with many of the $\pm L_i$ around it. This issue is usually handled by defining all the boundary pixels anew after each growth step and immediately recalculating all attributes for the set. Some computational savings could be realized, however, by only recalculating boundary-pixel attributes near the position of the last growth step, as this is the only region where the boundary changes significantly during that step.

Laplace Approximation

As discussed earlier in this chapter, the low Peclet number associated with snow crystal growth means that the particle diffusion equation turns into Laplace's equation, and growth modeling can be divided into separate diffusion and growth steps. This latter point, first made by Kelly and Boyer [2014Kel], provides a substantial simplification in CA modeling.

The first step is to assume a static crystal surface and iterate to a solution of Laplace's equation in the space surrounding the crystal. For a 3D model, the optimal propagation equation (see Equation 5.15) becomes

$$\sigma_a(k+1) = \frac{1}{9}\sum_{i=1}^{6} \sigma_i(k) + \frac{1}{6}\sum_{i=7}^{8} \sigma_i(k), \quad (5.29)$$

where σ_a is a vapor pixel, and the sum is over its eight nearest neighbors. As usual with CA models, this is the simple part, and the level of precision is mainly limited by the number of iterative steps computed.

Outer Boundary and Monopole Matching

The next-easiest part of the model is the outer boundary, and again we can use monopole matching to extend the outer boundary to infinity to a satisfactory approximation. This avoids complications associated with an adaptive grid, allowing the use of a constant grid spacing with the outer boundary that is fairly close to the growing crystal. How close depends on the overall accuracy desired, as a close outer boundary will distort the supersaturation field to some extent.

For our 3D grid, the outer boundary is defined as (see Equations 5.10 and 5.27)

$$\sigma_B(\rho_{far}) \rightarrow \sigma_\infty - \frac{dV/dt}{4\pi\rho_{far} X_0 v_{kin}}, \quad (5.30)$$

where σ_B refers to an outer boundary pixel, and ρ_{far} is the distance from the model's physical center to that outer boundary pixel location.

The rate of change of the total volume is given by (see Equation 5.26)

$$\frac{dV}{dt} = \sum \frac{G_1 \Delta x^3}{\delta t_b} \quad (5.31)$$

where the sum is over all surface boundary pixels, and $G_1 = 2/\sqrt{3}$ so the numerator is equal to the volume of a single pixel in our model, assuming Δx as defined in Figure 5.26 along with $\Delta z = \Delta x$. The δt_b are defined below.

Facet-Dominated Growth

As with the 2D model, most of the important physics in three dimensions rests in the surface boundary conditions, so it is important that we define these to reflect the correct underlying physical processes as accurately as possible. This is most easily done for a simple basal surface, where the continuum surface boundary condition

$$X_0 \left(\frac{\partial \sigma}{\partial n}\right)_{surf} = \alpha \sigma_{surf} \quad (5.32)$$

(see Chapter 3) becomes (on upper basal surfaces only)

$$\sigma_b = \sigma_{b+1}\left(1 + \alpha(\sigma_b)\frac{\Delta x}{X_0}\right)^{-1} \quad (5.33)$$

on our CA grid, where σ_b is the supersaturation in the basal boundary pixel, σ_{b+1} is the supersaturation in the vapor pixel just above the boundary pixel, and $\alpha(\sigma_b) = \alpha_{basal}$ is the attachment coefficient at the boundary pixel.

Note that this equation disagrees with the Kelly and Boyer boundary condition [2014Kel, Equation 11], as the latter is incorrect for a flat basal surface. Note also that Equation 5.33 should be used as part of the iterative process of defining the supersaturation field, as discussed in connection with Equation 5.7. When applied in this way, any functional form for $\alpha(\sigma_b)$ can be used, regardless of complexity. Both σ_b and $\alpha(\sigma_b)$ should converge smoothly to the correct result during this iterative process.

Unfortunately, the 3D grid geometry is such that Equation 5.33 cannot be used for other boundary pixels, even on a prism facet. We therefore define a generalized boundary condition that is valid for all boundary pixels

$$\sigma_b = \sigma_{opp}\left(1+\alpha(\sigma_b)\frac{G_b \Delta x}{X_0}\right)^{-1}, \tag{5.34}$$

where σ_b is any boundary pixel, σ_{opp} is the average supersaturation in all vapor pixels that "oppose" ice pixels for this boundary pixel (see Figure 5.29), and G_b is a dimensionless geometrical factor that must be defined for each boundary pixel. Conveniently, $G_b = 1$ for both [20] (prism facet) and [10] (basal facet) boundary pixels, which is why we defined Δx as we did in Figure 5.26.

Admittedly, calculating σ_{opp} is a chore for every boundary pixel in every iterative step of the relaxation process, but this could be done in a straightforward manner by using

$$\sigma_{opp} = \frac{1}{N_{weight}} \sum_{i=1}^{8} M_i \sigma_i, \tag{5.35}$$

where M_i is a weighting vector equal to 1 for an "oppose" pixel and 0 otherwise, and $N_{weight} = \Sigma M_i$. As with other boundary pixel attributes, the M_i can be calculated once after each growth step. This vector then allows for rapid calculation of σ_{opp} at each step in the Laplace iteration of the supersaturation field.

For facet-dominated growth, the values of G_b on nonfacet boundary pixels may not greatly change the outcome of a model run. It might be beneficial to do the geometry correctly for a [21] boundary pixel, however, much like it was with the 45-degree surface in the 2D model discussed earlier in the chapter. A few added touches like this can substantially reduce the intrinsic anisotropy in the CA model, which could improve how some familiar snow crystal features are reproduced in the model.

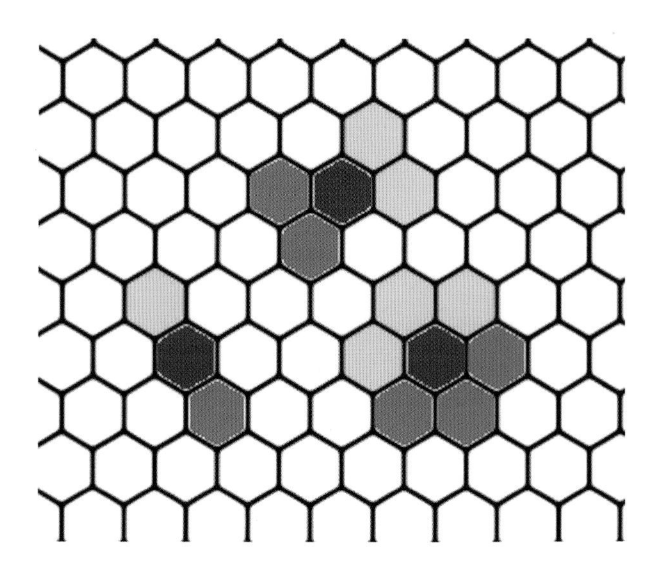

FIGURE 5.29. This diagram illustrates the calculation of σ_{opp}, the supersaturation averaged over all pixels that "oppose" a given boundary pixel. In each of these examples, the red cell is a boundary pixel, blue cells are ice pixels, and the yellow cells "oppose" the ice pixels. The value of σ_{opp} is calculated by averaging the supersaturations in the yellow cells. This 2D diagram does not show additional cells in the vertical direction.

Another possibility is to use $G_b = 1$ on [01] and [20] facet surfaces while leaving it as a constant, but adjustable, model parameter on all other surfaces to see what happens. It appears that some additional research is needed to determine the best course of action. Greater algorithmic complexity can reduce intrinsic model anisotropies, but the additional effort may not greatly influence the outcomes for facet-dominated growth. Most important at this point is simply to make sure that the surface boundary condition in Equation 5.34 reproduces the correct attachment physics on the facet surfaces with the highest possible precision.

Growth Steps

As with the 2D model described above, there are additional geometrical factors inherent in how we define a growth step. We begin with a generic functional form for the time required to "fill" the remainder of each boundary pixel, generalizing Equation 5.18 to

$$\delta t_b = \frac{H_b \Delta x}{v_n}(1 - f_b), \tag{5.36}$$

where f_b is the filling factor described above (equal to 0 when a boundary pixel first appears and 1 when a boundary pixel turns to ice),

$$v_n = \alpha v_{kin}\, \sigma_{surf} = \alpha(\sigma_b) v_{kin}\, \sigma_b \tag{5.37}$$

is the growth velocity normal to the ice surface, and H_b is a dimensionless geometrical factor near unity.

As with Equation 5.34, $H_b = 1$ for both [20] (prism facet) and [10] (basal facet) boundary pixels. This accurately describes facet growth partly because of our choice of Δx in Figure 5.26, along with $\Delta z = \Delta x$, and partly because the surface normals lie along the grid axes for both primary facets. As with the G_b factors, determining H_b for nonfaceted boundary pixels is difficult to do with high accuracy, as the correct surface normal is not easily determined in a CA model. I described this same problem for the 2D model, where the facet-vicinal model appears to achieve reasonably accurate geometrical factors for all flat vicinal surfaces, thus substantially reducing the built-in anisotropies in the CA model.

Extending the facet-vicinal ideas to a 3D model would be challenging, as one would have to deal with quite a bit of additional geometrical complexity. I can imagine various numerical strategies for calculating approximate G_b and H_b using the $\pm L_i$ data, but nothing that would be both simple and accurate. At this stage in our understanding of 3D CA snow crystal models, I suspect that it is premature to worry too much about the intrinsic anisotropies and geometrical precision associated with these geometrical factors. Perhaps simply using $G_b = H_b = 1$ for all boundary pixels is good enough to produce reasonable model results. However, it is valuable to recognize that these geometrical factors exist and that errors in their calculation result in intrinsic anisotropies in the model. Understanding the extent and importance of this problem is left for another day.

Attachment Coefficients on Faceted Surfaces

I believe that the most important challenge right now is to incorporate accurate attachment coefficients in snow crystal models, particularly on the primary facet surfaces. This will be something of an iterative scientific process, as detailed 3D modeling will help us better understand the attachment kinetics, which, in turn, will allow us to build better models, and so on.

As described in Chapter 4, targeted experiments measuring α_{basal} and α_{prism} on broad facets are necessary but not sufficient, as the attachment kinetics appear to depend on supersaturation, temperature, terrace width, and perhaps other factors yet unknown. Growing and modeling fully 3D structures at different background pressures will likely be needed to fully comprehend what is going on with the attachment kinetics. And we are only just beginning to execute this scientific program in earnest.

Given that our understanding is quite limited at present, a good first step is to parameterize α_{basal} and α_{prism} with $\alpha_{facet} = A\exp(-\sigma_0/\sigma_{surf})$, where A and σ_0 are model parameters that may depend on a variety of other factors. This functional form clearly applies to large-facet growth, and it is likely a good approximation for narrow facets as well. However, both A and σ_0 may depend on facet width to some degree, as described in detail in Chapter 4. Whatever the form chosen, the faceted α_{basal} and α_{prism} apply only to upper-terrace boundary pixels or to those far removed from any kink sites that can dramatically increase α via surface diffusion. In terms of the $\pm L_i$ data, all these parameters must be greater than zero, or the smallest negative L_i must be greater than some prescribed limit (which is a parameter in the model). It is an unfortunate reality in snow crystal modeling that we do not yet fully understand attachment kinetics, although the CAK model (see Chapter 4) can be used as a reasonable approximation.

Upper Terrace Effects

Sharp edges are somewhat common features in snow crystal growth, and it appears that these require some special attention in CA models. The edges of thin plates are especially prominent near −15°C, yielding narrow prism facets, but narrow basal facets can also be found on the edges of sheathlike hollow columns. In both cases, the radii of curvature of the edge surfaces appear to be few times X_0 in extreme cases, although the dimensions are not known with great precision.

We have regularly observed thin plates in normal air with overall thicknesses down to 1 micron using accurate interferometric measurements [2008Lib1, 2009Lib], and this thickness puts an upper limit on the edge curvature for these crystals. Sheathlike edges appear to be a few times thicker, but I am not aware of any accurate measurements for these crystals. In a curvature class all their own are the electric ice needles (see Chapter 8), where the tips of c-axis needles have exhibited radii of curvature down to 100 nm in the most extreme cases [2002Lib]. Incorporating electric-field effects in CA models is beyond the scope of this chapter, but suffice it to say that even 100 nm surface curvatures are not impossible to realize in snow crystal growth experiments.

There are two primary physical effects that one should consider when modeling these high surface curvature features: the Gibbs-Thomson effect (Chapter 2) and the SDAK effect (Chapter 4). The former is a well-understood phenomenon but appears to play a relatively small role in snow crystal growth. The latter appears to play a substantial role in the formation of thin plates near −15°C and hollow columns near −5°C. I believe that both these physical phenomena can be incorporated into a CA model, to a reasonable approximation anyway, by considering only the upper terraces of faceted surfaces. The upper terraces are easily identified for a given boundary pixel by requiring that all the L_i be positive, as described above. Moreover, the curvature of the surface at a given point can be estimated from the L_i data as well; for example, by the sum of opposing L_i. The precise algorithm is probably not especially important, and one can imagine different approaches. What likely does matter is consistently identifying large arrays of high-curvature surfaces, for example, the edges of thin plates that remain thin as the crystal grows.

In the SDAK effect, a high-curvature basal or prism surface would exhibit an attachment coefficient that is far larger than the usual α_{prism} associated with broad prism facets (see Chapter 4). This effect can be incorporated into the model by letting the facet attachment coefficients depend on the upper facet width, as extracted from the L_i boundary pixel attributes. This factors into $\alpha(\sigma_b)$ in Equation 5.34, and we see that $\alpha(\sigma_b)$ changes automatically with every update of the boundary pixel attributes. The same goes for the growth-step calculation in Equations 5.36 and 5.37.

For the Gibbs-Thomson effect, the surface curvature need only be used to adjust the growth step, so Equation 5.37 is replaced by the general form

$$v_n = \alpha(\sigma_b) v_{kin} (\sigma_b - d_{sv}\kappa) \tag{5.38}$$

where κ is again estimated from the L_i boundary pixel attributes. Because the Gibbs-Thomson effect is not strong in snow crystal growth, it is probably not necessary to provide an extremely accurate algorithm for estimating κ. Likewise, it is probably not necessary to include the Gibbs-Thomson effect when solving Laplace's equation to determine the supersaturation field around the growing crystal, as the perturbation arising from surface curvature is so small.

Throughout this discussion, we see that the key element for including the Gibbs-Thomson effect and investigating the SDAK phenomenon in a 3D CA model is the initial step of generating boundary pixel attributes that include nonlocal information via the $\pm L_i$ data. Only nearest-neighbor considerations were used in previous

3D CA models [2009Gra, 2014Kel], but it is becoming clear (in my opinion) that nonlocal surface structure will be absolutely necessary before CA models will be able to match experimental observations of snow crystal growth rates and morphologies.

Vicinal Surfaces and the FSD Approximation

While nonlocal information can be important on upper-terrace faceted surfaces, it is also relevant on vicinal surfaces, which include essentially all non-upper-terrace surfaces in the CA model. The basic idea here is that admolecules on normal terraces (not upper terraces) can diffuse along the surface until they reach kinks sites, where they are readily absorbed. Thus, surface diffusion increases the attachment kinetics to $\alpha \approx 1$ for surface locations closer than x_{diff}, the surface diffusion length, to the nearest kink sites. Moreover, as discussed above in connection with Figure 5.20, the important parameter for CA modeling is not x_{surf}, but x_{surf}/a, as surface diffusion effectively operates over x_{surf}/a cells in the CA model. So in a practical sense, $\alpha \approx 1$ should be assumed on most non-upper-terrace surfaces in the model. Upper-terrace surfaces are described by α_{facet}, as are surfaces where $|L_i| \gg x_{surf}/a$ for all negative L_i. But all other surfaces are best described by $\alpha \approx 1$. This is the essence of the FSD approximation described above. Because essentially all nonfaceted surfaces have $\alpha \approx 1$, the overall growth dynamics are largely determined by the behavior of the faceted surfaces.

Incorporating surface diffusion is relatively straightforward using the $\pm L_i$ boundary pixel attributes. If the distance to the nearest kink site is less than some prescribed value L_{sd} in the model, then one simply sets $\alpha \approx 1$ for that boundary pixel and proceeds. It is likely, given our current knowledge of surface diffusion, that this method will set $\alpha \approx 1$ over much of the surface for all but the most cleanly faceted crystals. Nevertheless, the existing measurements of x_{surf} are quite poor, and the value could be quite different on the basal and prism facets. As a result, surface diffusion is perhaps best left as two adjustable parameters (one for each facet) at present.

Note that even a leaky Ehrlich-Schwoebel barrier does not mean that $\alpha \approx 1$ on upper-terrace surfaces. Moreover, the x_{surf}/a logic described above does not apply to upper terrace surfaces like it does to other surfaces in CA models. In the case of a leaky Ehrlich-Schwoebel barrier (Chapter 4), admolecules near the edges of upper terraces could diffuse to an edge and be absorbed by a lower kink site. Physically, however, this only happens within a distance x_{surf} from the edge, meaning it does not affect the entirety of a large facet area. It is reasonable to assume, therefore, that α_{facet} on an upper terrace in our CA model is also negligibly affected by a leaky Ehrlich-Schwoebel barrier.

Quantitative Modeling

It is a telling statement that no 3D snow crystal models to date have ever been directly compared with quantitative experimental data. There is an opportunity here, as the CA techniques described in this chapter are beginning to yield realistic faceted + branched structures, and the corresponding snow crystal forms can now be grown reliably on the ends of electric ice needles (see Chapter 8). These are both quite recent developments, and many new insights may be forthcoming when they are brought together for detailed comparisons.

Our understanding of the underlying physics is incomplete at present, but that is what makes the scientific challenge especially interesting. The best way to proceed is likely just to plunge forward with both modeling and quantitative growth studies, comparing one with the other to see what works and what does not. A complex menagerie of morphological behaviors awaits explanation. This appears to be a fruitful path to follow in the ongoing quest to improve our understanding the physical dynamics of snow crystal growth.

FIGURE 6.1. This synthetic snow crystal measures about 3 mm from tip to tip and was grown by the author using the Plate-on-Pedestal (PoP) technique described in Chapter 9. The edges and faceted features are noticeably sharper than can be found in natural snow crystals, because this laboratory specimen was photographed as it was growing.

SIX
Laboratory Snow Crystals

A scientist does not study nature because it is useful; he studies it because he delights in it, and he delights in it because it is beautiful.

—HENRI POINCARÉ, *SCIENCE AND METHOD*, 1908

Creating synthetic snow crystals is an essential activity in our quest to understand the physical dynamics of ice growth. Laboratory investigations reveal how morphologies change with temperature, supersaturation, air pressure, and other parameters, while careful growth measurements allow us to develop models and test quantitative hypotheses. Comparing experimental data with theoretical predictions is the time-honored scientific method that allows us to gradually comprehend the underlying molecular processes that cause formless water vapor to spontaneously consolidate into intricately patterned snow crystals.

Beginning with Nakaya's pioneering work creating the first laboratory snowflakes, researchers have developed a variety of strategies for nucleating, preparing, imaging, and measuring ice crystals under a range of environmental conditions. In some cases, including the example shown in Figure 6.1, the primary motivation is an artistic one, and I describe how to grow specimens like this one in Chapter 9. More commonly, however, snow crystals are produced for specific scientific objectives, and I describe some detailed techniques along those lines in Chapters 7 and 8. Regardless of the underlying goals, it is useful to possess a toolbox of practical experimental skills that can be employed to create quality specimens. While the scientific literature includes a considerable list of hardware approaches, I survey here some broadly applicable techniques that I have found especially useful when studying the physical dynamics of snow crystal growth.

FREE-FALL SNOW CRYSTALS

One of the simplest ways to grow synthetic snow crystals (not the same as artificial snow, as seen in Figure 6.2) is by letting them fall through the air as they grow, essentially imitating the formation of atmospheric snowflakes. Of course, a laboratory growth chamber is vastly smaller than a winter cloud, so we can expect that freely falling synthetic snow crystals will be substantially smaller than the natural variety. But the size difference is not as great as one might naively expect, which we can show from an understanding

FIGURE 6.2. Artificial snow at ski resorts (left) is made from rapidly freezing liquid water droplets, which produces small globules of ice (right). Synthetic snow crystals, such as the example in Figure 6.1, are made from freezing water vapor, which can yield the same elaborate morphological structures seen in natural snow crystals.

of snow crystal growth and aerodynamics. From Chapter 3, the terminal velocity of a small spherical ice crystal of radius R is

$$u_{term} \approx \frac{2}{9} \frac{\rho_{ice} g}{\mu} R^2, \tag{6.1}$$

and assuming that the growth of the sphere is purely diffusion limited at a pressure of 1 atmosphere gives the growth velocity

$$v \approx \alpha_{diff} v_{kin} \sigma_\infty \approx \frac{X_0}{R} v_{kin} \sigma_\infty . \tag{6.2}$$

Integrating this expression gives the crystal radius as a function of time:

$$R \approx \sqrt{2 X_0 v_{kin} \sigma_\infty t}. \tag{6.3}$$

Using typical values of $v_{kin} \approx 300$ µm/sec and $\sigma_\infty \approx 0.01$, integrating the terminal velocity over time gives a fall distance of $h \approx 1$ meter after a fall time of $T \approx 140$ seconds, at which point the crystal radius is $R \approx 11$ µm. Importantly, these results scale as $R \sim h^{1/4}$ and $T \sim h^{1/2}$, so crystals falling 1 meter will only be about six times smaller than those falling a kilometer. The details will depend on the attachment kinetics, crystal morphology, and other factors, and thin plates will fall slower and grow larger than spheres. Thus, even with a modest fall distance in a laboratory free-fall chamber, crystal diameters of up to 100 microns are readily achievable in a relatively short time.

Freezer Snowflakes

A basic top-loading household freezer is perhaps the least expensive means of creating freely falling synthetic snow crystals. The set-up illustrated in Figure 6.3 was first conceived by Vincent Schaefer and colleagues in the 1940s as an easy demonstration of basic cloud physics [1981Sch]. Opening the top of the freezer and simply breathing down into it produces a visible cloud of water droplets, as water vapor from your breath condenses on dust particles in the cold air. The cold air sinks stably into the freezer, and the fog will float inside for many minutes before slowly turning into frost on the freezer walls. The supercooled liquid water droplets supersaturate the air in this environment, making an excellent nursery for growing snow crystals.

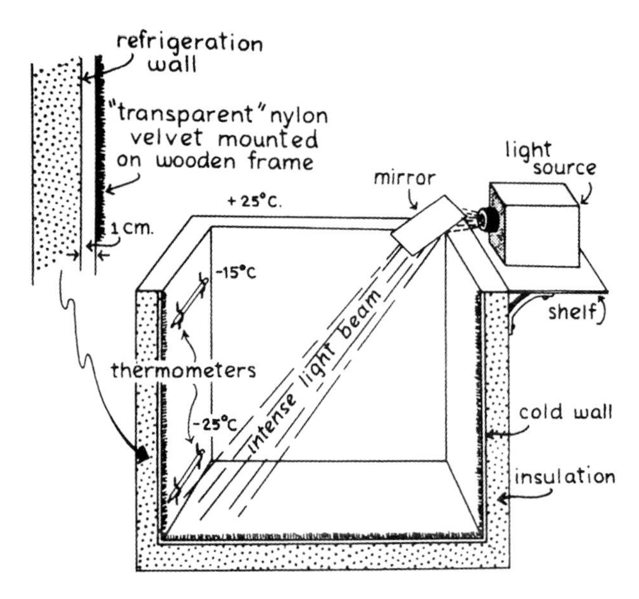

FIGURE 6.3. A top-loading household freezer can be used to demonstrate several meteorological processes involved in the formation of snow crystals. Image adapted from [1981Sch].

The next step in making freezer snowflakes is to provide a nucleation event that will stimulate the production of some microscopic ice particles. This step is necessary, because supercooled liquid droplets can remain unfrozen for many hours at −15°C, which is a typical freezer temperature. One effective nucleation method is to drop a small fleck of dry ice into the floating freezer cloud, because dry ice sublimates at a temperature of −78°C. The frigid particle produces an extremely high supersaturation in its vicinity that leaves a trail of nucleated ice crystals in its wake as it falls through the freezer. By repeatedly breathing into the freezer and dropping in flecks of dry ice, one can create a veritable flurry of tiny snow crystals. Popping individual bubble-pack cells in a syringe can nucleate crystals as well [1976Rya], although not as effectively as dry ice. Amusingly, Schaefer also described mixing iodine vapor with lead pollution from car exhaust to form lead-iodide ice nucleators [1981Sch], but that technique is no longer an option in a world of unleaded gasoline. Particles of silver-iodide smoke are also good ice nucleators [1982Gon, 1994Gon], and the expansion nucleator described below provides an especially simple and effective technique.

Household freezers are usually set to about −15°C, because this is cold enough to reliably freeze liquid water after an hour or two, but this temperature is also ideal for growing thin, hexagonal snow crystal plates. By shining a bright flashlight or laser pointer into a cloud of freezer snowflakes, sparkling reflections from the flat basal facets can be observed quite easily. Covering the walls of the freezer with dark cloth greatly enhances the view. A cloud of water droplets looks like a rather dull gray fog, but a cloud of platelike crystals in front of a black background makes a beautiful swirl of sparkling diamond dust.

The tiny freezer snowflakes are generally too small to resolve with the naked eye, so making out their morphology requires a microscope. As described in the calculation above, crystal sizes in a free-fall chamber are typically some tens of microns, which is smaller than the diameter of a human hair, so quite high magnification is needed just to see their overall hexagonal form. Moreover, the viewing is best at the bottom of the freezer, where the crystals fall, so a household freezer is perhaps not the ideal scientific tool for making detailed snow crystal observations.

Expansion Nucleator

An excellent general-purpose method for initiating ice crystal growth is the expansion nucleator shown in Figure 6.4. In this device, pressurized air fills the nucleator body via an air hose and a flow restrictor, supplied by a commercial oil-free air compressor. The nucleator body is typically 5–10 cm in length, the air pressure is about 15–30 psi, and I usually pass the compressed air through a column of activated charcoal grains to absorb any remaining chemical impurities. With the solenoid valve closed, pressurized air fills the nucleator in a few seconds. Abruptly opening the solenoid valve allows the compressed air to expand rapidly into the surrounding medium. The dynamics of the expansion is impossible to

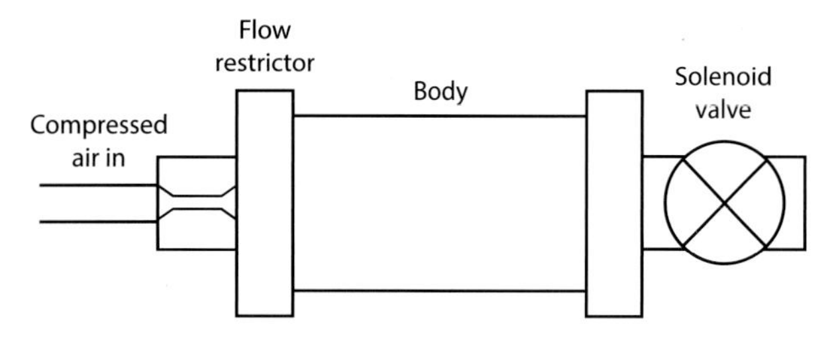

FIGURE 6.4. This expansion nucleator is a simple apparatus for creating small ice crystals. When the valve is pulsed opened, the rapid expansion nucleates numerous tiny crystals.

calculate precisely, as the flow is turbulent, nonadiabatic, and generally quite ill defined. Nevertheless, the rapid expansion cools the air, thus greatly increasing the supersaturation at localized positions in the outflow, which is sufficient to nucleate the formation of numerous tiny ice crystals.

The expansion nucleator relies on the fact that normal room air invariably contains a good amount of water vapor along with an ample supply of dust particles that act as ice nucleation sites. It is only necessary to increase the air pressure until ice crystals appear. A higher body temperature requires more cooling of the air, and thus higher air pressures, so I usually keep the nucleator at about −15°C. At this temperature, no crystals form if the pressure is much below 10 psi, while 30 psi is generally high enough to form some hundreds or thousands of minute crystals each time the valve is pulsed open.

The expansion nucleator creates ice particles through a process essentially like that occurring in a winter cloud, except much faster. The rapid expansion cools the air and causes water droplets to condense on suspended dust particles, which is basically how clouds form in the atmosphere. As in a cloud, the first nucleation step produces liquid water droplets rather than ice crystals, in accordance with Ostwald's step rule. But the tiny droplets only last a fraction of a second in the nucleator pulse, as continued expansion provides additional cooling, causing some of the droplets to freeze. The nascent ice particles quickly absorb water vapor around them, causing any remaining water droplets to evaporate (see Chapter 1). Thus, the process could be described as a miniature, high-speed snowfall. Each discharge of the nucleator results in a puff of microscopic crystals.

Note that the expansion nucleator in Figure 6.4 loses its effectiveness when the air discharges into a low-pressure environment. Before the valve opens, the water vapor density inside the nucleator body is roughly equal to c_{sat} at the body temperature. After expansion, the air fills a much larger volume at low pressure, thus reducing the water vapor density by roughly the expansion ratio. If the reduced vapor density is below c_{sat} at the ambient temperature inside the main growth chamber, then any nucleated crystals will quickly sublimate away. This dilution effect can be countered by using much higher pressures, a task that is best accomplished using a supersonic *de Laval* nozzle [1995Hua].

Convection Chambers

A relatively simple method for creating free-fall snow crystals is the convection chamber illustrated in Figure 6.5, which has become something of a workhorse device in my lab. In the chamber, a heated water reservoir provides a source of water vapor via evaporation, simultaneously driving convection that carries the water vapor upward. Turbulent convection mixes the air to yield a steady-state supersaturated environment that is reasonably uniform, although not as homogeneous as in a droplet-filled cloud chamber. An expansion nucleator generates seed crystals on demand, at a known time, and these crystals grow as they slowly descend in the supersaturated air. Typical fall times τ_{fall} are about 1–5 minutes, depending on supersaturation, at which point crystals begin to appear on the substrate for observation. By about $2\tau_{fall}$, most of the crystals have settled to the bottom of the chamber.

Increasing the temperature of the water reservoir increases the supersaturation in the chamber, giving the convection chamber added flexibility compared to a

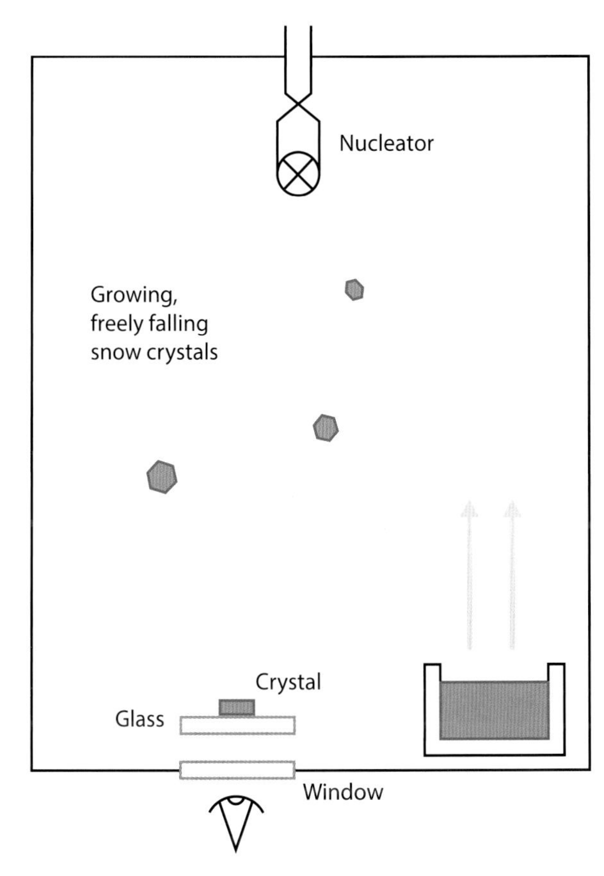

FIGURE 6.5. A free-fall convection chamber. The expansion nucleator near the top of the chamber produces microscopic seed crystals that subsequently grow and slowly fall in the supersaturated air. After a few minutes of growth, the crystals fall to the bottom of the chamber, where some of them land on a glass substrate for observation.

cloud chamber. However, while it is straightforward to produce supersaturations that range from zero to σ_{water}, measuring the precise value of the supersaturation is a challenging task. We presented a calibrated method using differential hygrometry measurements of air sampled from the chamber [2008Lib2], but the potential for systematic errors with this technique is significant. Another issue is that the growing crystals remove water vapor from the air, and its rate of replenishment from the water reservoir is not well known. Despite these problems, convection chambers can provide reasonably quantitative data over a broad range of temperatures and supersaturations in air [2008Lib1,2009Lib]. Figure 6.6 shows some photographic examples of snow crystals grown in a convection chamber at supersaturations close to σ_{water}, and Figures 6.7 and 6.8 show measurements of growth rates. The different experiments show generally good agreement with one another, but the large diffusion corrections in air at one bar make it difficult to determine σ_{surf} from σ_∞.

These measurements illustrate that a free-fall chamber is a simple and effective apparatus for growing small, simple snow crystals under reasonably controlled conditions. Growth times are limited, but many crystals can be observed from a single nucleation pulse. The growth morphologies are not encumbered by supporting substrates or filaments, and no complex levitation strategies are needed. While not ideal for every experimental goal, a free-fall chamber can be quite a useful tool when growing synthetic snow crystals.

A Seed-Crystal Generator

One of my favorite applications of the free-fall convection chamber is to produce an on-demand supply of copious ice seed crystals. The basic idea is a chamber like that shown in Figure 6.5 but perhaps smaller and without the observation hardware. By opening the nucleator valve about once every 10 seconds during operation, a continuous cloud of fresh ice crystals can be found floating in the chamber. The small crystals grow quickly at first, while larger crystals settle to the floor of the chamber. These two effects combine to yield a seed-crystal size distribution inside the chamber that typically peaks at around 20–50 microns. As seen in Figure 6.6, thin plates or slender columns can be produced, depending on the chamber temperature. Crystals with simple, faceted prism morphologies are readily produced if the supersaturation is kept substantially lower than σ_{water}. The experiments described in Chapters 7 and 9 began with seed crystals created using this method.

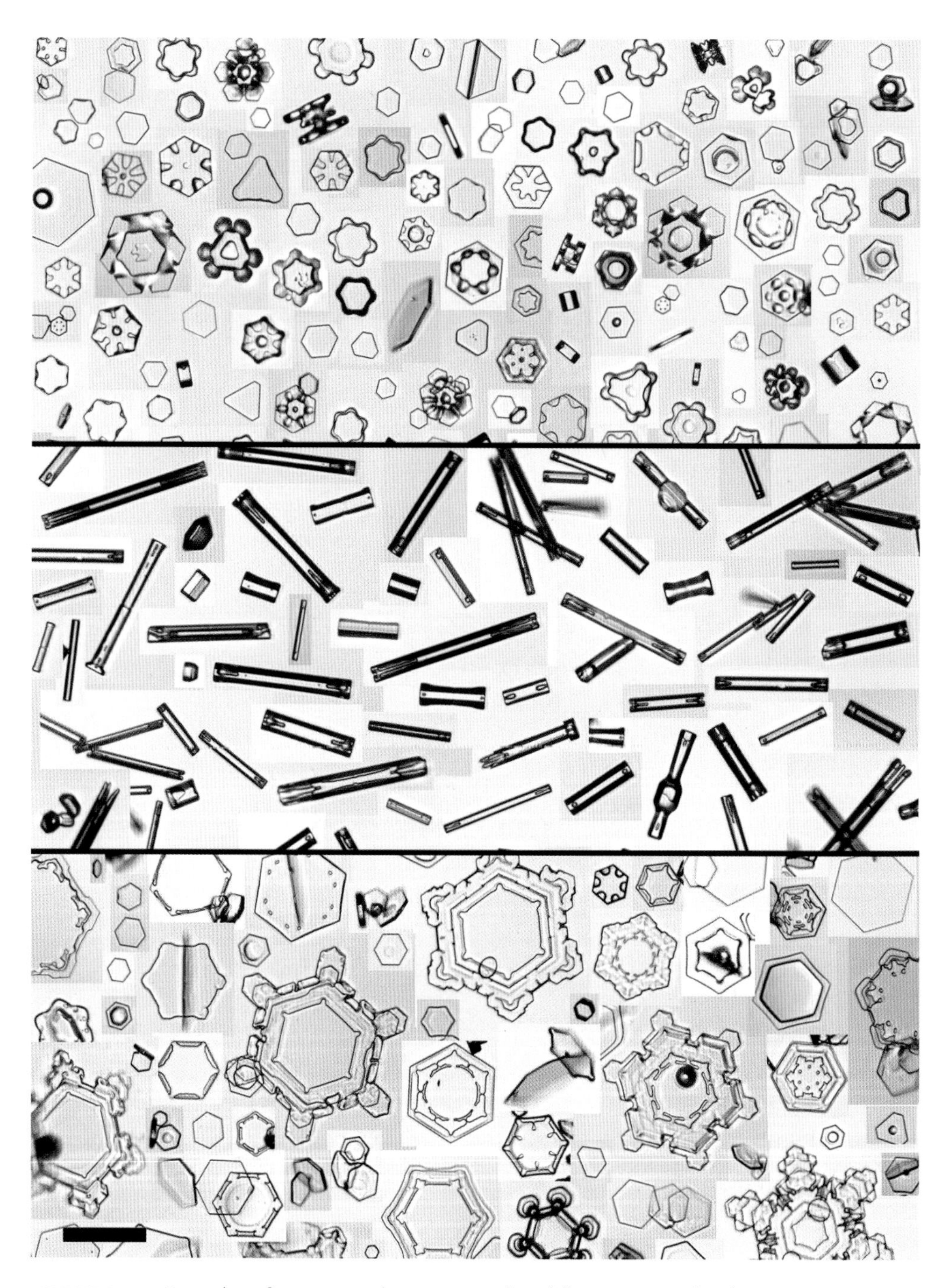

FIGURE 6.6. Examples of snow crystals grown in a free-fall convection chamber at temperatures of −2°C (top), −5°C (middle), and −15°C (bottom). The scale bar in the lower left corner of the image is 50 microns in length. The variation in crystal size and morphology at each temperature reflects inhomogeneities in temperature and supersaturation in the chamber. Overall, however, the morphologies are consistent with expectations from the Nakaya diagram.

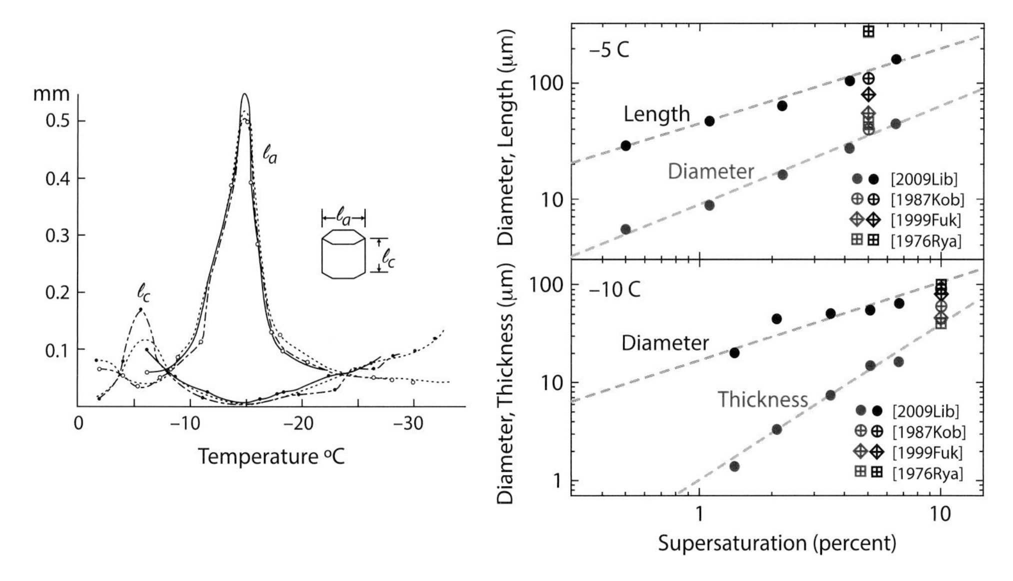

FIGURE 6.7. (Left) Measurements of the diameters and thicknesses of snow crystals after growing in air for 200 seconds in a free-fall cloud chamber. Note that the variation in crystal aspect ratio as a function of temperature matches expectations from the Nakaya diagram. Image adapted from [1987Kob]. (Right) Similar measurements at −5°C and −10°C as a function of background supersaturation σ_∞.

In typical operation, the seed-crystal generator operates continuously and is located adjacent to a separate experiment. When some seed crystals are needed, a bit of air is drawn out of the free-fall chamber and wafted over a waiting cold substrate. A few seed crystals drop onto the substrate, which can then be examined to select a suitable test crystal for subsequent observations. When a free-fall seed-crystal chamber is set up with a temperature near −15°C and modest supersaturation levels, roughly half of the crystals are well-formed simple hexagonal plates, the remainder being malformed to some degree, perhaps from mid-air collisions, polycrystalline nucleation, or other factors.

A highly desirable feature of this seed-crystal generator is its intrinsic self-cleaning nature. Ice crystals are created in large numbers, and they typically fall to the bottom of the chamber in just a few minutes. The growing crystals, both in the chamber and covering the walls, absorb chemical contaminants from the chamber air, thus rapidly and continuously cleaning it. In contrast, seed crystals that sit for long periods, require a lot of handing, or experience sublimation/regrowth cycling are apt to become contaminated with surface impurities. I believe that chemical contamination effects have been, and continue to be, underestimated in many snow crystal experiments, and beginning with fresh, clean seed crystals is a good first step in reducing potential contamination effects.

SUBSTRATE SUPPORT

Dropping a seed crystal onto a fixed, flat substrate and watching its subsequent development is a popular method for studying snow crystal formation. Sapphire is often the substrate material of choice, as it is transparent and has a high heat conductivity. The fact that sapphire is highly resistant to scratching is a big plus as well. Some advantages of growing snow crystals on substrates include:

1) Individual crystals can be observed for long periods of time.

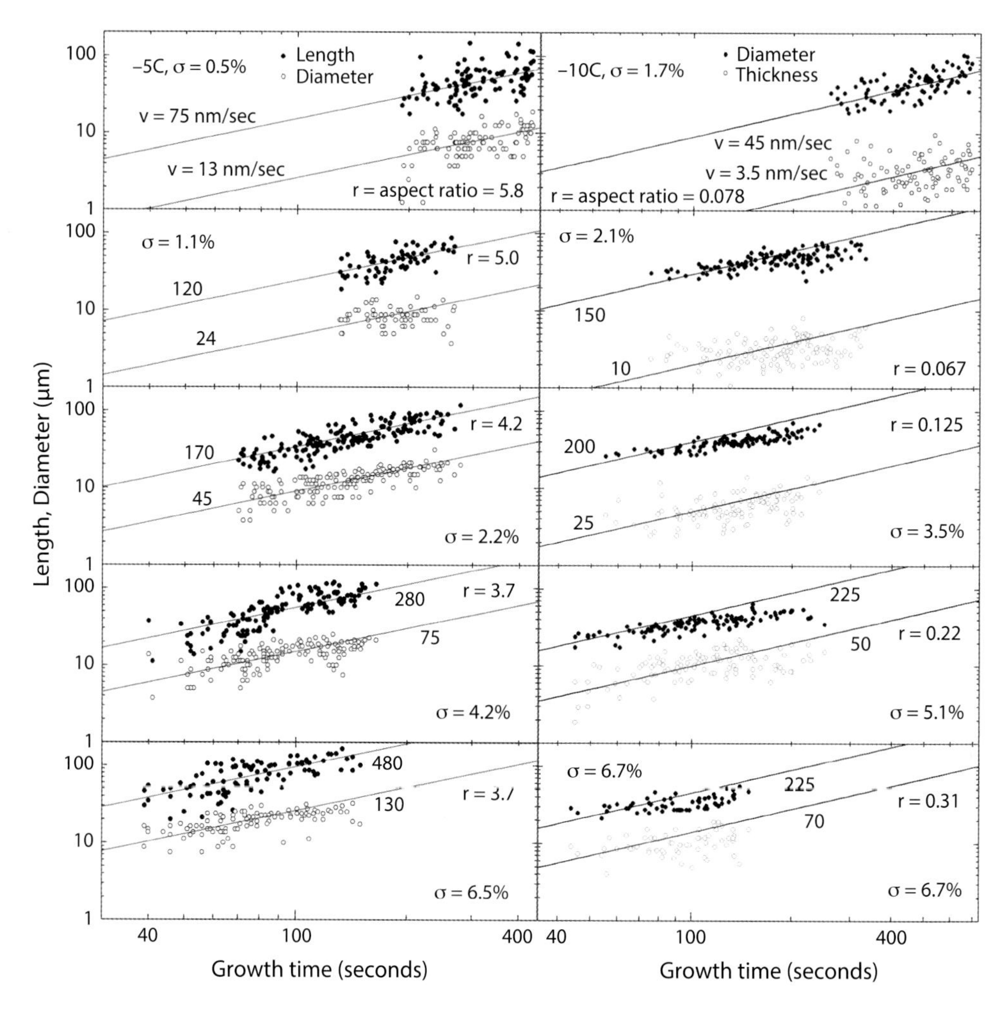

FIGURE 6.8. Free-fall growth data (from [2009Lib]), showing crystal sizes as a function of fall times for several different values of σ_∞, obtained at growth temperatures of −5°C (left) and −10°C (right). Each pair of points (length and diameter) represent one observed crystal. Lines show constant-velocity trajectories, and the aspect ratio is the ratio of these velocities. Note that the aspect ratio tends toward unity with increasing supersaturation.

2) The stationary crystals can be measured using high-resolution microscopy or interferometry, allowing precise measurements of sizes and growth velocities.
3) Supersaturations can be accurately modeled with a plane-parallel ice reservoir above a fixed substrate (see Chapter 7).
4) Crystals experience good thermal coupling to the substrate, greatly reducing heating effects that may be problematic in free-fall or levitation experiments.

However, substrates have some significant disadvantages as well, including:

1) Water droplets readily condense on substrate surfaces, generally limiting supersaturations to σ_{water} and below.
2) The additional boundary condition at the substrate surface can affect the ice growth behavior.
3) Heterogeneous nucleation at the ice/substrate interface can strongly affect the growth of faceted surfaces that intersect the substrate.

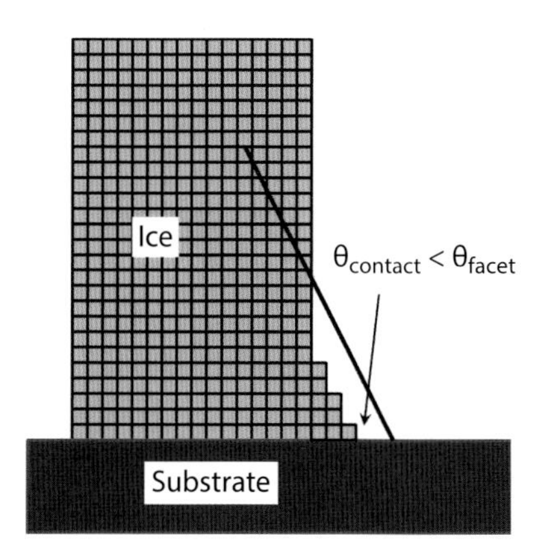

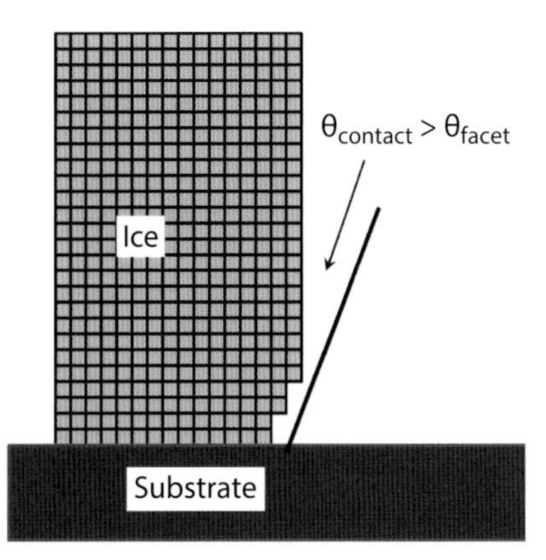

FIGURE 6.9. If the ice/substrate contact angle is less than the facet/substrate angle (left sketch), then new terraces will be nucleated at the ice/substrate contact line. With a larger contact angle (right sketch), new terrace nucleation can only occur on the free faceted surface, away from the contact region. The enhanced terrace nucleation in the first case can greatly increase the growth rates of facets contacting the substrate.

Regarding this last point, Figure 6.9 illustrates how substrate interactions can result in the nucleation of new terraces on faceted ice surfaces. The ice/substrate contact angle is like the often-discussed water/substrate contact angle, and both depend on how well water molecules bond to the substrate surface. Enhanced ice growth from this mechanism can be quite pronounced on hydrophilic surfaces (see below). The ice/substrate contact angle seems to be generally larger than with water, usually increasing somewhat at lower temperatures. Sapphire and most other surfaces are slightly *hydrophilic*, with $\theta_{contact} < 90$ degrees, and the contact angle is usually smaller if the surface is not clean. However, surface coatings can often be applied to yield robust *hydrophobic* surfaces with $\theta_{contact} > 90$ degrees. Creating robust *superhydrophobic* coatings with $\theta_{contact} > 150$ degrees is an active area of research. One concern with these advanced coatings, however, is they can be quite thick and have low thermal conductivities. This may reduce the thermal coupling to the substrate to the point that latent heating can significantly reduce the measured growth rates (see Chapter 3).

The advantages of using substrates generally outweigh the disadvantages to the point that most ice growth investigations have been conducted on a substrate of some kind. I discuss two such experiments involving the growth of simple ice prisms in Chapter 7, where precise growth rates were measured under well-controlled conditions for investigating the attachment kinetics. Chapter 9 presents another apparatus used for the artistic creation of large stellar crystals, again growing crystals on a sapphire substrate.

Thin Filaments

Filamentary support of snow crystals has been popular ever since Nakaya's first experiments using rabbit hair [1954Nak], and the current state-of-the-art has been the development of thin capillary tubes for this purpose [1996Nel]. Heating and drawing capillaries with tip diameters down to 10 μm has been a staple of laboratories for many decades, and suitable commercial micropipettes have recently become available. Creating an isolated ice crystal at the tip can be accomplished by freezing water in the capillary from below [1996Nel] or by using ice transfer from the top, as shown in Figure 6.10.

Filament support reduces some of the problems associated with using substrates, but not entirely. Substrate interactions can still cause unwanted heterogeneous nucleation, and droplet condensation on the filament will still be present at high supersaturations. The persistence of these substrate-related issues plus low experimental throughput in creating crystals have made capillary support generally less popular compared to other methods.

Negative Snow Crystals

Pulling a vacuum through a thin capillary tube embedded in a block of single-crystal ice can yield "negative" snow crystals like the one shown in Figure 6.11. Water

FIGURE 6.10. Growing an oriented snow crystal on a capillary tube with a 20-μm diameter. In the upper left image, a water-filled capillary is brought up to the basal surface of an ice crystal, freezing the water with the ice c-axis (nearly) collinear with the capillary axis. With the ice plate removed, the remaining images show subsequent development of the crystal in a diffusion chamber. The stellar snow crystal in the lower image measures 3 mm from tip to tip, but this crystal eventually grew to 12 mm from tip to tip. This capillary technique is well suited for growing large snow crystals at high supersaturations. Photos by the author.

vapor is extracted through the capillary, leaving behind a void in the ice. Facets appear during this process, because it is especially difficult to remove water molecules from a perfectly faceted surface, as each is tightly bound by neighbors on all sides. It is comparatively much easier to remove molecules from a terrace step, as step-edge molecules have fewer nearest neighbors and are thus less tightly bound. Beginning with an arbitrarily shaped void, molecules are preferentially removed from terrace edges, eventually leaving behind a faceted void in the shape of a hexagonal prism.

Note that the faceted shape in this image does not arise from surface energy effects, and the minimum energy shape is likely nearly spherical (see Chapter 2). As with snow crystal growth, the formation of a prismatic void is a dynamical effect resulting from sublimation kinetics, which are related to attachment kinetics. Moreover, the time needed for this prismatic void to relax to its equilibrium shape can be exceedingly long. If the capillary were somehow extracted to leave behind a clean faceted void, then relaxation to a spherical shape would mean removing molecules from the facet surfaces and depositing them in the corners. This process is strongly suppressed by the nucleation barrier, greatly increasing the relaxation time.

The formation of negative snow crystals is also affected by thermal diffusion, resulting in the peculiar protruding shapes seen in the void in Figure 6.11. Beginning with a perfectly faceted void, removing material via sublimation cools the ice, and the extracted heat must be replaced by thermal diffusion from the surrounding medium. The prism corners, sticking out farther into the ice, are more efficiently heated by diffusion, so the corners sublimate more quickly than the facet centers do.

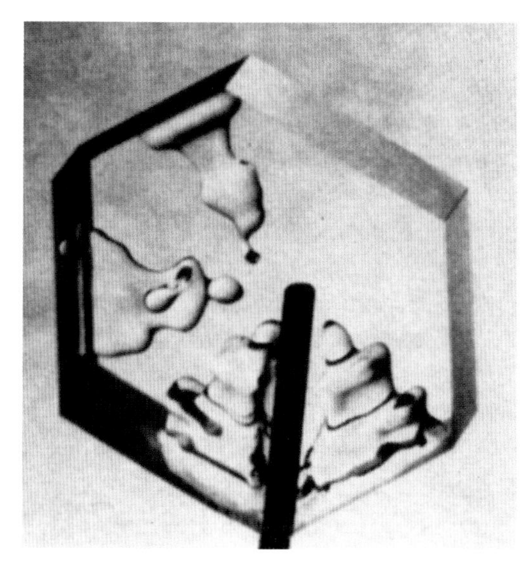

FIGURE 6.11. A "negative" snow crystal growing at −14°C in a block of single-crystal ice, as a vacuum pump removes water vapor via a 0.45-mm-diameter capillary tube. The morphology of the void is determined from thermal diffusion effects together with a strong nucleation barrier in the sublimation kinetics. As with normal snow crystal growth, surface energy effects are negligible compared to anisotropic sublimation kinetics. Image adapted from [1965Kni].

As this process continues and the void grows larger, ice protrusions extend from the facet centers into the void. This is a "negative" version of the Mullins-Sekerka instability applying to thermal-diffusion-limited sublimation (see Chapter 3).

Negative snow crystals have received relatively little study [1965Kni, 1993Fur], in part because they are somewhat difficult to grow under well-defined environmental conditions. Determining the undersaturation with useful accuracy is challenging, and negative crystals tend to be substantially larger than normal snow crystals. Quantitative growth measurements of negative crystals are subject the same kinds of systematic errors discussed in Chapter 7, and these errors are generally smaller with normal snow crystals. Nevertheless, one can imagine an array of microscopic holes in a substrate contacting a block of ice, allowing one to perform hundreds of simultaneous sublimation experiments. Nothing of this sort has been attempted to date, but it presents another avenue that might be explored someday.

Oriented Ice Crystals

It is often desirable to create large, single-crystal specimens of bulk ice in the lab for a variety of functions, and how one obtains such samples is not immediately obvious. A century ago, researchers would "mine" the Mendenhall glacier in Alaska for this purpose, as high-purity single-crystal ice specimens could be found there. Fortunately, a remarkably easy method for creating large single-crystal specimens in a top-loading household freezer was described by Knight [1996Kni], obviating the need for polar ice mining.

In Knight's method, one simply fills an open, insulated container with water, places it in a freezer, and waits. Water vapor first evaporates from the water's surface and deposits a bit of frost on the lid of the freezer above the water container. Over time, the water cools to below freezing, and at some point (if all goes well) a single crystal of frost will break off from the lid and fall into the water. If the water temperature is just below freezing, the ice seed will grow out as a thin disk crystal. As it floats on the surface of the water, buoyancy forces automatically orient the growing disk so that its c-axis points in the vertical direction as the edges of the disk grow outward. Soon the ice disk expands and covers the surface, preventing nucleation by subsequent falling frost crystals. Because the container is insulated, the rest of the water freezes slowly, its thickness increasing at a rate of about 1 cm per day.

This experimental procedure is not particularly well controlled, but usually there will be some sections of single-crystal ice on the surface, with the crystalline c-axis oriented vertically. Viewing the sheet between crossed polarizers verifies that no domain boundaries are present. This technique could be better developed, but I can verify that it works surprisingly well with little effort. Ice samples with higher purity, or with controlled chemical doping, can be grown using the

Czochralski method [2017Bru, 1973Bil], but this method requires a substantially greater investment in effort and equipment. Etch pits can also be used to determine the lattice orientation of single-crystal samples [2017Bru].

For smaller ice sheets, I have also created an on-demand version of Knight's method using an expansion nucleator. First a temperature-controlled glass window is treated with a hydrophilic coating and wetted with a film of water. The window is then placed, water side up, in a supersaturated chamber, and its temperature is set to just below 0°C. A small puff of air containing seed crystals from the nucleator is passed over the disk, such that a single crystal will randomly drop onto the water's surface. With a bit of luck, surface tension and buoyancy will align the crystal with respect to the water film, and a small disk of ice will slowly grow outward over the glass, as shown in Figure 6.12. This disk is a single crystal of ice oriented with its crystalline c-axis perpendicular to the glass surface, and it took just a few minutes to grow. I have used this technique in an experiment growing snow crystals on glass capillaries (described earlier in the chapter), but the method may have other applications as well.

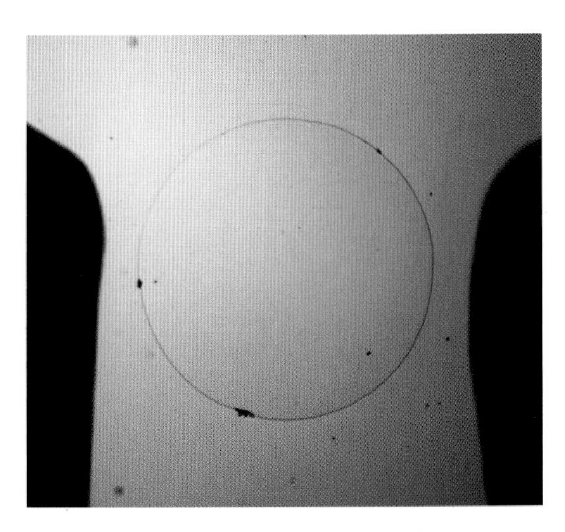

FIGURE 6.12. A 2-mm-diameter disk of ice grows outward on the surface of a thin film of slightly supercooled water covering a glass plate. The dark regions are copper arms supporting the glass. The c-axis of the oriented ice crystal is aligned perpendicular to the glass surface.

If oriented ice crystals in the size range of 10–200 μm are sufficient, the seed-crystal generator described earlier in this chapter can be used to drop prismatic crystals onto a waiting substrate. Thin hexagonal plates are produced when the generator temperature is set near −15°C, and these typically fall with one basal facet flat against the substrate. Prism faceting can be used to determine the orientation angle about the c-axis. Operating the seed-crystal generator near −7°C yields blocky crystals, and dropping these on a substrate produces a significant yield of crystals with one prism facet resting against the substrate. As illustrated in Figure 6.13, using a hydrophilic substrate stimulates preferential growth of the basal and prism surfaces that contact the substrate. After some growth in a near-vacuum environment, one obtains a large platelike crystal that includes a single prism facet oriented parallel to the substrate. Being freshly prepared on demand, these fully oriented crystal surfaces are relatively clean and uncontaminated compared to those obtained from other methods.

LEVITATION

Levitation appears to be an attractive method for investigating snow crystal growth, combining long observing times with none of the substrate-related issues described above. Of course, the technical challenges are substantial in comparison to simply dropping a seed crystal on a substrate, but two different levitation methods have been demonstrated for ice crystal growth with promising results.

Electrodynamic Ion Trapping

Electrodynamic trapping of charged aerosol particles was first developed in the 1950s, and this levitation technique was applied to snow crystal research by Brian Swanson and colleagues in 1999 [1999Swa]. In their apparatus, AC electric fields provide trapping forces together with DC electric fields that counter the downward force of gravity, yielding stably trapped ice crystals. Particle sizes

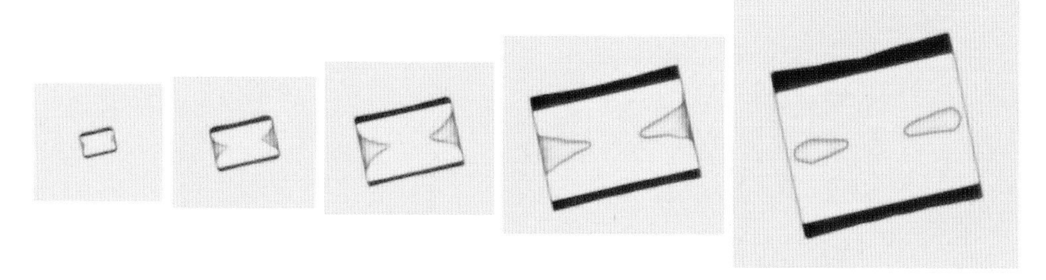

FIGURE 6.13. Large prism facets can be created by growing seed crystals on a hydrophilic substrate. This series of images shows a specimen growing at −7°C to a final size of 150 microns. The increased terrace nucleation at the ice/substrate interface enhances growth along the substrate, yielding an overall flat morphology with a large prism facet oriented parallel to the substrate. Note that small voids became trapped at the substrate surface, below the unblemished top prism facet. This technique allows the easy preparation of large, clean, faceted prism surfaces with known orientation.

were in the 10–100 μm range, carrying charges typically of 0.1–0.5 pC. The physics underlying microparticle ion trapping is beyond the scope of this book, but a summary can be found in Libbrecht and Black [2018Lib], along with techniques for building simple ion traps for laboratory demonstrations. Subsequent development of this technique has led to different electrode geometries and other improvements [2003Bac, 2016Har]. For example, Figure 6.14 shows an electrode geometry that traps ice crystals at the center of a plane-parallel diffusion chamber, allowing precise control of temperature and supersaturation surrounding the particle.

Electrodynamic trapping allows observations of single, isolated, levitated ice prisms growing under well-controlled environmental conditions, which is a somewhat ideal experimental setup. Moreover, particle mass changes can be precisely determined by monitoring the levitating electric fields. Some disadvantages include difficulties loading seed crystals, the generally slow throughput that comes with examining individual crystals, and difficulty maintaining levitation during substantial mass increases. When used at low pressures, latent heating can also become problematic in any levitation apparatus, as heat diffusion through the surrounding medium tends to limit the growth (see Chapter 2). The heating problem is especially severe near 0°C but becomes less of an issue as the temperature is reduced.

A Laminar-Flow Chamber

Tsuneya Takahashi and Norihiko Fukuta developed an ingenious system for levitating a falling snow crystal in a vertical flow of air [1988Tak, 1991Tak, 1999Fuk], allowing the growth of quite large crystals. A slightly tapered flow tube gently pushes a growing snow crystal toward the tube's central axis, and the laminar flow rate is continually adjusted to keep the crystal's vertical position fixed in an observation region as it grows. A fog of water droplets is added to the flowing air to keep the supersaturation at σ_{water}.

Using this laminar-flow chamber, these authors reported extensive observations of snow crystals growing in air for up to 30 minutes, covering a range of temperatures from −2°C to −24°C, including the examples shown in Figure 6.15. This unique data set provides the best record to date of laboratory snow crystals growing in controlled environmental conditions quite close to what can be found in dense clouds. If an adjustable humidification method could be worked out, the laminar-flow chamber could perhaps be used over a range of supersaturations as well. The electric-needle technique presented in Chapter 8 might be better suited overall for comparison with numerical modeling studies, but the laminar-flow chamber provides an attractive alternative for studying the growth of large levitated snow crystals in air.

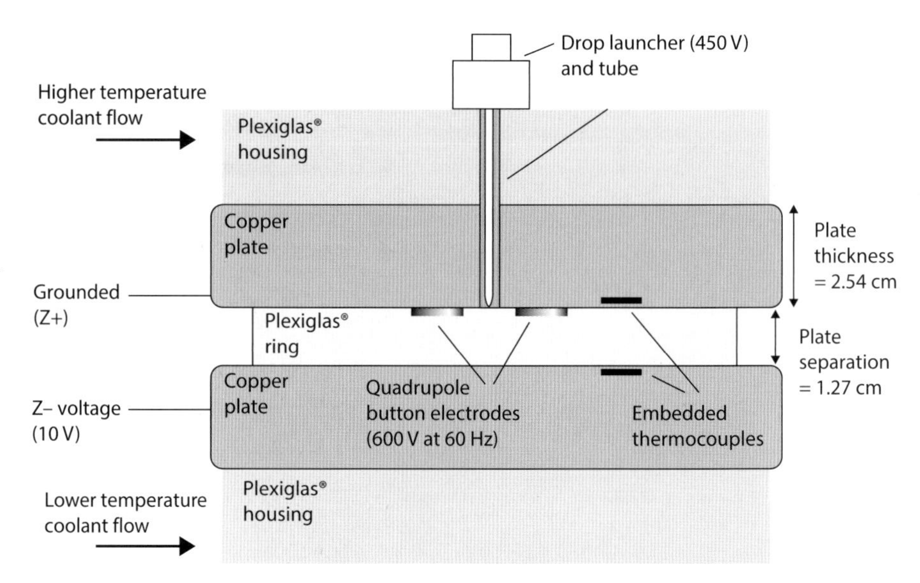

FIGURE 6.14. In this electrodynamic trap [2016Har], a trapping quadrupole electric field oscillating at 60 Hz is delivered via a set of "button" electrodes, while a constant vertical electric field balances gravity. The parallel-plate diffusion-chamber geometry allows for precise modeling of the supersaturation at the position of a trapped ice crystal. Image © American Meteorological Society, courtesy of Jerry Harrington.

CONTINUOUS DIFFUSION CHAMBERS

While several different experimental techniques can be employed to study slowly growing snow crystals, obtaining the high supersaturation levels needed to observe fast-growing crystals is best accomplished with a continuous diffusion chamber. The basic layout consists of an insulated box with a strong vertical temperature gradient, as illustrated in Figure 6.16. Water vapor evaporates from the liquid water reservoir at the top of the chamber and diffuses downward into the colder air below, causing the air to become supersaturated. If any kind of substrate is placed in this region, snow crystals will nucleate on its surface and grow. Because the diffusion process happens continuously, the supersaturation remains roughly constant in time unless perturbed by convection or growing snow crystals. The supersaturation reaches its peak value near the center of the chamber and drops to zero at the chamber walls, as the walls soon become covered with water droplets or frost crystals.

The basic physical concepts being applied in diffusion chambers have been known since the cloud-chamber work of Charles Wilson and others in the early twentieth century. Continuous snow crystal diffusion chambers like the one shown in Figure 6.16 were used by Schaefer [1952Sch] and by Hallett and Mason [1958Hal] in numerous studies of snow crystal growth and morphologies. One clever trick used in snow crystal diffusion chambers is to hang a vertical filament down the

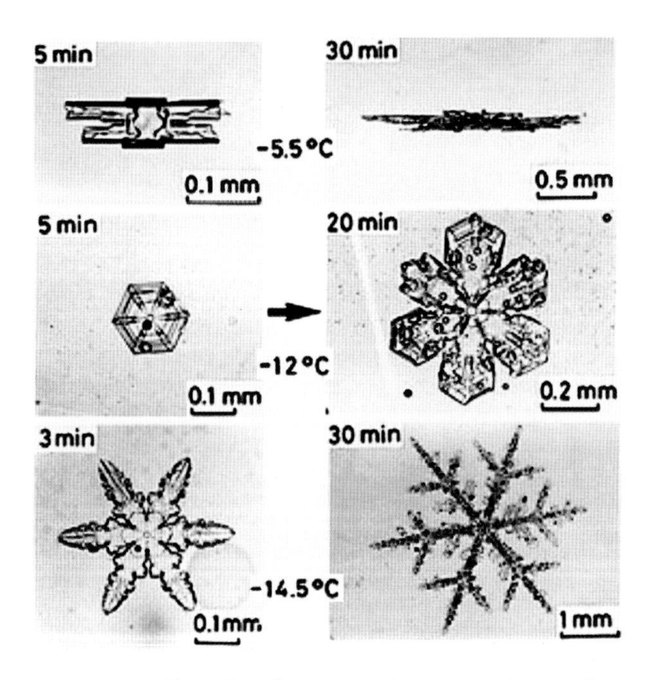

FIGURE 6.15. Examples of snow crystals grown in a laminar-flow chamber. Image adapted from [1999Fuk].

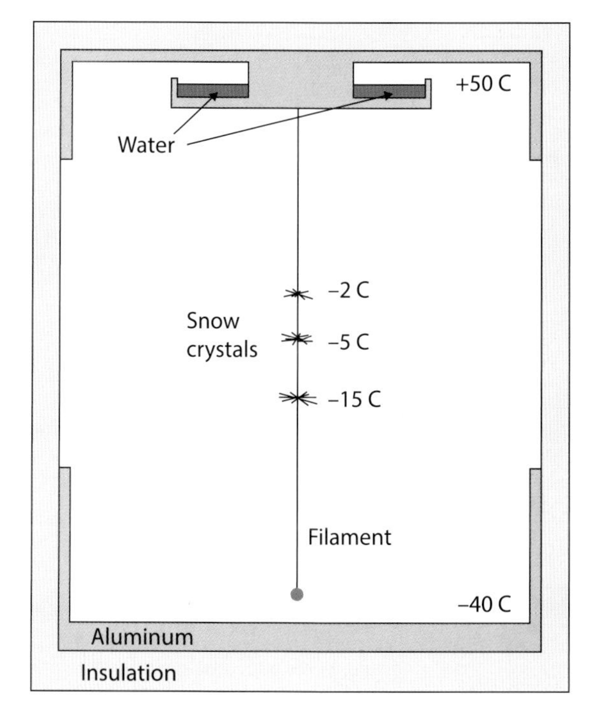

FIGURE 6.16. (Left) This snow crystal diffusion chamber creates a steep vertical temperature gradient with the top warmer than the bottom. Water vapor evaporates from the warm water reservoir and diffuses downward into the colder air below, yielding highly supersaturated air in the interior of the chamber. In this illustration, snow crystals grow at different temperatures on a vertical filament. (Right) A filament supports snow crystals growing at high supersaturation in a diffusion chamber like that illustrated. The fastest-growing dendritic crystals cluster at −2°, −5°, and −15°C. At other temperatures, the crystals tend to be blocky in form with slower growth rates.

central axis of the chamber, as illustrated in Figures 6.16 and 6.17. Snow crystals will grow all along the filament, allowing one to view a variety of different growth morphologies as a function of temperature, reproducing many aspects of the Nakaya diagram simultaneously [1958Hal]. When the supersaturation is high, fast-growing dendritic crystals tend to grow out faster than blocky crystals do, often yielding three distinct clusters along the filament at −2°, −5°, and −15°C.

The aluminum side walls at the top and bottom of the apparatus shown in Figure 6.16 are there to conduct heat and tailor the temperature profile in the chamber. The supersaturation generally increases as the vertical temperature gradient dT/dz increases, and "clamshell" geometries like this are often used to create especially high supersaturation levels. Figure 6.18 shows an extreme example of a clamshell diffusion chamber design that will create an exceptionally high water vapor supersaturation level at its central point. It is straightforward to numerically solve the static heat diffusion equation in this chamber, giving naïve constant-temperature surfaces like those illustrated in the sketch. I call these surfaces "naïve," because this clamshell design is unstable to the weak convection air currents illustrated, so the static diffusion equation does not apply in this situation. For this reason, it is not easily possible to accurately calculate either the temperature or the supersaturation profile in a clamshell diffusion chamber.

Figure 6.18 also illustrates a linear-gradient diffusion chamber that is amenable to accurate thermal and supersaturation modeling. In this design, the stainless-steel plates have a modest thermal conductivity that results in a linear temperature gradient with a constant dT/dz along the walls. The aluminum clamshell plates are still present to reduce heating from the outer boundary, and some engineering is required to obtain an accurately linear temperature gradient. Solving the heat diffusion equation then yields a constant dT/dz in interior region as well, and this can be verified by direct measurements. With entirely horizontal isothermal surfaces, this ther-

FIGURE 6.17. Shown are two strings of snow crystals like that shown in Figure 6.16, but at higher resolution. The cluster at −5°C is made of fishbone dendrites (see Chapter 3), while the cluster at −15°C is made from individual branches of fernlike stellar dendrites. As soon as these fast-growing crystals extend out away from the filament, they tend to shield water vapor from reaching smaller crystals nearer the filament, stunting their growth.

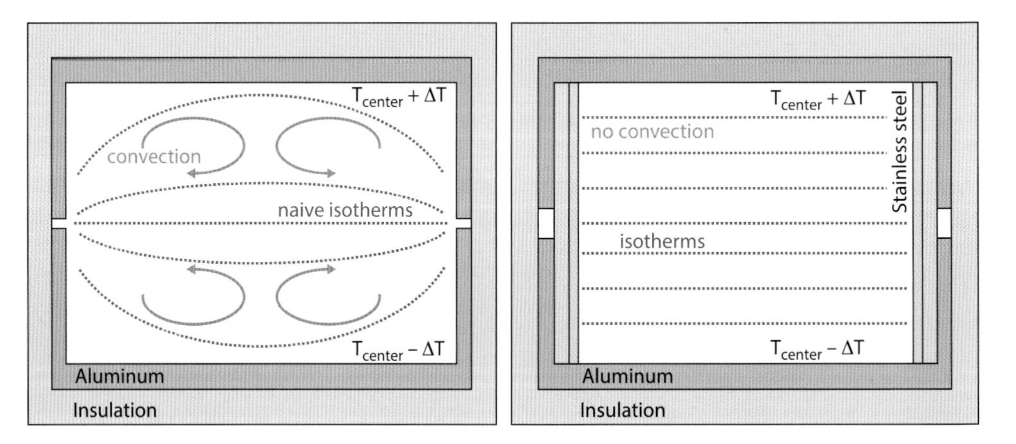

FIGURE 6.18. (Left) A clamshell diffusion chamber designed to create a high supersaturation level at the center of the chamber. This system is unstable to convection, however, making it quite difficult to accurately characterize the temperature and supersaturation in the chamber. (Right) A linear-gradient diffusion chamber designed using moderately conducting stainless-steel walls to produce a linear temperature profile. With this simpler thermal structure, it becomes possible to model the supersaturation in the chamber with quite high accuracy.

mal profile is stable against convection, so the static heat diffusion equation will describe the correct 3D temperature field $T(\vec{x})$ within in the chamber.

Once the temperature profile is stable and well characterized, it then also becomes possible to solve the particle diffusion equation to obtain the water vapor field $c(\vec{x})$ throughout the chamber. In a nutshell, the diffusion equation becomes Laplace's equation (because $\partial c/\partial t=0$), with the boundary conditions $c=c_{sat}(T_{surf})$ on all surfaces. Solving this diffusion problem yields the supersaturation field $c(\vec{x})$ and thus the interior supersaturation

$$\sigma(\vec{x})=\frac{c(\vec{x})-c_{sat}(T(\vec{x}))}{c_{sat}(T(\vec{x}))}. \tag{6.4}$$

If we move the side walls of the linear diffusion chamber out to infinity, giving a 1D parallel-plate geometry, then the diffusion equation yields a simple linear gradient for $c(\vec{x})$, giving

$$\begin{aligned}\sigma_{cent} &\approx \frac{1}{2}\frac{1}{c_{sat}(T_{cent})}\frac{d^2c_{sat}}{dT^2}(T_{cent})\cdot(\Delta T)^2 \\ &\approx C_{diff}(\Delta T)^2\end{aligned} \tag{6.5}$$

at the center point midway between the two plates, where $\Delta T=(T_{top}-T_{bottom})/2$. The last column in Table 2.1 in Chapter 2 lists some C_{diff} values. With the side walls not at infinity, as sketched in Figure 6.18, a numerical solution of Laplace's equation is required to determine $c(\vec{x})$. Further investigations of snow crystals growing in a linear diffusion chamber are described in Chapter 8.

SNOW CRYSTAL IMAGING

The most straightforward approach to measuring ice crystal morphology and growth is via optical microscopy. Growth velocity can be derived from observations of the crystal size as a function of time, and velocity measurements can be combined with a knowledge of the supersaturation to examine the attachment kinetics. Moreover, optical images are ideal for investigating morphologies and morphological transitions. High-quality microscope objectives are readily available, and modern digital cameras produce excellent images even in low light levels.

Commercial bench microscopes typically use a two-step approach to imaging. First the microscope objective focuses an image onto an intermediate aperture, the purpose of which is to block all light except for that coming from the field of view around the desired subject. This first image is then reimaged onto the camera sensor by a second lens group. This technique greatly reduces problems arising from scattered light in the microscope body, but it requires a second high-quality lens for reimaging. A more direct approach is to use a microscope objective on the end of a long extension tube. The objective then images the subject directly onto the camera sensor, without reimaging, which is a substantial gain in optical simplicity. This approach also facilitates mounting the microscope objective inside the cold chamber, near the growing snow crystals, which is often necessary for high-resolution imaging. The direct-imaging method does tend to increase problems associated with scattered light, but these can usually be mitigated using any or all of the following methods:

1) Use a field stop near the test crystal to block extraneous light. Then only the area around the test crystal is brightly illuminated.
2) If a field stop is impractical, image the incident light onto the subject, again only illuminating a small area around the test crystal. This can generally be done using inexpensive optics that are not part of the imaging optical system.
3) Add baffles inside the extension tube.
4) Coat the inside of the extension tube with light-absorbing material

The first two methods on this list reduce the amount of light that does not strike the test crystal but might still make its way into the microscope. This light contributes nothing to the desired image but will add scattered light that reduces contrast in the final image. The last two methods help absorb scattered light in the extension tube before it can strike the imaging sensor.

Once the diffraction limit is reached in any optical systems, one always faces the question of optical resolution versus depth of focus. For example, I often use a Mitutoyo long-working-distance 10X objective for crystals with sizes in the range 5–50 microns, as it has a 0.28 numerical aperture, 1.0 micron resolving power, and a depth of focus of 3.5 microns. The low depth-of-field value can be problematic with this objective, however, as crystals thicker than 3.5 microns will not focus well, which tends to interfere with achieving the rated resolving power. For somewhat larger crystals, or for imaging flat, platelike crystals, I like the Mitutoyo long-working-distance 5X objective. This lens has a 0.14 numerical aperture, 2.0 micron resolving power, and a depth of focus of 14 microns, producing bitingly sharp images of stellar plate snow crystals. Even for quite small crystals, the depth-of-field issue means that the 10X objective is not substantially better than the 5X in many circumstances. When lower resolution is sufficient, I often build in a 3X Mitutoyo compact objective. This lens has a working distance of 78 mm when imaging at infinity, becoming longer when imaging to shorter distances without a secondary lens. The resolving power is 3.6 microns, which is adequate for larger snow crystals, especially those with complex morphologies, and the depth of focus is a comfortable 50 microns.

In nearly all my work, I often use focus stacking to achieve a higher effective depth of focus for a given resolution. Focusing by hand is adequate for stacking just a few images, and this can make a surprisingly large difference in overall image quality. For better consistency when acquiring a greater number of images, a StackShot automated focus-stacking rail (or something similar) can make the image acquisition process quite simple. For postprocessing, there are a variety of software tools available (for example, Helicon Focus) for combining images. I discuss optical microscopy techniques further in Chapter 11.

Optical Interferometry

When additional measurement precision is desired, optical interferometer can be applied to snow crystal observations [1990Gon1, 1993Fur1, 1994Gon]. My favorite example is determining the thickness of thin, platelike crystals using a white-light interferometer, as illustrated in Figure 6.19. The essential idea is to interfere a reflection from the ice/substrate interface with a second reflection from the nearby ice/air interface, as illustrated in the upper-right corner of the figure. Both reflections have roughly the same amplitude, because the index of refraction jumps at the two interfaces are similar. The interference of the two reflections will depend on the wavelength of the incident light, ranging from constructive to nearly fully destructive. A transmission grating disperses the reflected light to reveal a pattern of fringes. Note that the dark line segments in Figure 6.19 are images of the slit after destructive interference. The ends of the line segments indicate the edges of the thin ice prism.

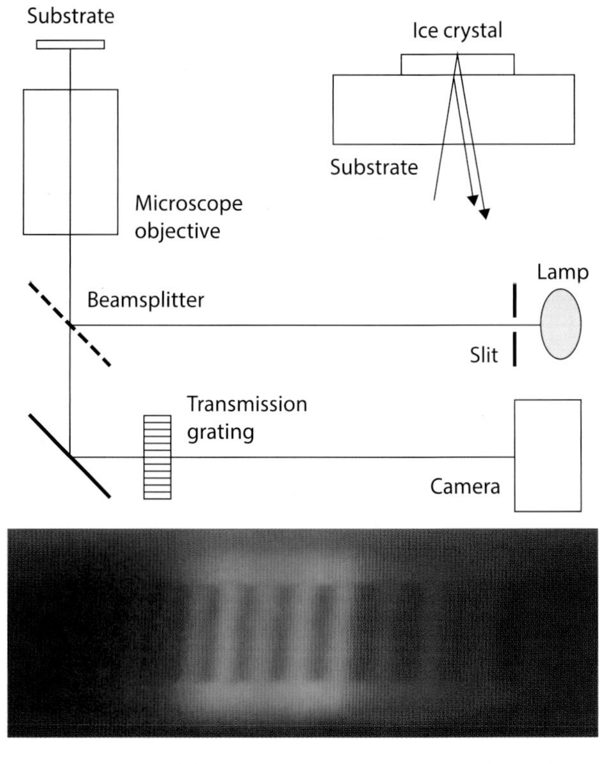

FIGURE 6.19. (Top) An optical setup for measuring the thickness of thin ice crystal plates using white-light interferometry, described further in the text. The upper-right sketch shows the interfering reflections from the substrate/ice and ice/air interfaces. (Bottom) An example of a white-light spectrum, in true color, showing a series of dark interference fringes. The absolute thickness of the ice crystal can be derived from the measured spacing between the optical fringes [2013Lib].

The spacing between the fringes depends on the thickness of the ice crystal, with thicker crystals producing more closely spaced fringes. Working through the math, the ice crystal thickness h is given by

$$h = \frac{\lambda^2}{2n\Delta\lambda}, \tag{6.6}$$

where λ is the wavelength of light, $\Delta\lambda$ is the fringe spacing, and n is the index of refraction of ice. Note that the use of white-light interferometry allows for absolute thickness measurements, and one can achieve sub-micron precision with careful calibration. As a crystal grows thicker, the fringes move laterally across the image, and the fringe spacing becomes smaller. The fringe motion can be determined with especially high accuracy, allowing velocity measurements down to 1 nm/sec in many cases.

The astute reader may note that the ice crystal at the focus of the microscope objective acts as a "cat's eye" reflector. A nearly collimated light beam enters the objective and reflects from the flat surfaces at the microscope's focus, then exits through the objective as a nearly collimated beam. Reflections from other surfaces, such as the bottom surface of the substrate, do not take place in the focal plane and thus do not yield collimated exit beams. The cat's-eye effect explains why only the interference pattern from the ice crystal reflections appears on the camera sensor; interference patterns involving other reflections in the optical system are suppressed. However, this same effect means that especially thick crystals produce poor fringe patterns. In practice, it becomes difficult to discern fringe patterns when the crystal thickness is more than a few times the depth-of-focus of the microscope objective.

For thicker crystals, white-light interferometry becomes difficult as the fringe contrast becomes too low to observe. In this case, one can use direct laser interferometry by removing the transmission grating and replacing the lamp and slit in Figure 6.19 with a collimated laser beam. Simple imaging then shows a bright spot superimposed on the crystal, as shown in Figure 6.20. As the crystal thickness increases, the laser intensity rises and falls, again from interference between reflections from the ice/substrate and ice/air interfaces. This technique does not yield absolute thickness measurements, and velocity measurements are less precise than using white-light interferometry. Note that the laser intensity is far too low to significantly affect the crystal temperature or growth dynamics.

For crystals smaller than about 10 microns, the cyclical brightness oscillations from laser interferometry can be difficult to interpret. I have found that multiple reflections in the ice, especially for slender columnar crystals (substantially smaller than that shown in Figure 6.20), can produce a puzzling variety of brightness patterns. But once the reflecting surfaces become larger than the laser spot size, the oscillating brightness signal matches expectations from basic plane-wave interferometry theory. When applied to simple ice prisms with clean faceted surfaces, white-light and laser interferometry are quite valuable in precision growth experiments, and I describe this technique in more detail in Chapter 7, along with resulting measurements of the attachment coefficients over a broad range of conditions.

Laser interferometry can also be applied to larger snow crystals with complex surface structures [1993Fur1, 1994Gon, 2016Shi]. In Figure 6.21, for example, narrow-band light reflecting off the surface of a natural dendritic snow crystal interferes with a reference beam reflecting off a flat mirror in a standard Michelson interferometer arrangement. The interference fringes create a pattern much like a topographic map, mapping the overall surface structure of the observed crystal. With care, both surfaces of a platelike crystal can be observed simply by changing the focus of the imaging system. This photograph nicely reveals the contours of the central ridges on the branches of this crystal, which can be compared to the discussion of ridge formation in Chapter 3. In a laboratory setting, this technique would allow, for example, many interesting in situ studies of the PoP snow crystals described in Chapter 9.

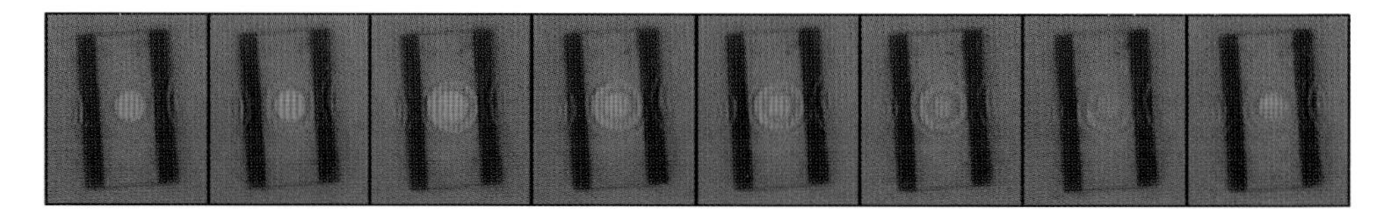

FIGURE 6.20. A series of images of a columnar ice prism showing a reflected spot from a (extremely low power) helium-neon laser beam. As the crystal grows (left to right), the laser spot first increases and then decreases in brightness, owing to interference between reflections from the ice/substrate and ice/air interfaces.

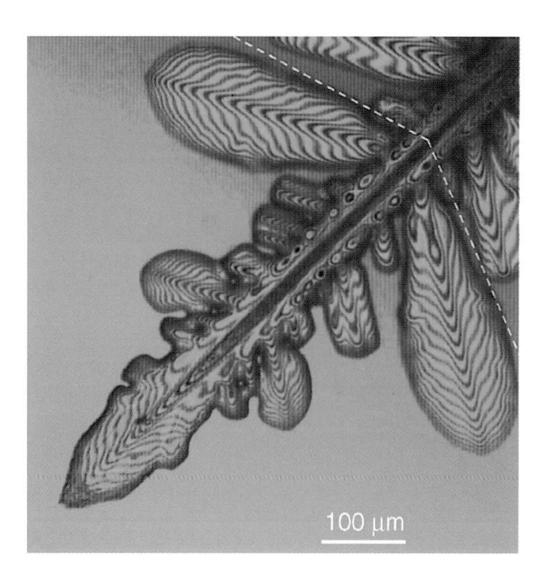

FIGURE 6.21. Optical tomography of one branch of a natural stellar dendrite snow crystal [2016Shi]. The interferometric fringe pattern produces a contour map of one surface of the crystal with a contour interval of 273 nm [2016Shi]. Note the contours of the ridge structures on the large sidebranches, which can be compared with the ridge formation mechanism described in Chapter 3. Image courtesy of Wataru Shimada.

Electron Microscopy

Beyond optical imaging and measurement techniques, electron microscopy has also been adopted to examine ice crystal structure and growth. In a series of papers beginning in the 1990s, William Wergin and collaborators examined natural snow crystals that were collected, transported in liquid nitrogen, sputter coated with a several-nanometer thickness of platinum (to provide a conductive surface), and imaged using a low-temperature scanning electron microscope [1995Wer, 1996Ran, 2002Wer]. More recently, researchers have developed environmental scanning electron microscopy (ESEM) that allows direct imaging of uncoated snow crystals in humid, low-pressure environments [2014Mag], making it possible to observe in situ growth and sublimation on the ESEM stage, as shown in Figure 6.22 [2010Pfa]. Several authors also observed the development of peculiar "trans-prismatic strands" during sublimation [2014Mag]. Clearly, ESEM investigations have great potential for examining small-scale snow crystal surface structures.

Molecular Imaging and Surface Probes

Optical imaging of individual terrace steps on faceted mineral crystals was first demonstrated in the early 1950s using precision interferometry [1950Gri, 1951Ver], but related techniques have been applied to ice crystal surfaces only recently [2010Saz]. The image shown in Figure 6.23 was obtained using laser confocal microscopy combined with differential interference contrast microscopy (LCM-DIM), and the authors clearly verified that the observed features were one-terrace-high molecular steps.

Subsequent studies using LCM-DIM have investigated the surface premelting structure near the triple point [2016Mur] and ice wetting by liquid water (including the water/ice wetting contact angle) [2015Asa, 2016Asa]. Measurements of step velocities have also been used to determine the surface diffusion length for admolecules on faceted surfaces [2014Asa, 2018Ino], although these measurements may have been affected by a lower-than-estimated supersaturation near the ice surface [2015Lib].

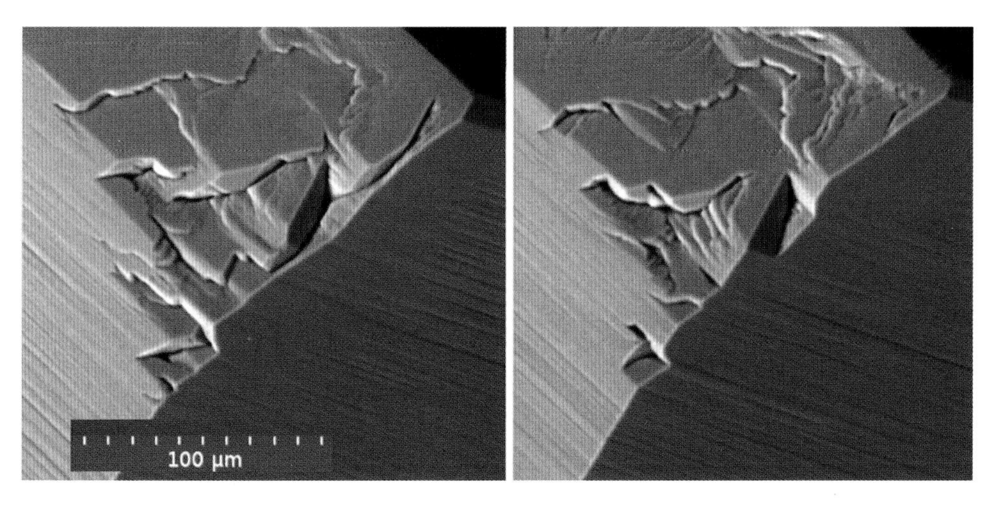

FIGURE 6.22. ESEM images before (left) and after (right) a period of ice growth that filled in part of the unfaceted corner [2010Pfa]. Images courtesy of Steven Neshyba.

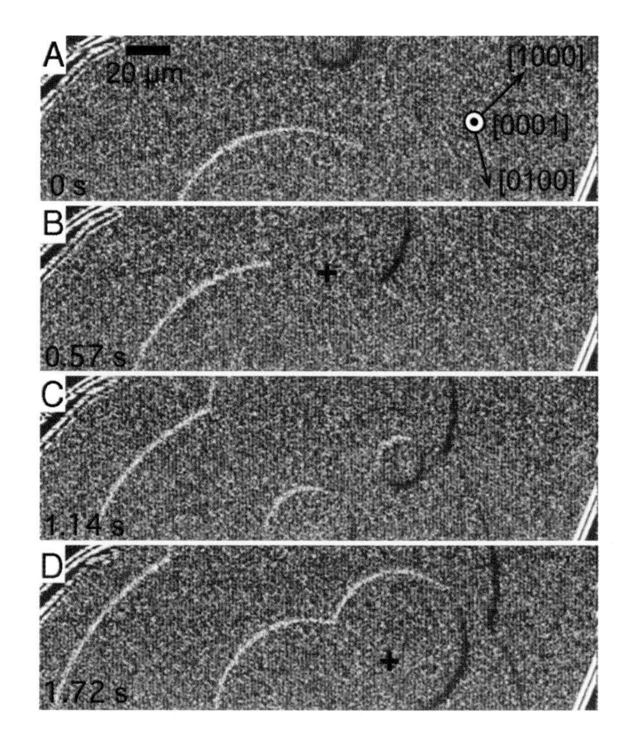

FIGURE 6.23. Photomicrographs of terrace steps on the basal face of an ice crystal that display the nucleation and growth of new terraces [2010Saz]. Image courtesy of Gen Sazaki.

Scanning probe microscopy has also been used to image terrace steps on many faceted crystalline surfaces, but so far, not on ice under normal environmental conditions [1997Pet, 1998Dop, 1998Pit, 2001Zep, 2018Con]. Surface premelting appears to interfere with molecular-scale resolution, perhaps by surface tension forces, but the details are not yet well known. Molecular imaging of single ice bilayers on metal surfaces has recently been demonstrated [2013Thu, 2020Ma], but only at temperatures below −100°C.

Beyond imaging, a plethora of surface probes have been applied to ice, including Brewster reflectometry [1993Elb], ellipsometry [1980Bea], Fourier-transform infrared spectroscopy [2002Sad], interfacial force microscopy [2009Goe], photoelectron spectroscopy [2002Blu], proton scattering [1977Gol], grazing-incidence X-ray diffraction [1995Dos], and second harmonic generation spectroscopy [2015Abd]. These techniques are generally beyond the scope of this book, as they have focused on aspects of the materials science of ice to date [1970Fle, 1974Hob, 1999Pet]. Nevertheless, one can certainly imagine future connections to snow crystal growth as we better develop our understanding of the molecular attachment kinetics (see Chapter 4).

FIGURE 7.1. A collection of simple ice prisms growing in near vacuum at −7°C on a sapphire substrate. The size of typical crystal in this image is about 30 μm.

SEVEN

Simple Ice Prisms

To see a World in a Grain of Sand,
And a Heaven in a Wild Flower.
Hold Infinity in the Palm of your Hand,
And Eternity in an Hour.

—WILLIAM BLAKE, *AUGURIES OF INNOCENCE*, CA. 1803

Understanding the physics of snow crystal formation begins with a careful consideration of the simple ice prism. With its minimalist shape, bounded by two basal facets and six prism facets, the hexagonal ice prism is like the hydrogen atom of snowflakes—the most elemental form possible, with none of the complicating embellishments one finds on stellar dendrites and other elaborate structures. Following the well-traveled path of scientific reductionism, our goal in this chapter is to observe the growth of these archetypal snowflakes with the highest possible precision, so we can better understand the physical processes that govern their development. Doing so builds the foundation for further progress, as much of our understanding of the attachment kinetics stems from measurements and models of microscopic crystals grown under well-controlled conditions. Before we can develop algorithms to fashion complex computational snow crystals that faithfully reproduce the full morphological menagerie of laboratory and natural specimens, we must first thoroughly understand the formation of simple ice prisms.

I begin this story with the conclusion, as Figures 7.2 and 7.3 encapsulate what are (in my opinion) the best measurements to date of the ice/vapor attachment kinetics on large faceted surfaces. Much of this chapter is devoted to a discussion of how simple ice prisms are central to such measurements, including descriptions of the two experiments used to acquire these specific data. I will demonstrate that the functional form $\alpha(\sigma_{surf}) = A \exp(-\sigma_0/\sigma_{surf})$ provides an excellent fit to ice growth data, thus supporting the terrace nucleation theory described in Chapter 4. The theory also connects the terrace step energies $\beta(T)$ to the measured $\sigma_0(T)$, thus relating the data in Figures 7.2 and 7.3. Considerable attention is also given to a critical analysis of systematic errors that can corrupt ice growth measurements and analysis, as identifying and controlling these errors has proven to be quite important for obtaining consistent results.

The principal methodology for determining the attachment coefficients is straightforward enough—measure the growth velocity v_n normal to the surface of a faceted surface, determine the water vapor supersatura-

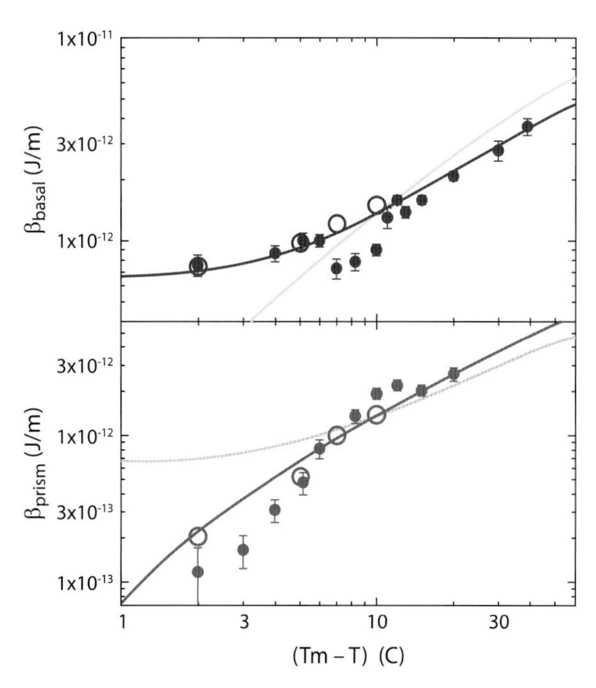

FIGURE 7.2. Measurements of the basal (top) and prism (bottom) step energies from the VIG (solid points) and VPG (open circles) experiments described in this chapter. Lines are from the CAK model (see Chapter 4).

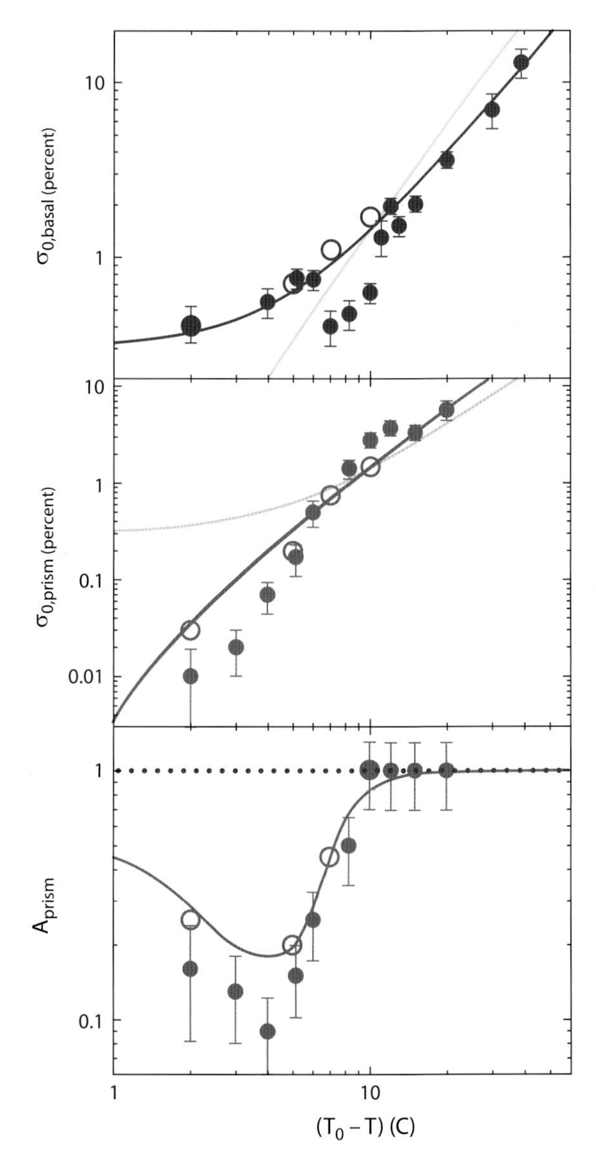

FIGURE 7.3. Measurements of the basal (top) and prism (middle) nucleation parameters from the VIG and VPG experiments, using the functional form $\alpha_{basal} = \exp(-\sigma_{0,basal}/\sigma_{surf})$ and $\alpha_{prism} = A_{prism} \exp(-\sigma_{0,prism}/\sigma_{surf})$. The bottom plot shows A_{prism} measurements, and all lines are from the CAK model (see Chapter 4).

tion σ_{surf} just above the growing surface, and use that information to extract the attachment coefficient via the usual growth equation $v_n = \alpha v_{kin} \sigma_{surf}$. And then repeat this procedure to map out α_{basal} and α_{prism} as a function of temperature T, σ_{surf}, and perhaps other factors. In the ideal case—infinite faceted surfaces growing in pure water vapor—there are no other factors, so $\alpha(T, \sigma_{surf})$ must be a single-valued function. Surface diffusion, surface premelting, and other surface properties are automatically incorporated into the attachment coefficient. Adding a background gas of air could conceivably change $\alpha(T, \sigma_{surf})$, but the data presented below suggest that such effects are perhaps small enough to be considered negligible. SDAK effects are quite important on small facets (see Chapter 4), but our focus in this chapter will be on large facets only.

Given how easy it is to make and grow simple ice prisms (see Chapter 6), one might expect that measuring their attachment kinetics under controlled conditions would be straightforward as well. Indeed, measuring growth velocities to sufficient accuracy is not particularly difficult. But precisely determining the growth conditions, especially σ_{surf}, has long been problematic.

Taking a close look at some of the earlier ice growth measurements at temperatures above −25°C [1969Fuk, 1972Lam, 1976Rya, 1982Bec, 1983Bec, 1984Kurl, 1989Sei, 1998Nel, 2003Lib], it soon becomes apparent that the measurement uncertainties are not just statistical in nature; they also exhibit a significant level of systematic errors associated with diffusion corrections, substrate interactions, and other factors. Moreover, because the potential for systematic errors was not always fully appreciated in the early experiments, one cannot tell from published reports the degree to which these unwanted effects influenced the resulting data and conclusions. All one knows for sure is that substantial unexplained discrepancies exist between the different data sets, as evidenced by the large amount of scatter seen when they are all plotted together [2019Har]. My goal here will not be to give all the published results equal weight, but to focus on producing better measurements with a level of accuracy and precision that supersedes earlier efforts. The observed scatter in the data in Figures 7.2 and 7.3 is roughly five times smaller than a similar plot including all published data [2019Har], and this fivefold reduction is what made it possible to develop the CAK model.

I begin by briefly examining several ice growth strategies and by performing an in-depth analysis of a generic substrate-based experiment. Having laid this foundation, I then take a close look at the Libbrecht and Rickerby experiment [2013Lib], which I believe is the best comprehensive ice growth data set above −25°C at the time of this writing. From there I continue with a description of a new, improved experiment also designed to explore the growth of simple prisms in the same temperature region [2019Lib2]. Although the two experiments use substantially different measurement strategies and completely different hardware, they have produced remarkably similar results so far, lending credence to the notion that we are indeed converging on a set of reliable and accurate measurements of the attachment coefficients $\alpha(T, \sigma_{surf})$, at least for simple ice prisms over a limited range of environmental conditions. My primary focus is on the temperature range from 0°C to −30°C, but I also briefly examine lower temperatures at the end of the chapter.

PRECISION ICE GROWTH MEASUREMENTS

As described in Chapter 6, many possible experimental techniques can be employed to observe and measure snow crystal growth, including simple ice prisms. No single apparatus is appropriate for all measurements, as each has its own advantages and disadvantages. As technology develops and general observations become replaced by targeted experimental investigations, the tools for measuring ice growth become ever more varied and sophisticated. Designing an experimental plan and apparatus thus involves many choices.

Air or Vacuum? Growing ice crystals in ordinary air is certainly convenient and may be desirable if one wants to simulate a meteorological environment. But if the goal is to better understand the attachment kinetics, measurements in air can be problematic. The spherical growth analysis described in Chapter 3 tells us that ice growth rates will be determined primarily by particle diffusion whenever $\alpha_{diff} \ll \alpha$, and α_{diff} is inversely proportional to both the background gas pressure and crystal radius. If $\alpha_{diff} \ll \alpha$, then it becomes nearly impossible to extract information about the attachment kinetics using direct growth measurements. Even with a small crystal (for example, one having a radius of 5 microns), we find $\alpha_{diff} \approx 0.03$ in air at a pressure of 1 bar, which is quite small. From the discussion of the CAK model in Chapter 4, we see that restricting a set of measurements to $\alpha < 0.03$ places serious limitations on the available physics scope, as this regime describes mainly the slow growth of large facets. Much of the interesting physics is clearly at higher α, so one would like to probe the fast-growth region as much as possible using simple ice prisms. These considerations provide an incentive to make measurements using the smallest possible ice prisms at low background gas pressures.

Reducing the background gas pressure to zero gets rid of the particle diffusion problem entirely, but working in a pure water vapor environment can present some quirky experimental problems of its own. If the growth chamber is not perfectly isothermal (which it never is), then water vapor will tend to quickly deposit as frost on the coldest point. Pressure generally equilibrates rapidly inside a sealed chamber, so temperature inhomogeneities can drive some rather large and unpredictable water vapor flows, especially if the point-to-point temperature variations are not constant in time. Analyzing all this is difficult in practice, so a pure water vapor environment is perhaps best avoided if possible.

There is a broad experimental sweet spot when the background gas pressure is several times greater than the saturated water vapor pressure, meaning residual air pressures of around 5–50 mbar, depending on temperature. In this environment, water vapor will still want to migrate to the coldest point in the chamber, but particle diffusion through the background gas creates a substantial hindrance to water vapor flows. And while the total pressure will still equilibrate quickly, the partial pressure of water vapor need not be constant throughout the chamber. Without getting into a serious diffusion analysis, the bottom line is that a bit of inert background gas pressure greatly reduces large-scale water vapor flows, thus yielding a generally more benign and stable growth environment.

Substrate or Levitate? Growing a simple ice prism levitating in nearly empty space has a certain intuitive appeal, providing something of an ideal crystal growth experiment. But the strategy has an important downside in that latent heat generated by crystal growth is only slowly dissipated by conduction through the surrounding medium. Again using the spherical analysis discussed in Chapter 3, we see that heating will limit a levitation experiment to the region of phase space with $\alpha < \alpha_{heat}$, which is only marginally better than working in air at 1 bar, although this situation improves rapidly at low temperatures. This analysis shows that a levitation apparatus like that described by Harrison et al. [2016Har] is close to ideal for observing cirrus cloud particles in the lab at low temperatures, but it is not so well suited for examining ice prism attachment kinetics at low pressures and temperatures above −20°C.

Observing crystals growing on a substrate largely eliminates the latent heating problem, owing to the relatively high thermal conductivity of ice (see Chapter 3), but this strategy also introduces the possibility of substrate interactions. Hydrophobic surface coatings can help (see Chapter 6), but the problem can never be eliminated entirely, and it is not always obvious whether substrate interactions are introducing significant systematic errors or not. Nevertheless, substrates have additional advantages over levitation, including experimental simplicity, fully stable crystals, and issues arising from mass changes during growth in levitation. For these and other practical reasons, it is hard to beat watching crystals grow as they simply rest on a substrate.

Optical or Other? Ordinary optical imaging is a standard method for making basic growth measurements, but the optical resolution limit makes it challenging to observe especially small ice prisms or those growing especially slowly. Various forms of optical interferometry can provide improved velocity sensitivity, and electrodynamic levitation has what is essentially a built-in, extremely sensitive system for mass measurement. Electron microscopy can provide ultra-high spatial resolution, as can scanning probe microscopy, but these techniques present significant experimental challenges. Some of these advanced techniques are described in Chapter 6, and all have found use in modern ice growth experiments. Over a substantial range of parameter space, however, the biggest experimental limiting factor is determining σ_{surf}, not achieving optical resolution. In this case, increasing the velocity sensitivity may not provide a large improvement in the resulting data quality.

Diffusion Modeling? There is no real need for a question mark after this heading—diffusion modeling is essential when using ice growth data to probe the physical processes underlying snow crystal formation. It is likely that much of the scatter in the early measurements of attachment kinetics resulted from imperfect diffusion modeling that yielded inaccurate determinations of σ_{surf} around growing test crystals. This subject deserves especially careful attention, and it should be an important consideration when designing any ice growth experiment. Even the most sophisticated measurement apparatus may yield scientifically uninteresting results if σ_{surf} is not well determined. Before delving into actual experiments and results, therefore, we next consider a diffusion analysis of a basic apparatus for measuring the growth of simple ice prisms.

A Generic Substrate Experiment

Figure 7.4 sketches a simple substrate-based ice growth experiment, showing a test crystal resting on a substrate at temperature $T_{substrate}$ accompanied by a nearby ice reservoir at temperature $T_{reservoir}$. The test crystal will grow as long as $T_{reservoir} > T_{substrate}$, and the goal of the experiment is to determine the attachment coefficient $\alpha(\sigma_{surf})$. If we remove all the substrate crystals for a moment, then it is straightforward to determine the temperature $T(\vec{x})$ and water vapor number density $c(\vec{x})$ as a function of position $\vec{x}$ throughout the chamber. Solving the heat diffusion equation yields a simple linear temperature gradient profile

$$T(\vec{x}) = T_{substrate} + \Delta T \cdot \left(\frac{z}{L} \right), \quad (7.1)$$

where z is vertical distance above the substrate, L is the chamber inner height, and $\Delta T = T_{reservoir} - T_{substrate}$. This simple solution is valid in the presence of nearby walls, provided their heat conductivity is relatively low. Likewise, solving the particle diffusion equation yields the trivial solution $c(\vec{x}) = c_{sat}(T_{reservoir})$ throughout the chamber. Putting these two solutions together yields the supersaturation

$$\sigma_{subst,0} = \frac{c_{sat}(T_{reservoir}) - c_{sat}(T_{substrate})}{c_{sat}(T_{substrate})} \approx \frac{1}{c_{sat}} \frac{dc_{sat}}{dT} \Delta T \approx \eta \Delta T \quad (7.2)$$

at the surface of the substrate, while $\sigma = 0$ at the surface of the reservoir. Note that this solution only applies when no ice crystals are present on the substrate or on the walls of the chamber. If we then place a single small ($R \ll L$) test crystal on the substrate with no additional "extraneous" crystals, then it will begin

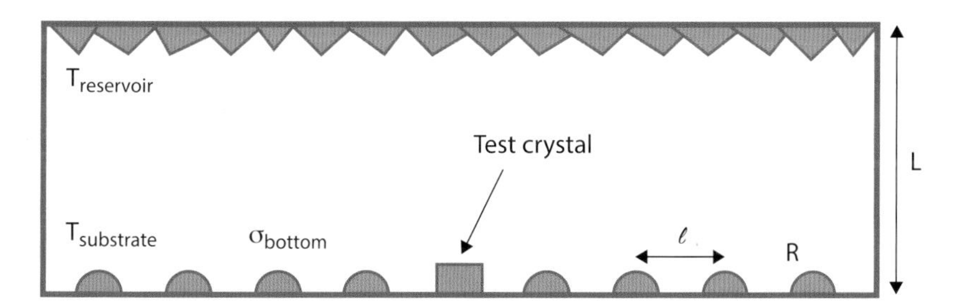

FIGURE 7.4. A schematic layout of a basic ice growth experiment. The test crystal being measured rests on an isothermal substrate at temperature $T_{substrate}$, surrounded by some additional "extraneous" ice crystals. Water vapor is supplied by a large collection of frost crystals on the top of the chamber at temperature $T_{reservoir}$. The walls are free of ice crystals and have a relatively low thermal conductivity.

growing as if surrounded by a faraway boundary condition $\sigma_\infty = \sigma_{subst,0}$. Thus, for this ideal experimental geometry, the growth conditions seen by the test crystal are quite well known.

In a real experiment, however, it is not always easy to drop a single, isolated test crystal on the bottom of a growth chamber. More often, several additional crystals end up on the substrate, as illustrated Figure 7.4, and these may even be treated as additional test crystals, allowing numerous measurements in parallel. In this case, we need to consider how the entire set of crystals affects our calculation of the supersaturation, because Equation 7.2 may no longer be a good approximation for the supersaturation near the substrate.

Large-Scale Diffusion Correction

To simplify the analysis with multiple crystals on the substrate, I have drawn Figure 7.4 with a simple square array of ice crystals, all hemispherical in shape with a uniform radius R, each spaced a distance ℓ from its nearest neighbors. The additional crystals will not affect the temperature solution given by Equation 7.1, but they will alter the water vapor field $c(\vec{x})$, so our first task is to calculate the supersaturation σ_{subst} at the bottom of the chamber, for which I will assume $R \ll \ell \ll L$.

From the spherical solution for diffusion-limited growth, the substrate crystals will all grow at a rate

$$v = \varepsilon \alpha_{diff} v_{kin} \sigma_{subst}, \tag{7.3}$$

where

$$\varepsilon = \frac{\alpha}{\alpha + \alpha_{diff}} \tag{7.4}$$

and α is the usual attachment coefficient. For mainly diffusion-limited growth, $\varepsilon \approx 1$. This growth implies an overall downward flux of water molecules equal to

$$F = \frac{2\pi R D c_{sat} \varepsilon \sigma_{subst}}{\ell^2} \tag{7.5}$$

from mass conservation, where $c_{sat} = c_{sat}(T_{substrate})$. Equating this to the downward diffusion flux $F = D\nabla c$ then gives the result

$$\begin{aligned} \sigma_{subst} &\approx \left(1 + \frac{2\varepsilon\xi L}{R}\right)^{-1} \sigma_{subst,0} \\ &\approx \left(1 + \frac{2\pi\varepsilon R L}{\ell^2}\right)^{-1} \sigma_{subst,0}, \end{aligned} \tag{7.6}$$

where

$$\xi = \frac{\pi R^2}{\ell^2} \tag{7.7}$$

is the substrate filling factor, equal to the fraction of the substrate surface area that is covered with ice crystals. If we put in some typical numbers, with a filling factor of $\xi = 0.01$, a chamber height $L = 1$ cm, a crystal radius $R = 10$ μm (giving $\ell \approx 180$ μm), and simple diffusion-limited growth with $\varepsilon = 1$, we find $\sigma_{subst} \approx 0.05\sigma_{subst,0}$. Because the filling factor is quite small and $R \ll \ell$, one might naively think that $\sigma_{subst} \approx \sigma_{subst,0}$ would be reasonably accurate. But this naive assumption would overestimate σ_{subst} by a factor of 20! Lowering the pressure reduces ε, but often the correction factor is substantial even at quite low pressures. In practice, I have found that the large-scale diffusion correction arising from having extraneous ice crystals on the substrate is nearly always an important consideration.

Several ice growth measurements described in the literature have used growth chambers with geometries roughly like the one illustrated in Figure 7.5, which turns out to be a somewhat poor choice, especially if operated at normal atmospheric pressure. Topologically, this geometry is like that shown in Figure 7.4, but the effective L is quite large because the ice reservoir is located far from the growing crystals. Even if there are only a few extraneous ice crystals near the test crystal, an assumption of $\sigma_{surf} \approx \sigma_{subst,0}$ could be quite inaccurate. Moreover, if the extraneous crystals are not visible inside the chamber, then it will not even be clear how large the diffusion correction might be. It appears that such systematic

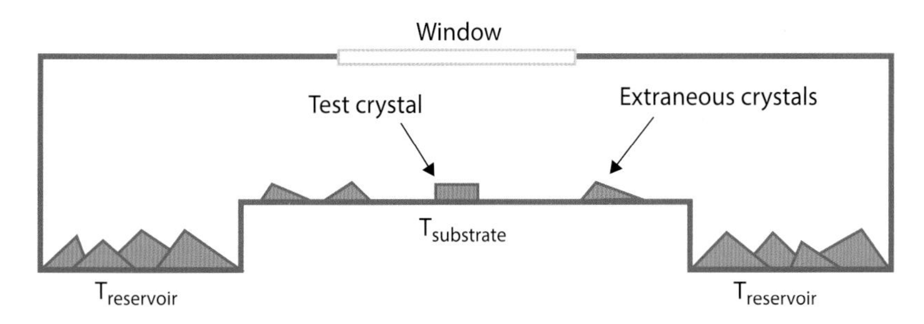

FIGURE 7.5. An example of a somewhat poor chamber design for making precision ice growth measurements. The presence of any extraneous ice crystals on the substrate, coupled with the ice reservoir being far from the test crystal (making L large in Equation 7.6), make it quite difficult to determine the supersaturation near the test crystal with high accuracy.

errors can explain at least some of the discrepancies seen in published ice growth data.

Going back to Figure 7.4, a related analysis shows that one can use the substrate growth observations directly to estimate the diffusion correction in determining σ_{subst}, provided one can view the extraneous crystals near the test crystal [2019Lib2]. The basic idea is to watch the crystals growing on the substrate and use the measured crystal sizes over time to estimate $\dot{V}$, which is equal to the change in ice volume per unit time and per unit area on the substrate. Because this ice is supplied by a downward flux of water vapor from the ice reservoir, a straightforward analysis yields a measurable correction factor

$$\sigma_{subst} \approx \sigma_{subst,0} - \frac{c_{ice}}{c_{sat}} \frac{L}{D} \dot{V}. \tag{7.8}$$

Moreover, one can also estimate an uncertainty in $\dot{V}$, from measurements of the growing crystals, so Equation 7.8 can be used to estimate both σ_{subst} and the experimental uncertainty in σ_{subst}.

The takeaway message from this exercise is that it is difficult to produce precise measurements of α without limiting the number of crystals on the substrate to quite low values, lower than one might initially expect. Achieving $\sigma_{subst} \approx \sigma_{subst,0}$ requires that the filling factor satisfy

$$\xi \ll \frac{R}{2\varepsilon L} \tag{7.9}$$

or, equivalently, having a spacing between test crystals of

$$\ell \gg \sqrt{2\pi\varepsilon RL}. \tag{7.10}$$

Reducing the background gas pressure reduces this diffusion issue, but it may introduce the temperature inhomogeneity problem discussed earlier in the chapter. I have found that the best way to manage these large-scale diffusion problems is to make careful measurements and do a throughout diffusion analysis. As a rule of thumb, if σ_{subst} is less than about $\sigma_{subst,0}/2$, then the large-scale diffusion correction becomes so large that it may be difficult to obtain meaningful results from the data.

Small-Scale Diffusion Correction

To proceed with the next step in our diffusion analysis, first assume that we have managed to place a single, isolated test crystal on the substrate, thus achieving the experimental geometry illustrated in Figure 7.4, but with no extraneous crystals. Assuming that the test crystal size is much less than L, the crystal is surrounded by a supersaturation given by $\sigma_{subst} = \sigma_{subst,0}$ in Equation 7.2. Put another way, because $R \ll L$, the isolated test crystal will grow as if it were immersed in a uniform supersaturation environment with $\sigma_\infty \approx \sigma_{subst}$. Starting with this faraway boundary condition, we must then determine σ_{surf}, the supersaturation at the crystal surface, in order to extract α using the growth equation $v_n = \alpha v_{kin}\, \sigma_{surf}$.

For a faceted test crystal, the supersaturation will vary with position on the crystal surface, so numerical modeling of the diffusion equation is necessary to determine $\sigma_{surf}(\vec{x})$ over the entire surface with extremely high accuracy. But the monopole matching approximation (see Chapter 5) is often sufficient if the diffusion correction is not too large, and it requires only the analytic solution for the growth of a spherical ice crystal. The solution for a hemispherical crystal on a substrate is similar to the spherical case, as we can apply reflection boundary conditions to solve the diffusion equation. The result is the supersaturation field $\sigma(r)$ at all radii, including on the surface of the hemisphere, which we write as

$$\sigma(r) = \sigma_\infty - \frac{\dot{V}_s}{2\pi r X_0 v_{kin}}$$
$$\sigma(R) = \sigma_\infty - \frac{R}{X_0}\frac{v_R}{v_{kin}}, \tag{7.11}$$

where $\dot{V}_s = 2\pi R^2 v_R$ is the time derivative of the volume of the hemispherical test crystal growing at velocity v_R.

For a nonspherical crystal, Equation 7.11 will still give a good approximation for the supersaturation field $\sigma(r)$ when r is much larger than the size of the crystal, as this is the nature of the monopole approximation. For a nearly isometric test crystal, the monopole solution is likely a good representation of the real solution nearly all the way down to the crystal surface. Thus, Equation 7.11 provides a useful estimate of the surface supersaturation, which we can write as

$$\sigma_{surf} \approx \sigma_\infty - \frac{\dot{V}_s}{2\pi R_{eff} X_0 v_{kin}}, \tag{7.12}$$

where $\dot{V}_s$ is the time derivative of the volume of the nonspherical test crystal, and R_{eff} is an effective radius estimated from the size and morphology of the crystal. If R_{eff} is not precisely known, this translates into an experimental uncertainty in the estimate of σ_{surf}. Of course, Equation 7.12 is less accurate for highly anisometric crystals, but it can be used as a reasonable first approximation in many cases. If we have a collection of crystals growing on the substrate, then σ_∞ in Equation 7.12 is replaced by σ_{subst}. And, once again, if σ_{surf} is less than about $\sigma_{subst}/2$, then the small-scale diffusion correction becomes so substantial that it may be difficult to obtain meaningful results from the data.

The overarching theme in this exercise is that a careful diffusion analysis of the entire growth chamber is necessary to accurately determine σ_{surf} around a growing ice prism. While a full-blown 3D solution to the diffusion equation would be ideal, this level of accuracy is unnecessary if the crystal morphology is simple and the diffusion corrections are reasonably small. Importantly, if the growth chamber is specifically designed to minimize diffusion corrections and make their calculation feasible, then relatively simple analytic solutions to the diffusion equation can yield good estimates for σ_{surf} in many practical situations, along with realistic uncertainty estimates as well. Without this careful attention to detail from the outset, however, it is remarkably easy for systematic errors to creep in and seriously degrade the results from an otherwise excellent ice growth experiment.

A TALE OF TWO EXPERIMENTS

The set of ice growth measurements described in the Libbrecht and Rickerby paper [2013Lib] initiated something of a turning point in my investigations, as this was the first data set that I felt was extensive enough and accurate enough to allow a detailed investigation of the attachment kinetics. Compared to earlier measurements, including some of my own, these data stood out as finally providing a reliable sample. These data played a central role in developing the CAK model (Chapter 4), resulting in what I believe is a clearer picture of the growth of large basal and prism facets. Some years later, I developed another precision ice growth experiment with similar, but somewhat complementary characteristics, which was again able to measure attachment kinetics over a broad range of growth conditions [2019Lib2]. Although it will take several additional years to fully analyze the newer experiment and collect a comprehensive set of data, so far it is giving results that are in good agreement with the 2013 results. After many years slogging through a careful examination of systematic errors of every ilk, the measurements are finally beginning to provide (in my opinion) a convincing physical picture of the snow crystal attachment kinetics.

In this section, I examine these two experiments in detail, in each case focusing on the apparatus design, ice crystal handling, measurement strategy, and data analysis. Special attention is given to the identification and minimization of potential systematic errors. My motivation for this section is threefold: 1) these experiments present substantial improvements over earlier efforts and so deserve some attention and scrutiny, 2) the discussion and comparison of techniques may be useful for designing future precision ice growth experiments, and 3) the resulting data played a foundational role in the development of the CAK model described in Chapter 4.

Case Study I: The Vacuum Ice Growth Experiment

Figure 7.6 illustrates the vacuum ice growth (VIG) apparatus that we used [2013Lib], which was designed to approximate the ideal growth-chamber geometry shown in Figure 7.4. The easiest way to describe this device is to walk through the steps used during its operation. This somewhat pedagogical approach will both explain the apparatus details and examine various design choices and their consequences.

Vacuum Chamber. The pancake-shaped outer box in Figure 7.6 depicts a short cylindrical vacuum chamber with an outer diameter of 7.5 cm, machined out of aluminum. The chamber is black anodized to seal the aluminum surfaces, and it includes a lid that bolts to the lower box, sealed using a silicone O-ring. The high thermal conductivity of aluminum keeps the chamber walls at a uniform temperature, and the silicone O-ring retains its pliability (and vacuum seal) at low temperatures. A digital temperature controller maintains a constant chamber temperature by means of a thermistor sensor embedded in the aluminum using thermally conducting epoxy together with thermoelectric modules on the bottom of the outer box. The chamber is opened, cleaned, and baked between runs to minimize chemical vapor contaminants. The sapphire substrate is also removed and thoroughly cleaned between runs to remove dirt and chemical residues. The substrate is given a final rinse with deionized water before being installed in the chamber, again to reduce remaining solvent residues.

The Subchamber. Figure 7.7 shows a close-up of the subchamber where the test crystals grow. A key feature

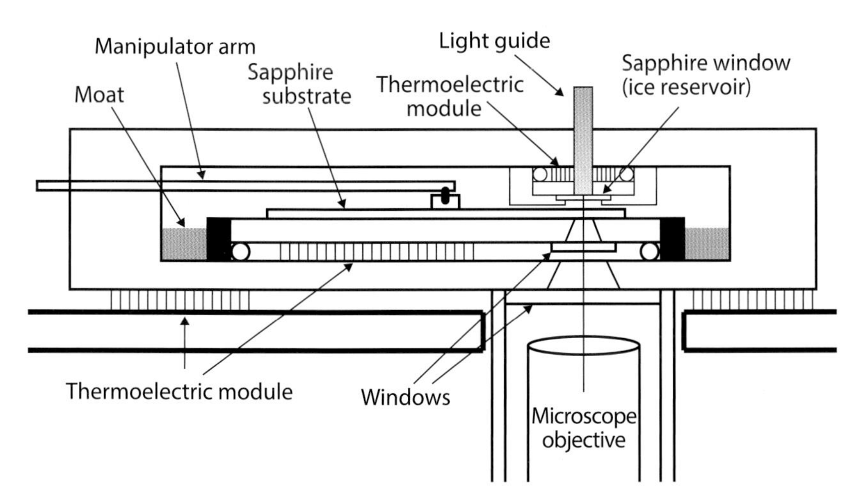

FIGURE 7.6. Schematic diagram of the VIG chamber used to measure the growth of small ice crystals in near vacuum [2013Lib].

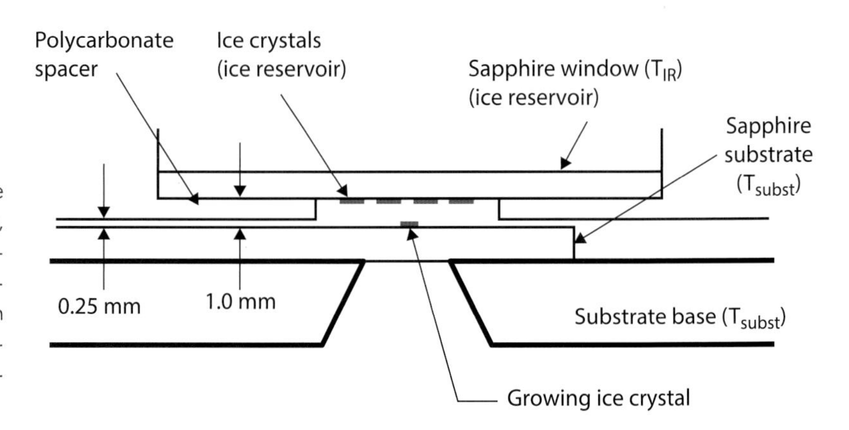

FIGURE 7.7. A close-up view of the ice growth region of the apparatus in Figure 7.6, showing the ice reservoir and single test crystal. Note the similarities to the ideal growth-chamber geometry shown in Figure 7.4, with a close spacing (L = 1 mm) between the reservoir and substrate to reduce large-scale diffusion corrections.

in this region is the spacing $L = 1$ mm between the ice reservoir and the test crystal. This spacing is as small as practical to minimize the difference between σ_{subst} and $\sigma_{subst,0}$, as given in Equation 7.6. The 0.25 mm gap between the substrate and inner walls is large enough to allow free movement of the substrate but small enough to isolate the subchamber somewhat from the main vacuum chamber, reducing perturbations to the supersaturation in the subchamber. The overall geometry of the subchamber, together with the procedure for placing a single test crystal in it (described below), are quite important for creating a well-defined supersaturation in the vicinity of the test crystal.

Seed-Crystal Generator. The VIG vacuum chamber rests on the bottom of a much larger refrigerated chamber filled with ordinary air, about 1 meter in height, shown in Figure 7.8. The larger chamber serves as a continuous seed-crystal generator (see Chapter 6), producing a constantly replenished cloud of small ice crystals that slowly drift down all around the small vacuum chamber. These crystals grow as they free fall in air, and a typical crystal grows for just a few minutes before settling out. New crystals are nucleated every 10 seconds using an expansion nucleator, thus yielding a steady supply of pristine seed crystals. The seed-crystal chamber is somewhat self-cleaning, because the cloud of fresh ice crystals tends to continuously remove residual chemical impurities from the air in the chamber.

Temperature Control. The aluminum vacuum chamber, substrate base, and ice reservoir are all independently temperature regulated using thermistor sensors and thermoelectric heating/cooling. The sapphire window that defines the ice reservoir is bonded using thermal epoxy to a small copper plate that contains a small thermistor for temperature sensing. The high thermal conductivity of copper, sapphire, and thermal epoxy makes for a well-defined ice-reservoir temperature. The sapphire substrate can slide freely over the anodized aluminum substrate base, and only the latter is temperature regulated. However, the large-area flat-on-flat contact between the substrate and base, along with the fact that the substrate has almost no heat load, keeps the substrate and base at essentially the same temperature.

The ice-reservoir servo uses a home-built controller that does not regulate $T_{reservoir}$ directly but instead regulates $\Delta T = T_{reservoir} - T_{substrate}$. Thus ΔT is controlled with high accuracy, minimizing effects from any substrate temperature drifts. During normal operation, ΔT is quite close to zero, yielding a nearly isothermal environment in the vacuum chamber.

Temperature Calibration. The thermistor response is known from the manufacturer's specifications, and the chamber and substrate temperatures are set using calibrated temperature controllers. The value of ΔT is especially critical, so the $\sigma = 0$ point is measured for each test

FIGURE 7.8. The VIG experiment in the lab. The aluminum vacuum chamber is a small black package inside the much larger copper-walled seed-crystal chamber. A heat lamp inside the large chamber is baking the system in this photo. The optics are covered in black panels below the large chamber, and the high-resolution camera is contained in the white styrofoam box at the lower left.

crystal by adjusting ΔT (changing $T_{reservoir}$) until the crystal is neither growing nor sublimating. This can be accomplished with especially high accuracy at low pressures, when the response to temperature change is swift. With care, one can locate the $\sigma = 0$ point with an absolute uncertainty of about $\delta\sigma \approx 0.001$, which is equivalent a temperature accuracy for ΔT of about 0.01°C.

Supersaturation Verification. Observing the condensation of water droplets on the substrate provides an excellent method for verifying that the supersaturation is equal to that given by Equation 7.2. With the chamber evacuated and no test crystals present, one can increase ΔT until water droplets appear on the substrate and then slowly adjust ΔT until the droplets are neither growing nor evaporating. The value ΔT_{stable} when the droplets are just stable must be producing a supersaturation equal to σ_{water} at the substrate surface. Calculating σ_{water} from the measured ΔT_{stable} yields excellent agreement between theory and measurements with no adjustable parameters or fits to the data [2013Lib].

Preparing the Ice Reservoir. At the beginning of a run, after the system has reached its operating temperature and the vacuum chamber is stably temperature controlled, a butterfly vacuum valve on top of the aluminum vacuum chamber (not shown in Figure 7.6) is opened, allowing some seed crystals to fall onto the large sapphire substrate. A small electric motor rotates the substrate about a central pivot point so that seed crystals land at all points around the circumference of the substrate. The vacuum pump, controlled by a variable needle valve, slowly draws air and seed crystals from the large seed-crystal chamber into the small vacuum chamber, and the crystal density on the substrate is monitored using the imaging system during this process.

Once an ample supply of ice crystals has landed on the substrate, the ice-reservoir temperature is set lower than the substrate temperature. The butterfly valve is closed, and a vacuum is drawn inside the chamber while the substrate slowly rotates. Some ice on the substrate then sublimates and deposits on the sapphire ice-reservoir window. The microscope is focused on the latter surface to verify that a thick coating of frost appears on the window, forming the ice reservoir. After this loading process is complete, the substrate is further warmed to drive off any remaining ice, and air is let back into the chamber. With the ice reservoir thus prepared, $T_{substrate}$ and $T_{chamber}$ are set to the desired operating temperature, while ΔT is set to zero. The chamber is then ready for the main experimental session to commence. Additional ice can be added to the ice reservoir during the run as needed.

Positioning an Isolated Test Crystal. When the substrate is ice free and a test crystal is desired, the butterfly vacuum valve is opened once again, allowing some seed crystals to fall onto the rotating sapphire substrate. The needle valve to the vacuum pump is again opened slightly to draw air down from the seed-crystal chamber, facilitating the transfer of seed crystals onto the substrate. This process is continued for some tens of seconds to yield a low density of seed crystals on the substrate. With the butterfly valve closed, a live video view through the microscope objective is scrutinized while the substrate slowly rotates to search for a suitable seed crystal. The substrate's central pivot can also be translated using the manipulator arm shown in Figure 7.6, allowing a 2D sweep of the substrate surface for test crystals. The crystal density is low, and not every seed crystal has an ideal prism morphology, so it often takes some searching to locate a suitable test crystal. Typically, one looks for a well-formed ice prism with either a basal or prism facet lying flat on the substrate. Polycrystalline forms, malformed crystals, or poorly oriented crystals are all rejected. Crystals with nearby neighbors are also rejected. If need be, the substrate can be heated to remove all the crystals, so it can be reloaded with new seed crystals for another attempt. Finding high-quality, isolated seed crystals can be a laborious process, sometimes taking 10–20 minutes to locate a suitable specimen.

Pumping Out the Chamber. Once a high-quality test crystal has been positioned at the center of the microscope field of view, the variable needle valve is opened slowly to begin the pump-down. This is a somewhat tricky step when the pressure becomes low, as it is relatively easy to sublimate away the test crystal with overzealous pumping. The operator carefully watches the test crystal and adjusts ΔT and the pump-out speed appropriately to make sure that the test crystal neither grows nor sublimates appreciably during the pump-down. Once the pressure has been reduced to about 30 mbar, the test crystal is ready for a growth measurement.

The Optical System. A white-light interferometer system is used to measure the thickness of test crystals, using the optical layout described in Chapter 6. The main illumination is from above, with light passing through the ice reservoir. Calculations show that this light and all other relevant light sources are far too weak to affect the temperature of the test crystals or the ice reservoir. A low-resolution camera is used for finding and positioning a suitable test crystal and for verifying that the test crystal has no nearby neighbors. A second image is projected onto the high-resolution camera for direct imaging of the test crystal along with its interferometer fringe pattern. Figure 7.9 shows two typical images from the high-resolution sensor.

Basal and Prism Growth. As implemented in this experiment, the white-light interferometric measurement of the crystal thickness works only for thin, platelike crystals, as demonstrated in Figure 7.9. As the crystal thickness increases, the fringes become closely spaced, and the fringe contrast diminishes. Thus, the technique is well suited for measuring growth of basal facet surfaces using thin plates, but it cannot be used when measuring

FIGURE 7.9. Two still images from a video showing the growth of a test crystal. The top parts of both images show interferometer fringes in true colors, while the bottom parts show a direct image of the crystal including illumination from the slit. The numerical sub-images show a projected voltage from which ΔT can be determined. The left image was taken near the beginning of the growth cycle. The right image shows the same crystal after it had grown larger and thicker. The bright slit light is periodically blocked to provide a better direct image of the crystal.

the growth of prism facet surfaces using columnar crystals, as the prism facet-to-facet spacing is too large.

Therefore, when performing runs with columnar test crystals, the interferometer lamp and slit are replaced by a helium-neon laser giving direct images like those shown in Figure 7.10. As with light from the slit, the He-Ne beam reflects off both the substrate/ice and ice/vacuum interfaces, and these two reflections interfere with each other. The brightness of the reflected spot then depends on crystal thickness, oscillating between bright and dark as the crystal grows. This method is not as precise as white-light interferometry, nor does it yield an absolute measurement of the crystal thickness. With care, however, it can yield acceptable growth measurements of faceted prism surfaces.

A Growth Sequence. Once a suitable test crystal has been found and positioned, and the chamber has been pumped down to about 20–40 mbar, the $\sigma = 0$ point is determined by adjusting the ice-reservoir temperature until the test crystal is neither growing nor sublimating. This procedure also gives the chamber a few minutes to equilibrate near $\sigma = 0$. A growth sequence then commences by slowly increasing σ by changing $T_{reservoir}$ while monitoring ΔT. The substrate and chamber temperatures remain constant as ΔT increases. Figure 7.11 shows typical data during a growth sequence, where σ_{subst} here is defined from ΔT using Equation 7.2. The small size of the ice reservoir, and the high thermal conductivity of copper and sapphire, give the reservoir a fast temperature response while keeping the overall temperature equal to that indicated by its thermistor sensor. Thus, the measured voltage gives an accurate indication of ΔT and thus σ_{subst} during the sequence. The temperatures of the substrate and vacuum chamber are kept fixed during this time, as they have longer thermal response times.

At the end of the growth sequence, the test crystal is discarded by heating the substrate. We have found that "recycling" a test crystal—sublimating it back down to a smaller size and growing it out again—generally leads to unreliable results. Surface impurities become concentrated during large-scale sublimation, which often seems to corrupt the subsequent regrowth of the crystal. Sometimes a recycled crystal behaves normally, but this is not always the case. To avoid any potential problems from crystal recycling, we typically use each test crystal only once.

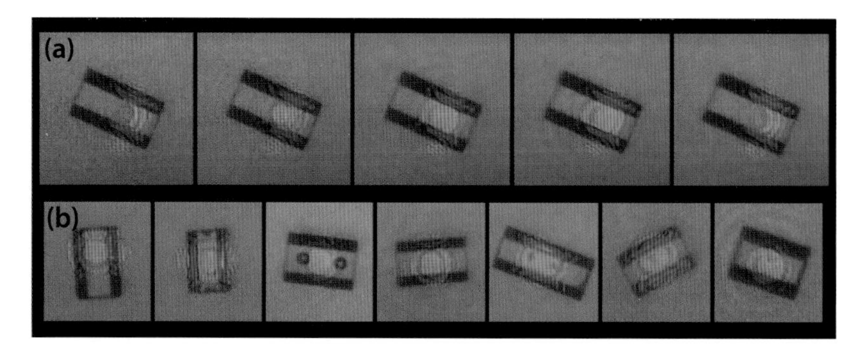

FIGURE 7.10. (Top) A series of five images showing the interference of two reflections from an incident helium-neon laser beam, one from the substrate/ice interface and one from the ice/vacuum interface. As the columnar crystal grows (left to right), the spot brightness oscillates from dim (first image) to bright (middle image) and back to dim (last image). (Bottom) Snapshots from several other example crystals.

Chemical Contamination Tests. Contamination of ice surfaces from unwanted chemical vapors is always a concern in any ice growth experiment, as one is never sure how clean is clean enough. Opening the entire system and baking it between runs is a first line of defense against chemical contaminants, as they tend to bake out after numerous thermal cycles. Another plus is using a continuous seed-crystal generator (see Chapter 6), as fresh seed crystals keep the air in the seed-crystal generator chamber quite clean during a run. As a test crystal is being grown, the growing ice surface is also somewhat self-cleaning with respect to chemical contaminants. As a crystal grows, each expanding ice terrace edge tends to push surface chemicals ahead of it, as few chemicals are readily incorporated into the ice lattice. Surface contaminants are thus swept aside as a crystal grows, cleaning the faceted surface in the process.

One can test this process in a single growth sequence by first increasing ΔT with time and then decreasing it back to zero. After a short period of rapid growth, when ΔT is high, the fresh ice surface should be especially free of contaminants. Quickly bringing ΔT back down then allows a growth measurement of this pristine surface. If the growth velocity $v(\sigma_{surf})$ looks the same whether ΔT is increasing or decreasing, then it suggests that the ice surface is reasonably clean.

Data Analysis

To analyze a growth sequence, the video is first transcribed to produce time-dependent measurements of H, R, and ΔT as a function of time, as shown in Figure 7.11. Then the H and R data are used to extract growth velocities (also seen in the figure), while σ_{subst} as a function of time is derived from the ΔT voltage measurement. Because the chamber design keeps the correction in Equation 7.6 quite small, ΔT gives σ_{subst} directly from Equation 7.2. This gives the velocities V_H and V_R as a function of σ_{subst} as the test crystal grew. Note that the growth sequence of a single test crystal yields growth velocities over a range of supersaturations. Thus, every test crystal growth sequence can be used to extract $\alpha(\sigma_{surf})$ for σ_{surf} ranging from zero to some maximum value.

Substrate Interactions. As described in Chapter 6, substrate interactions can have detrimental effects on ice crystal growth measurements. Especially important is that a low ice/substrate contact angle may cause spurious nucle-

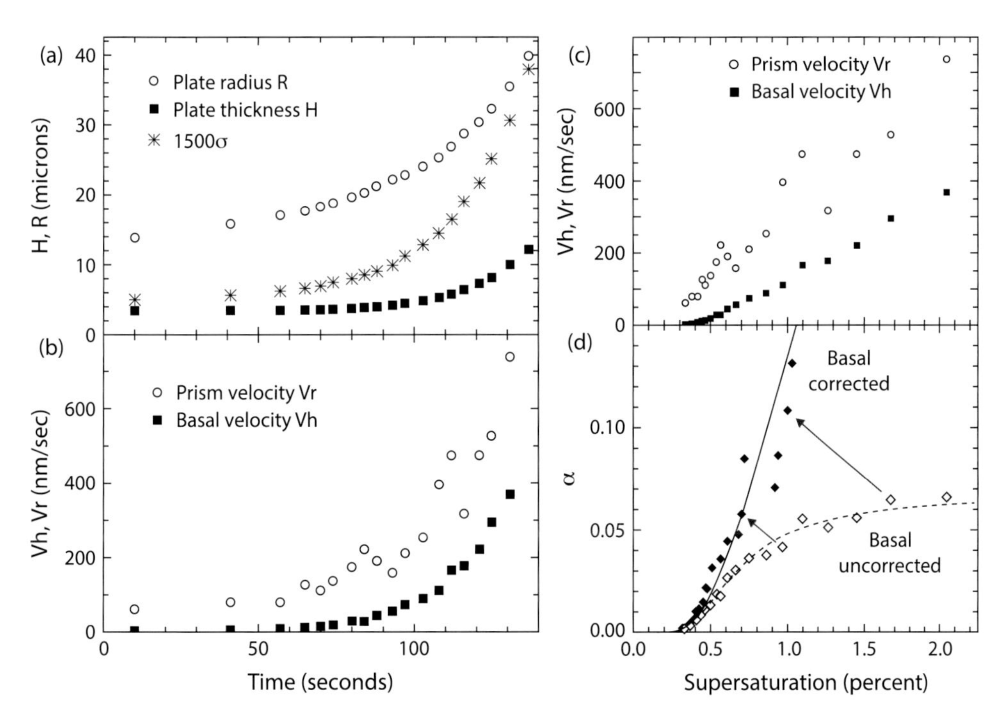

FIGURE 7.11. (a) A typical growth sequence for a single thin-plate crystal at −12°C, like the crystal shown in Figure 7.9. As the supersaturation is slowly increased by increasing ΔT, the plate thickness H is measured using white-light interferometry while the plate "radius" R (here defined as half the distance between opposing prism facets) is measured from direct imaging. Note that $R \gg H$, indicating a thin, platelike crystal. (b) Growth velocities of the basal and prism facets derived from the size data in (a). The higher accuracy of the interferometric measurements yields lower noise in the basal thickness and velocity data. (c) The growth sequence in (b) after being converted to velocity versus supersaturation σ_{subst} just above the substrate surface. Note that σ_{subst} is not generally equal to the supersaturation σ_{surf} at the crystal surface. (d) The basal growth data converted to the basal attachment coefficient α as a function of supersaturation. The open points show $\alpha_{uncorrected} = v/v_{kin}\,\sigma_{subst}$ plotted as a function of σ_{subst}, while the filled points show the corrected $\alpha = v/v_{kin}\,\sigma_{surf}$ plotted as a function of σ_{surf}. The arrows show how two individual points transformed from uncorrected to corrected. A nucleation-limited growth model $\alpha = A\exp(-\sigma_0/\sigma_{surf})$ (solid line) provides a good fit to the corrected data using $A = 1$ and $\sigma_0 = 2$ percent. Including diffusion gives the dashed curve, equal to $\alpha\alpha_{diff}/(\alpha + \alpha_{diff})$ with $\alpha_{diff} = 0.075$ (Chapter 3). At low σ_{surf}, the basal growth is strongly limited by a nucleation barrier, so $\alpha \ll \alpha_{diff}$ and the diffusion correction is small. At higher σ_{surf}, $\alpha \gg \alpha_{diff}$ and the diffusion correction is quite large.

ation of new terraces, thereby increasing growth rates compared to surfaces that do not contact the substrate. This effect can be seen directly in Figure 7.11, as the prism growth velocity V_R is much larger than the basal growth V_H at early times, when the supersaturation is low. In these data, the basal growth is suppressed by a large nucleation barrier, while the prism growth is aided by substrate-mediated terrace nucleation. We have found that substrate interactions can be both significant and somewhat unpredictable on bare sapphire surfaces. For example, the ice/substrate contact angle is sensitive to surface chemical residues, which may vary with position on the substrate.

Therefore in the VIG experiment, we treat growth velocity data for all surfaces that contact the substrate directly as suspect. In Figure 7.11, for example, we mostly discard the V_R data for this platelike crystal, but retain the V_H data, as the latter came from measurements of the top basal surface, which is parallel to the underlying substrate. For columnar crystals like those in Figure 7.10, we retain only growth data from the upper prism surface for the same reason. Interestingly, Beckmann et al. [1983Bec] observed similar substrate interactions to those just described, but they chose to discard data from the facets that were not contacting the substrate, keeping data from the those that did, rather than the other way around.

Diffusion Correction. The next step in the data analysis is to recognize that while the ΔT data give $\sigma_{subst} \approx \sigma_{subst,0}$ with good accuracy (for a suitably isolated test crystal), σ_{subst} is not generally equal to σ_{surf} at the surface of the growing crystal. Even at a low pressure of 30 mbar, the diffusion correction is substantial at high growth rates and must be calculated. The monopole approximation described earlier in the chapter is adequate to convert σ_{subst} to σ_{surf}, using the volume derivative $\dot{V}_s$ calculated from the crystal size and velocity data [2012Lib], and using both the V_H and V_R data, so the $\delta\sigma$ correction contains no adjustable parameters. The correction is thus based entirely on measured quantities, and the underlying physics is well understood.

Figure 7.11(d) shows an example of this diffusion correction being applied to data from a single test crystal. Because the correction changes σ_{subst} to the lower value σ_{surf}, a given point moves to the left in the diagram. The same point also moves vertically upward as $\alpha_{uncorrected} = v/v_{kin}\sigma_{subst}$ changes to $\alpha = v/v_{kin}\sigma_{surf}$. The filled points in Figure 7.11(d) then give the desired function $\alpha(\sigma_{surf})$ for this test crystal. Clearly the correction is quite large for the highest-velocity points, as σ_{surf} is only about half as large as σ_{subst} for the final point measured. We find that diffusion corrections of this magnitude are acceptable, but barely. Pushing the experiment to substantially higher growth rates, however, becomes problematic.

Nucleation-Limited Growth. Once corrected for diffusion effects, the VIG data strongly suggest that the attachment kinetics are primarily limited by terrace nucleation. Figure 7.12, for example, shows that the data are well fit by a nucleation model, while a spiral-dislocation model does not fit the data (Chapter 4). This conclusion

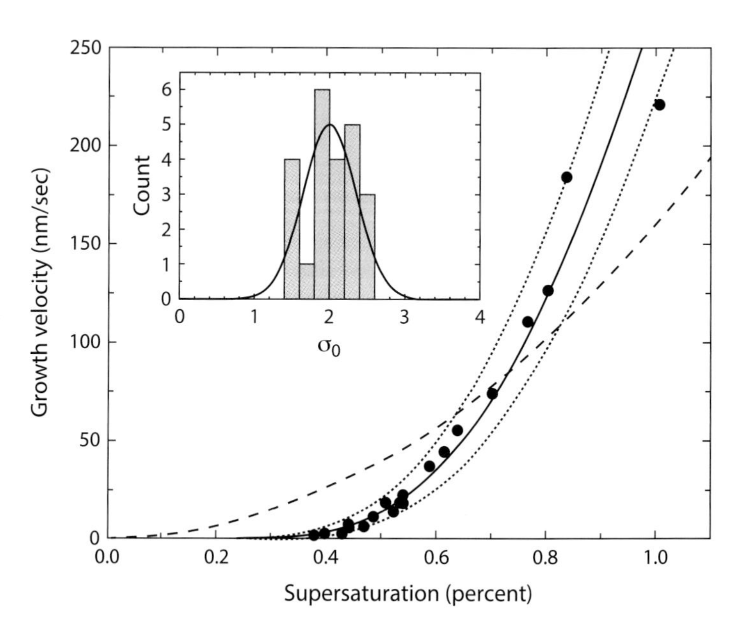

FIGURE 7.12. An example showing corrected measurements (data points) of the growth velocity of the basal surface for a single ice crystal as a function of σ_{surf} at −12°C. The solid line through the points gives the model $v_{basal} = \alpha_{basal} v_{kin} \sigma_{surf}$ with $\alpha_{basal}(\sigma_{surf}) = \exp(-\sigma_0/\sigma_{surf})$ and $\sigma_0 = 2.3 \pm 0.2$ percent, flanked by dotted curves using $\sigma_0 = 2.1$ and $\sigma_0 = 2.5$ percent. The dashed line shows a spiral-dislocation model with $v \sim \sigma_{surf}^2$, which is a poor fit to the data. The inset graph shows an unweighted histogram of measured σ_0 values for 23 crystals. A weighted fit to these data gives an estimated mean of $\sigma_0 = 1.95 \pm 0.15$ percent.

applies broadly to both basal and prism growth data, and Figure 7.13 shows additional measurements of α_{basal} plotted as a function of $1/\sigma_{surf}$ for a variety of temperatures. Note that displaying α versus $1/\sigma_{surf}$ in a semi-log plot provides an especially good view of the nucleation-limited growth behavior seen in the data, as the functional form $\alpha(\sigma_{surf}) = Ae^{-\sigma_0/\sigma_{surf}}$ appears as a straight line in such a graph. Moreover, the convergence of these data on $\alpha_{basal} \to 1$ as $\sigma_{surf} \to \infty$ was not imposed by the analysis in any way. Once the proper (independently determined) diffusion corrections were applied to the data, this behavior appeared, suggesting that $A \approx 1$ is an accurate description of the basal growth data over the entire temperature range measured. The same cannot be said for the prism data, however, as the measurements indicate $A < 1$ at temperatures above $-10°$C.

Precision Measurements. One important lesson from the VIG experiment is that obtaining high-quality measurements of the attachment kinetics requires a great deal of attention to detail regarding apparatus design, systematic errors, and data analysis. With the VIG experiment, these details included:

1) The chamber was designed specifically to produce a well-defined supersaturation near the test crystal, following a careful diffusion analysis. The various correction factors are important and must be examined carefully. Even at low pressures and with a small filling factor of crystals on the substrate, diffusion effects can still be quite significant.
2) Several measures were taken to avoid chemical vapor contamination. The entire system was baked between runs, a self-cleaning seed-crystal generator was used, and the vacuum chamber was purged with fresh air throughout each run.
3) We were careful to select test crystals with the highest visual quality. Only near-perfect ice prisms with no nearby neighbors on the substrate were chosen.
4) Crystals were not sublimated and then regrown. A new test crystal was selected for each growth sequence.
5) We spent a great deal of time perfecting the apparatus, data acquisition procedures, and analysis methods. More than 200 crystals were grown and analyzed, which allowed many consistency checks and redundancies.

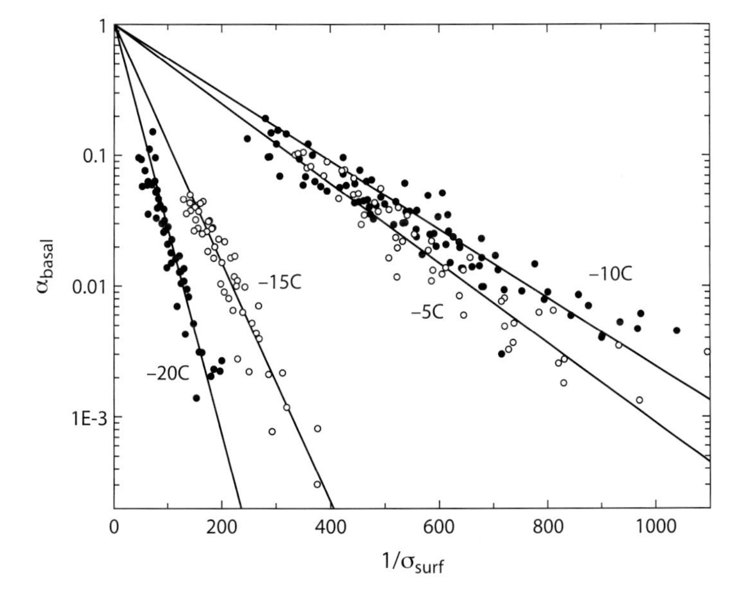

FIGURE 7.13. Experimental data showing the attachment coefficient α_{basal} plotted as a function of $1/\sigma_{surf}$, including measurements taken at four different temperatures. Plotted this way, data exhibiting a nucleation-limited growth behavior with $\alpha_{basal}(\sigma_{surf}) = A\exp(-\sigma_0/\sigma_{surf})$ appear as straight lines. The data at each temperature extrapolate to $\alpha \approx 1$ at large σ_{surf}, indicating rapid kinetics in the absence of a nucleation barrier. The values of σ_0 at different temperatures can be extracted from the slopes of these lines [2013Lib, 2017Lib].

Attachment Coefficients

We found that all VIG data could be well represented using attachment coefficients having the functional form $\alpha(\sigma_{surf}) = A\exp(-\sigma_0/\sigma_{surf})$, thus reducing the entire large-facet attachment kinetics data to the functions $\sigma_0(T)$ and $A(T)$ shown in Figure 7.14. These data clearly favor a terrace-nucleation model for ice crystal growth from water vapor over a broad range of environmental conditions, and I discussed the physical implications of this result in Chapter 4. Remarkably, the data suggest that ice crystal growth rates on the basal and prism facets are largely determined by the terrace step energies as a function of temperature, which are fundamental equilibrium properties of the ice crystal lattice (see Chapter 2).

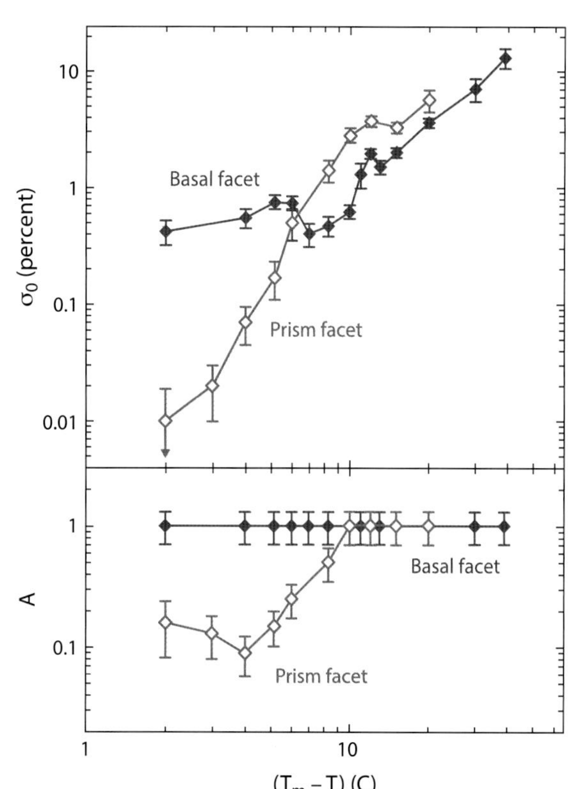

FIGURE 7.14. Experimental data showing the attachment coefficient fit parameters $\sigma_0(T)$ and $A(T)$ for the basal and prism facets, assuming a functional form $\alpha(\sigma_{surf}) = A\exp(-\sigma_0/\sigma_{surf})$ for both α_{basal} and α_{prism} [2013Lib].

If these step energies can be independently determined by molecular dynamics simulations, this would provide a major step forward in solving the full problem of snow crystal growth dynamics.

Case Study II: The Variable Pressure Growth Experiment

The VIG experiment yielded some excellent ice growth data, as just described, and it also provided us with valuable insights that we soon put to use in developing an even better apparatus. With the specific goal of measuring the attachment kinetics on small ice prisms, we found the following:

1) Interferometric velocity measurements provided an unnecessarily high level of velocity sensitivity. Higher is always better, of course, but extracting the attachment coefficients from velocity data was limited primarily by our ability to determine σ_{surf} with precision. Except at the lowest growth rates, extracting growth velocities from direct imaging is good enough and much simpler. Moreover, with direct imaging, one can examine substantially smaller crystals, thus reducing the diffusion corrections and allowing better σ_{surf} determinations.
2) Substrate interactions were generally not as problematic as expected, especially at low temperatures, provided the sapphire substrate was adequately cleaned. Moreover, we subsequently found suitable hydrophobic coatings that reliably yield water contact angles on sapphire that are greater than 90 degrees, reducing this problem further. Thus, with some care, we could use direct imaging to extract velocity measurements without suffering large systematic errors caused by substrate interactions.
3) Placing a single test crystal in the VIG apparatus was quite laborious, which reduced overall throughput to about one usable crystal per hour. As camera sensors improved rapidly in terms of pixel count and sensitivity after the VIG experiment, it became possible

to drop 100 or more small crystals onto a substrate field-of-view without excessive crowding effects. This allows doing multiple measurements in parallel to greatly increase throughput, plus this strategy provides a quick visual check on the uniformity of overall crystal size and morphology that is useful for managing systematic errors.

4) Determining α_{basal} and α_{prism} with overall measurement uncertainties of about a factor of two is acceptable (for now), because these quantities vary by orders of magnitude with temperature and supersaturation. Thus, systematic errors are typically acceptable if below that factor-of-two level. More important is to measure α_{basal} and α_{prism} over a broad range of environmental conditions, as an extensive exploration of parameter space is the best way to gain information about the attachment kinetics.

Based on these considerations, we developed the variable-pressure growth (VPG) apparatus illustrated in Figures 7.15 and 7.16. Once again, it is useful to walk through this experiment as a case study in how one can measure the growth behavior of small ice prisms.

Vacuum Chamber. The VPG vacuum envelope is a stainless-steel Conflat chamber with the usual array of ports, including one high-quality, coated optical window at the bottom of the chamber for optical imaging. This window used an indium vacuum seal to avoid warping the overall figure or damaging the optical coatings. The chamber is wrapped on all sides with copper cooling plates covered with soldered copper pipe for cooling using a programmable recirculating chiller. The chamber is opened and baked between runs to reduce chemical contaminants as much as possible.

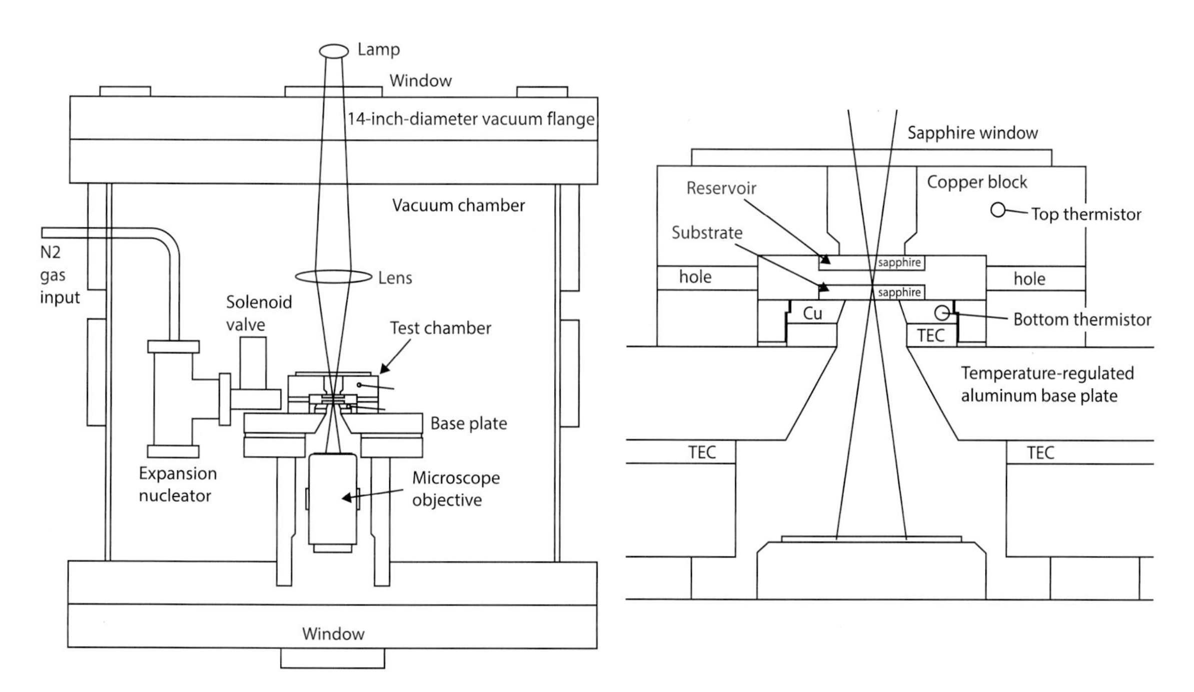

FIGURE 7.15. A schematic drawing showing the VPG vacuum chamber (left) and the test chamber (right). The expansion nucleator creates a collection of minute ice crystals, and some of these make it into the test chamber and onto the waiting substrate. The microscope objective creates an image of these crystals on a camera sensor outside the chamber. A thin film of ice on the upper sapphire window serves as a water vapor reservoir to grow small crystals on the substrate below. The supersaturation is set by adjusting the temperature difference between the two plates, which are separated by a 2-mm gap.

Test Chamber. The inner test chamber consists of a thick-walled copper block resting on a temperature-controlled base plate, illustrated in Figure 7.15. Thermal joint compound provides a good thermal connection to keep the copper block at a fixed temperature. This design supports a pair of sapphire windows in the parallel-plate configuration, as this allows for accurate modeling of the supersaturation in the test region. The top window is wet with a thick water film before each run, which subsequently freezes to form the ice reservoir. The status of the reservoir can be viewed on the camera by refocusing the optical system. The lower window, on which the test crystals rest, is attached to a small thermoelectric module for temperature control. The absolute temperature of the copper block is maintained using a temperature controller operating on the base place, and a second controller sets the temperature difference ΔT between the ice reservoir and substrate. As in the VIG experiment, ΔT determines the supersaturation $\sigma_{subst,0}$ at the substrate surface. The copper block is also lifted off and cleaned between runs to minimize chemical contaminants. Clean air is also cycled into the inner chamber periodically during each run to maintain cleanliness. Because the system temperature is quite low during operation, outgassing from the thermal joint compound and other surfaces does not appear to cause serious problems.

Substrate Interactions. Before each run, the sapphire substrate is thoroughly cleaned, and a hydrophobic surface coating is applied to reduce possible substrate inter-

FIGURE 7.16. The VPG experiment in the lab.

actions (see Chapter 6). Hendlex Nano Glass Pro and Glass Prepare Cleaner give good results, as treatment with these products yields water contact angles that are consistently greater than 90 degrees at room temperature in air, whereas bare sapphire surfaces generally yield contact angles slightly below 90 degrees. Moreover, the prescribed surface preparation seems to give quite robust coatings that are insensitive to imperfect surface cleaning. The ice/substrate contact angle at temperatures below 0°C seems to be generally somewhat higher than the water/substrate contact angle at room temperature, so this bodes well that a suitable coating will largely eliminate significant substrate interactions. However, we have not tested this statement conclusively. Coating technology is rapidly advancing on many fronts at present, so it is likely that new and improved coatings will become available in the near future.

Unfortunately, there is no clear test that would guarantee the absence of significant substrate interactions in an experiment of this kind. Indeed, we have some preliminary evidence that growth rates at especially low σ_{surf} are somewhat higher than expected from a terrace nucleation model, as the model predicts extremely low growth rates when $\sigma_{surf} \ll \sigma_0$ on a faceted surface. Thus there may be some low level of heterogeneous terrace nucleation that arises from substrate interactions under even the best conditions. This is one area where the VIG experiment was superior, as it measured ice surfaces that did not contact the substrate.

Test Crystal Preparation. Once the vacuum pressure, base temperature, and substrate temperature are all set and stable, a growth set is initiated by pulsing the expansion nucleator. This produces a small cloud of nascent test crystals that spray onto the outside of the test chamber, and some enter it through a small hole in the copper block, depositing an assortment of submicron ice crystals on the substrate surface. These crystals are imaged by the microscope objective onto a camera sensor via a direct optical path. For a typical growth set, the camera data acquisition is set to a fixed cadence before nucleation, as the crystals are too small to be easily seen until after the growth cycle is over. With experience, the nucleator can be set to produce a suitable density of crystals on the substrate.

The desired spacing between test crystals is set by an experimental trade-off: on one hand, a larger sample makes it easier to find well-formed specimens for analysis, plus one can examine the distribution of crystal sizes to avoid outliers; on the other hand, the large-scale diffusion correction becomes detrimental if there are too many crystals on the substrate, as shown in the diffusion analysis above. For example, with the 4.9 mm × 3.3 mm field of view of the VPG, a set of 100 test crystals means a typical spacing of $\ell \approx 400$ μm between adjacent neighbors, which results in a fairly small diffusion correction under low-pressure conditions. Figure 7.17 shows a representative set of crystals after growing large enough for analysis. In most runs, the overall crystal sizes are kept below 10–20 microns, which is large enough for making satisfactory growth measurements but small enough to avoid large diffusion corrections.

Crystal Selection. Having dozens of crystals on the substrate during a growth set makes it especially easy to examine the crystal uniformity and avoid outliers. The eye is quite good at rapidly discerning the overall consistency on a field, even when the crystal orientations are somewhat random. Figure 7.18 shows a measurement of the size distribution of crystals on a single field (more crowded than that in Figure 7.17), yielding an overall width of about ±20 percent for both the basal and prism dimensions. In most cases, it is possible to determine the approximate center of the size distribution simply by a judicious visual examination of a single substrate image showing the crystals when their overall sizes are reasonably large. In this way, one can use the entire set of crystals to select a "typical" specimen for detailed analysis. In contrast, other experimental strategies involving the measurement of one crystal at a time (such as the VIG experiment and levitation experiments) are subject to troublesome inconsistencies if a crystal chosen for de-

FIGURE 7.17. (Left) A sample image of the substrate after a set of crystals had grown to a fairly large size at −7°C, showing the entire 4.9×3.3 mm field of view of the camera. (Right) Several individual crystals from this set are shown, with each small frame measuring 37×31 microns. The crystals fell onto the substrate with random positions and orientations, but most grew to similar sizes with a simple prism morphology. A few fast-growing outliers can also be found in this field as well, probably affected by crystal dislocations or other defects.

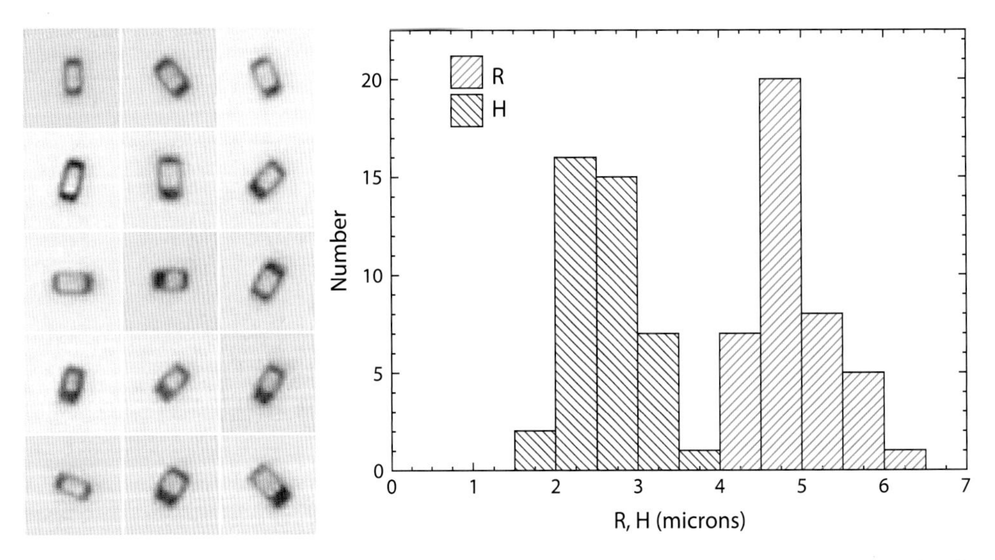

FIGURE 7.18. (Left) Several ice crystals selected from a single image of the VPG substrate after a brief growth period at −5°C. (Right) The histogram shows measured effective radii R and half-thicknesses H from these crystals along with numerous others, including all crystals that exhibited simple prismatic morphologies and were oriented such that both R and H could be measured.

tailed study happens to be a significant outlier for some reason. Recognizing outliers necessarily requires a large sample size, and the VPG experiment supplies this without having to make laborious measurements of many individual crystals.

Supersaturation Management. Before nucleation, the supersaturation is set to a predetermined value by means of the temperature difference ΔT, so the crystals begin growing as soon as they land on the substrate. The nucleation pulse introduces a brief temperature perturbation that initially disturbs the growth conditions by a small amount, but this perturbation settles after about 10 sec, with the time scale being set by the temperature servo response. In a typical growth set, the crystals are allowed to grow for some fixed period under constant conditions, then ΔT is reduced linearly with time to slowly reduce the supersaturation and slow the growth. The ramp is continued until the supersaturation drops below zero and the crystals begin sublimating. Observing this transition from growth to sublimation allows a precise determination of the $\sigma = 0$ point, better than what can be extracted from the thermistor data alone.

Modeling Tiny Prisms

A growth set in the VPG experiment consists of a series of images of the substrate together with a simultaneous recording of ΔT from the temperature controller. In a typical set, the substrate contains a collection of crystals that have landed at random positions and orientations, with a spacing of at least several hundred microns between crystals on average. With roughly isometric prisms, some specimens will be oriented with one prism facet resting flat on the substrate, allowing a simultaneous measurement of $R(t)$ and $H(t)$, these being respectively the crystal "radius" (approximating the hexagonal prism as a circular prism) and half-thickness as a function of time. Choosing one of these crystals for analysis yields a time series like that shown in Figure 7.19. With up to 100 growing crystals on the substrate, it is usually straightforward to visually inspect the field to choose a suitable specimen for detailed analysis. Or, if the crystal uniformity is observed to be poor, the whole collection can be discarded.

Once a suitable test crystal is chosen and the measurements of $R(t)$ and $H(t)$ are obtained from the image series, the data are fit using a fast "1.5-dimensional" numerical modeling program [2019Lib2]. Fixed program inputs include: 1) the $R(t)$ and $H(t)$ data, 2) the corresponding time series ΔT from the temperature controller, 3) the measured average crystal spacing ℓ on the substrate, and 4) a variety of physical parameters. Adjustable program inputs include: 1) the ΔT value of the $\sigma = 0$ point, which is when $\sigma_{subst,0}$ goes to zero, and 2) a guess for the attachment coefficients $\alpha_{basal}(\sigma_{surf})$ and $\alpha_{prism}(\sigma_{surf})$. Using these inputs, the program integrates the crystal growth to determine a model $R(t)$ and $H(t)$ that it plots along with the data, as illustrated in Figure 7.19. The inputs are then adjusted to give a good fit to the data, which takes little time, because the 1.5-dimensional code runs in a matter of seconds. In most cases, it is straightforward to converge on an unambiguous set of parameters that give a reasonable fit to the data.

As the code runs, it first uses the calculated test crystal size and growth velocities from the previous time step to calculate σ_{subst} from $\sigma_{subst,0}$ using the large-scale diffusion correction described earlier in the chapter. This section of the code uses the measured crystal spacing ℓ on the substrate, and it assumes a square array of crystals all identical to the modeled test crystal. Clearly, this result is only an approximation of the actual distribution of crystals on the substrate, but it is adequate if the correction is not too large. The code further calculates σ_{surf} from σ_{subst} using the small-scale diffusion correction described above along with a volume-conserving estimate for R_{eff} in Equation 7.12. Once σ_{surf} is suitably determined, it is straightforward to calculate the growth

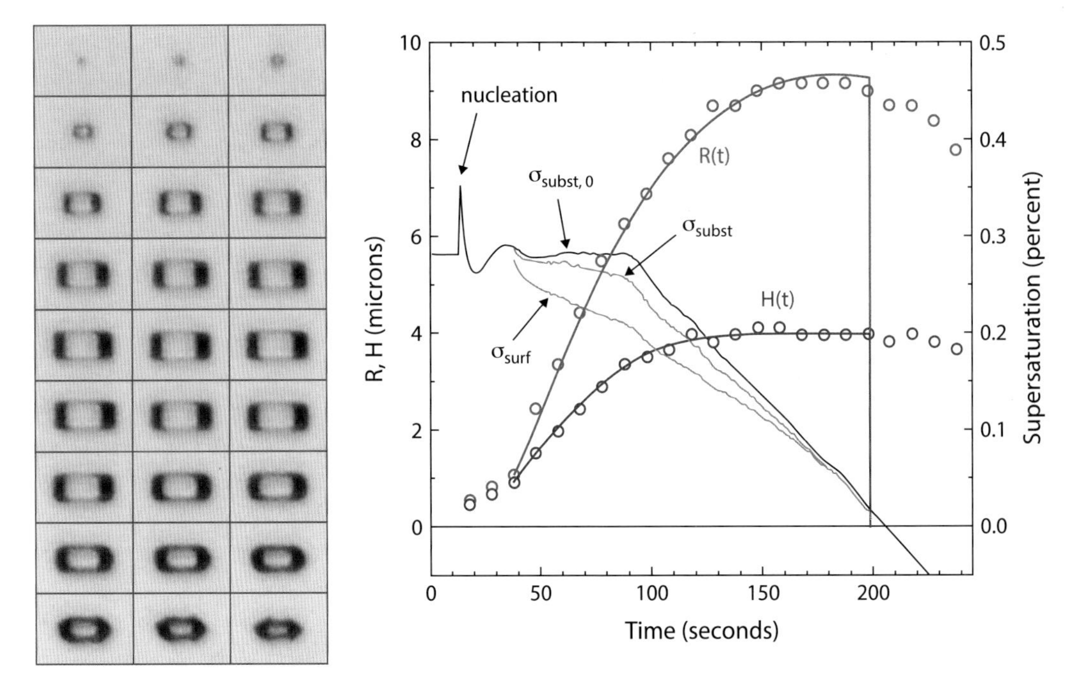

FIGURE 7.19. The panel on the left shows a series of images of a single ice prism growing and then sublimating at −5°C with an air pressure of 60 mbar. The effective radius $R(t)$ and half-thickness $H(t)$ were obtained from each image to yield the data points shown in the accompanying graph (left scale). Lines through the points show a model fit to the data, as described in the text. Additional lines show the calculated supersaturation (right scale) in the absence of any test crystals ($\sigma_{subst,0}$), after a large-scale diffusion correction (σ_{subst}), and after an additional small-scale diffusion correction to yield the near-surface supersaturation (σ_{surf}).

velocities and advance the crystal by a time step. As is something of a theme in this chapter, the difficult part of the analysis is always determining σ_{surf} accurately around the growing crystal. Figure 7.19 illustrates how $\sigma_{subst,0}$, σ_{subst}, and σ_{surf} evolve with time as ΔT was slowly ramped down in this growth set, and additional modeling details can be found in the literature [2019Lib2].

Note that the 1.5-dimensional analysis also makes it straightforward to quantify and reduce systematic errors in the VPG experiment, simply by adjusting the various parameters to see how this changes the fit to the data. The vacuum pressure, $\sigma = 0$ point, crystal spacing ℓ, the attachment kinetics parameterization, and other factors can all be examined and adjusted to estimate the measurement uncertainties with reasonable confidence. This information can further be used to change various experimental parameters in subsequent runs to obtain better measurements under a variety of growth conditions.

The largest uncertainty in the VPG data usually comes from the position of the $\sigma = 0$ point, which is shown as the vertical line in Figure 7.19. The flattening of the growth rate prior to this time arises from terrace nucleation, because $\alpha(\sigma_{surf}) = A\exp(-\sigma_0/\sigma_{surf})$ drops precipitously once $\sigma_{surf} \ll \sigma_0$. After the $\sigma = 0$ point, the crystal slowly begins sublimating away, which happens first at the corners and later on the broad facets. Modeling the sublimation behavior has proven somewhat

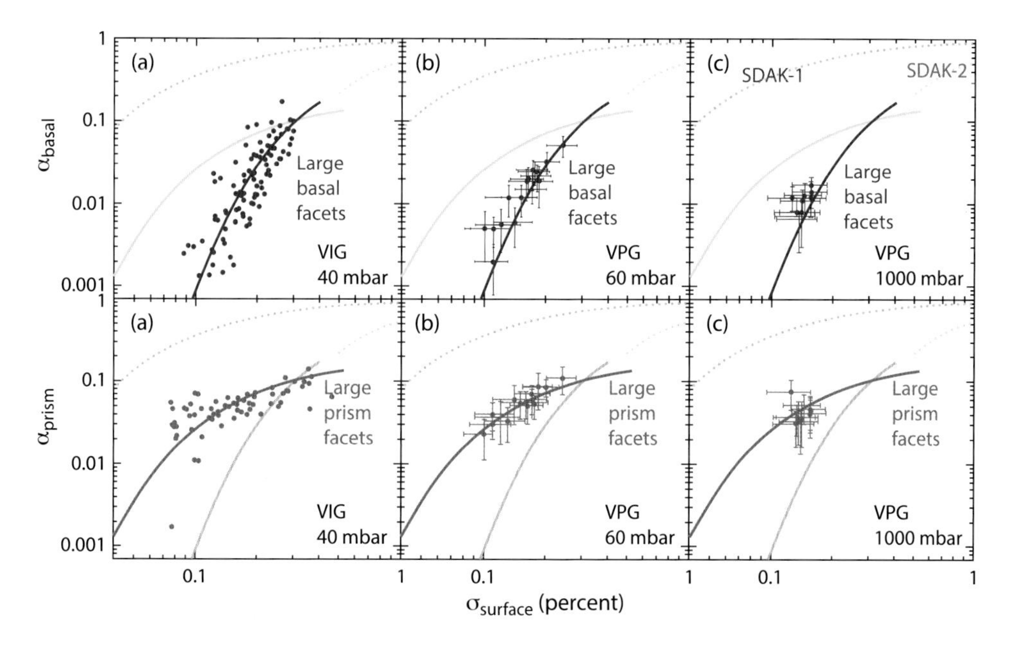

FIGURE 7.20. Measurements of $\alpha_{basal}(\sigma_{surf})$ and $\alpha_{prism}(\sigma_{surf})$ from (a) the VIG experiment at low pressure, (b) the VPG experiment at low pressure, and (c) the VPG experiment at a pressure of one bar [2019Lib1]. All measurements were corrected for diffusion effects to yield the attachment kinetics on the faceted surfaces of simple-prism crystals. Lines are from the CAK model (Chapter 4) for −5°C, and the additional CAK curves are shown faded in the different panels. Excellent agreement is seen between the two experiments, and there is no apparent pressure dependence in the attachment kinetics on either facet at −5°C.

difficult, so it has not yet been possible to fit to the full growth/sublimation curve in detail. Note also that the modeled $\sigma = 0$ position is not precisely at $\sigma = 0$, being shifted slightly by the Gibbs-Thomson effect on these small crystals. Although the uncertainty in determining this point is not large, the fitted growth parameters are quite sensitive to its location. Thus, as usual, the most difficult part of the VPG experiment is accurately determining σ_{surf} around each growing crystal.

SIMPLE-PRISM CONVERGENCE

At the time of this writing, we have not yet obtained a comprehensive set of measurements over a broad range of temperatures with the VPG experiment, but early data show good agreement with the VIG experiment, and one example of this is shown in Figure 7.20. Several immediate conclusions can be reached from these data:

1) The VIG and VPG experiments show excellent agreement at −5°C, even though they use significantly different measurement strategies and completely different hardware. Moreover, the basal and prism facets were measured separately in the VIG experiment, using different crystals in different runs, and even different types of interferometry on the basal and prism facets. In contrast, both facets were measured simultaneously on smaller ice prisms in the VPG experiment. Obtaining such clear agreement

from quite different experiments is itself a noteworthy achievement, given the considerable measurement uncertainties and discrepancies found in earlier ice growth experiments.

2) There is no obvious air-pressure dependence in either α_{basal} or α_{prism} as a function of σ_{surf} in the VPG experiment, at least over the limited range that was measured. This supports our assumption in the CAK model that air at a pressure of 1 bar has little effect on the attachment kinetics.
3) The data all strongly support the terrace-nucleation mechanism for the large-facet attachment kinetics in the CAK model. Specifically, the function forms for both $\alpha_{basal}(\sigma_{surf})$ and $\alpha_{prism}(\sigma_{surf})$ are both well described by the terrace-nucleation model.
4) The data support $A_{basal} \approx 1$ and $A_{prism} < 1$ at this temperature, as described in Chapter 4. Constraining the model to have $A_{prism} = 1$ is not excluded completely by the measurements shown here, but both data sets show quite similar trends that prefer $A_{prism} < 1$. The same is true for the data at −2°C shown in Chapter 4. In all four cases (VIG and VPG data at −2°C and −5°C), the individual data sets each separately support $A_{prism} < 1$.
5) As described in Chapter 4, substantial evidence from a variety of experiments now indicates that platelike crystals are the norm at −5°C when the overall morphology is that of a simple hexagonal prism. Columnar forms appear at higher supersaturations in air when the basal facets are narrow, which is a manifestation of the SDAK phenomenon.

It will take some years to complete our experimental program with the VPG experiment, examining the full available temperature range and further investigating whether the attachment kinetics are significantly affected by air pressure. Figure 7.20 shows results at −5°C [2019Lib1], and similar data at −2°C are shown in Chapter 4 [2020Lib]. Combining all the VIG and VPG data yields the measurements shown in Figures 7.2 and 7.3. Although there is still much work to be done, the data we have acquired to date provide a self-consistent picture of the ice/vapor attachment kinetics on large faceted surfaces, providing the foundation for the CAK model described in Chapter 4.

At temperatures below −25°C, there are limited data, and the overall picture is not so clear. Measurements by Bailey and Hallett [2004Bai, 2009Bai, 2012Bai] indicate that columnar forms predominate at temperatures between −40°C and −70°C at relatively high growth rates, with typical aspect ratios (length/diameter) of roughly 10. Converting the observations to a characteristic supersaturation (a surrogate for σ_0 in a terrace-nucleation model) [2019Har] shows essentially a power-law behavior that extends the $\sigma_0(T)$ trends seen in Figure 7.3, with $\sigma_{0,basal}$ and $\sigma_{0,prism}$ differing by no more than about a factor of 2 below −30°C. Arrowhead crystals at −40°C (see Chapter 2) are consistent with an aspect ratio of about 10.

The CAK model predicts that there should be almost no SDAK effects at these low temperatures, so the observed columnar growth at −40°C should result from the terrace-nucleation model. Assuming this model, one expects a crystal length/diameter aspect ratio of approximately $\exp(\Delta\sigma_0/\sigma_{surf})$, where $\Delta\sigma_0 = \sigma_{0,prism} - \sigma_{0,basal}$. At −40°C, the CAK model was chosen to have $\sigma_{0,prism} \approx 23$ percent, $\sigma_{0,basal} \approx 14$ percent, and thus $\Delta\sigma_0 \approx 9$ percent, so an aspect ratio of 10 requires $\sigma_{surf} \approx 4$ percent, which is not unreasonable for growth in air at this temperature, and far below $\sigma_{water} \approx 47$ percent. This quick calculation shows that a simple terrace nucleation model is sufficient to yield columnar growth at low temperatures, while the SDAK effect is necessary to produce thin plates and slender columns at higher temperatures, as described in Chapter 4.

In summary, the focus in this chapter has been on experimental studies of simple ice prisms, because these studies provide the foundation for our understanding of the attachment kinetics. Over the temperature range from −2°C to −30°C, data from both the

VIG and VPG experiments are well described by a terrace nucleation model with $\alpha(\sigma_{surf}) = A\exp(-\sigma_0/\sigma_{surf})$. The basal growth data additionally indicate $A_{basal} \approx 1$ over this entire temperature range, while the prism data indicate $A_{prism} < 1$ at temperatures above $-10°C$. The two experiments yield remarkably similar results, and both appear to be substantial improvements relative to earlier experiments. In this realm, therefore, the data appear to be converging on a consistent picture of simple prism growth.

FIGURE 8.1. Stellar snow crystals grow on the ends of slender ice needles in this laboratory photo. The c-axis needles were created using high electric fields to accelerate normal crystal growth by using the techniques presented in this chapter.

EIGHT

Electric Ice Needles

The universe is full of magical things patiently waiting for our wits to grow sharper.

—EDEN PHILLPOTTS, *A SHADOW PASSES*, 1919

Experimental hardware features prominently in this book, because laboratory investigations are so important in our attempts to understand the physical dynamics of snow crystal growth. Theory provides a necessary mathematical framework for the discussion, along with new ideas regarding the underlying molecular processes, but experiments are essential to test those ideas and suggest additional theoretical directions. Progress happens when the two are applied in concert. While Chapter 7 looked at growing simple ice prisms for investigating the molecular attachment kinetics on broad facets, this chapter examines a particularly valuable technique for studying larger snow crystals exhibiting thin edges and complex morphologies. By creating slender "electric" ice needles and cultivating ordinary snow crystals on their ends, it is possible to explore growth behaviors that arise from structure-dependent attachment kinetics (SDAK), which plays such a major role in defining the Nakaya diagram (see Chapter 4). As I describe in this chapter, the electric-needle method allows the reproducible growth of complex structures over a broad range of temperatures and supersaturations, starting from well-defined seed crystals. These synthetic snowflakes are especially well suited for analysis alongside their 3D computational counterparts, thus facilitating direct quantitative comparisons between theoretical models and experimental measurements.

SNOWFLAKE ON A STICK

Figure 8.2 shows a typical set of c-axis electric ice needles, which I also call *e-needles*. One begins with a thin metal wire exposed to highly supersaturated air in a diffusion chamber, so the wire quickly becomes covered with frost. The wire is seen at the bottom of the photograph, supporting with an assortment of small ice crystals growing on its surface. In this example, the temperature surrounding the wire tip was near −6°C, so the frost crystals grew in random orientations with a generally columnar morphology, as expected from the Nakaya diagram (see Chapter 1). The wire extends down to the bottom of the growth chamber where it exits and is connected to a high-voltage power supply.

When a high voltage (typically +2,000 volts DC) is applied to the wire, slender ice e-needle crystals grow outward from the tips of some frost crystals in just a few seconds, provided that the conditions are right. When the temperature at the wire tip is close to −6°C, the water vapor supersaturation is near 100 percent, and the air contains trace quantities of acetic acid vapor, then the e-needles will typically emerge, growing along the crystalline c-axis with tip velocities of about 100–150 microns/sec [2002Lib]. In these conditions, 3-mm-long e-needles can be grown in less than a minute.

While electrically modified needle growth is an intriguing phenomenon in its own right, c-axis e-needles are also quite useful as seed crystals for a broad range of snow crystal investigations. Ordinary (nonelectrified) growth commences as soon as the applied high voltage is removed, yielding well-formed single-crystal specimens growing on the tips of the e-needles. In Figure 8.3, for example, a set of e-needles was transferred to a second growth chamber near −15°C, causing platelike crystals to grow on the ends of the e-needles. As the plates became larger, they soon shielded the ice needles on which

FIGURE 8.2. A set of slender e-needles grows on the end of a frost-covered wire. The e-needle centered in the image is about 3 mm in length. The e-needles formed when +2,000 volts was applied to the frost-covered wire at the bottom of the photo, which was simultaneously exposed to highly supersaturated air at a temperature near −6°C. The e-needle phenomenon is the result of an electrically induced growth instability described in this chapter. The radius of curvature of the tip of a fast-growing e-needle can be as small as 100 nm.

FIGURE 8.3. Thin, platelike snow crystals growing on the ends of c-axis e-needles. This example illustrates how e-needles can be used to cleanly support isolated snow crystals as they growth and develop. After the thin plates had grown out, the supersaturation was raised to stimulate branching from the corners of the plates. Figure 8.1 shows a similar example after additional growth.

they grew, so the plate structures were only moderately perturbed by the presence of their supporting needles.

The e-needle method is an especially versatile tool for studying snow crystal growth in the dual-chamber apparatus described in detail below. The first chamber is optimized for rapidly and reproducibly growing high quality c-axis electric needles, while the second chamber is designed to provide a carefully controlled growth environment that can produce a broad range of temperatures and supersaturations. Once created in the first chamber, the e-needles are quickly transported to the second growth chamber, so their subsequent normal growth can be observed and measured. With such a dual-chamber setup, the entire morphology diagram can be explored using quantitative growth measurements. The challenge then becomes creating realistic computational models that can reproduce both the observed growth rates and morphologies. Figure 8.4 illustrates just a few of the possibilities.

If there is one thing I have learned with great certainty in this field, it is that snow crystal growth exhibits a mystifying variety of complex growth behaviors that will require much additional study to appreciate. While the simple ice prisms discussed in Chapter 7 generally fit a basic terrace-nucleation model, the formation of thin plates and hollow columns requires the SDAK and ESI phenomena (see Chapter 4), and dendritic structures arise from a convoluted mix of diffusion-limited growth

FIGURE 8.4. E-needles make excellent seed crystals for studying the development of complex snow crystal morphologies at high supersaturations, as shown in these examples. Once a c-axis e-needle has grown to a desired length, the high voltage is removed, and normal growth commences on the needle tip. Each single-crystal structure shown was grown in air at a constant temperature and supersaturation, and time-lapse photography can record the full growth history if desired. Observations like these are wonderfully suited for comparing quantitative growth measurements with detailed computation models, hopefully leading to a better understanding of the physical dynamics of snow crystal formation.

and anisotropic attachment kinetics. Understanding how all this works will require a variety of experimental tools that can be used in conjunction with sophisticated 3D computational models. This undertaking is nontrivial, and I believe that experiments using electric ice needles will play an important role in advancing the science of snow crystal formation.

Advantages . . .

At this point, I believe it is instructive to examine what makes the e-needle method especially useful as a scientific tool and to compare it with other experimental techniques for creating snow crystals for quantitative study.

Single-Crystal Specimens. One normally thinks of a seed crystal as being a tiny crystalline speck, a minute hexagonal prism in the case of ice. A single-crystal specimen is almost always desirable, as polycrystalline samples are unnecessarily complicated and less well suited for investigating the underlying crystal growth dynamics. But being small in all three dimensions is not an essential requirement, and a slender e-needle, small in two dimensions only, can still be considered to be a seed crystal. Moreover, with a well-defined crystal orientation and a sub-micron tip radius (at least during its initial growth phase), a c-axis e-needle embodies many qualities one seeks in an ideal seed crystal.

Versatile Support. Once a seed crystal has been created, the question of supporting it while it grows must be addressed. In the case of e-needles, this question answers itself, as the support is already provided by the wire from which the needle originally grew. This support is robust and is easily manipulated, so the e-needle can be moved to a second growth chamber, positioned in front of a camera lens, and rotated to a desired orientation with relative ease. Moreover, a complex snow crystal growing on the end of a long, slender e-needle is well isolated from other parts of the apparatus, so its surrounding environment can be carefully manipulated and controlled.

No Substrate Interactions. When a seed crystal is supported by a non-ice surface (for example, a supporting surface, filament, or capillary tube), there is a good chance that the substrate will influence the ice growth rates. Flat substrates and capillary supports all suffer from this problem to some extent. I discuss the topic of substrate interactions in some detail in Chapters 6 and 7, because this has been a significant issue in many of my own ice growth experiments. Moreover, the substrate interaction problem was often not fully appreciated in the past, sometimes yielding growth data of dubious quality. With crystals growing on the tips of e-needles, however, no non-ice surfaces are present, so substrate interactions are completely absent. When the entire structure is made of ice, it all becomes part of the same computational modeling problem.

No Water Condensation. Another problem with non-ice surfaces in a supersaturated environment is water condensation. If the supersaturation is above the dew point, then water droplets tend to condense on all available surfaces, greatly perturbing the surrounding supersaturation field. For example, with small seed crystals placed on a planar substrate, droplet condensation readily occurs around the crystals (see Chapter 9), which effectively precludes any useful studies at high supersaturation levels. E-needles avoid this issue, as there are no non-ice surfaces for water to condense on. Thus with e-needle support, one is free to explore quite high supersaturations with relative ease compared to other techniques.

Witness Surfaces. The supersaturation level is difficult to know with high accuracy in ice growth experiments, and e-needles provide a way around this problem, at least partially. In many circumstances, the columnar body of the needle can serve a "witness surface" for determining the surrounding supersaturation. If the columnar growth satisfies $\alpha_{diffcyl} \ll \alpha_{prism}$ (see Chapter 3), then the radial growth of the needle is determined to a good approximation by just the needle radius and the faraway supersaturation. In practice, then, the measured growth of the body of the needle can often be used to effectively measure the faraway supersaturation.

Rapid Turnaround. To make any real progress toward understanding the physics of snow crystal growth, one must measure a lot of crystals. The underlying molecular processes are complex and difficult to isolate, plus everything changes substantially with temperature, supersaturation, and a variety of other factors. Therefore, one of my favorite features of the e-needle method is that it is possible to grow a lot of crystals in a short time, while still examining each one as it grows. Because several e-needles typically form simultaneously on a wire tip (for example, as shown in Figure 8.3), the observer can select the best of several specimens, while examining the others to gauge the overall variability in growth and morphological development. This is useful for avoiding crystal outliers that can distort one's scientific conclusions. The wide spacing between needle tips results in only minor interactions between the growth of the different crystals in a cluster. Equally important, a new cluster of e-needles can be created in about a minute's time, allowing many observations in a single observing session. By comparison, techniques with lower turnaround often look good in a first demonstration experiment but then lose their luster when it ends up taking all day to produce just two or three useful measurements.

The Double-Plate Problem. Another surprisingly beneficial feature of the e-needle method is that a needle tip contains only one exposed basal facet, compared to two basal facets on a small hexagonal prism. As illustrated in Figure 8.5, when a thin plate grows out from an e-needle, there will only be one plate. But two plates may emerge from the two ends of a stout columnar crystal, and the double plate often presents a substantially more difficult analysis problem. In the lab, these two plates are subject to a growth instability, in that they will compete until one dominates over the other. In computational models, however, the double-plate symmetry is typically built into the code, so both plates grow equally. This difference can make it problematic to compare models and experiments. This "double-plate" problem does not exist with the e-needle case, as there is only one plate from the outset.

Moreover, the geometry of an e-needle tip and its surroundings provides a degree of symmetry breaking that can be valuable when examining morphologies and growth rates in detail. In many circumstances, a plate-

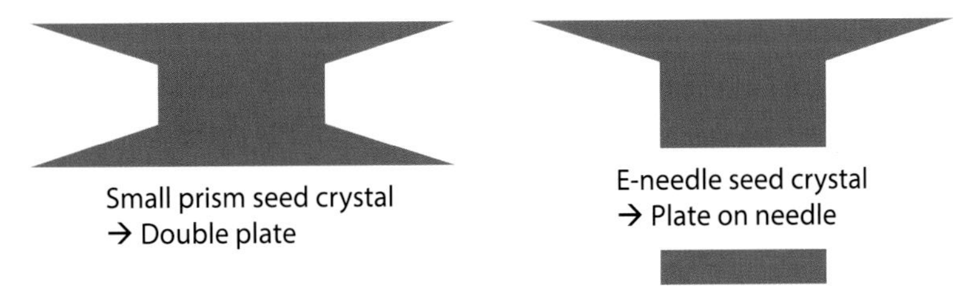

FIGURE 8.5. Using e-needles avoids the "double-plate problem" in snow crystal modeling. Simple-prism seed crystals often grow into double-plate structures (left), both in natural snow crystals and in computational models. This added complication can be problematic when comparing models with laboratory experiments, as the two plates tend to interfere with one another. In similar conditions, however, only a single plate emerges from an e-needle seed crystal (right). Thus, snow crystals on e-needles are often easier to grow and to analyze than crystals growing from small seed crystals, especially at high supersaturations when morphologies are complex.

like crystal growing on the end of an e-needle will exhibit a slightly concave upper basal surface and a slightly convex lower basal surface (depending on the detailed growth conditions). This built-in asymmetry means that ridge structures form only on the lower basal surface, while inwardly propagating macrosteps are found only on the upper basal surface (see Chapter 3). Several other morphological features are similarly isolated when using an e-needle seed crystal, which turns out to be surprisingly helpful when trying to decipher the characteristics and physical origins of these features, especially when they appear together.

. . . and Disadvantages

Although the use of e-needles as seed crystals has numerous experimental advantages when examining complex snow crystal structures, there are some disadvantages associated with the technique as well.

Larger Diffusion Effects. Because $\alpha_{diffcyl} < \alpha_{diff}$ at fixed R (see Chapter 3), the diffusion effects arising from an e-needle of radius R are substantially larger than those of a hexagonal prism of overall size R. This is bad news for the e-needle method, as it means that e-needles are not especially well suited for making quantitative measurements of the attachment kinetics. As presented in Chapter 7, the smallest possible seed crystals are needed for this purpose. I have tried to use simple columnar e-needles for making measurements of attachment kinetics, but the diffusion corrections are always large and problematic. In general, the e-needle method shines best when growing crystals with complex morphologies, at high supersaturations, for comparison with corresponding computational growth models.

Weight Restrictions. Although e-needles can support their own weight along with some build-up of material on their tips, there are limits. The contact point at the base of an e-needle is especially weak, and the needle will fall if it becomes sufficiently top heavy. Oddly enough, e-needles almost never crack and break the way one might expect from a crystalline structure. Instead, an e-needle tends to rotate slowly downward when too much weight accumulates at its tip. The e-needle itself behaves like a rigid structure, but its base support does not. When a growing crystal on the tip becomes sufficiently heavy, the entire needle structure usually pivots about its support point, slowly falling like a stick with its bottom end held in chewing gum.

Growth at Room Air Pressure. The apparatus described in this chapter operates in air at 1 bar, and modifying it to operate in a vacuum environment would be difficult. Thus, while exploring snow crystal growth as a function of background gas pressure is desirable, achieving this with e-needles presents a significant experimental challenge.

Complicated Construction. A final disadvantage with the e-needle method is that the apparatus is complex and therefore nontrivial to construct. I describe my dual-chamber setup in some detail in this chapter and in [2014Lib1], but it is impractical to list every nuance of its construction and operation in any publication. A great deal of trial-and-error experimentation was necessary to produce satisfactory crystals and growth measurements, and much of this effort would likely have to be repeated (to some degree) with a new e-needle venture. Although this disadvantage is nontrivial, I certainly hope that the results presented in this chapter stimulate at least some interest in developing the e-needle technology to greater heights.

E-NEEDLE FORMATION

Electrically enhanced ice growth was discovered in 1963 by Bartlett, van den Heuvel, and Mason [1963Bar], who observed the spontaneous formation of fast-growing e-needles when large, positive DC voltages were applied to ice crystals growing at high supersaturations. Libbrecht and Tanusheva explained the underlying physical cause as an electrically induced growth instability 35 years later [1998Lib, 1999Lib1, 1999Lib2]. The importance of chemical influences on the crystalline orientation of e-needle growth was discovered soon thereafter [2002Lib], leading to the reliable technique for growing high quality c-axis e-needles described in this chapter.

Basic Theory

The physical mechanism that produces electric ice needles can be understood by first considering the equilibrium vapor pressure of a charged ice sphere. From basic electrostatics, if the sphere has some nonzero conductance, then the static charge must all reside on the surface of the sphere, while the electric field inside the sphere is zero. Pulling a neutral water molecule off the sphere reduces its radius but not its charge, and this brings the surface charges closer together than they were before the water molecule was removed. Because like charges repel, it requires some energy to reduce the size of the sphere and pull the surface charges closer together. It follows that pulling a water molecule off a charged sphere requires slightly more energy than pulling a water molecule off an uncharged sphere. For this reason, the equilibrium vapor pressure of a charged sphere is slightly lower than that of an uncharged sphere. A high voltage applied to the sphere has the same effect.

The argument is essentially the same as for the Gibbs-Thomson effect presented in Chapter 2, and the math is also similar. Adding in the electrostatic self-energy term, the equilibrium vapor pressure of a charged sphere of radius R becomes

$$c_{eq}(R) \approx c_{sat}\left(1+\frac{2d_{sv}}{R}-\frac{R_{es}^2}{R^2}\right), \tag{8.1}$$

where

$$R_{es}^2 \approx \frac{\varepsilon_0 \varphi_0^2}{2c_{ice}kT}, \tag{8.2}$$

φ_0 is the applied electrical potential, and ε_0 is the vacuum permittivity in SI units [2002Lib].

When the growth of the sphere is mainly diffusion limited, we can add the electrostatic term to the same perturbation expansion described in Chapter 3 to obtain

$$v \approx \frac{X_0}{R} v_{kin}\left(\sigma_\infty - \frac{\sigma_\infty}{\alpha}\frac{X_0}{R}+\frac{R_{es}^2}{R^2}\right), \tag{8.3}$$

where here we have neglected the small effect from surface tension. The corresponding equation for the tip velocity of a growing Ivantsov parabolic crystal becomes

$$v_{tip} \approx \frac{2X_0 v_{kin}}{BR_{tip}}\left(\sigma_{far} - \frac{\sigma_{far}}{\alpha}\frac{2X_0}{BR_{tip}}+\frac{GR_{es}^2}{R_{tip}^2}\right), \tag{8.4}$$

where R_{tip} is the radius of curvature of the parabola at the tip, and G is a dimensionless geometrical factor. This equation describes the growth of the parabolic crystal illustrated in Figure 8.6.

Examining the individual terms in Equation 8.4 gives us a picture of the essential physics underlying the e-needle growth instability. The first term in the parentheses gives the constant tip velocity of a parabolic crystal when its growth is entirely diffusion limited. This is the Ivantsov solution discussed in Chapter 3, arising solely from the solution to the particle diffusion equation. The second term is rather small compared to the first, but it reduces the growth velocity as R_{tip} becomes smaller. This term, albeit small, plays an essential role in stabilizing the normal growth of a parabolic crystal, as it "selects" the final R_{tip} for the growing Ivantsov parabola via solvability theory (see Chapter 3). Together, these first two terms describe the growth of a normal ice needle or dendrite that has an approximately parabolic shape near its tip.

The third term in Equation 8.4 tends to destabilize the normal parabolic growth, and this is the term that drives the growth of e-needles. As R_{tip} becomes smaller, this term increases the tip velocity relative to the normal growth, and the R_{tip}^{-2} dependence means that this term eventually overwhelms the kinetic term in the equation as the tip sharpens. Including this third term in an extension of solvability theory and following the algebra

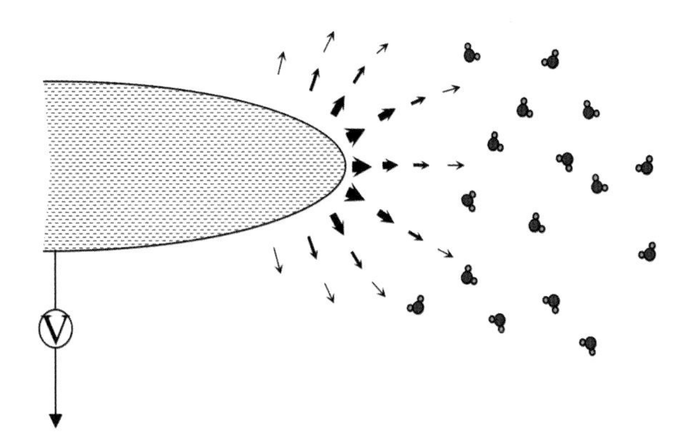

FIGURE 8.6. When a high voltage is applied to an ice needle, strong electric fields (arrows) are concentrated at the tip. These fields decrease the equilibrium water vapor pressure near the tip and increase its growth rate, yielding fast-growing e-needles.

through, the resulting equation for the parabola tip radius can be written in the form of a quadratic equation [2002Lib]

$$R_{tip}^2 - R_0 R_{tip} + A R_{es}^2 \approx 0, \tag{8.5}$$

where R_0 is the tip radius in the absence of an applied electrical potential (the normal solvability-theory result), and A is a dimensionless constant.

Solving the quadratic equation gives $R_{tip} = R_0$ when there is no applied potential, which is the normal solvability result. As the potential is turned on slowly, at first the solution yields a tip velocity that is only slightly larger than the normal velocity. In this regime, the normal solvability solution is only slightly perturbed by the applied potential, decreasing R_{tip} and increasing v_{tip} as φ_0 becomes larger. Thus, there is no dramatic effect when a small voltage is applied, as one would expect.

This "perturbative" regime remains in effect as long as $R_{tip} < 2R_0$, or, equivalently, as long as v_{tip} is no greater than about twice its normal growth result. Under typical ice growth conditions, the perturbative regime holds as long as the applied voltage is less than about 1,000 volts. Beyond that point, the quadratic equation no longer has any real roots, meaning that the second term can no longer stabilize the growth as described by solvability theory.

Physically, the destabilizing electrostatic term eventually brings about a full-blown growth instability. Above a threshold voltage of about $\varphi_{thresh} \approx 1{,}000$ volts, the physical influence of the third term in Equation 8.4 exceeds that of the second term, providing a positive feedback effect that leads to runaway growth. Reducing R_{tip} makes the tip electric fields higher, which turns up the growth rate and reduces R_{tip} still more, further increasing the tip electric fields. All this quickly leads to an abrupt increase in v_{tip} and the formation of an electric ice needle.

Figure 8.7 shows a direct comparison of experiment measurements with the theory described above [1998Lib], illustrating the initial perturbation of the solvability solution followed by a runaway instability that leads to the formation of an electric needle. Consistent with the solvability model, the dendrite tip growth increases in a well-behaved fashion until reaching about twice its normal value, at which point the e-needle forms, and the tip velocity increases abruptly.

Figure 8.8 shows measurements of tip velocities near −5°C for normal needle growth, e-needle growth along axes other than the c-axis, and c-axis e-needles. Although the tip radius R_{tip} was often too small to measure optically, the value of R_{tip} can be accurately estimated from the Ivantsov solution and the measured v_{tip}. This solution dictates that R_{tip} is proportional to v_{tip}^{-1}, indicating that the inferred tip radius R_{tip} falls to values below 100 nm for the fastest growing c-axis e-needles.

This electrostatic theory is a natural extension of solvability theory, and it seems to fit the observations reasonably well. But it only describes the tip behavior in the perturbative regime when the tip radius is still stabilized by the kinetic term in Equation 8.4. Above the voltage threshold, there must be some other stabilization mechanism that selects the final e-needle tip velocity. This necessary stabilization cannot be provided by vapor-pressure effects stemming from either surface energy or the attachment kinetics, as these effects both go only as R_{tip}^{-1}, as seen in the second term in Equation 8.4. Therefore, these terms

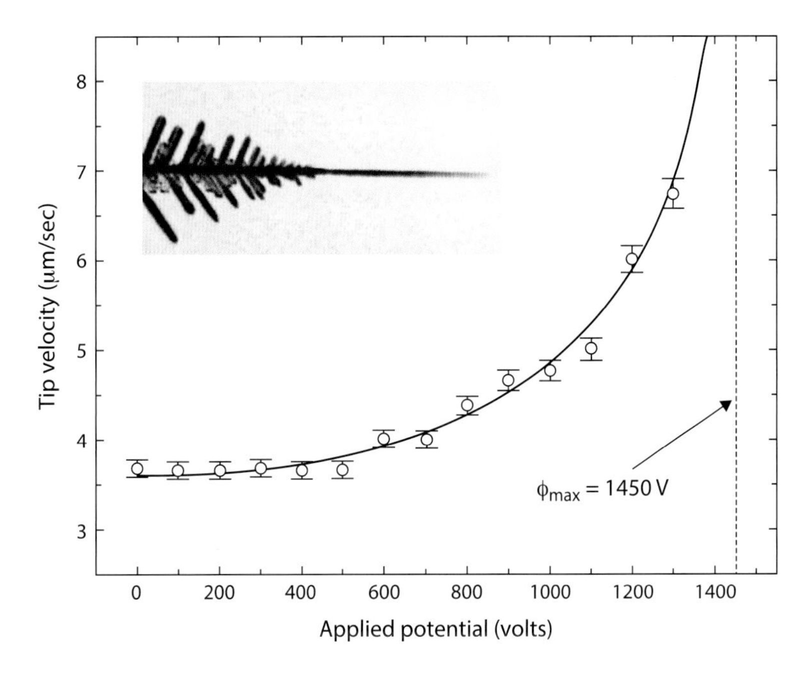

FIGURE 8.7. Experimental measurements (data points) showing a gradual increase in the tip growth velocity of an ice dendrite at −15°C as the applied voltage is increased. The curve through the data points comes from the theory contained in Equation 8.5, with *A* adjusted to fit the data. Note that the dendrite morphology continued to exhibit sidebranching as the voltage was initially increased. Once a threshold voltage was exceeded, however, normal dendrite growth gave way to the formation of an e-needle. The inset image shows a similar run in which the normal dendrite growth transformed into a fast-growing a-axis e-needle above threshold [1998Lib].

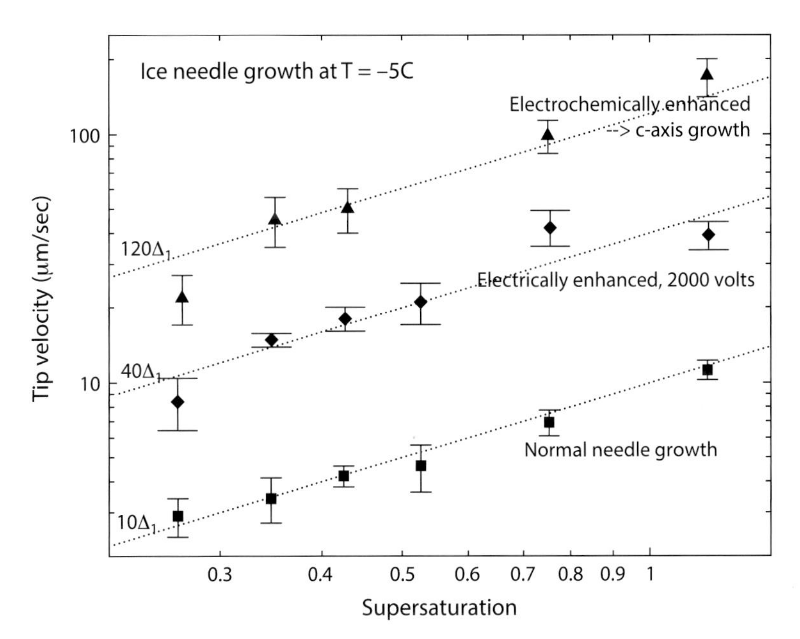

FIGURE 8.8. Experimental measurements (data points) showing the tip velocities for normal needle growth (lower points), e-needles growing along crystalline axes other than the c-axis (middle points), and c-axis e-needles (upper points) [2002Lib]. The data were all taken at a temperature of −5°C, and in all cases, the tip growth velocity is linearly proportional to the supersaturation, as predicted by solvability theory (see Chapter 3). Growth along the c-axis was stimulated using chemical vapor additives, as described in this chapter.

cannot compete with the R_{tip}^{-2} term once the tip radius becomes small. At the 100 nm scale, surface tension may provide sufficient mechanical force to halt additional tip sharpening, but this is just a guess. As of this writing, the e-needle tip stabilization mechanism is not known.

Polarizability Effects

The alert reader will note that the above theory does not involve the polarizability of the highly polar water molecule in the electric field near the needle tip. It turns out that molecular polarizability brings about two electrical effects that nearly cancel each other in the theory. First, the vapor pressure of a charged sphere is increased by the polarizability, because removing a molecule from the zero-field region inside the sphere to the high-field region outside releases energy. Second, the water vapor density in the high-field region is increased as polarized water molecules are preferentially drawn into this region. The theory becomes somewhat complicated at this point, but the result is that the polarizability of the water molecule can be ignored to first order [1999Bre, 2002Lib]. It appears that molecular polarizability contributes somewhat to the energetics, but it is not as important as the electrostatic effect for creating e-needles.

The electrostatic effect is somewhat universal in that it does not depend much on the characteristics of the vapor molecules in the problem, including the polarizability. Because of this universality, one expects that the e-needle phenomenon should be observable in other high-vapor-pressure material systems besides ice. Indeed, Libbrecht, Crosby, and Swanson [2002Lib] demonstrated a similar e-needle effect in iodine crystal growth, even though this simple dipole molecule has a quite low molecular polarizability.

E-Needle Crystal Orientation

E-needles can be persuaded to grow with a variety of crystalline orientations, depending on growth conditions, as is demonstrated in Figure 8.9. Near −15°C, e-needles often prefer to grow along the a-axis of the ice crystal, but sometimes they grow preferentially along the $[1\bar{1}00]$ axis, as shown in the middle image in the figure. I have not fully explored the causes of the different e-needle orientations, and the precise conditions needed to produce growth along the a-axis or the $[1\bar{1}00]$ axis are not presently known. The preference for growth along the $[1\bar{1}00]$ axis is especially puzzling, as this axis seems to play little role in other aspects of ice crystal growth, and the phenomenon shown in Figure 8.10 adds another twist to the mystery.

FIGURE 8.9. Under different experimental conditions, e-needles can grow preferentially along the a-axis (top photo), the $[1\bar{1}00]$ axis (middle photo), or the c-axis (bottom photo). In many conditions (not shown), e-needle growth is not along a well-defined crystalline axis but appears to be somewhat random and is likely dependent on the orientation of the seed crystal from which the e-needle formed. In each of these photos, the e-needle growth was followed by a period of normal growth, during which no voltage was applied. The orientation of the e-needle could then be determined by the orientation of the subsequent normal growth.

At temperatures near the needle peak around −6°C in the Nakaya diagram, e-needle growth is usually not along a well-defined crystal axis. Instead the growth axis appears to be roughly the same as the growth direction of fishbone dendrites, which is somewhat temperature and supersaturation dependent. However, we found that c-axis electric needles can be reliably produced near −6°C by adding trace quantities of vaporous chemical additives to the air in which the needles grow [2002Lib].

Various chemicals were found to promote c-axis e-needles near −6°C, including hydrocarbons (such as gasoline vapor), various alcohols, and other solvent vapors. After some trial-and-error investigations, we found that acidic acid vapor is especially effective, with concentrations as low as 1 ppm readily promoting needle growth along the c-axis. However, the best vapor we have found for promoting c-axis e-needles is that emitted from GE Silicone II caulk. Acetic acid is the primary solvent used in this caulk, but the vapor appears to include additional proprietary volatile organic compounds at low concentrations. This chemical-vapor trick can be used to produce c-axis e-needles with nearly 100 percent efficiency. Why trace chemical impurities promote c-axis e-needle growth so effectively remains a mystery.

FIGURE 8.10. The preferential growth of e-needles along the $[1\bar{1}00]$ axis sometimes yields a peculiar tip splitting phenomenon seen in dendrite growth near −15°C, illustrated in this photo. When the applied voltage is just slightly above the e-needle threshold, an $[1\bar{1}00]$ axis e-needle apparently begins to form, but the growth halts before the structure can turn into a full-fledged e-needle. Instead the reorientation of the crystal axis brings about a tip splitting that yields two primary dendrite branches and two secondary branches, the latter perpendicular to the original branch axis. This split-tip structure lowers the electric fields sufficiently (at constant applied voltage) so that near-normal growth commences from the split tip. In this example, the two primary branches then grew farther apart until the electric fields passed through threshold again, so the two branch tips each underwent an additional splitting. The applied potential was not changed after the first tip splitting occurred.

FIGURE 8.11. The serendipitous discovery of chemically induced c-axis e-needles (1997). The appearance of five perpendicular stars on five needles in this photo initiated our search for a chemical "recipe" for reliably producing c-axis needles.

A Serendipitous Discovery

In 1997, summer student Victoria Tanusheva captured the photograph shown in Figure 8.11 during our early studies of electric ice needles. The image shows five beautifully formed stellar snow crystals growing on the tips of the five e-needles. Notably, all five stars are perpendicular to their respective e-needles, indicating that all five e-needles had grown along the crystalline c-axis. We had witnessed c-axis e-needles previously, but these were rare occurrences. Most of the time, the e-needle axes were somewhat randomly oriented with respect to the crystal axes, which was not ideal for growing snow crystals on the needle tips. Turning off the voltage and setting the growth conditions to produce stellar crystals on these e-needles yielded mostly lopsided, rather malformed stars. Seeing this image, with all five e-needles growing along the c-axis, we realized that there must be some recipe for reliably making c-axis e-needles.

During the weeks and months after seeing this lone photo, we were unable to reproduce the high yield of c-axis e-needles. We carefully explored growing e-needles at different temperatures, supersaturations, voltages, and other parameters in our apparatus, but nothing worked, and the desired recipe eluded us during many frustrating tests and trials. Having excluded many other possibilities, we began to think that unwanted chemical vapors in our apparatus may have been affecting our results. The diffusion chamber was constructed from aluminum, styrofoam, glass, and other materials, and much of it was held together with silicone caulk, which does emit a characteristic odor.

Removing the contaminating vapors entirely was impossible, but baking the chamber would slowly reduce the contaminant levels. So, we heated the chamber to about 50°C and left it alone for several weeks, focusing our attention on other projects for the duration. When the time seemed right, and the odors had clearly subsided substantially, we turned off the bake and tried our luck once more. And, lo and behold, now we saw no c-axis e-needles whatsoever; the bake had reduced the yield to effectively zero. At that point, the light bulb turned on, and we realized that vapor contaminants were not the problem but instead were an essential part of our desired recipe. Adding a bit of caulk vapor back into the chamber was straightforward, and in short order, we were producing superb c-axis e-needles reproducibly with a nearly 100 percent yield.

After additional tests, we found that many chemical vapor additives could bring about the formation of c-axis e-needles. Just about anything with a significant odor seemed to do the trick. Acetic acid (a.k.a. vinegar) worked especially well, and it was entirely fortuitous that this was a main constituent in the caulk we had been using all along (G.E. Silicone II caulk). In fact, the caulk vapor ended up being slightly better than pure acetic acid, and better than any other chemical additive we tested. We soon developed a highly reproducible procedure for creating copious c-axis e-needles, which is presented in this chapter. As with many empirical recipes, however, we still do not understand why it works!

Some Remaining Questions

Although the above theory is probably correct at a basic level, the e-needle phenomenon is nevertheless largely unexplored, both experimentally and theoretically. Some remaining questions and ideas for further research include the following.

- What stabilizes the e-needle growth above the instability threshold? Solvability theory is no longer adequate for this problem, so some new theoretical ideas seem to be required.
- Why do c-axis e-needles grow about four times faster than e-needles growing along less preferred lattice directions?
- What mechanism is responsible for chemical impurities promoting the growth of c-axis e-needles?
- How would the e-needle phenomenon change in different gases or as a function of gas pressure? No experiments along these lines have ever been performed, to my knowledge.
- Is it possible to grow individual c-axis e-needles in a more controlled fashion, for example, producing c-axis e-needles that are oriented perpendicular to a fixed substrate? To date, I have only been able to grow c-axis e-needles from a frost-covered wire, which leads to somewhat random spatial orientations.
- Why do e-needles sometimes prefer growth along the $[1\bar{1}00]$ axis near −15°C, while at other times selecting the a-axis at this same temperature?
- Are there other preferred e-needle orientations under different growth conditions? Much of the available parameter space remains largely unexplored, leaving open the possibility of new discoveries.
- What other materials exhibit the e-needle growth instability? Iodine exhibited some e-needle-like behavior, but little work has been done exploring this phenomenon in other materials that grow from the vapor phase.

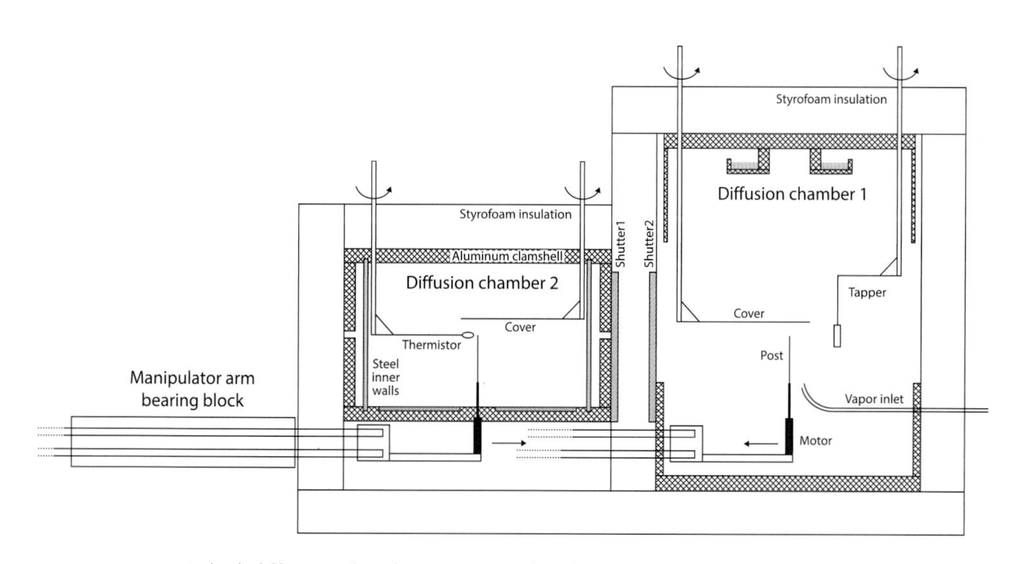

FIGURE 8.12. A dual-diffusion-chamber apparatus for observing snow crystal growth on e-needles. Diffusion Chamber 1 (DC1, on the right) provides the necessary conditions for creating c-axis e-needles easily and quickly. The e-needles are then transported to Diffusion Chamber 2 (DC2, on the left), which provides a well-controlled environment that can achieve a broad range of temperatures and supersaturation levels. The inside height of DC2 is 10 centimeters [2014Lib1].

AN E-NEEDLE DUAL DIFFUSION CHAMBER

To exploit e-needles for additional studies of snow crystal growth, I constructed the dual-diffusion-chamber apparatus shown in Figures 8.12 and 8.13 [2014Libl]. The basic idea here is to grow c-axis e-needles in one diffusion chamber and then move the e-needles to a second diffusion chamber, where their subsequent ordinary growth can be observed over a broad range of conditions. The first diffusion chamber can then be optimized for the task of creating c-axis e-needles quickly and reliably, while the second chamber can be separately designed to produce a well-controlled environment with a precisely known temperature and supersaturation level.

In many respects, this is a next logical step in the morphological studies begun by Nakaya [1954Nak] and advanced by Mason [1958Hal, 1963Mas], Kobayashi [1961Kob], Bailey and Hallett [2004Bai, 2009Bai], and others. The main difference is that we can now make precise measurements of single, isolated crystals growing in situ, avoiding the complications of substrate effects and crystal crowding. In turn, this enables the next substantial phase in the scientific progression—making detailed, quantitative comparisons with modern computational models that can yield new insights into the underlying physical processes that govern snow crystal growth.

Diffusion Chamber 1

Referring to Figure 8.12, DC1 was designed to produce c-axis electric needles quickly, reliably, and easily. Its basic construction is a partial clamshell diffusion chamber (see Chapter 6) with a top temperature of +60°C and a bottom temperature of −35°C [2014Libl]. These temperatures, along with the dimensions of the aluminum clamshell walls, were adjusted (somewhat by trial-and-error) to yield a high supersaturation ($\sigma \approx 100$ percent) and a temperature of −6°C at the location of the wire tip of the support post.

FIGURE 8.13. A laboratory photograph of the dual-diffusion-chamber apparatus depicted in Figure 8.11. The recirculating chiller that cools the chambers, as well as several temperature controllers and other pieces of electronic hardware, are not visible in this picture.

The base of the apparatus is cooled using a recirculating chiller that circulates methanol at −35°C to a 12 × 18-inch aluminum base plate, on which the rest of the hardware is assembled. Four 1 × 1-inch copper bars thermally connect this base plate to the top surface in DC2. Three sets of thermoelectric modules provide adjustable temperature control for the bases of both DC1 and DC2, and the top of DC2, set by independent electronic temperature controllers.

A key feature in DC1 is that the supersaturation must be high enough to reliably produce c-axis e-needles, which do not readily form when when $\sigma < 100$ percent. The clamshell diffusion-chamber design is well suited to this task, although it is difficult to calculate the supersaturation a priori from the design parameters. Some trial-and-error reckoning was necessary, therefore, to achieve the desired environmental conditions in DC1.

The temperature profile in a clamshell diffusion chamber is typically nonlinear, and Figure 8.14 shows the vertical profile along the centerline in DC1. This profile varies with horizontal distance from the walls, so the air in the chamber is not stable against weak convection currents. The resulting slow air circulation in the chamber complicates any attempt to calculate the supersaturation using diffusion modeling. Even quite slow air currents are important to consider, as the time necessary to establish the final supersaturation profile is on the order of the diffusion time $\tau = L^2/D$, which is about 10 minutes in this chamber. It is not necessary, however, that the air be perfectly still or that the temperature profile be precisely known. All that really matters in DC1 is that the temperature be near −6°C and the supersaturation be at or above 100 percent at the position of the wire tip.

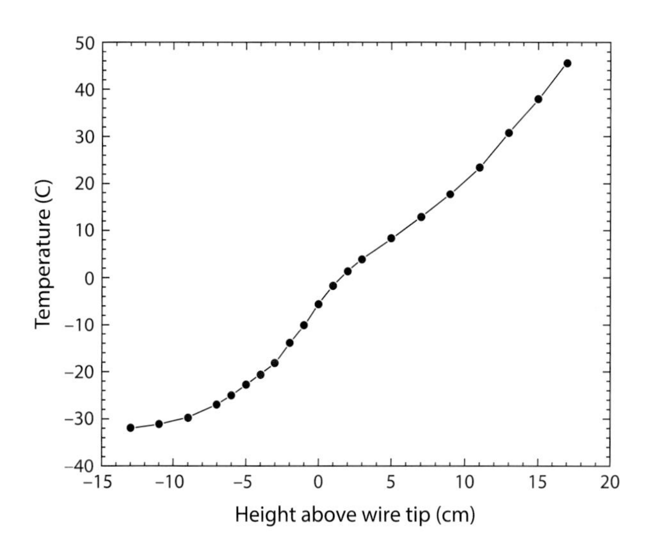

FIGURE 8.14. A measurement of the temperature profile along the central vertical axis in DC1.

The top plate in DC1 includes a reservoir that holds about 100 mL of water in a shallow pan to facilitate evaporation. The plate, clamshell walls, and water reservoir are all made from copper, soldered together for good heat conduction. The top-plate assembly is heated using a sealed resistive heating element with a digital temperature controller to maintain a well-defined plate temperature throughout the duration of an experimental run.

At the beginning of a run, and for other testing purposes, it is often convenient to hang a weighted length of thin monofilament fishing line down the center of DC1 to observe the resulting ice growth at the position where the wire tip will be placed. Two cylindrical observing ports are included in the DC1 walls for this purpose (one for viewing, one for back illumination), and the long-distance viewing microscope can be seen in Figure 8.13. When DC1 is operating correctly, fishbone dendrites (see Chapter 2) appear at temperatures near −5°C on the filament, as shown in Figure 8.15. These fast-growing dendrites provide an effective temperature measurement inside the chamber, and they indicate that a high supersaturation level has been achieved. This filament is removed during normal operation of the chamber, so as not to interfere with the e-needle growth on the wire tip (at the end of the post in Figure 8.12).

A small amount of chemical vapor is needed to produce c-axis electric ice needles, as they do not readily form in clean air. At the same time, too much chemical vapor might contaminate the subsequent growth in DC2, which would clearly be undesirable for quantitative

FIGURE 8.15. An image showing ice crystals growing on a segment of 200-micon-diameter nylon fishing line hanging in the center of DC1. The scale bar is 1 mm long, and the crystal growing time was 19 minutes [2014Lib]. The crystal morphology is strongly temperature dependent, with the fast-growing "fishbone" dendrites appearing near −5°C. The best wire tip location for producing c-axis electric needles is just below the fishbone peak, where the temperature is near −6°C. Hollow columnar crystals appear at this position.

analysis. Therefore, the vapor inlet tube shown in Figure 8.12 is included in DC1 so only a minute amount of chemical vapor need be used to create c-axis e-needles.

The following methodology was found to work quite well in practice. First, a good amount of GE Silicone II caulk is deposited into an empty half-liter soda bottle. The caulk has quite a strong odor, mostly from acetic acid, and capping the bottle nicely traps this vapor for weeks at a time. When the wire tip is in place in DC1, a syringe is inserted into the bottle through a small hole in the cap, and 2 ml of odoriferous air is drawn into the syringe. Note that the caulk itself remains untouched at the bottom of the bottle; only the air above it is used. The air in the syringe is then ejected through the inlet tube into the region surrounding the wire tip. A high voltage (typically +2,000 volts DC) is quickly applied to produce c-axis e-needles. The injected air quickly disperses in DC1, and the quantity is too low to significantly contaminate the air in DC2.

Remarkably little caulk vapor is needed to stimulate the formation of c-axis e-needles. It is often sufficient to draw 2 ml of caulk-bottle air into the syringe, eject this air out of the syringe (not into DC1), and then draw another 2 ml of normal lab air into the syringe and eject that air into DC1. The residual caulk vapor coming from whatever comes off the inner walls of the syringe is enough to produce c-axis e-needles in DC1 with nearly 100 percent efficiency. However, completing these same steps with no initial caulk-bottle draw does not work, yielding almost entirely non-c-axis e-needles. Why this chemical vapor is needed to produce c-axis e-needles remains a mystery, and this procedure was developed almost entirely from an initial serendipitous observation followed by trial-and-error experimentation. As discussed in Chapter 4, our overall understanding of chemical vapor influences on snow crystal growth is extremely poor.

The swing-in cover in DC1 (shown in Figure 8.12) is made from a strip of 0.1-mm-thick plastic sheet mounted horizontally, rigid enough to maintain its flat shape, about 4 cm in width. The cover quickly becomes covered with frost crystals, and it is normally kept near the chamber walls, where it does not perturb the supersaturation to a great extent. Swinging the cover into place, so it is positioned above the wire tip, quickly reduces the supersaturation at the tip by a substantial (but not well measured) factor. With this cover in place, the voltage can be turned off, and the simple e-needle structure remains stable for some tens of seconds. Without the cover over the e-needles, turning off the high voltage produces normal fishbone growth that greatly broadens the tip structure in just a few seconds.

I have also found that judicious use of the swing-in cover greatly improves the transfer of e-needles from DC1 to DC2. The best procedure is to first get the e-needles started with the cover removed, as this requires the highest available supersaturation. After the needles grow to about 1 mm long, swing the cover into place while leaving the high voltage on. This slows the e-needle growth by about a factor of 2–3, but their sharp morphology remains. In this state, let the e-needles growth another 1–2 mm, then turn off the high voltage (leaving the cover in place) and pull the wire tip into DC2. Following this procedure, the transfer efficiency is nearly 100 percent; the e-needles mostly survive the journey into DC2. Without the slower growth step with the cover in place, however, the e-needles frequently break off from the wire tip before they make it into DC2, which can be quite frustrating in practice.

The "tapper" shown in Figure 8.12 is a convenient tool for removing built-up frost from the wire tip. The tapper consists of a small cylindrical weight on a string that swings into place like the cover plate. The weight swings from its string and effectively knocks crystals off the wire, readying it for making more e-needles. Over time, the frost buildup on the wire tip becomes so high that it cannot be removed effectively with the tapper. When this happens, the wire tip can be cleaned by inserting a long plastic rod into the top of the chamber. A plastic rod with a 2-cm-long wire end seems to work best. Touching the thicker, room-temperature wire to the thin wire tip immediately melts and removes the ice buildup. The frost-covered tapper is then brought in to tap the wire tip a few times to nucleate freezing of the remaining water. (Without this nucleation step, only liquid water condenses on the wire tip, as the temperature is not cold enough to quickly nucleate spontaneous freezing.)

After some trial-and-error development of this apparatus, the result is an efficient electric-needle "factory" that works remarkably well and nearly always yields a set of c-axis e-needles in DC2 like those shown in Figure 8.2. Moreover, the dual chamber has a rapid sample turnaround that is especially important for turning a proof-of-principle demonstration into a workhorse experiment that produces valuable scientific data.

Diffusion Chamber 2

DC2 was designed to be a linear-gradient diffusion chamber, optimized to allow accurate modeling of the interior supersaturation. As shown in Figure 8.12, a pair of aluminum clamshells provide a cold barrier around the chamber, while stainless-steel inner walls conduct heat vertically to establish a linear vertical temperature gradient along the walls and throughout the chamber interior (see Chapter 6). The thickness of the stainless-steel walls (1.6 mm) was chosen to provide sufficient conduction to define a linear temperature gradient, but not so thick that the resulting heat conduction is difficult to sustain. Figure 8.16 shows a measured temperature profile at the center of the chamber.

With a linear temperature gradient and frost-covered walls to produce the boundary condition $c_{wall}(z) = c_{sat}(T_{wall}(z))$ at the chamber walls, it becomes possible to accurately model the interior temperature and supersaturation. In the limit that the width of DC2 is much greater than the height, the heat and particle diffusion equations yield linear profiles for both the temperature and particle density inside the chamber. From this, one obtains the supersaturation at the chamber center

$$\sigma_{center} \approx \frac{1}{2} \frac{1}{c_{sat}(T_{cent})} \frac{d^2 c_{sat}}{dT^2}(T_{cent})(\Delta T)^2 = C_{diff}(\Delta T)^2, \tag{8.6}$$

where $T_{cent} = (T_{top} + T_{bottom})/2$, $\Delta T = (T_{top} - T_{bottom})/2$, and Table 2.1 lists C_{diff} as a function of temperature.

This faraway-walls approximation is not too far off for DC2, but the walls do reduce the supersaturation somewhat. Using a finite-element diffusion analysis to calculate the effects of the walls and the post supporting the crystals yields the supersaturation model shown in Figure 8.17 [2016Lib]. This model predicts that, over a

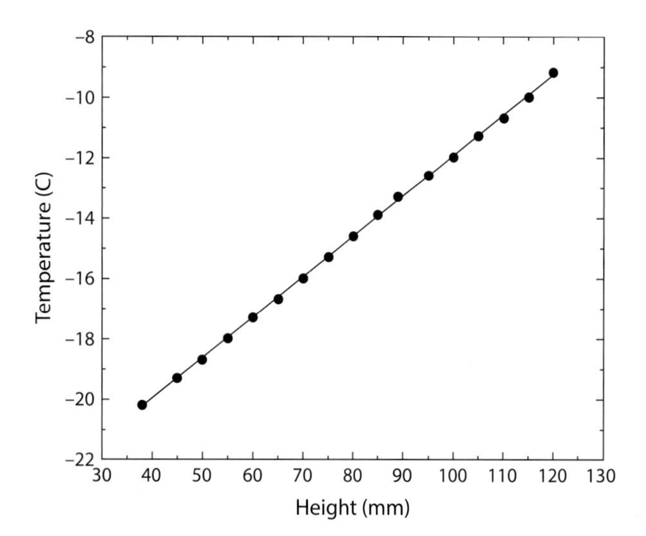

FIGURE 8.16. Measurements of the linear vertical temperature profile in DC2.

broad range of growth conditions, the center supersaturation is well described by

$$\sigma_{center} \approx G_{mod}\, C_{diff} (\Delta T)^2, \tag{8.7}$$

where $G_{mod} \approx 0.72$ is a correction factor arising from the walls and post.

DC2 also includes a swing-in thermistor to monitor the temperature at the chamber center and a swing-in cover to lower the supersaturation if desired. The calibrated thermistor has an absolute accuracy of $\pm 0.1°C$ and can be rotated in to measure the temperature at the location of the growing crystals. In principle, the center temperature is the average of the top and bottom temperatures, but small perturbations arise from the imaging optics, optical viewports, and other factors.

The DC2 cover consists of a 0.1-mm-thick, 1-cm-wide plastic strip mounted horizontally. Swinging the cover into position just above the center of the chamber greatly reduces the supersaturation seen by the growing crystals. The cover is typically put in place right before the e-needle transfer from DC1, which then allows time to position the needles and focus the camera under conditions of low supersaturation. Swinging the cover away then restores the normal supersaturation in a time of roughly $\tau \approx L^2/D_{air} \approx 5$ sec, where $L \approx 1$ cm and $D_{air} \approx 2 \times 10^{-5}$ m^2/sec. The impact of the cover on the supersaturation can be verified by direct measurements of the crystal growth.

During the several-hour-long cool-down of the system, the bottom surface of DC2 is heated to produce

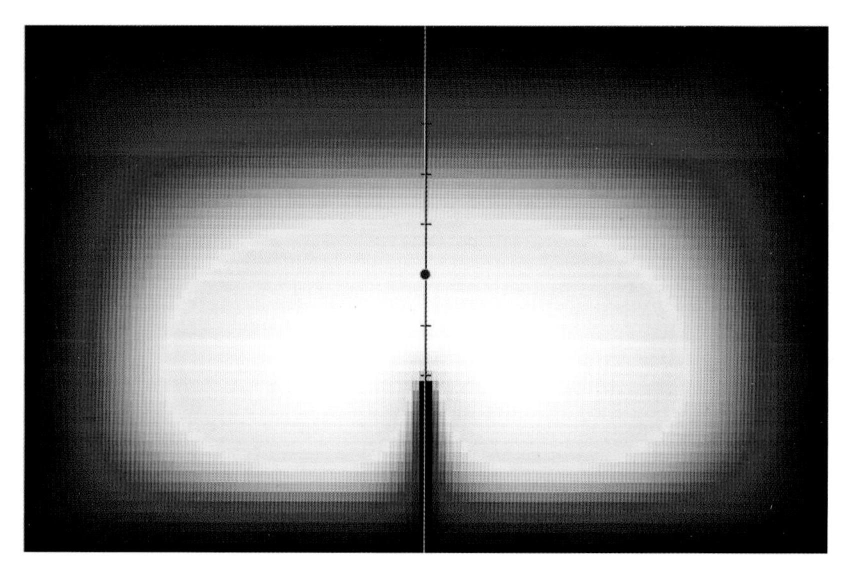

FIGURE 8.17. A finite-element diffusion model of the supersaturation field in DC2 at a fixed ΔT. The supersaturation goes to zero (black) at the frost-covered walls and exhibits a broadly peaked maximum (yellow/white) below the chamber center. The black dot indicates the center of the chamber, which is the location of the growing e-needles. By running the model with different top and bottom temperatures, one can determine the supersaturation at the center of the chamber as a function of ΔT.

strong convection within the chamber. Evaporation from water reservoirs at the bottom of DC2 (see Figure 8.12) produces water vapor that is then transported upward to deposit as frost on the walls and upper surface of the chamber, and this frost provides the water vapor source for normal operation of the diffusion chamber. At the start of cooldown, the DC2 bottom temperature is set to +35°C, and it remains at this temperature for about an hour, producing a substantial evaporation rate for icing the upper surfaces. The DC2 bottom temperature drops as the chiller cools the base plate during cooldown. This temporary inverted temperature profile during cooldown provides an ample supply of ice on the upper surface and walls of DC2, thus maintaining the assumed boundary condition that all DC2 surfaces are covered with ice.

Beginning during cooldown, clean air is introduced slowly into the top of DC2 to reduce effects from chemical contamination. Air is first sent through an activated charcoal filter to remove residual chemical contaminants, and the clean air is then injected into the top of DC2 at a rate of 60 cc/min via a flow meter. This slow trickle of clean air continually replaces the air in DC2 about once per hour without significantly affecting the temperature or supersaturation profiles. The clean-air purge is not essential for operating DC2, but it does seem to improve the overall reproducibility of the crystal growth observations.

The Manipulator Arm

Moving the e-needles reliably from DC1 to DC2 is a nontrivial challenge, and Figure 8.12 shows the manipulator arm that was constructed for this task. The lateral motion is guided by a pair of precision-polished stainless-steel rods moving through linear-motion bearings, providing a smooth ride, so that the e-needles are not shaken off the frost-covered wire tip on which they formed. The drawing shows the post assembly placed in both DC1 and DC2, but there is only one post that shuttles back and forth between these two chambers.

The post assembly consists of a set of telescoping stainless-steel capillary tubes that produce minimal perturbation of the supersaturation field while still providing the necessary support and rigidity. The top of the post, extending out from the smallest stainless-steel capillary tube, is a sharpened, 120-micron-diameter stainless-steel acupuncture needle (J type). The base of the post is connected, via an insulating coupler, to a 6-mm-diameter DC motor that rotates the entire post assembly about its vertical axis. Wires for the motor and the high-voltage brush connection to the post pass through a tube that runs along the entire length of the manipulator arm and out the back end.

A pair of insulating sliding-plate shutters (see Figure 8.12) are used to open and close a keyhole-shaped passage between DC1 and DC2. These shutters are normally kept closed to maintain the temperature profiles in the two chambers, and they are slid open horizontally only briefly to allow passage of the post assembly. A narrow slot at the base of DC2 is not shuttered, as the temperature below the DC2 bottom plate is colder, so the air below does not mix with the air in DC2.

Optical Microscopy

Imaging of the growing e-needles in DC2 is done using a 3X Mitutoyo Compact Objective, with 0.07 Numerical Aperture and a 2.5-micron resolving power. The objective is built into the back wall of DC2, and the front surface of the objective has a fixed distance of 69 mm from the chamber center. A short plastic tube placed over the front of the objective keeps frost from forming on the optical surface. As with all optical microscopy, this objective was chosen as a compromise between resolution, depth-of-field, and working distance (see Chapter 11).

A full-frame digital camera at room temperature is positioned behind the objective, separated by a three-window cylindrical viewing port. No additional optics are placed between the microscope objective and the camera sensor, while extension tubes minimize scattered light from the room lighting. Focusing is done by moving

FIGURE 8.18. A thin platelike crystal was first grown on this e-needle, after which the supersaturation was increased to produce dendritic branches.

the manipulator arm slightly (perpendicular to its main line of travel) and by sliding the camera back and forth on an optical rail.

Illumination is provided by an LED lamp positioned outside the chamber, and another viewing port near the lamp completes the optical path. The manipulator arm is held only by the bearing block, so the wire tip exhibits several microns of shake when in normal operation, brought about largely from unavoidable coupling to vibrations from the recirculating chiller. A camera shutter speed of 1/8000 sec effectively freezes the crystal motion to provide sharp imaging. This dual-chamber apparatus has become something of a workhorse for my ongoing investigations of snow crystal growth. The hardware has evolved to where it has a nearly turn-key operation, able to churn out observations of a broad range of single-crystal structures on the tips of e-needles.

THE NAKAYA DIAGRAM ON E-NEEDLES

Using the dual-diffusion-chamber apparatus just described, one can examine ice growth over a substantial range of temperatures and supersaturations, thus exploring a broad parameter space in the snow crystal morphology diagram. Figure 8.18 shows one example, and Figure 8.19 shows an array of photographs of normal growth on e-needles as a function of temperature from −0.5°C to −21°C and as a function of supersaturation from 8 to 128 percent.

Each tile in this collection of photos shows a representative example using fixed values of temperature and supersaturation that remained constant as the crystals grew. In these images, the smallest needlelike structures have diameters of about 30 microns, while the largest dendritic plates have diameters of about 1.5 mm. The image scale and cropping were separately adjusted for each image, and growth times ranged from about 5 minutes at the highest supersaturations to 30 minutes at lower supersaturations. This set of photos serves mainly to convey a qualitative sense of the morphological changes that occur as a function of temperature and supersaturation.

Earlier versions of the Nakaya diagram were created from observations of snow crystals growing on filaments, and this technique has disadvantages relative to e-needle support. Filamentary materials can influence the observed growth behaviors to some degree, as can competition from closely spaced crystals. With filament experiments, therefore, it can be difficult to observe sharp morphological boundaries and other features, and detailed comparisons with numerical models are not practical. In contrast, e-needles can be used to grow isolated single crystals with no substrate interactions and in well-defined environmental conditions. The resulting crystals exhibit nearly flawless morphologies with excellent sixfold symmetry. Because each crystal begins as a simple ice needle, the subsequent tip growth behavior is remarkably reproducible over the entire observed range of environmental conditions. E-needle observations are thus nearly ideal for examining detailed morphological features.

Importantly, each of the photos in Figure 8.19 presents an opportunity for quantitative comparisons between observed growth behaviors and numerical models. Properly calibrated images of snow crystals growing on e-needles allow a broad investigation into the detailed physical processes that determine all the various growth behaviors. Thus e-needle observations open up new opportunities for better understanding of the physical processes underlying the broad diversity of snow crystal growth behaviors.

Robust Features

The e-needle observations shown in Figure 8.19 demonstrate a rich variety of robust morphological features that are characteristic of snow crystal growth. In this context, I use "robust" to indicate a specific behavior that is easily generated in the lab, distinctive, and can be reliably found over a well-characterized range of environmental conditions. Numerical models that cannot readily reproduce these robust features are clearly incomplete or incorrect in some way.

A first robust feature is one that has long been part of the Nakaya diagram, namely, the increased degree of complexity in crystals grown at higher supersaturations. E-needles grown at low supersaturations often develop as simple columns, simple blocky structures, or perhaps thick plates on stout columns. At even lower supersaturations than those shown in Figure 8.19, e-needles generally grow slowly to become simple hexagonal columns. As the supersaturation increases, branching often begins with six primary branches exhibiting little or no sidebranching. Sidebranching eventually develops at the highest supersaturations shown, although some temperatures are more prone to copious sidebranching than others.

Near −15°C, dendritic branching is mainly confined to a nearly planar structure, as growth outside the plane is limited by strong basal faceting. On e-needles, these dendritic plates are typically slightly conical in overall shape, as the top basal surface grows faster than the bottom surface. The resulting cone angle depends rather strongly on growth conditions, and these morphological trends tell a story about how $\alpha_{prism}/\alpha_{basal}$ varies with temperature and supersaturation.

The six primary branches develop into dendrite structures at the highest supersaturations, with the distinctive fishbone dendrites appearing near −5°C, while fernlike dendrites develop near −15°C. Over the entire temperature range in Figure 8.19, the morphology of a single dendritic branch at high supersaturation (see Chapter 3) defines the shapes of the six primary branches. Unfortunately, reaching high supersaturations at low temperatures is experimentally difficult, so this region of phase space has yet to be explored. At −0.5°C, melting prevented the formation of snow crystals at $\sigma = 128$ percent.

Simple stars are a robust feature near $(T, \sigma) = (-14°C$, 32 percent), and the spikelike primary branches are

FIGURE 8.19(A). The Nakaya diagram illustrated by snow crystals growing on the ends of slender ice needles in air. The horizontal and vertical axes indicate temperature and faraway supersaturation, as labeled. Platelike growth is common at high temperatures, with greater morphological complexity as the supersaturation increases. Plates transition to somewhat blockier forms as the temperature falls from −0.5° to −3°C. Dendritic sidebranching is weak in this temperature range, and the dendrite growth direction varies with temperature and supersaturation. Prism and basal faceting are both strong even at −0.5°C. The upper-left panel is missing, because the fast growth rate at those conditions causes melting.

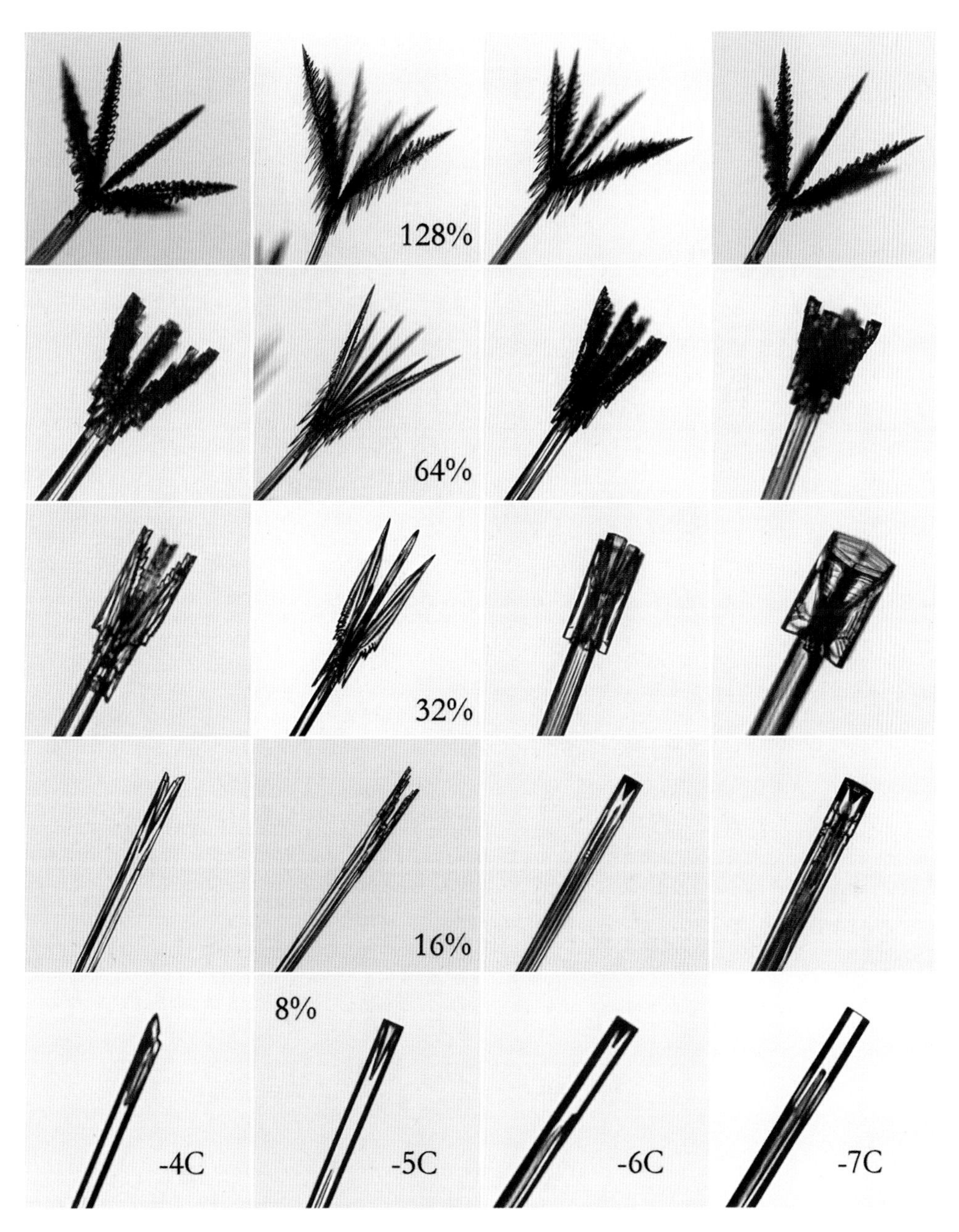

FIGURE 8.19(B). The Nakaya diagram illustrated by snow crystals growing on the ends of slender ice needles in air. The horizontal and vertical axes indicate temperature and faraway supersaturation, as labeled. Columns and needles are common near −5°C, turning into distinctive "fishbone" dendrites at the highest supersaturations. Hollow columns appear at (−5°C, 8 percent), branching into needles at (−5°C, 16 percent). Exceptionally thin-walled cups can be seen at (−7°C, 32 percent). The "tridents" at (−5°C, 32 percent) result from competition between neighboring branches (see Chapter 3).

FIGURE 8.19(C). The Nakaya diagram illustrated by snow crystals growing on the ends of slender ice needles in air. The horizontal and vertical axes indicate temperature and faraway supersaturation, as labeled. The basal and prism growth rates are nearly identical at −8°C, yielding blocky forms and weak sidebranching at high supersaturations. Thin plates emerge as temperatures drop just a few degrees below −8°C. Deep ridging is seen at (−8°C, 32 percent), developing into I-beam structures at −9° and −10°C (see Chapter 3). Ridges generally tend to become thinner as the temperature moves toward −15°C.

FIGURE 8.19(D). The Nakaya diagram illustrated by snow crystals growing on the ends of slender ice needles in air. The horizontal and vertical axes indicate temperature and faraway supersaturation, as labeled. Exceptionally thin plates appear in a narrow temperature range near −14°C, accompanied by nearly flat fernlike dendrites at high supersaturations, exhibiting exceptionally well-developed sidebranching. Simple starts are common around (−14°C, 32 percent), quickly transitioning to dendrites at higher supersaturations.

FIGURE 8.19(E). The Nakaya diagram illustrated by snow crystals growing on the ends of slender ice needles in air. The horizontal and vertical axes indicate temperature and faraway supersaturation as labeled. Thin platelike crystals transition to blockier forms as the temperature drops from −15°C to −21°C, with I-beam structures appearing during the transition. Hollow plates form at (−16°C, 16 percent), while much thinner plates appear at (−16°C, 32 percent). Panels on the upper right are missing, because it is difficult to reach high supersaturations at low temperatures in a linear diffusion chamber.

observed to grow stably to substantial lengths with no sidebranches. I suspect that this morphology will be difficult to reproduce in 3D modeling without the ESI, but that question remains to be investigated. Hollow columnar structures appear on the ends of e-needles near −5°C, but they occupy a rather small region of parameter space. Hollows do not readily form if the supersaturation is too low, and the walls break up into a cluster of needle-like structures if the supersaturation is too high. Deep cups form near (−7°C, 32 percent), with the cup opening angle depending strongly on temperature. When growing on e-needles, the cups are typically flanked by straight "fins" on the outside edges.

Ridges on the six corners of hexagonal plates appear over a broad range of growth conditions, making these features especially robust. Their morphology depends quite strongly on the cone angle of the growing plates, and often the ridges develop an "I-beam" structure, for example, seen clearly at (T, σ) = (−9°C, 32 percent). Ridges generally become more pronounced and more structured as the cone angle of the plates increases, eventually yielding the fins described in the previous paragraph. The diversity and widespread appearance of ridgelike structures is quite remarkable in e-needle growth. In part, this is because the supersaturation gradient around an e-needle tip yields slightly conical plates that exhibit especially distinctive ridgelike structures. As mentioned above, this built-in supersaturation gradient can be quite beneficial in that it accentuates these morphological features and facilitates their detailed investigation.

The formation of exceedingly thin plates on e-needle tips near −15°C is another noteworthy feature of the Nakaya diagram. As seen in Figure 8.19, remarkably thin, nearly featureless hexagonal plates form at several locations in the (T, σ) plane near −15°C. Ridges are sometimes absent at low supersaturations, but delicate ridging is present in the largest, thinnest plates. The CAK model in Chapter 4 provides a possible explanation for why such thin plates form so dramatically near −15°C, over just a narrow temperature region so far from the freezing point.

The Next Grand Challenge

The obvious next step in this scientific progression is to quantify the e-needle observations over a broad range of conditions and then compare the results with 3D numerical models. At least the first part of this statement is relatively straightforward, as the technology for creating and exploiting e-needles is already quite mature. Unfortunately, the theory side of this research program now substantially lags the experimental side, as described in Chapter 5.

Full 3D cellular-automata models have yielded structures that nicely resemble many aspects of snow crystals, and this puts them ahead of other numerical modeling techniques. Nevertheless, the models to date have not progressed much beyond their demonstration phases. They have not yet incorporated realistic parameterizations of the attachment kinetics, so their morphological successes do not always reflect a good understanding of the underlying crystal growth physics. Plus, the 3D models have not yet reached the kind of turnkey operation needed for churning out dozens or hundreds of models for direct comparison with observations of growth rates as well as morphologies.

There appear to be no obvious roadblocks to developing suitable numerical models at this point, however, so it appears likely that researchers will begin making quantitative 3D comparisons between observations and numerical models in the not-too-distant future. When this happens, I expect it will lead to rapid progress as the attachment kinetics and surface diffusion effects are studied and adjusted to provide good quantitative agreement between observations and theory over a broad range of environmental conditions. And this, hopefully, will spark new theoretical insights into the molecular processes that underlie the best-fit model parameters. At some point, we may finally achieve a fundamental understanding of how snow crystals form.

SIMPLEST E-NEEDLE GROWTH

Examining the most basic columnar growth of e-needles provides a good validation of the supersaturation model for DC2, and it provides a basic confirmation of our overall understanding of ice growth from water vapor [2016Lib]. This example tests the accuracy of Equation 8.7, confirms our understanding of the 1D cylindrical growth model (see Chapter 3), and generally supplies a "reality check" that our basic picture of diffusion-limited growth is quantitatively correct. In my opinion, it is important to perform these kinds of model-validation experiments if one expects to realize an accurate quantitative understanding of complex ice growth phenomena.

The general idea in this basic experiment is to start with a set of e-needles in DC2, like that shown in Figure 8.2, rotate the support post to bring a single e-needle into focus perpendicular to the imaging axis, and then measure the radial growth of the chosen needle as a function of time. The growth is subsequently compared with the 1D cylindrical model for different temperatures and supersaturations. If everything is working properly, then we should find good quantitative agreement between theory and measurements.

Figure 8.20 shows some sample data of the growth of a relatively simple e-needle tip structure. At later times, the formation of the blocky crystal at the needle tip likely affects the overall needle growth somewhat, but the morphology is quite simple at early times, being that of a slightly tapered hexagonal column. At these early times, it is reasonable to approximate the shape as a simple cylinder, and the radius $R(t)$ can be extracted from the image data. As shown in the graph, a simple power-law functional form fits the data quite well even at later times.

Drawing a line through these data, I then use the fitted line to determine the growth velocity dR/dt when the needle radius was with $R = 5\ \mu\text{m}$, before the blocky structure appeared at the tip of the needle. Analyzed in this way, producing a single dR/dt value at early times and at a position quite far from the needle tip, this value can be taken as a reasonable proxy for the analogous measurement of the growth velocity of an infinite cylinder with $R = 5\ \mu\text{m}$. Figure 8.21 shows a series of growth measurements reduced in this way from measurements of $R(t)$ on individual needles at different supersaturations.

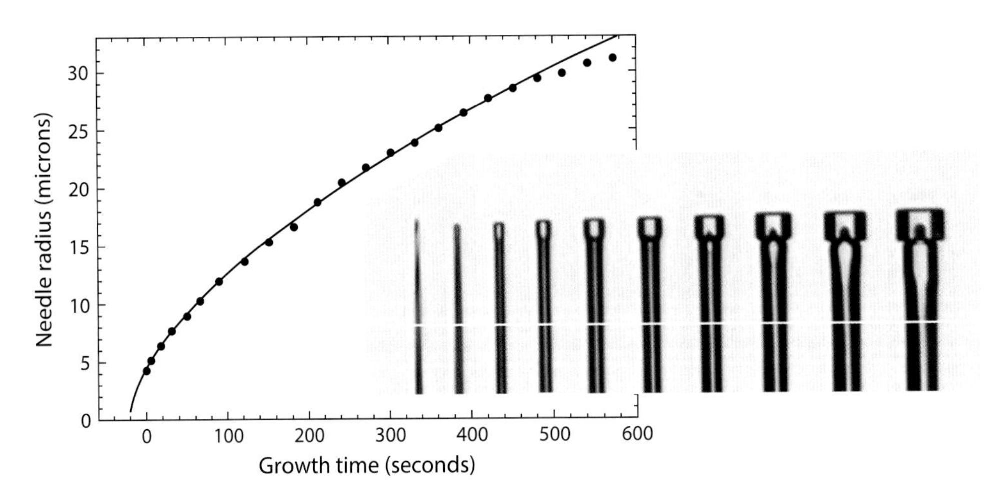

FIGURE 8.20. Example observations at −2°C of the growth of a single electric ice needle after being transported into DC2. The composite image shows the needle at several different times, and the graph shows measurements of the needle radius as a function of time, measured at the position of the white horizontal line in the image, which is approximately 100 μm below the needle tip. The images correspond to the first ten data points in the graph [2016Lib].

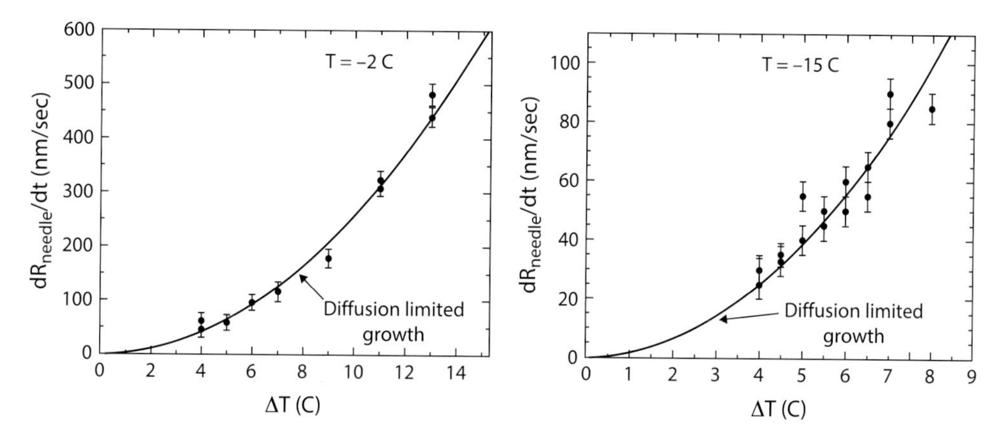

FIGURE 8.21. Example measurements at −2°C and −15°C showing the radial growth velocity of cylindrical ice needles when the needle radius was equal to 5 µm. The data are shown as a function of $\Delta T = (T_{top} - T_{bottom})/2$ in DC2, and Equation 8.7 indicates that the supersaturation is proportional to ΔT^2. Fitting the data reduces it to a single parameter $A(T)$, with the growth velocity given by $dR/dt = A(t)\,\Delta T^2$ when the needle radius is equal to 5 µm [2016Lib].

We see that dR/dt (when $R = 5$ µm) is proportional to ΔT^2 to reasonable accuracy, meaning that the growth velocity is proportional to supersaturation at the growth region in DC2, as is expected from theory.

The theory of cylindrical growth is well understood for the limiting case of an infinitely long cylinder, as described in Chapter 3. In most circumstances of interest in this section, $\alpha_{diffcyl} \ll \alpha_{prism}$ is a good approximation, so the growth is not substantially limited by attachment kinetics, but instead is determined solely by particle and heat diffusion around the cylinder. The theory then depends only on well-known physics, and, importantly, it is independent of the not-so-well-known molecular attachment kinetics. Assuming Equation 8.7 provides an accurate model of the supersaturation around the growing ice needles in DC2, the radial growth velocity of an infinite ice cylinder is then given by

$$v \approx \frac{1}{1+\chi_0} \frac{G_{mod}}{B} \frac{X_0}{R} C_{diff} (\Delta T)^2, \qquad (8.8)$$

where R is the radius of the cylinder, $B = \log(R_{out}/R)$ derives from the cylindrical boundary conditions of the diffusion problem, and χ_0 is a thermal parameter that derives from latent heating of the crystal during growth.

Writing $v = A(T)(\Delta T)^2$, the proportionality constant $A(T)$ depends only on the growth temperature, and it is known directly from the analytical theory describing cylindrical growth. The only unknown parameter in the theory is B, which varies only logarithmically with R_{out} and can be estimated to an accuracy of about 20 percent [2016Lib]. Putting all this together yields the result shown in Chapter 3 (Figure 3.17), where we derived how thermal diffusion affects snow crystal growth. As discussed in that chapter, latent heating is mostly important at higher temperatures, and its effects can be approximated by changing the effective supersaturation around the crystal.

This relatively simple experiment provides a nice demonstration that our basic understanding of particle and heat diffusion in snow crystal growth is indeed correct. To my knowledge, this is the first experiment that has clearly observed the simultaneous effects of both particle and heat diffusion in snow crystal growth. As such, it a step forward in the quest to make quantitative observations that can be compared directly with theoretical models of snow crystal growth. The experiment also demonstrates the inherent accuracy that can be obtained with careful modeling of linear diffusion chambers.

AN ANALYSIS EXAMPLE: E-NEEDLES AT −15°C

The formation of thin, platelike crystals at temperatures near −15°C is an especially intriguing aspect of snow crystal growth that is also quite amenable to analysis using e-needles. At the lower supersaturations shown in Figure 8.19, we see that the thinnest plates appear only in a narrow temperature range between about −13° and −15°C. In Chapter 4, this behavior is attributed to the SDAK phenomenon, specifically, the localized reduction in $\sigma_{0,prism}$ in this temperature range that I call the "SDAK dip." In this section, I describe a focused investigation that helps quantify this distinctive feature in the attachment kinetics [2015Lib2].

A notable element of this investigation is that the phenomenon of plates growing on e-needles is relatively straightforward to analyze, as the overall morphology of a thin hexagonal plate growing on a slender hexagonal e-needle is quite simple in its structure. As described in Chapter 5, basic columnar and platelike morphologies can be modeled with reasonable accuracy using a cylindrically symmetric 2D numerical model that is substantially easier to create and run than a fully 3D model. With a 2D model, one can quickly run dozens of configurations using a wide range of parameters for comparison with experimental measurements, greatly speeding up the analysis. Thus, this exercise presents a good example of how a quantitative analysis of e-needle growth can provide valuable information about the SDAK effect as well as potentially many other aspects of snow crystal growth.

On the experimental side, I have found that selecting an especially well-formed, representative crystal specimen and analyzing it in detail tends to give better results than trying to form averages over many crystals. The present observations begin with creating e-needles in the dual-chamber apparatus described earlier in this chapter and photographing the normal growth of a single crystal in the linear diffusion chamber (DC2). Figure 8.22 shows a sample image selected from a series of similar photographs. By rotating the support post (see Figure 8.12), the entire needle is brought into focus in the frame, allowing a measurement of the needle radius R_{needle}, the plate radius R_{plate}, and the overall needle

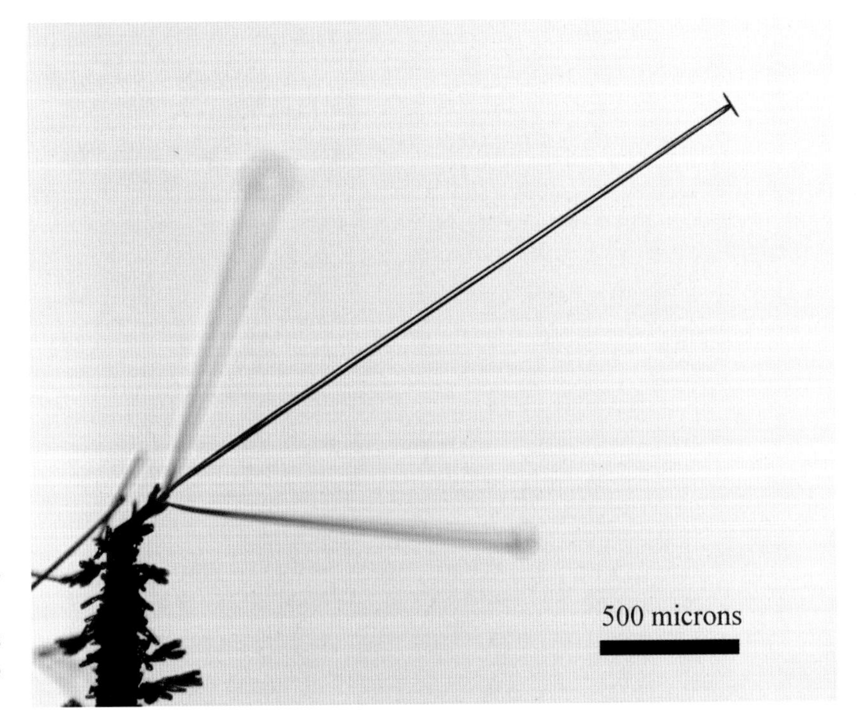

FIGURE 8.22. Example of a thin snow crystal plate growing on the end of a slender ice needle. The supporting wire was rotated so that the entire needle is in the focal plane of the image.

length H. The length is determined relative to the frost crystals growing at the base of the needle, which do not provide an ideal reference point. As a result, the measurement uncertainty in H is generally greater than with the radius measurements.

In a typical run, the overhead cover in DC2 is rotated into place to reduce the supersaturation during the e-needle transfer from DC1. After orienting the ice needle and bringing it into focus on the camera, the cover is then rotated away, restoring the supersaturation to its normal level in a few seconds. The growing crystal is then photographed at regular intervals, and the images are subsequently analyzed to give quantitative measurements. Figure 8.23 shows three separate sets of data taken this way, with additional details presented in Libbrecht et al. [2015Lib2].

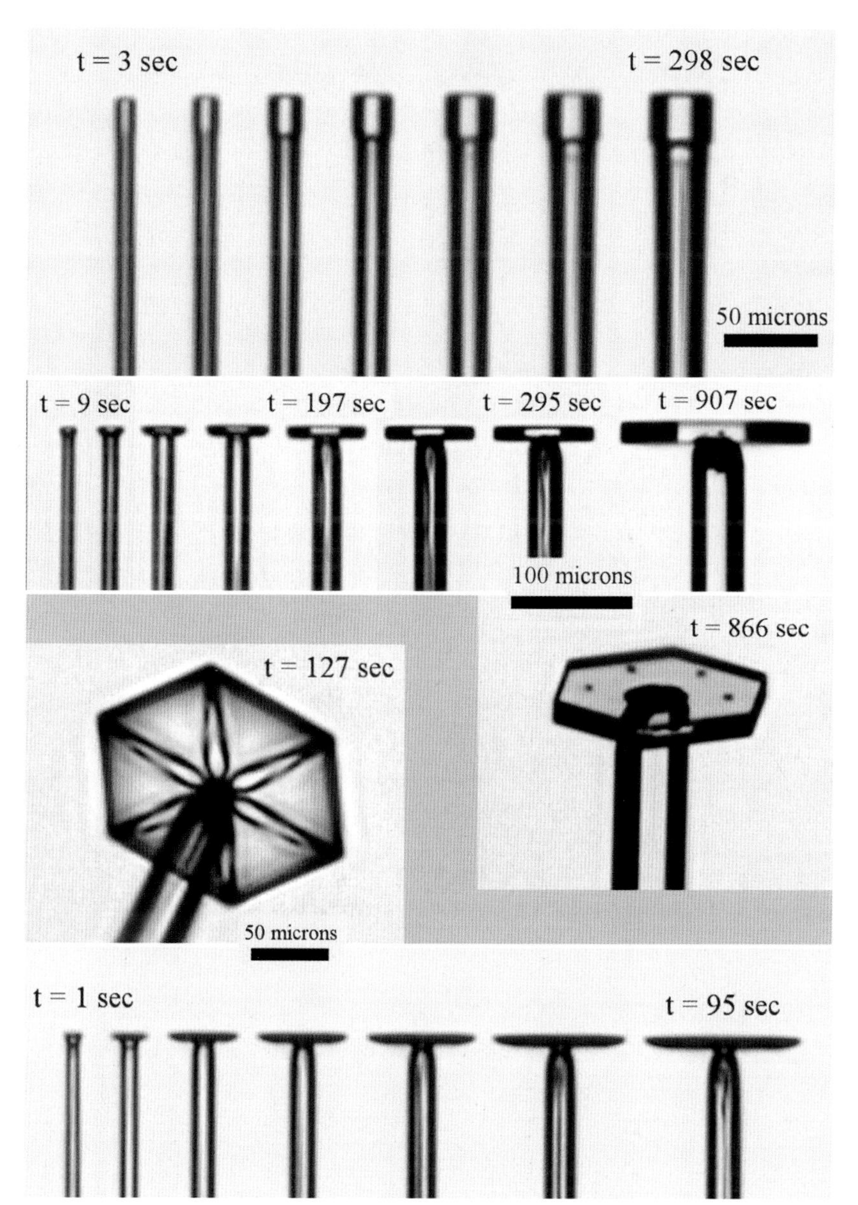

FIGURE 8.23. Growth of e-needles in air at −15°C when the supersaturation was set to 4.6 percent (top), 7 percent (middle), and 11 percent (bottom), illustrating the transition from blocky to platelike growth. Because these morphologies are relatively undeveloped, a simpler 2D cylindrically symmetric model (see Chapter 5) can be used to analyze the growth [2015Lib2].

Figure 8.24 shows an example comparing the 11 percent data in Figure 8.23 (including other images not shown) with a set of 2D computational models. The data points in the figure show measurements of $H(t)$, $R_{plate}(t)$, and $R_{needle}(t)$ extracted from the set of e-needle images, where the plotted height is after subtracting a constant value, as the overall needle length is quite long (as illustrated in Figure 8.22). The lines through the data points came from 2D numerical models, and the right side of Figure 8.24 shows model cross-sections at the times given in the figure. Here the brightness outside the crystal is proportional to the supersaturation (Chapter 5).

An interesting feature in this example is that many of the model parameters are quite strongly constrained from the outset. We know from Chapter 4 that assuming α—$Aexp(-\sigma_0/\sigma_{surf})$ with $A=1$ at $-15°C$ is reasonable for both the basal and prism facets, and doing so reduces the attachment kinetics parameters to just $\sigma_{0,basal}$ and $\sigma_{0,prism}$. Moreover, the basal growth is so slow that the data do little to constrain our choice of $\sigma_{0,basal}$, and assuming the large-facet value of $\sigma_{0,basal} \approx 0.02$ seems to give good results.

While the value of σ_{far} (at the outer boundary of the model space) is constrained by our knowledge of the supersaturation in DC2 from the diffusion modeling discussed above, (yielding $\sigma_\infty \approx 0.11$), we can also use the measured $R_{needle}(t)$ to fix σ_{far} with higher accuracy. Running models with different parameter choices quickly reveals that $R_{needle}(t)$ is mainly sensitive to σ_{far}, being quite insensitive to reasonable choices for $\sigma_{0,basal}$ and $\sigma_{0,prism}$. The reason for this is simply that $\alpha_{diffcyl} \ll \alpha_{prism}$ over a broad range of parameters, which means that $R_{needle}(t)$ is essentially entirely diffusion limited. A value of $\sigma_{far} \approx 5 \pm 1$ percent fits the data well, which is consistent with expectations for the size of the modeling space.

After determining σ_{far} directly from $R_{needle}(t)$, the only remaining free parameter that matters much in the model is $\sigma_{0,prism}$, so Figure 8.24 shows four models run with different values of this parameter. And we see that quite a low value of $\sigma_{0,prism} \approx 0.15$ is needed to fit the data. This is not surprising, because we also know that the thin plate morphology in this crystal is close to sprouting branches, as this transition does take place at moderately higher values of σ_∞, as seen in Figure 8.19. The formation of prism facets on this crystal indicate that $\alpha_{prism} < 1$, but the observation of branching at slightly higher supersaturations indicates that α_{prism} is not much below unity.

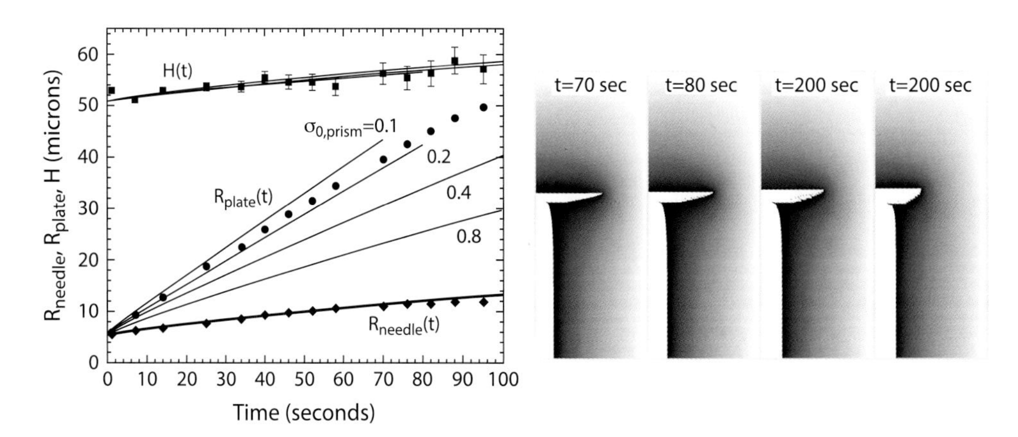

FIGURE 8.24. (Left) Measurements of the plate radius, needle radius, and axial growth as a function of time, from the 11 percent data shown in Figure 8.23 (plus other images not shown). Lines through the data are from numerical models of the growing crystals, as described in the text. (Right) Computer-generated cross sections of the four model crystals shown in the data graph [2015Lib2].

The main takeaway message from this exercise is that the only way to provide a quantitative fit to these data is with a choice of model parameters that gives a value of α_{prism} that is close to unity. This fits our overall understanding of thin-plate growth, as this morphology requires $\alpha_{prism} \gg \alpha_{basal}$, and additionally having α_{prism} near unity is needed to reproduce the formation of branching at slightly higher supersaturations. This all follows rather simply from the observation of the thin-plate morphology. At the same time, the appearance of nearly isometric simple prisms at −15°C (see Chapter 7) requires $\alpha_{prism} \approx \alpha_{basal}$ in that growth regime. These two experiments yield conflicting results, yet both results are quite robust. As discussed in Chapter 4, the CAK model provides one way to reconcile these facts via the SDAK mechanism, and I believe that there are no viable alternative explanations at this time.

The story at −15°C develops further when one starts varying the supersaturation and thereby growing different crystal morphologies on the ends of e-needles, as shown in Figure 8.23. Modeling these data independently following a similar procedure, we quickly find that $\sigma_{0,prism} \approx 0.6$ percent is needed to fit the 7 percent data and $\sigma_{0,prism} \approx 1.5$ percent is needed for the 4.6 percent data [2015Lib2]. Surprisingly, the models also indicate that the value of surface supersaturation remains at roughly $\sigma_{surf} \approx 0.5$ percent at the outermost prism edges of all three crystals. It is a somewhat fascinating result when you think carefully about it, but it all fits with the CAK model presented in Chapter 4. It further suggests that there is perhaps much more to be learned by investigating snow crystal growth over a wider range of conditions using computational models. Finally, this exercise illustrates the usefulness of e-needles for quantitative investigations of the physical dynamics of snow crystal growth over a broad range of temperatures and supersaturations.

E-NEEDLE VIGNETTES

This final section presents a collection of different growth behaviors observed on the ends of e-needles. This is by no means a complete accounting of morphological features that can be found on e-needles, but rather a sampling of phenomena we have observed to date. As with any scientific endeavor, this investigation of e-needles is a work in progress, presenting considerable opportunity for further investigation and discovery.

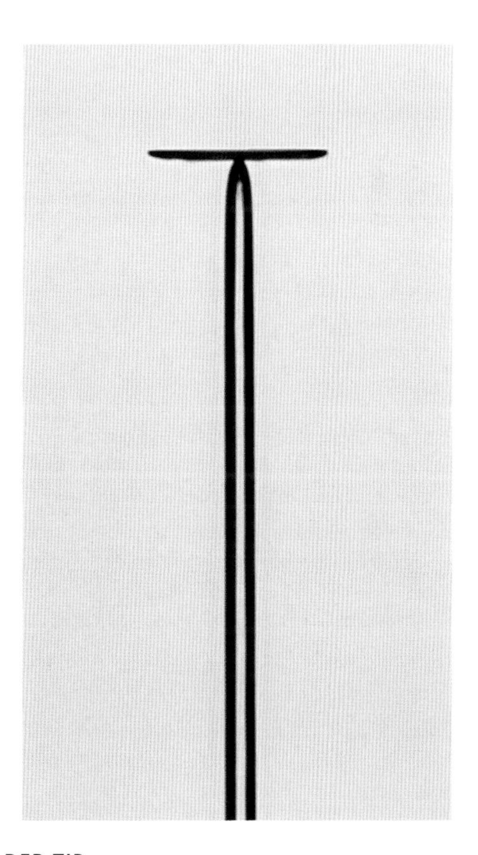

SHIELDED TIP
When a platelike crystal grows on an e-needle, it shields the needle tip, resulting in a small point of contact.

PRECARIOUS SUPPORT (ABOVE)
This platelike crystal is supported by a remarkably small ice tip. The plate has a slightly conical shape overall, reflecting a slight gradient in the surrounding supersaturation.

ICY BROOMSTICK (RIGHT)
This image shows a set of six fishbone dendrites (Chapter 3) growing out from an e-needle.

CRYSTAL BOUQUET

Thin, near-perfect hexagonal plates readily grow on the ends of c-axis e-needles. The front plate is almost perfectly flat on its top surface but has a slightly convex shape on the underside, the latter supporting a set of ridges (Chapter 3).

COLUMNAR SIMPLICITY (LEFT)
When the supersaturation is especially low at any temperature, e-needles typically just thicken into simple columnar forms.

MULTIPLE MACROSTEPS (BELOW)
This series of images shows a platelike crystal exhibiting multiple macrosteps, propagating both inward and outward, on both the top and bottom surfaces of the plate. Deciphering the growth of each is a nontrivial problem.

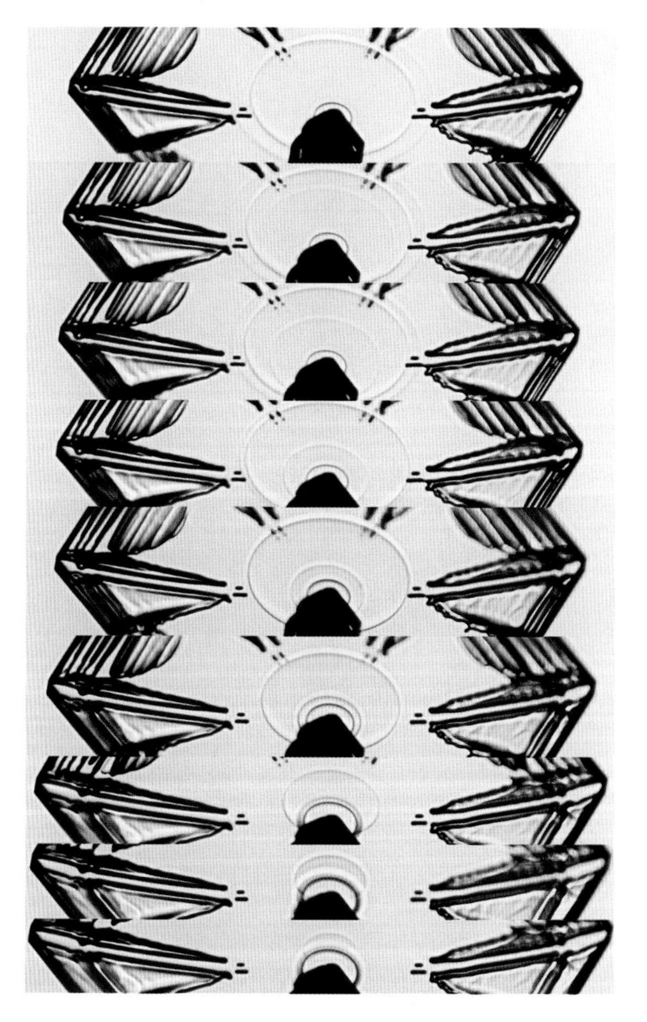

CURVED BRANCHES (ABOVE)
The spikelike branches on this crystal grew up and out at high supersaturation, then mainly out at lower supersaturation, giving them an overall curved appearance.

STELLAR DENDRITE (BELOW)
Spontaneous sidebranching is a characteristic feature of a fernlike stellar dendrite.

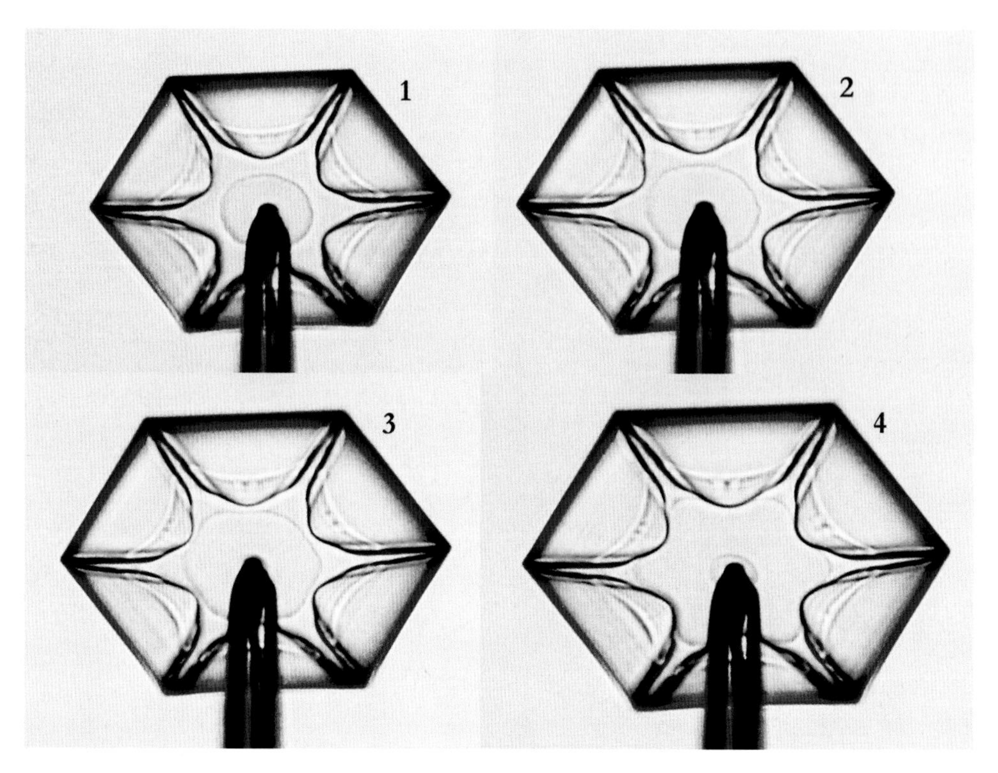

OUTWARDLY PROPAGATING RINGS

A ringlike pattern (image 1) formed on the underside of this platelike crystal at $(T, \sigma) = (-15°C, 16 \text{ percent})$, and the ring slowly expanded outward with time (2-3). The Gibbs-Thomson effect pulled water vapor onto the needle contact point, owing to its sharply concave shape there. This vapor was supplied by evaporation from the nearby plate, resulting in the expanding ring. Once the ring had grown large (3), and new small ring appeared (4).

CRYSTAL BATTLEMENTS

This unusual structure emerged when a cup-with-fins crystal (Chapter 3) grew near $(T, \sigma) = (-7°C, 32 \text{ percent})$ and was subsequently exposed to a higher supersaturation, causing the rim of the cup to break up into six sections.

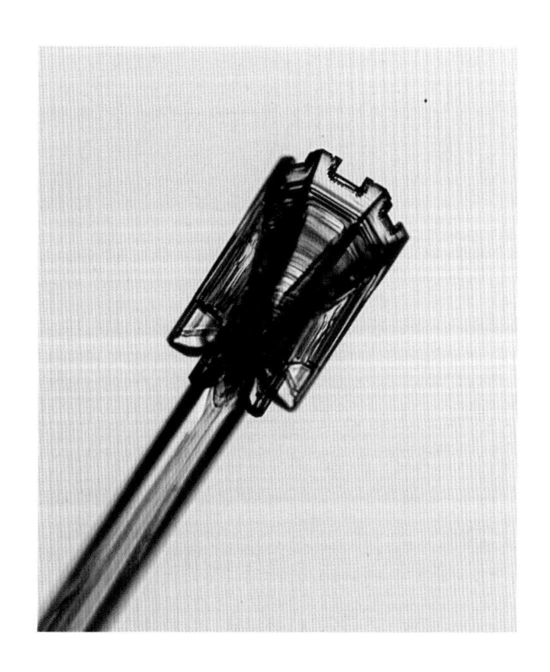

SHEATH ON A STICK

This image shows a thin-walled sheath crystal growing on the end of an electric needle crystal at −7°C. Sheathlike morphologies are rarely found in natural snow crystals, as they require nearly constant growth conditions in quite narrow ranges of temperature and supersaturation.

TIP SPLITTING

This photograph shows a crystal grown on the end of an e-needle at $(T, \sigma) = (-15°C, 128$ percent$)$. Early in this crystal's development, when the crystal was small and the near-surface supersaturation especially high, all six branches underwent tip splitting, yielding six sets of split branches. As discussed in Chapter 3, tip splitting indicates nearly isotropic prism attachment kinetics, in this case because α_{prism} was close to unity. As the crystal grew larger, diffusion lowered the surface supersaturation, which in turn lowered α_{prism}, so the branches experienced no additional tip splitting. Tip splitting is common at these extreme conditions around −15°C, although the nearly symmetrical tip splitting seen in this example is unusual.

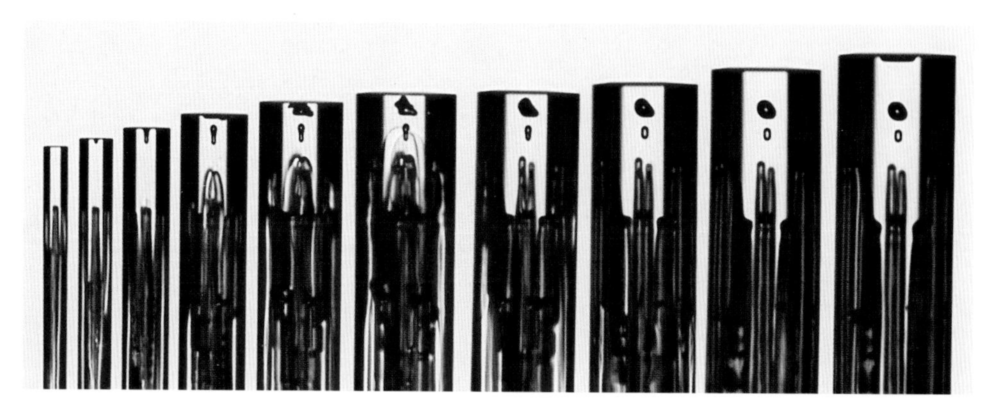

BUBBLE EVOLUTION
This series of images shows the formation and evolution of two bubbles inside a columnar crystal grown over several hours near −7°C. The large markings running down the length of the column are surface features, while the bubbles grew from hollows at the end of the column.

CROWNS ON NEEDLES
This frame shows a set of e-needles that were made to branch into arrays of additional e-needles. After that, fishbone-dendrite crowns were grown on the ends of the needles.

CRYSTAL CHANDELIER (ABOVE)
After creating a snow crystal star on an e-needle, columns were added on the stellar tips, and then plates were grown on the columnar ends.

PLATE WITH THICK RIDGES
This plate exhibits thick ridges on its underside, while the top surface is quite flat.

FIGURE 9.1. A laboratory-grown Plate-on-Pedestal (PoP) snow crystal, measuring 3.3 mm from tip to tip, grown by the author. Remarkably, most of this stellar plate is not in contact with the transparent substrate, but lies above it, balanced atop the tiny flower-shaped nub seen at the center of the crystal.

NINE

Designer Snow Crystals

Build a better snowflake, and the world will shovel a path to your door.

—KGL

In this chapter, I set aside the science and consider the snowflake as an objet d'art—a diminutive sliver of ice worthy of admiration for its intricate beauty and complex symmetry. In search of new artistic vistas, I go beyond photographing natural specimens falling from the clouds and describe a technique for crafting synthetic snow crystals that are even more spectacular than the best nature has to offer. I like to call these *designer snowflakes*, as one can create a desired growth morphology simply by adjusting the applied temperature, humidity, and other environmental factors as a function of time. Moreover, one can photograph a designer snowflake as it forms, allowing a time-lapse recording of the entire process, thereby adding a temporal dimension that is unavailable when viewing natural crystals. The activity becomes a novel type of emergent ice sculpture, discarding the chisel and using molecular self-assembly and the laws of crystal growth to create beautiful, symmetrical crystalline structures. While creating synthetic snow crystals is an essential endeavor for investigating the underlying physical dynamics of ice solidification, the occupation also presents an excellent opportunity for artistic expression.

Designer snowflakes have lagged their natural counterparts as photographic subjects, in part because it had not previously been possible to grow synthetic snow crystals of a quality that compared to the best natural specimens. Just as synthetic diamonds have only recently begun to rival their quarried counterparts in size and quality, designer snowflakes are only now surpassing those found in nature, exhibiting sharper facets and more precise symmetries.

This chapter focuses on a Plate-on-Pedestal (PoP) technique that I developed for creating well-formed stellar snow crystals in the laboratory. The PoP apparatus is relatively straightforward to construct and operate, plus it is specifically engineered for capturing high-resolution photographs of designer snow crystals as they form and develop. Both the growth temperature and supersaturation can be separately adjusted as a function of time, allowing the manufacture of a nearly infinite variety of complex morphologies. And as an added bonus, one need not wait for a suitable snowfall, or brave the frigid weather, to photograph these icy creations. All the cold parts of the apparatus are built into an insulated enclo-

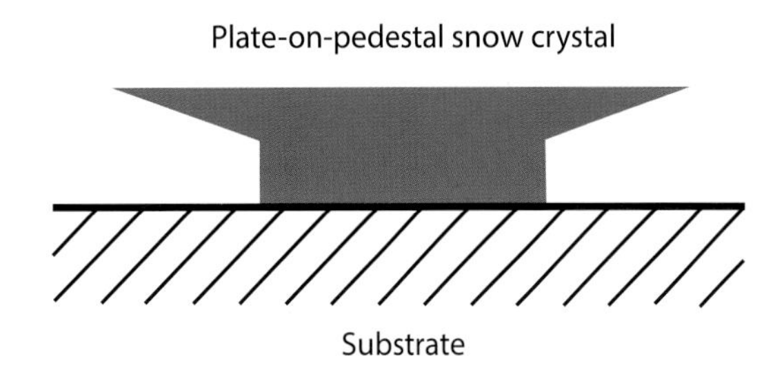

FIGURE 9.2. A side view of the basic PoP snow crystal geometry. A thin platelike crystal grows outward from the top edge of a small hexagonal prism, while the lower basal face of the prism rests on a transparent substrate.

sure, allowing one to explore the artistic side of snow crystal formation on one's own schedule while working in room-temperature comfort.

THE PLATE-ON-PEDESTAL METHOD

Creating a PoP snow crystal begins by producing a cloud of small ice prisms in a free-fall growth chamber (see Chapter 6) and letting a few of these crystals fall onto a transparent substrate held at a temperature near −12°C. Some of the prisms will land with one basal facet resting flat against the substrate surface, as illustrated in Figures 9.2 and 9.3. Select one of these crystals, expose it to a moderately high supersaturation in air, and a thin ice plate will commence growing out horizontally from the top edge of the prism. Because the upper plate is supported above the substrate by the central ice prism, I refer to this as a Plate-on-Pedestal geometry.

The physics underlying the formation of the PoP geometry arises from diffusion-limited growth and the edge-sharpening instability (ESI) described in Chapter 4, which itself arises from the phenomenon of structure-dependent attachment kinetics (SDAK). The underlying molecular physics responsible for this growth behavior is not completely understood at present, as the attachment kinetics are generally a subject of current research (see Chapters 4 and 7). Nevertheless, like many aspects of engineering, one need not understand a phe-

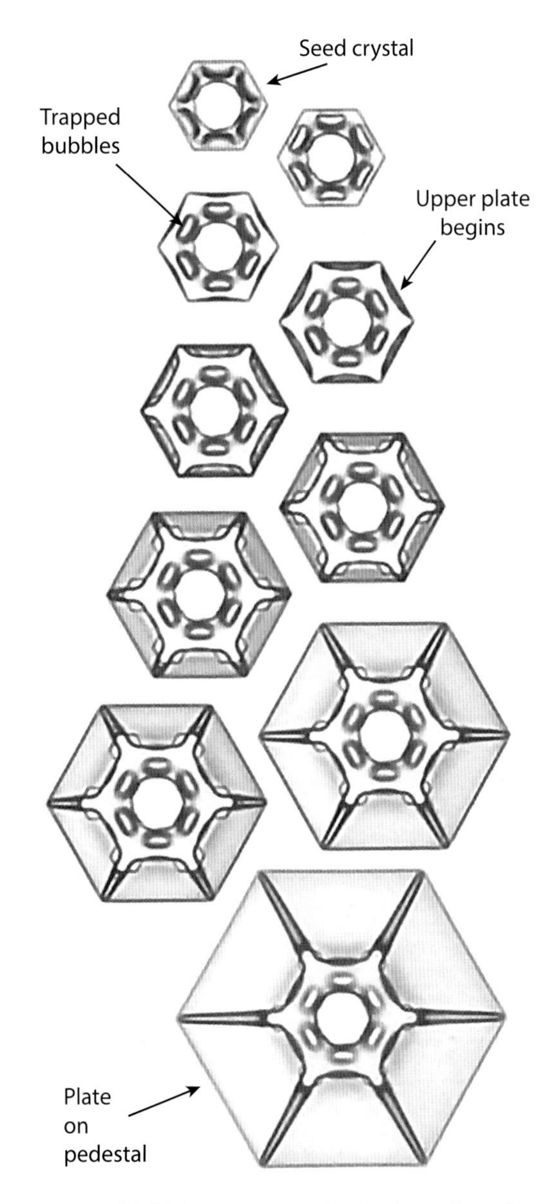

FIGURE 9.3. Multiple images showing the formation of a small PoP snow crystal from a seed crystal, here growing from about 50 μm to 170 μm in diameter. The initial seed crystal exhibited some nonfaceted structure, and its early growth trapped six small bubbles between the ice and the substrate surface. Once formed, these isolated bubbles did not evolve substantially over time. Soon the upper plate began growing from the top basal/prism corner, as sketched in Figure 9.2. Note that the thin hexagonal plate in the last photo is not touching the substrate but is supported above it by a small ice pedestal.

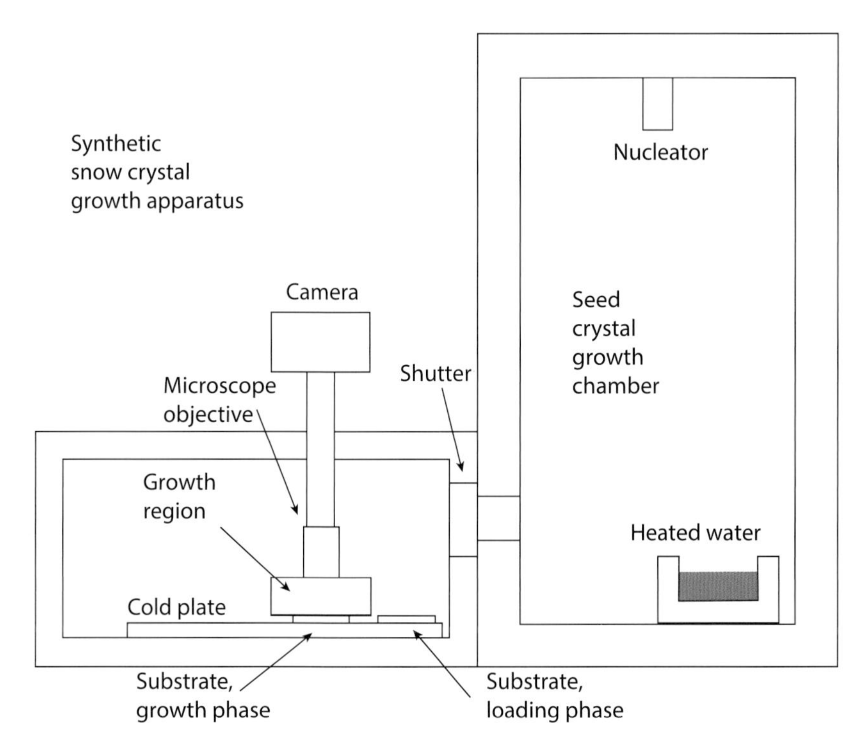

FIGURE 9.4. The apparatus used to grow and photograph PoP snow crystals. Small, freely falling ice prisms in the seed-crystal chamber first pass through a shutter, and some of them fall randomly onto a waiting sapphire substrate (loading phase). The substrate is then moved to the observation region (growth phase) for subsequent PoP crystal development.

nomenon perfectly to put it to good use. Figure 9.3 shows the formation of a small PoP crystal, which is remarkably easy to achieve in practice.

Once the PoP structure has been established, the upper plate will continue to grow outward, provided the temperature remains within a few degrees of −15°C. Even as the ice plate becomes quite large, it grows entirely above the substrate, balanced atop the initial small prism. I have used this technique to create a variety of intricately patterned stellar snow crystals, many of which are shown at the end of this chapter. Being stationary and supported above a transparent substrate, PoP snow crystals can be photographed easily as they form and develop, allowing high-resolution imaging and striking time-lapse videos of their emerging structure.

It appears that Gonda, Nakahara, and Sei grew snow crystals using a similar technique in the 1990s [1990Gon, 1997Gon], although these papers do not explicitly describe the PoP structure. However, little subsequent work appeared after these initial results, perhaps because the PoP technique is not especially well suited for scientific investigations. After recognizing the PoP geometry in my own studies, I continued its development mainly to grow snow crystals in a more artistic realm.

PoP Hardware

Figures 9.4 and 9.5 illustrate the apparatus I constructed for creating and photographing PoP snow crystals [2015Lib3]. The tall chamber is essentially the seed-crystal generator described in Chapter 6, producing a continuous cloud of small ice prisms that slowly fall through the chamber as they grow. Upon opening a shutter connecting the seed chamber to the adjoined lower chamber, some of the seed crystals waft through the opening and fall randomly onto a waiting sapphire substrate, shown in its loading position in Figure 9.4. After loading crystals for a few seconds, the substrate is moved over to its growth position under a photomicroscope, shown in greater detail in Figure 9.5. While watching the

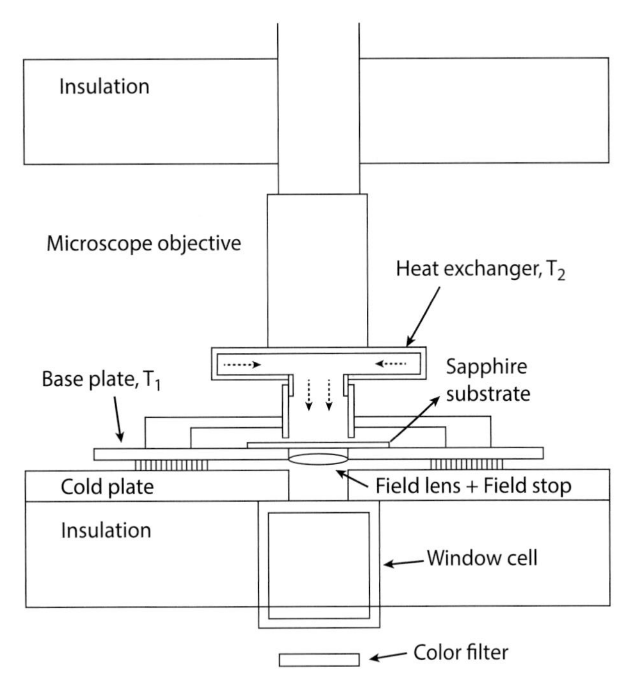

FIGURE 9.5. An expanded schematic view of the PoP growth region shown in Figure 9.4.

microscope image displayed on a TV monitor, the substrate is moved around using a pair of manipulator arms to search for a well-formed, isolated hexagonal prism that can be grown into a PoP snow crystal. Once a suitable ice prism has been positioned in the microscope field, moist air is blown gently down onto the crystal in the growth region. When the proper temperature and supersaturation conditions are applied to the seed crystal, the early growth produces the PoP geometry, which can then be grown further into a large stellar plate.

When growing the PoP crystal, air passes through the heat exchanger to become saturated with water vapor at temperature T_2. This air blows gently down onto the substrate at temperature $T_1 < T_2$ to grow the crystal. Not shown in Figure 9.5 are the microscope camera and a white-light LED lamp placed underneath the color filter. It typically takes 20–60 minutes to produce a large PoP snow crystal, during which time one can monitor its progress using the photomicroscope. Changing the substrate temperature T_1 changes the growth temperature of the crystal, while changing the heat-exchanger temperature T_2 adjusts the effective supersaturation around the crystal. Increasing the air flow though the heat exchanger increases the supersaturation as well. Changing the growth conditions frequently and abruptly tends to produce especially complex growth morphologies with a high degree of sixfold symmetry (see Chapter 3).

With a bit of experience, one can, at least to some degree, plan the structural features of a PoP snow crystal in advance or improvise its morphological development in real time as it grows. Each change in the temperature and supersaturation alters the growth behavior, and these parameters become the tools needed to create a wide variety of snow crystal forms. As one begins to develop a set of empirical rules for guiding the crystal development, the PoP process becomes a unique and quite satisfying form of additive ice sculpture. In working with this apparatus to date, I have typically observed the crystal formation in real time and made temperature and air-flow adjustments without a great deal of pre-planning. However, it would be straightforward to add computer control to these inputs and develop specific algorithms for producing a variety of morphological outcomes.

The Seed-Crystal Generator

The seed-crystal growth chamber in Figure 9.4 has inside dimensions of approximately 40 × 40 × 100 cm and is made from a frame of aluminum T-rail covered with 1/8" thick aluminum panels. Methanol coolant from a recirculating chiller flows through central holes in the four vertical T-rails, and heat conduction through the aluminum rails and panels is sufficient to cool the remainder of the chamber, which is well insulated from the room by Styrofoam panels.

An insulated reservoir containing 1 liter of ordinary tap water rests on the bottom of the seed-crystal chamber, and the water temperature is kept constant by an electronic regulator using an immersed water heating element and water temperature sensor. The top of the container is open to allow evaporation, and the resulting water vapor is carried by convection throughout the rest of the chamber. The continuous evaporation and convection maintain a steady-state supersaturation in the seed-crystal chamber that can be adjusted by changing the water temperature.

Water vapor is continually removed from the air by the growing ice crystals and by frost depositing on the walls of the chamber, and this water vapor is continually replenished by evaporation from the water reservoir. The air temperature is typically kept near −15°C, as measured by a thermistor near the center of the chamber, as this temperature yields small platelike seed crystals. The supersaturation is difficult to determine precisely but can be inferred to some degree by the morphology of the growing crystals. The chiller temperature is typically set to −19°C (giving a −15°C air temperature) and the water temperature to 17°C, as this yields a continuous supply of thin, hexagonal platelike seed crystals with diameters in the 20–50 micron range. Higher water temperatures yield somewhat branched morphologies, which is not desirable for PoP seed crystals.

The expansion nucleator (see Chapter 6) at the top of the seed-crystal chamber consists of a standard 1.33-inch Conflat vacuum nipple with an interior volume of about 25 cc that is connected to a solenoid valve on the output side. Pressurized room air flows into the nucleator through a needle valve that constricts the rate of input air flow. The nucleator assembly is placed inside the growth chamber, so its temperature is approximately −15°C during operation. The overall flow rate is slow enough that the air temperature becomes roughly equilibrated inside the nucleator with a water vapor content near the saturated value.

Every 10 seconds the solenoid valve is pulsed open, causing the pressurized air to rapidly expand into the growth chamber. The expansion cools the air (at least in some localized regions) sufficiently to nucleate ice crystals. Air pressures as low as 15 psi will usually nucleate some crystals, while 30 psi produces many thousands per pulse. Water buildup inside the nucleator is removed after each run by operating it for several hours when the chamber is at room temperature. With no initial ice buildup, the nucleator can run continuously for at least 10 hours without difficulty.

The nucleated ice crystals float freely as they grow, until they eventually settle to the bottom of the chamber. The fall times are typically a few minutes, depending on temperature and supersaturation. Pulsing the nucleator valve open every 10 seconds thus produces a steady state in which roughly a million seed crystals are growing inside the chamber at any given time (this number being determined by a visual estimate of the typical spacing between crystals floating inside the chamber during operation). Shining a bright flashlight into the chamber reveals sparkles caused by reflections off the crystal facets, and this is a convenient way to verify that seed crystals are present.

Compressed air for both the nucleator and the crystal growth region is supplied by an ordinary oil-free workshop air compressor with a built-in storage tank and regulator, which automatically maintains the required 30 psi air pressure. The compressed air is passed through an oil filter and then an activated charcoal filter (containing coconut-husk charcoal) to remove remaining chemical contaminants from the air, and then it passes through a fine-pore fiber filter to remove any remaining charcoal dust.

A 50-mm-diameter hole in the side of the seed-crystal growth chamber connects it to the adjoining main growth chamber seen in Figure 9.4. The cold plate at the bottom of the growth chamber is cooled using the same circulating coolant that flows through the walls of the seed chamber. To grow a PoP crystal, the ice-free substrate is first moved to its loading position, and a simple plate shutter is then slid open between the two chambers. The convective air currents in the seed chamber cause a slight air flow between the chambers that carries a small number of seed crystals into the growth chamber, and some of these crystals fall onto the substrate within a few seconds. The shutter is then closed, and the substrate is moved to a covered region in the main growth chamber. The substrate is positioned using a pair of manipulator arms to locate a suitably isolated seed crystal and center it under the microscope for subsequent growth and observation.

The Growth Chamber

As illustrated in Figure 9.5, the substrate is an uncoated sapphire disk, 50 mm in diameter and 1 mm thick, with the sapphire c-axis perpendicular to the disk surface. The principal advantages of using sapphire in this application are its high thermal conductivity and its resistance to scratching. Using c-axis sapphire avoids birefringence issues that can interfere with optical imaging. The substrate slides on a smooth anodized aluminum plate with its temperature T_1 maintained by a digital temperature controller using a thermistor temperature sensor in the aluminum plate, together with thermoelectric heating/cooling modules beneath it. The thermistors have an absolute accuracy of better than ±0.1°C, and the temperature regulation is stable to better than ±0.01°C under normal operation. However, the substrate and the air immediately above it will not be at precisely the same temperature as the aluminum plate, which adds some uncertainty to the ice crystal growth temperature. Moreover, the PoP geometry itself provides a rather poor thermal coupling between the snow crystal and the substrate.

The primary heat exchanger above the substrate is an aluminum plate at a temperature T_2 maintained by a separate temperature controller. Filtered room air from the compressor first passes through a baffled precooler kept near $T_{precool} = 0°C$ to lower the air temperature and remove a large fraction of the water vapor it contains. This step reduces the thermal load on the primary heat exchanger and prevents it from clogging with ice during long observing runs. The precooled air then passes through a series of serpentine channels in the primary heat exchanger before blowing down onto the substrate and the growing snow crystal. The air flow rate F is typically 200–300 cubic centimeters per minute (cc/min), measured using a tapered-tube flow meter and controlled with a simple needle valve. This flow rate replaces air in the guide tube (between the heat exchanger and the substrate, shown in Figure 9.5) about once per second, which is comparable to the time needed to equilibrate the air temperature to that of the guide tube. The interior diameter of the guide tube is 1.6 cm, and its overall length is approximately 2.3 cm. Air flows into the guide tube via four channels in the heat-exchanger plate, arranged symmetrically around the circumference of the top of the guide tube. The equal flow rates through the four input channels, along with the cylindrical geometry of the guide tube assembly, were engineered to produce a nearly cylindrically symmetric downward flow pattern in the guide tube, with the flow axis centered on the growing crystal. The guide tube temperature is kept near the substrate temperature T_1, and the guide tube is thermally isolated from the heat exchanger by a short section of thin-walled plastic tube.

An important consideration in this heat exchanger design is the uniformity and symmetry of the air-flow pattern around the growing snow crystal. If the temperature, supersaturation, or air flow are substantially nonuniform across the face of a stellar crystal, this will compromise the symmetry of its final morphology, creating a lopsided crystal. Thus, in pursuit of artistic snow crystal perfection, the apparatus was designed to produce quite uniform environmental conditions around a growing crystal.

This growth chamber design provides three adjustable parameters that can be used to control the crystal growth behavior: T_1, T_2, and F. The crystal temperature is nearly equal to T_1, which is typically kept within a few degrees of −15°C to grow stellar-plate snow crystals. The quantity $\Delta T = T_2 - T_1$ mainly determines the supersaturation, which can be at most

$$\sigma_{max} = [c_{sat}(T_2) - c_{sat}(T_1)]/c_{sat}(T_1) \approx \eta \Delta T.$$

During a cooldown of the apparatus, air is passed through the heat exchanger for 30 minutes to deposit ice on its inner surfaces. The temperature is set to $T_2 < -20°C$ during this time to make sure ice (and not supercooled water) is deposited inside the heat exchanger. Once the heat exchanger has been preconditioned in this way, air passing through it will exit at temperature T_2 and be saturated with water vapor relative to ice at T_2. As it approaches the substrate, this air cools to near $T_1 < T_2$ and thus becomes supersaturated.

Modeling the temperature and supersaturation at the growing ice surface is problematic with this apparatus for several reasons. The Reynolds number of the airflow in the guide tube is approximately 10, so the flow is probably not perfectly laminar, and the timescale for the air in the guide tube to become equilibrated with the guide-tube walls via diffusion is comparable to the time it takes air to flow through the tube. Moreover, a stagnation point in the flow occurs where the flow axis of the system intercepts the substrate surface, which is at the position of the growing crystal, and this further complicates the air-flow and thermal analysis. In general, however, a higher ΔT and a higher F produce a higher supersaturation around the growing snow crystal. To complicate the supersaturation analysis even further, water droplets often condense on the substrate near the crystal, as I describe below. The presence of liquid water substantially alters the supersaturation field, and the amount of water condensation changes substantially with changes in ΔT and F. The thermal connection between the edge of a growing PoP crystal and the underlying substrate is also difficult to determine accurately, given the pedestal geometry.

For all these reasons, I do not expect that the apparatus described here, in its current form, will ever find much application in performing precision measurements of ice growth rates under known conditions. It is simply too difficult to determine the temperature and supersaturation at the crystal surface to high accuracy. Nevertheless, the technique is quite suitable for qualitative studies examining snow crystal morphologies and growth behaviors, as well as for creating and photographing snow crystals purely for artistic purposes.

Optical Imaging

The microscope objective shown in Figure 9.5 is part of the heat exchanger assembly, but it is kept a few degrees warmer than T_2 by using a heater dissipating 1–2 watts into the objective body. This elevated temperature is necessary to keep fog from condensing on the glass face of the objective, which would interfere with optical imaging. A Mitutoyo 5X Plan Apo objective with a 250-mm focal length achromatic reimaging lens immediately behind it works well for single-frame imaging. The infinity-corrected objective has a working distance of 34 mm, a numerical aperture of 0.14, resolution of 2.0 μm, and a depth of focus of 14 μm, yielding excellent image quality. Focusing is done by moving the camera body on a StackShot linear positioning stage, and some amount of focus stacking (see Chapter 11) is typically needed for optimal imaging of large crystals, owing to the shallow depth of focus. The image projects to about 1 μm per pixel of the 36 × 25 mm, 5616 × 3744 pixel sensor in a Canon EOS 5D camera.

Small glass beads are a convenient tool for testing the focus quality of the optical system (see Chapter 11), as even a slight tilt of the substrate relative to the focal plane can degrade the image sharpness across the field of view, owing to the small depth of focus. This can be nicely corrected using focus stacking, which is also quite helpful when photographing PoP crystals that are slightly conical in shape (a common overall morphology for larger specimens). Obtaining high-quality microscope imaging is always a challenge with a custom-built apparatus, so some image testing like this is quite useful for characterizing and improving the optical system.

When shooting still images for making time-lapse videos of growing PoP crystals, I prefer a 3X Mitutoyo Compact Objective, which has a resolution of 4 microns and a depth of focus of 50 microns (see Chapter 11). The 3X images are noticeably less sharp, but focus stacking is no longer needed in most cases, greatly simplifying video production from the series of time-lapse stills. Also, the reduction in image sharpness is hardly noticeable in videos of growing snow crystals, as the edges are constantly moving.

The field lens shown in Figure 9.5 reimages a color filter onto a pupil in the objective for achieving a variety of illumination effects, which are discussed below. Figure 9.6 shows a ray diagram that illustrates how the microscope objective creates an image of the snow crystal on the camera sensor while the field lens images the color

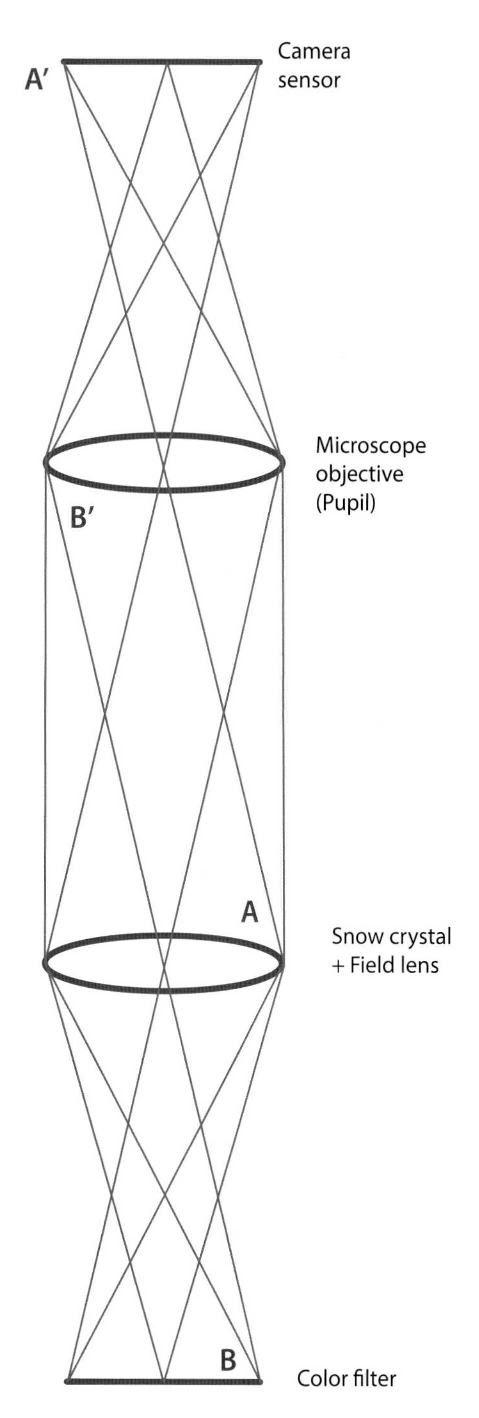

FIGURE 9.6. A simplified ray diagram of the optical layout used in the PoP apparatus. Note that the microscope objective images the snow crystal onto the camera sensor ($A \rightarrow A'$), while the field lens images the color filter onto a pupil inside the microscope objective ($B \rightarrow B'$).

filter onto the pupil inside the microscope objective. Understanding this ray diagram, particularly the importance of the pupil plane, is quite useful for creating desirable illumination effects that yield especially eye-catching color photographs. The topic of illumination effects is discussed in more detail in Chapter 11.

A rectangular field stop placed over the field lens blocks all incident light that does not transmit to the field of view of the camera, thus reducing the amount of unwanted scattered light in the optical system. The lens tube above the objective is also baffled and lined on its interior with a light-absorbing flocking material to further reduce scattered light. The window cell provides thermal insulation between the room and the cold plate.

Often a pellicle beamsplitter is placed right after the microscope objective to send an additional image to a second camera not shown in Figure 9.4. This is useful when collecting images rapidly from the main camera, as its live view seen on the TV monitor experiences a substantial dead time each time an image is being recorded. During times when the crystal growth behavior is changing rapidly, the second camera live view (seen on a second TV monitor) can be used to inform choices of T_1, T_2, and F that will achieve desired morphological effects.

Choosing a Seed Crystal

Finding a well-formed, isolated seed crystal on the substrate is perhaps the most difficult step in using this apparatus. Seed crystals fall randomly during loading, and their surface density on the substrate is adjusted by how long the shutter remains open with the substrate in the loading position. Also, many seed crystals are malformed or do not lie flat on the substrate, and these are not suitable candidates for further growth. If the surface density of loaded seed crystals is too low, it may not be possible to locate a well-formed specimen. If the density is too high, then it may not be possible to obtain a well-formed crystal that is sufficiently isolated. Often several loading attempts are needed to find a suitable specimen, with the substrate being heated between attempts to evaporate the existing crystals. The search process can be laborious and may end up taking anywhere from 1 to 30 minutes.

When growing large, platelike ice crystals, the ideal seed crystal is a simple, well-formed hexagonal plate with one basal surface flat against the substrate and with no additional ice within at least 1–2 millimeters from the chosen crystal. The subsequent growth phase typically lasts 20–60 minutes and is recorded via the imaging system. The temperatures T_1 and T_2 are adjusted with time (it requires about a minute for each to stabilize), along with the flow rate, to obtain the desired growth behaviors. At the end of the growth phase, the substrate is heated to just below 0°C so that the ice crystals sublimate away, at which point the cycle can be started once again.

After a typical day-long run growing crystals, the entire system is warmed to room temperature and baked to remove water. The base plate is typically heated to 40°C via its temperature controller, while the seed-crystal chamber is heated by means of an internal heat lamp, after emptying the water reservoir. Following about a day of baking, the entire system is clean and dry, reducing the presence of residual chemical contaminant vapors. Because a growing crystal is surrounded by air that has passed through the heat exchanger, special care is taken to reduce chemical contaminants in that air. The charcoal filter in the air stream removes contaminants coming from the air compressor, and the fiber filter downstream from the charcoal filter contributes little odor emission. Moreover, the heat exchanger is baked at 40°C overnight while clean air is passed through it before each run to remove residual contaminants. Seeing thin plates growing readily near −15°C is a good indication that the air flowing into the growth region is quite clean, as chemical contaminants readily inhibit thin-plate growth at this temperature (see Chapter 4).

ILLUMINATION AND POSTPROCESSING

A variety of illumination techniques can be explored using the relatively simple optical imaging system illustrated in Figure 9.6. The fixed microscope position and orientation are mainly optimized for photographing platelike stellar snow crystals, achieving a high imaging resolution while capturing crystals as they grow and develop. For most crystal morphologies, however, the type of illumination used can greatly affect the overall character of a resulting photograph. Ice is an almost perfectly clear material, so the choice of illumination is substantially more important than one might be accustomed to from photographing opaque subjects. Digital postprocessing can also be used for creating a variety of novel artistic effects, and, as with illumination techniques, there is considerable opportunity for enriching the overall look of a PoP snow crystal photograph.

Uniform Illumination

The most straightforward illumination method is to apply uniform white light from behind the crystal, replacing the color filter in Figure 9.5 with a simple round aperture. Uniform illumination tends to produce the sharpest microscopic details, and Figure 9.7 gives one example. This photo shows off the bitingly sharp facets and corners that are a special characteristic of most PoP snow crystals.

Natural snow crystals usually experience some sublimation after they fall out of the clouds, which gives their features a generally softer, "travel-worn" appearance (Chapter 11). This is not the case with PoP snow crystals, because they are being photographed as they grow, so their facets and corners are especially sharp. As a result, photographs of PoP crystals often reveal features that are rarely, if ever, seen in natural snow crystals. You need to be something of a snowflake connoisseur to notice the difference (or perhaps to care), but it does give PoP snow crystals a unique crispness in their finer details.

The specific appearance of the PoP snow crystal in Figure 9.7 can be understood from how light is transmitted through the clear ice. A flat pane of ice reflects some light incident on its surfaces, just as a flat pane of glass reflects some light. But not much is reflected, so the overall appearance of the flat areas of the crystal in this

FIGURE 9.7. Plain white-light illumination from behind yields a particularly high-resolution image, revealing exceptionally sharp facets and fine details in the surface structures. This crystal measures 2.7 mm from tip to tip, and five images were combined using focus stacking to improve the resolution slightly. Minimal postprocessing was applied, mainly just adjusting to background to full bright white and applying a slightly bluish hue to the dark contours.

photo are quite bright, almost as bright as the background. In these flat areas, most of the incident light simply passes through the clear ice, as it would a piece of glass. In contrast, the edges of a crystal refract transmitted light, diverting it away from its initial path. If the curvature of the edge is high, some of this light is diverted to such large angles that it does not enter the microscope objective, which gives the edges a darker appearance. More generally, the clear ice acts like a complex lens that refracts the transmitted light through a variety of angles. Edges tend to refract light to large angles, so the edges appear darkest in the image. Flat panes of ice refract the light less, so they appear brighter. No light at all is reflected or refracted where there is no ice, so the background is the brightest part of this photograph. While this type of plain white-light illumination is good for showing off the detailed structure in a PoP snow crystal, the resulting images tend to be a bit "flat" in character, as they give little sense of the snowflake's rich 3D structure.

When using a simple round aperture to produce white-light illumination, one soon finds that the size of the aperture affects the character of the resulting photograph. When the aperture is small, the resolution of the image is decreased, yielding generally fuzzier edges. The reason stems from how the field lens images the aperture onto the microscope objective, specifically onto the pupil plane, as shown in Figure 9.6. A close look at this ray diagram reveals that a smaller illumination aperture is essentially equivalent to reducing the aperture of the objective, because now no light enters the outer part of the objective. This reduces the resolution, because imaging is generally diffraction limited in microscopy. A smaller input aperture means a lower resolution in the diffraction limit, so using a small illumination aperture produces the same effect.

At the opposite extreme, if the illumination aperture is especially large, then the edges will not be as dark as they would otherwise be, which reduces the contrast of the final image. In the limit that white light is incident from behind the crystal at all possible angles, covering a full 2π steradians, the contrast would drop to nearly zero. In this extreme case, even large-angle refraction from the edges would not reduce the amount of light entering the objective, yielding a low-contrast image. This concept is best seen by running the rays in reverse, from the camera sensor to the illumination source. With wide-angle illumination, rays starting from any point on the image will run backward into the light source. Thus, every pixel on the camera will be exposed to light, which is the same as saying that the entire image will be washed out. This feature is what makes it especially tricky to photograph clear objects.

With plain white-light illumination, the optimal illumination aperture size is that for which the image of the aperture just fills the entrance of the microscope objective. This gives the maximum resolution, as the full objective is being used. And it produces a high contrast as well, as even small-angle scattering from the crystal edges will reduce the amount of light entering the objective. As can be seen from this exercise, a good understanding of the principles of optics is quite helpful for taking photographs of snow crystals. Commercial cameras and microscopes are not optimized for this purpose (as snow crystal photography is certainly not their primary market), so some DIY design effort and a fair bit of trial and error are required for obtaining high-quality photographic results.

Figure 9.8 shows this same photo after applying a "Bentley blocker" that digitally sets the background color to black. Wilson Bentley modified nearly all his photographs this way (see Chapter 1), although he did it the hard way by scraping the background emulsion off his glass photographic plates with a razor blade. Digital image processing reduces this task to a few clicks, but I am not a fan of the flat, high-contrast look. The main advantage of the Bentley blocker is that the crystal now appears white on a dark background, which many people feel is a more natural look for snow, even though individual snow crystals are actually clear, not white (see Chapter 11).

Figure 9.9 shows some additional image modifications that can be applied to a simple white-light PoP image to yield colorful effects. The various adaptations

FIGURE 9.8. Setting the background to black in Figure 9.7 yields this image, which looks much like the snow crystal photographs taken by Wilson Bentley (see Chapter 1).

FIGURE 9.9. Beginning with a single PoP snow crystal illuminated with white light, these modified images show a blue-on-white version from minimal postprocessing (left), an inverted image after color modification to give a white-on-blue appearance (center), and another inverted image modified to give a red-on-dark look (right).

FIGURE 9.10. An image of a PoP snow crystal using dark-field illumination. In the absence of any ice, no light enters the microscope objective, so the background is dark. The crystal appears bright, because the ice refracts some light from oblique angles toward the objective. In this case, the central light was blocked using a slightly off-center opaque spot in place of the color filter, giving an asymmetry to the overall illumination of the crystal.

do not change the underlying snow crystal structure, but simply present it in different ways. Given the amount of time and effort required to grow a single PoP snow crystal, it is often beneficial to explore a multitude of lighting and postprocessing effects like these.

Dark-Field Illumination

Another approach using white light is to replace the color filter in Figure 9.6 with a simple annulus that blocks the central light while letting a ring of light illuminate the crystal from an oblique angle. Figure 9.10 gives one example of this use of dark-field illumination in PoP snow crystal photography. Here the color filter has been replaced with an opaque disk on a clear glass holder. As shown in Figure 9.6, the field lens images the disk onto the microscope objective, so no light enters the microscope if no ice is present, giving the image a dark background.

When a snow crystal is present, it again acts like a clear, complex lens, this time refracting some of the light coming from outside the disk in such a way that it does enter the objective. For the photograph in Figure 9.10, the opaque disk filter was moved around and placed slightly off-center, producing different amounts of refraction on different sides of the crystal. Comparing Figures 9.8 and 9.10, one can see that off-center dark-field illumination gives the image a pleasing sense of depth along with an overall "glassy" look, as the brightness variations accentuate the 3D structure of the snow crystal. The "flat" image in Figure 9.8 is much less vibrant by comparison, and it gives the viewer no sense of the full 3D crystal structure. Using illumination to create a realistic sense of depth is one of the tricks of snow crystal photography (seen Chapter 11) that applies as well to PoP crystals.

Rheinberg Illumination

The technique of placing a patterned color filter in the pupil plane is a variation of dark-field illumination that was first described by microscopist Julius Rheinberg in 1896 and is now called *Rheinberg illumination*. Figure 9.11 shows one example of a PoP crystal photographed using this method. Rheinberg illumination produces an excellent sense of depth in snow crystal photographs, accentuating the full 3D structure better than other types of illumination do. Surface features remain sharp with high contrast, plus this technique adds a dimension of color to snowflake photography. This illumination method is described further in Chapter 11.

Because the Rheinberg color filter is placed at a pupil in the optical system, the pattern in the filter itself is not seen in the background image. One way to think about this illumination method is that different colors of light are shining onto the crystal from different angles. Because each point on the focal plane receives equal amounts of all the colors, the background image has a uniform average color. In this way, a uniform background color appears regardless of the color variations in the filter used. This is typical of optical systems, as patterns in the pupil plane have little effect on what is seen in the image plane.

Note also that the colors seen using Rheinberg illumination do *not* result from any dispersion effects, like the rainbow colors you see from a glass prism. Color dispersion is negligible in snow crystals, because the ice is simply too small and thin. It would take quite a large, thick ice prism to produce much color separation, and it

FIGURE 9.11. Rheinberg illumination was used in this PoP photo to accentuate the crystal surface structure while maintaining high resolution and adding some color to the image. Note that the background color is quite uniform across the image, while the snow crystal shows some red highlights introduced by the colorful Rheinberg filter (see Chapter 11).

FIGURE 9.12. Rheinberg illumination gives an especially pleasing sense of depth to thicker PoP snow crystals like this one, accentuating their surface features.

would only be noticeable with careful lighting and large, flat prismatic surfaces. Ordinary glass objects like cups and plates also show negligible color dispersion for the same reason.

With Rheinberg illumination, the colors all come from the color filter being used. As the light passes through the snow crystal, the ice can be thought of as a complex lens that refracts the light and changes its direction of travel, and this process is how color variations are produced. For example, if a bit of red light is shining on the image plane from an oblique angle, none of this light will normally enter the microscope objective, so none will make it onto the camera sensor. But if the ice bends some of that red light and sends it into the objective, and thus onto the camera sensor, then some red highlights will appear on the snow crystal image. And this is how the subtle red highlights in Figure 9.11 were created. Rheinberg illumination can also yield images with colorful shading that adds a sense of depth and accentuates the 3D structure in a thick snow crystal, as illustrated in Figure 9.12.

When photographing a PoP crystal, it is straightforward to swap color filters or move a patterned filter around while observing the live view on the TV monitor to produce a variety of pleasing color effects. If the crystal is growing slowly, as is often the case, one has plenty of time to experiment with different illumination techniques. Moreover, a PoP crystal is constantly changing as it grows, and each new morphological development provides what is essentially a new subject to record. With natural snowflakes, a crystal falls to earth and that is what you have. But photographing a PoP snow crystal is something of a continuous process, as each stage of its development presents a new photographic opportunity. After attaining some level of proficiency with the hardware, a single day of crystal growing can yield a bounty of excellent photographs.

POP GROWTH BEHAVIORS

Having described the PoP hardware, optics, and photography, I now step back from the photography aspect and discuss how one goes about growing a PoP snow crystal. Having worked with this apparatus for some years, I have developed some strategies for producing different morphological features under different growth conditions, and these have basically become a set of "recipes" for designing and fabricating different types of PoP snow crystals. As we will see, this hardware is quite versatile in that it can be used to grow a great variety of highly symmetrical stellar-plate snow crystals.

Establishing the PoP Geometry

Loading and positioning a seed crystal is usually done by setting the substrate temperature to −12.2°C, the heat-exchanger temperature to −12.0°C, and the air flow at about 250 cc/min. This establishes a slight supersaturation and a correspondingly modest degree of ice growth on the substrate, allowing some time to search for a suitable seed crystal. On opening the shutter, the substrate is moved into its loading position, where it stays for just a few seconds before being pulled back to the growth region. The substrate is then moved around in a 2D raster pattern to search for a well-formed seed crystal, one that exhibits a clean hexagonal prism morphology and has no nearby neighboring crystals. If none can be found, then the substrate is heated to −6°C for several minutes to drive off the seed crystals and then back to −12.2°C for another attempt. Once a suitable seed is in position at the center of the microscope field, the substrate temperature is lowered slightly to −12.5°C to commence the initial growth of a hexagonal PoP geometry.

The initial formation of the PoP geometry requires a careful selection of growth temperature and supersaturation. Thin plates grow out most readily from small hexagonal prisms when the temperature is near −15°C, which can be seen from the pictorial Nakaya diagram shown in Figure 8.19. The CAK model in Chapter 4 provides an explanation for why platelike growth is most prevalent at this temperature, but it can equally well just be taken as an empirical fact for this discussion.

Setting the initial temperature to near −15°C would quickly result in a PoP geometry, but growth at this temperature is also highly susceptible to branching. Even at relatively low supersaturations, the attachment coefficient on the plate edge will be so close to unity that the plate becomes unstable to branching even when the crystal is quite small. For this reason, −15°C is too cold if one desires a simple hexagonal plate like the one shown in Figure 9.3. However, the PoP geometry may not develop at all if the temperature is higher than −10°C. Blocky growth is common at this temperature, and the ESI (see Chapter 4) is often too weak to yield an upper plate at −10°C. With −15°C being too cold and −10°C too warm, a temperature of −12.5°C has been found to be about right for initiating the PoP geometry, and hexagonal sectored plates will grow out to around 0.5 mm in size at this temperature before branching occurs.

For even larger hexagonal plates, a good strategy is to move the temperature to near −10°C after the PoP geometry has been well established. A thin upper plate may not readily grow out from a small seed crystal at such a high temperature, but it will likely continue growing once the plate has formed. This type of hysteresis appears to be a feature of the ESI. After some amount of trial and error, I found that beginning near −12.5°C and slowly transitioning to −10°C provides a good recipe for growing a large, simple hexagonal plates. The plate growth rate decreases with plate size, however, so growing exceptionally large plates requires quite a lot of patience. The largest simple PoP plates I have made following this recipe have measured about 1.5 mm from facet to facet.

Fog Droplets

If the supersaturation at the substrate exceeds σ_{water}, then water droplets will readily condense on its surface, as illustrated in Figure 9.13. As these small droplets continue growing, they will slowly coalesce into larger droplets

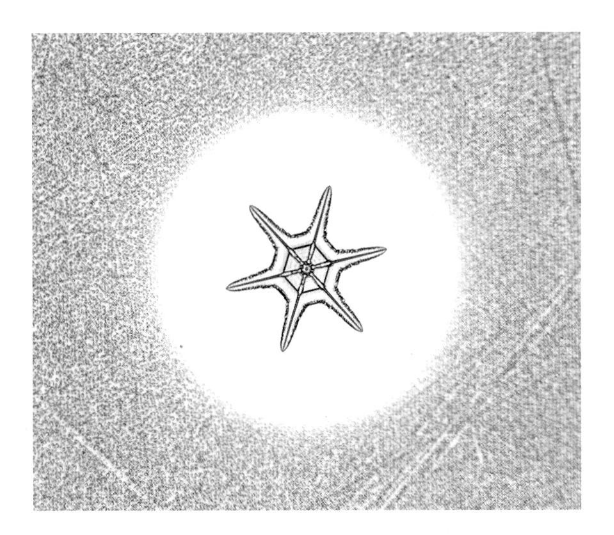

FIGURE 9.13. When the supersaturation exceeds σ_{water} near the substrate, water droplets will condense on it, here appearing as a fog around a small PoP crystal. The growing crystal absorbs water vapor in its vicinity, however, keeping the supersaturation below σ_{water} nearby. The circular transition region indicates where $\sigma \approx \sigma_{water}$ at the substrate.

through a process called *Ostwald ripening*. The process takes quite a bit of time, however, so often the droplet coalescence is not noticeable next to the rapid crystal growth. Moreover, fog droplets may evaporate when next to a growing snow crystal (see Chapter 1), because the vapor pressure of liquid water is higher than the vapor pressure of ice at the same temperature (Chapter 2). The left image in Figure 9.14 shows a small PoP crystal surrounded by an array of individual droplets that had recently condensed, while the right image shows the same crystal after substantial additional growth. As the crystal grows outward, it "pushes" the $\sigma \approx \sigma_{water}$ perimeter out in front of it, as nearby droplets evaporate to provide water vapor for the growing crystal. Within this perimeter, the nearby ice pulls the supersaturation down below σ_{water}, causing droplets to evaporate in that region.

Note that the supersaturation at the surface of a liquid droplet is essentially clamped at σ_{water}, making it nearly impossible to achieve supersaturations substantially above this level using the PoP method. This is true whenever snow crystals are supported on a large substrate, as water droplets will nucleate and grow on nearly any available surface when the supersaturation is above σ_{water}. As a result, one cannot grow true fernlike stellar dendrites using this apparatus. The electric-needle technique presented in Chapter 8 is better suited for investigating stellar dendrites and other forms that require supersaturations substantially above σ_{water}.

FIGURE 9.14. At high magnification (left), one can see individual droplets that recently nucleated on the substrate in this photo. At lower magnification (right), the droplets take on the appearance of a continuous fog. In this case, the $\sigma \approx \sigma_{water}$ contour is not circular but follows the overall shape of the PoP crystal.

It is possible to remove the condensed droplets around a crystal by reducing the supersaturation below σ_{water}, but the procedure can be quite time consuming in many circumstances. For a relatively large crystal, it may take 5–10 minutes to remove the droplets, and the crystal continues growing slowly during this time. The water evaporation is largely diffusion limited and therefore quite slow for a field of large droplets.

Because the substrate temperature is invariably below 0°C, eventually the liquid water droplets will freeze. However, they can remain in a metastable unfrozen state for quite some time, often for more than an hour even at temperatures down to −20°C. In some cases, a rapidly growing ice branch will approach a droplet field faster than the droplets evaporate away, causing the ice to grow into the supercooled water. The first droplet touched by the ice freezes instantly, and the solidification front slowly migrates outward as the frozen droplets each grow toward isolated liquid droplets in their immediate vicinity. Typically within several minutes, the entire droplet field becomes interconnected and frozen. Once this happens, the supersaturation will be clamped near zero by the ice-covered surface, thereby greatly reducing the supersaturation over the entire substrate. The PoP growth is thus greatly slowed when the droplet field freezes, as now the PoP crystal must compete for water vapor with all the surrounding ice. Increasing ΔT and the air flow rate will cause the whole ice field to continue growing slowly, along with the PoP crystal, as shown in Figure 9.15. But the supersaturation is generally too low to allow significant branching, and the PoP symmetry will also degrade with additional growth. In some of my PoP photographs (Figure 9.7 being one example), I have digitally removed the droplet field surrounding the crystal, simply to declutter the image background.

Branches, Wrinkles, and Spikes

Figure 9.16 illustrates the branching instability being initiated on the corners of a thin, faceted hexagonal plate. I use the word "initiated," because the branching process was brought about in this case by lowering the temperature of this substrate. The crystal first grew at −12.5°C with a low supersaturation, resulting in a simple hexagonal plate with sharp prism facets. Then I abruptly reduced the substrate temperature to −15°C, which both changed the growth temperature of the crystal and increased ΔT. This rapidly increased the supersaturation and initiated the sharp spikes seen in the figure.

The branching transition near −15°C is one of the most remarkable aspects of snow crystal formation, as the overall growth behavior is extremely sensitive to temperature in this region. The CAK model (see Chapter 4) provides a possible explanation of this phenomenon as a narrow "SDAK dip" on the prism attachment kinetics, but the underlying molecular dynamics responsible for this phenomenon is not yet well understood. Nevertheless, the high propensity for branching near −15°C is quite apparent when growing PoP snow crystals.

As the branches developed in Figure 9.16, they left behind a set of concave plate edges that experienced a new kind of "wrinkling" instability, which Figure 9.17 shows more closely. Close to the branch tip, α_{prism} is near unity and the growth is limited mainly by diffusion, resulting in a roughly parabolic tip shape that is essentially the Ivantsov solution to the diffusion equation (see Chapter 3), except with an added complication coming from strong basal faceting. Farther from the tip, the supersaturation is lower, so α_{prism} is reduced, and prism faceting becomes relatively more important. In this region, the faceting process turns the smooth concave edges into a series of faceted segments, yielding the overall serrated contours observed. Similar features can be seen in natural snow crystals, albeit not as clearly.

Figure 9.18 shows a slower branching transition that occurred at a temperature near −13°C. Before taking the first picture in this set, the crystal grew with a set of spiky branches that formed at −15°C. But the temperature was increased to −13°C soon after the branches formed, and the warmer conditions turned the narrow spikes into the broad-branched sectored plates seen in the top photo. The crystal continued growing at this temperature until

FIGURE 9.15. Soon after taking the top photograph, the field of water droplets became frozen. A long period of additional slow growth yielded the bottom scene.

the branching instability slowly took hold, breaking up the cleanly faceted prism surfaces and promoting the growth of narrower branches, faceted just at the tips, with each tip flanked by wrinkled concave surfaces. This kind of growth is typical when the branching instability is driven only weakly, resulting in broad branches with no sidebranches.

Figure 9.19 shows another example where a hexagonal plate grew out near −12.5°C, followed by an abrupt branching transition near −15°C. In this case, branches

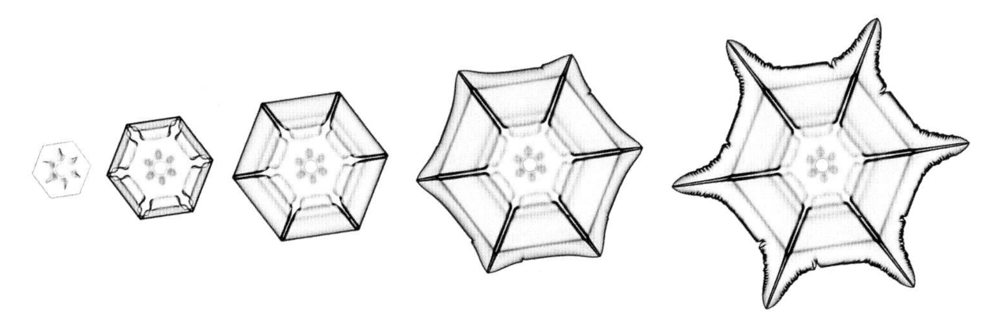

FIGURE 9.16. An example of the onset of branching on a simple hexagonal plate, brought about by lowering the temperature to near –15°C and increasing the supersaturation. First the edge of the plate first grew thinner, then the prism facet started to become curved, and finally the branching instability kicked in. The small "petals" at the center of the crystal are bubbles trapped at the base of the pedestal. As described in Chapter 3, the remaining surface structure is located on the underside of the plate, while the upper basal surface is almost perfectly flat.

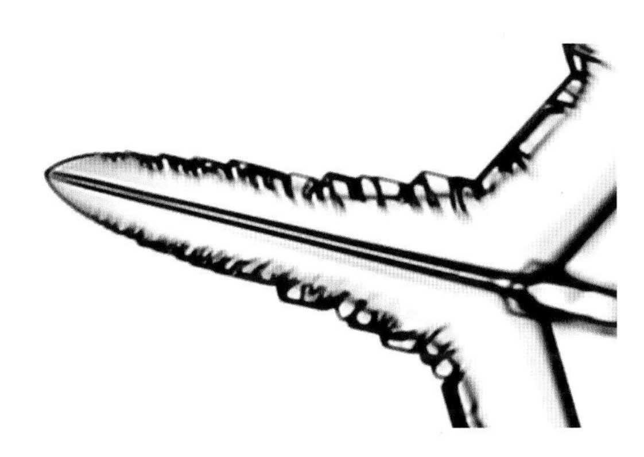

FIGURE 9.17. When the formation of branches results in a concave plate edge, the initially smooth perimeter is unstable to localized faceting, which produces a serrated edge and a series of "wrinkled" surface features.

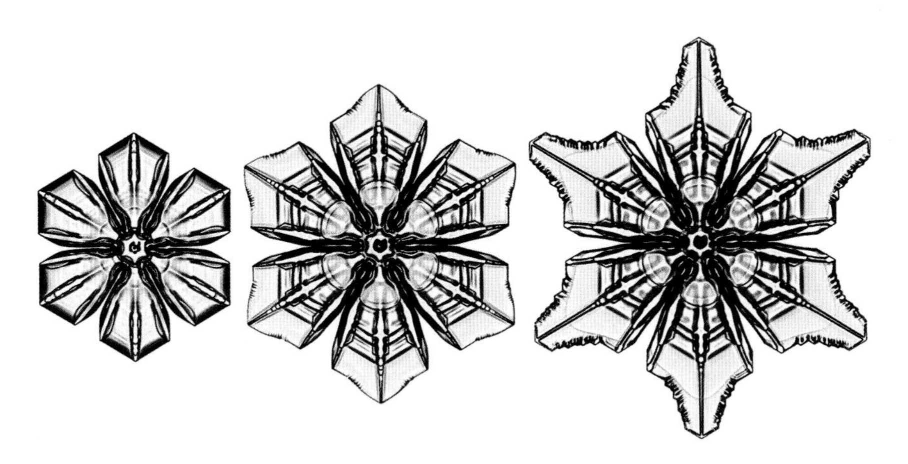

FIGURE 9.18. Starting as a sectored plate with strong prism faceting (left), this PoP crystal experienced a period of weak branching, resulting in broad outgrowths with faceted tips flanked by serrated edges (middle and right). Weak branching transitions like this are commonly seen in natural snow crystals.

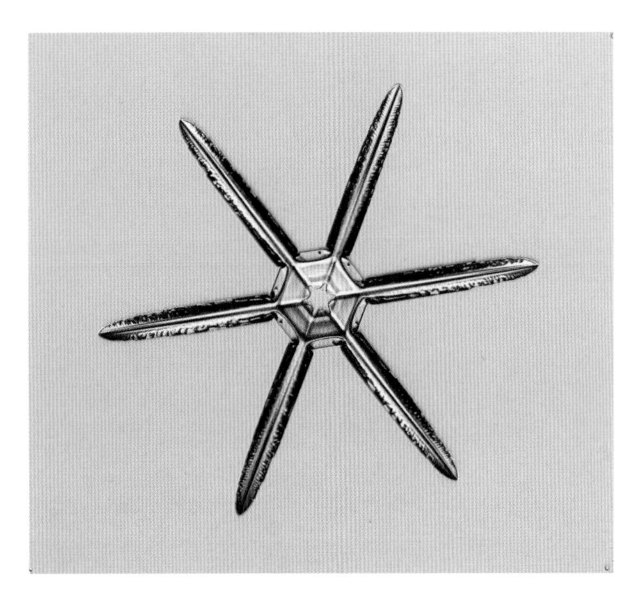

FIGURE 9.19. Spiky branches growing out from a hexagonal plate at −15°C. At constant growth conditions, these spikes will grow out indefinitely.

grew stably outward and would continue growing out for quite some time, resulting in long spikes with roughly parabolic tips and serrated sides. I have never seen spontaneous sidebranching in the PoP apparatus, however, because droplet formation generally limits the supersaturation to not much above σ_{water}, which is too low for spontaneous sidebranching even near −15°C. After growing out these spiky branches a bit more, I changed the temperature from −15° to −13°C to make a set of sectored plates, as seen in Figure 9.20, and then back to −15°C to put spikes on the sectored plates.

Sectored Plates

When grown at a temperature near −12°C with a constant supersaturation, PoP snow crystals often develop into simple sectored plates with the overall structure shown in Figure 9.21. Because the supersaturation is higher on the top basal surface than underneath, the plate edge grows up as well as out, resulting in a slightly conical shape overall. Basal terrace nucleation occurs near the plate edges, and the terraces subsequently propagate toward the center. If the basal growth is slow, then the top terrace may grow to cover the entire basal surface before another terrace nucleates, yielding what is called layer-by-layer growth. It is more likely, however, that there will be a continuous train of propagating terraces, as illustrated in Figure 9.21(a) and (b). If the number of unfinished terraces is small, the basal facet may look perfectly flat. But often the terrace steps bunch together to form a series of inwardly propagating rings, as described in Chapter 3.

While the upper basal surface of a PoP crystal typically exhibits a slightly concave shape, the lower surface is slightly convex, as illustrated in Figure 9.21. As the upper prism edges grow outward, terrace steps are left behind on the lower basal surface, and these steps subsequently propagate outward. Because the water vapor supply is highest near the corners, the lower basal steps tend to sprout 2D step branches that develop into ridges. This process is described in Chapter 3, being one of many manifestations of the Mullins-Sekerka instability in snow crystal growth. As a result, all the sector-plate ridges seen in PoP snow crystals form on lower basal surfaces.

Figure 9.22 illustrates a simple sectored-plate PoP snow crystal with several characteristic surface markings. First, the series of faint circular rings indicate step bunching on the top basal surface, and a time series of images reveals that these rings propagate slowly inward with time. Second, on the underside of the plate, a set of sectored-plate ridges divide the plate into six sectors, like equal slices of a hexagonal pie. In this example, however, the plate did not grow with perfect hexagonal symmetry, so the ridges are a bit curved. The ridges always terminate at the faceted plate corners, so nonlinearity of the ridges indicates a slight variation of the growth rates of the different prism facets. Third, this crystal also exhibits a few faint hexagonal surface markings I call "ribs". Unlike the rings and ridges, rib formation requires changes in the supersaturation with time, and the process is described in detail in Chapter 3.

Finally, four dislocation lines can be seen on this crystal, appearing as faint evaporation grooves on the

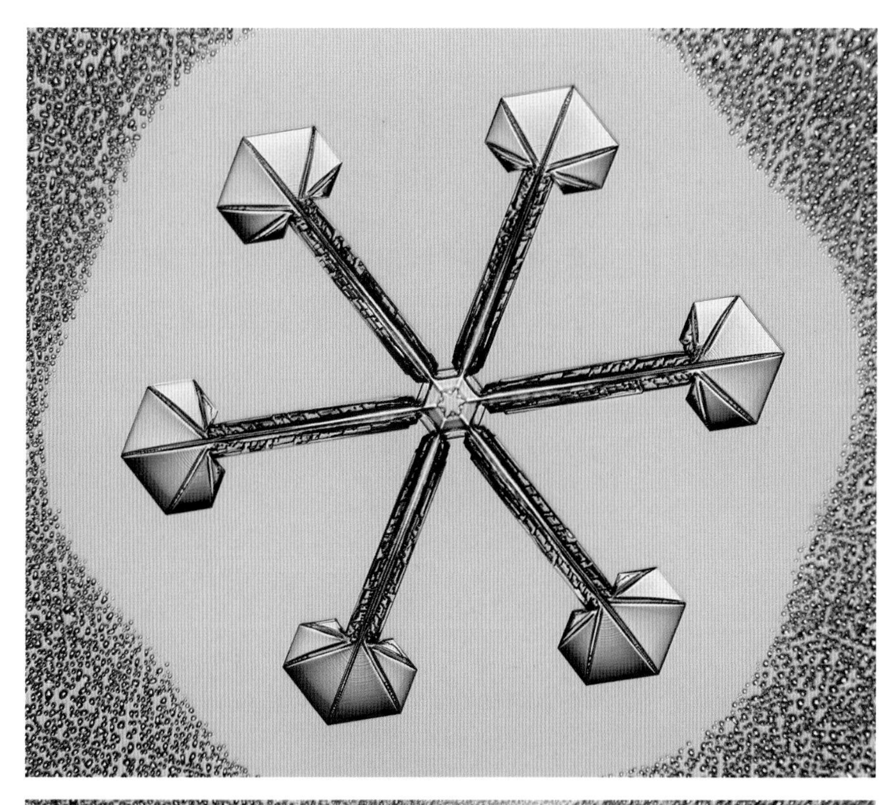

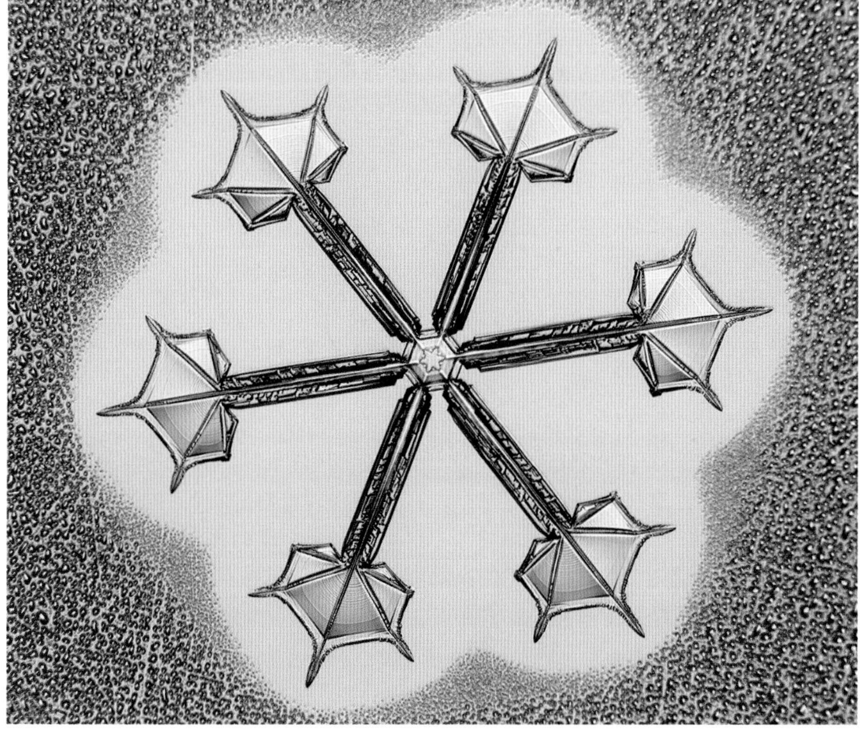

FIGURE 9.20. After the spikes in Figure 9.19 had grown longer (note the fixed size of the central hexagonal plate), the temperature was increased to −13°C to slowly grow a set of sectored plates on the ends of the spikes (top). Then the temperature was returned to −15°C and a new set of spikes grew out from the sectored plates (bottom).

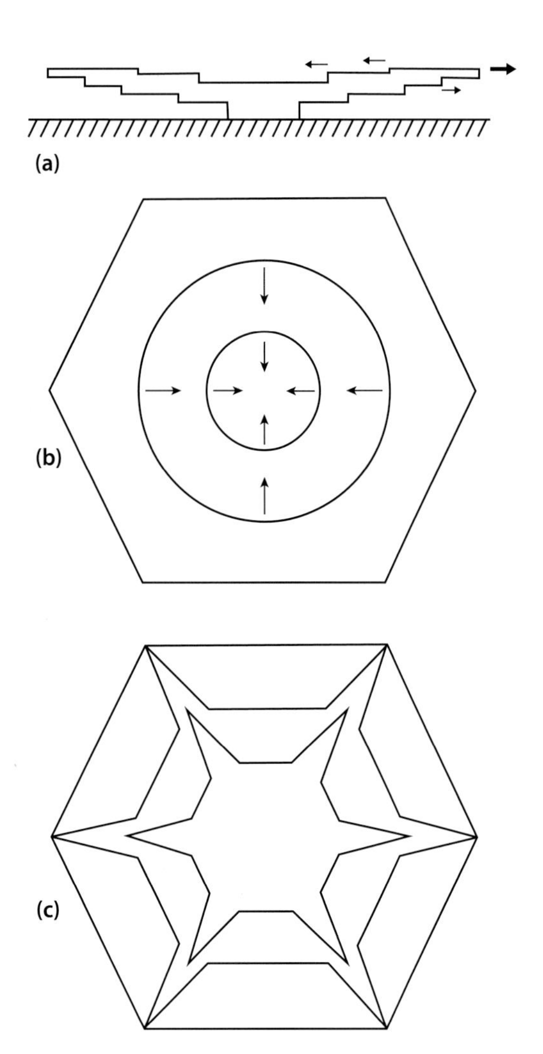

FIGURE 9.21. (a) The overall geometry of a PoP crystal, as seen from the side. The steps represent molecular terraces on the basal surfaces. (b) Growth on the concave top surface yields nearly circular inward-propagating steps. Step bunching sometimes turns a series of one-molecule-high steps into a coarser series of macrosteps that can be seen using optical microscopy. (c) Faster step advancement near the hexagonal corners yields ridges on the lower convex surface.

surface (Chapter 2). These originated at early times, when the outer prism facets began to break up at the beginning of a weak branching event (inset image). Seeing this happen, I lowered the supersaturation in the apparatus, which caused the prism edges to "heal" back to a simple hexagonal plate. The healing process did not happen with full molecular precision, however, and the remaining molecular mismatches turned into dislocation lines on the crystal. Unlike screw dislocations (see Chapter 4), these lattice errors had little effect on either the basal or prism attachment coefficients, so the overall crystal shape was not much affected by the dislocations.

Figure 9.23 shows another example of ridge formation on a PoP snow crystal, this time forming a broad-branched stellar plate. After a set of six branches grew out from the corners of this crystal, the conditions were set to promote the formation of flat plates near −12°C. With constant growth conditions, the sectored plate extensions are nearly featureless except for the distinctive ridging. The ridges are curved, because the plates interfere with one another as they develop, which stunts the sideways growth while maintaining fast outward growth. The curved ridges thus indicate the positions of the prism corners at different times. Similar morphological features can be seen in natural snow crystals, but large, simple plates like this are quite rare. They require constant growth conditions to form, and such conditions are unlikely in the turbulent atmosphere.

Inwardly propagating rings are also quite common in sectored-plate extensions on PoP crystals, and Figure 9.24 shows two examples. Similar structures can be found in natural snow crystals as well, but the asymmetry between the top and bottom plates in the PoP geometry facilitates their construction. Also, ring formation requires a significant growth time at nearly constant conditions, which again favors their observation in a laboratory setting. Ring formation is quite sensitive to the flatness of the top basal surface, with the most pronounced rings occurring on surfaces with larger cone angles. At −15°C, the prism/basal growth ratio is especially high, and the resulting plates are especially flat, so ring structures are typically absent, or nearly so. When the temperature is a few degrees higher or lower, the top plate surface becomes more concave and ring structures become more apparent.

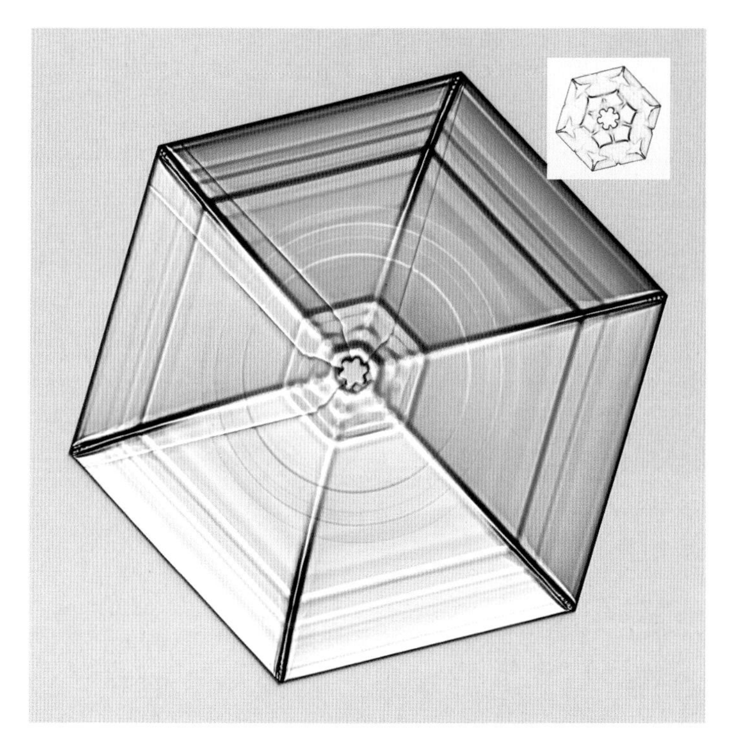

FIGURE 9.22. A simple hexagonal plate PoP crystal illustrating several common snow crystal morphological features, including ridges, inwardly propagating ring, and ribs. The four curved dislocation lines are not common, reflecting this crystal's unusual early growth history. The plate is supported by the small star-shaped pedestal seen at its center. The top-right inset shows the crystal at an earlier stage in its growth, at the same image scale as the main image.

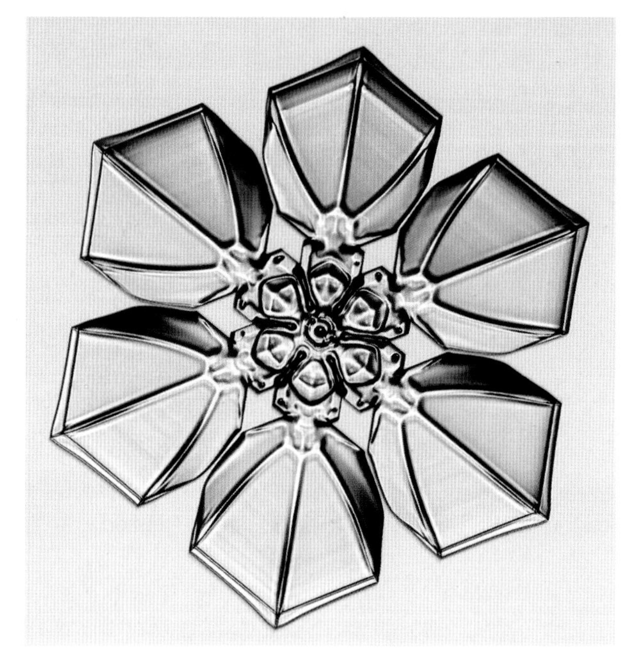

FIGURE 9.23. Sectored plate extensions. Curved ridges often form on broad-branched sectored plates, looking a bit like "duck's feet" in this PoP example. Because ridge formation originates at the faceted corners of a plate, the ridges trace out the location of the corners as a function of time as the crystal developed.

Stellar Dendrites and Imposed Symmetry

The PoP growth technique is especially adept a creating complex, symmetrical stellar dendrites using the process of *induced sidebranching* described in Chapter 3. The basic idea is to first form a set of six primary branches, then lower the supersaturation so the branch tips become faceted, and then increase the supersaturation abruptly to a high value, stimulating the formation of branches at all three of the exposed prism corners. Sidebranch formation is thus "induced" by first creating a faceted tip geometry and then quickly exposing it to a high supersaturation. Figure 9.20 shows one example of induced sidebranching, although this is something of an odd case. More typically, the supersaturation is lowered just long enough to develop clear prism faceting before abruptly increasing the supersaturation again to stimulate a set of sidebranches. Figure 9.25 shows a large PoP crystal that experienced many induced-sidebranching events.

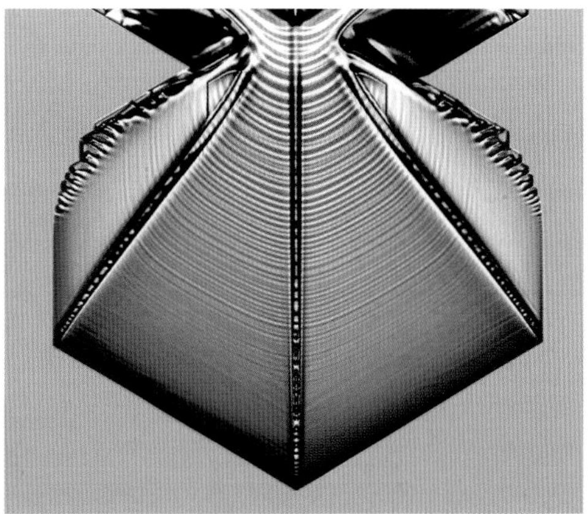

FIGURE 9.24. Large sectored-plate extensions on PoP snow crystals often exhibit concentric-ring patterns like those seen in these two examples. The rings slowly propagate inward as the plates grow, and they are located on the top basal surfaces, as sketched in Figure 9.21. Ridges and faint ribs can also be seen, and both these features are located on the lower plate surfaces.

With induced sidebranching, the formation of complex structure is coordinated by the timing of the events that caused it. Sidebranches form on all six primary branches simultaneously, and on both sides of each primary, yielding a symmetrical morphology. For both PoP and natural snow crystals, we see that the large-scale, complex symmetry does not result from any communication or other internal coordination between the primary branches. Instead it arises simply from the time-dependent environmental conditions being applied to the growing crystal, as described in Chapter 1. In the PoP apparatus, of course, the process of induced sidebranching can be applied repeatedly at varying temperatures and supersaturations, and with varying wait times between events. The example in Figure 9.25 shows how a series of growth transitions can be used to fabricate an extraordinarily complex, yet symmetrical, PoP snow crystal.

Natural snow crystals essentially never exhibit this degree of complexity accompanied by such near-perfect symmetry. However, the feat can be accomplished relatively easily in the laboratory, because the environment is carefully engineered and electronically controlled. It is straightforward to apply sudden and quite large changes to the growth conditions, thus stimulating sidebranching and other morphological features with considerable precision. It takes about an hour to grow a large snow crystal like this, but the process is quite reliable if the apparatus is working well.

In contrast to stimulated sidebranching, spontaneous sidebranching generally does not yield symmetrical stellar snow crystals. As described in Chapter 2, fernlike stellar dendrites exhibit copious sidebranching, but the sidebranch spacing is quite irregular over the entire crystal. The primary branch tips are roughly parabolic in shape, and the rounded profile makes it difficult for sidebranches to appear. High supersaturation is needed to produce sidebranches in this circumstance, and the process is often so haphazard that even the opposite sides of a single primary branch exhibit sidebranches at different locations. While stimulated sidebranching is thus somewhat chaotic, induced sidebranching is a controlled process that first prepares the tip by giving it faceted corners. The sharp corners are then more susceptible to the branching instability, so a lower supersaturation level is needed to make branching happen.

FIGURE 9.25. This elaborate PoP snow crystal underwent a series of induced sidebranching events, each producing a set of symmetrical sidebranches flanking the primary branches. The final crystal displays a decidedly complex overall structure with a remarkable degree of sixfold symmetry, substantially better than what one finds in natural snow crystals. Just before this photograph was taken, the supersaturation was increased to condense a fog of water droplets quite close to the outer perimeter of the crystal. This crystal measures 3.8 mm from tip to tip.

Controlling the applied growth conditions is what makes growing PoP snow crystals something of an artistic process. The entire temporal history of the temperature and supersaturation can be planned and manipulated to create a unique snow crystal morphology. Of course, there are limits to what snow crystal designs can be created, as one is bound by the underlying rules of snow crystal growth. And we cannot yet predict what patterns will appear in all circumstances, as our understanding the rules is somewhat incomplete. Nevertheless, the PoP technique provides an opportunity for artistic expression in the novel medium of snow crystal fabrication.

Other Forms

While the PoP apparatus was designed for creating stellar-plate snow crystals, columnar seed crystals are often present, and these can land with one prism facet resting on the substrate. Growing platelike crystals from these columnar seeds yields crystals like those shown in Figure 9.26, which are essentially double plates seen edge-on. Technically, these might be called "half-double plates," as the crystals can only grow up from the substrate. As seen in Figure 9.27, these crystals often grow side-by-side with the usual PoP geometry.

IDENTICAL-TWIN SNOW CRYSTALS

That familiar maxim that no two snowflakes are exactly alike appears to have had its origin with the photographs taken by Wilson Bentley. In his book with William Humphreys [1931Ben], Bentley presented pictures of nearly 2,000 stellar snow crystals, selected for their beauty and symmetry, and each was clearly different from all the others. Since then, the notion of snowflake uniqueness seems to be something we all learn at a young age (at least in North America), possibly in elementary school while participating in the nearly universal craft of cutting snowflakes out of paper.

There is good reason to believe that no two complex, natural snow crystals will ever look quite the same, as I described in Chapter 1. Because each snowflake follows a different path through the clouds, guided by the motions of a turbulent atmosphere, each one experiences different growth conditions during its journey. The number of possible variations is vast, so

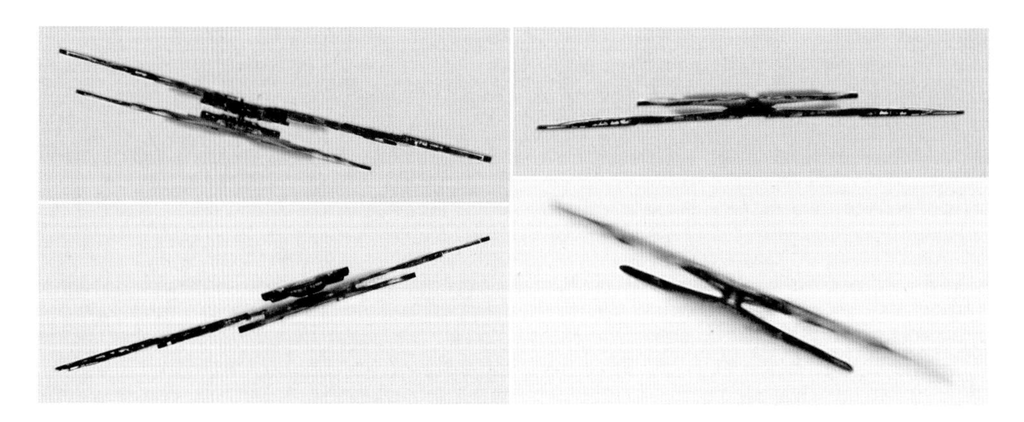

FIGURE 9.26. When a small columnar seed crystal on the PoP substrate is exposed to temperatures near −15°C, plates often grow out from the ends of the column. These images show some of the resulting double plates seen edge-on.

FIGURE 9.27. The double-plate crystal on the upper left grew out from a columnar seed crystal, while the stellar crystal followed the usual PoP geometry. As both seed crystals were exposed to the same growth conditions near −15°C, both developed thin-plate morphologies.

the probability of finding two identical snowflakes is vanishingly small. But this discussion changes when you consider growing PoP snow crystals. Now the growth conditions are not determined by random paths through a turbulent atmosphere. Instead, the temperature and supersaturation are controlled by precision temperature controllers that can be set and changed according to a prescribed schedule. In principle, one could engineer a precision snow crystal factory that would produce a continuous flurry of essentially identical snowflakes. (Of course, one must be careful about how you define the word "identical," as discussed at some length in Chapter 1.) I have not had any great urge to create such a factory, but it would not be an impossible task.

Instead of growing identical PoP snowflakes one after another, however, a much easier approach is to grow two at the same time, side by side, as illustrated in

Figure 9.28. This photo, unmodified except for cropping and small global brightness/contrast/color adjustments, shows two snow crystals that I grew simultaneously in the PoP apparatus. I like to call these "identical-twin" snow crystals, because they are not perfectly identical, and small differences can easily be found if you look carefully at the photo. But like human identical twins, they are clearly much more similar than one might generally expect to see. In the spirit of full disclosure, Figure 9.29 shows the original, unmodified image of the full 5.6×3.7 mm field of view of the camera.

To create this pair of identical-twin snow crystals, I first had to locate a pair of well-formed seed crystals, close to one another, but not too close, and reasonably isolated from any additional crystals. This is a high bar to clear, and Figure 9.29 shows that there was quite a bit of unwanted activity going on over the entire substrate, a fact that the cropped image does not convey. Pretty much every photo in this chapter was cropped in this way, to focus the viewer's attention on the primary subject.

With a suitable pair of seed prisms in place, I then proceeded to grow the dual PoP crystals by applying

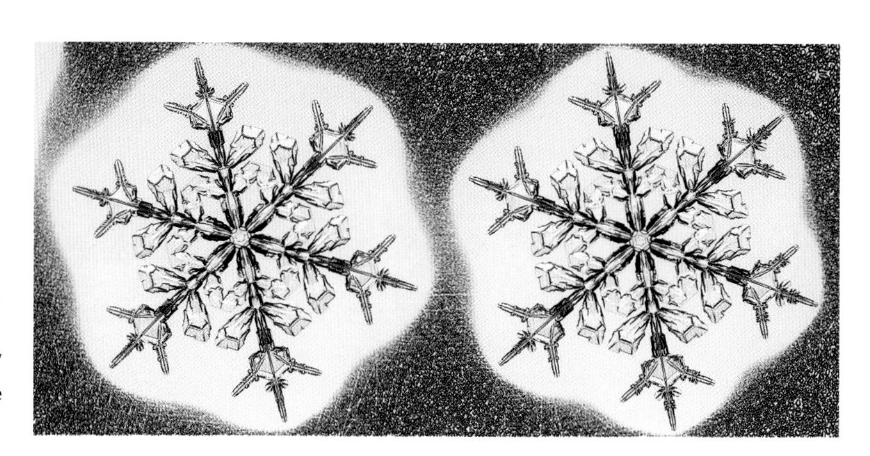

FIGURE 9.28. A pair of "identical-twin" snow crystals, grown side-by-side in the PoP apparatus.

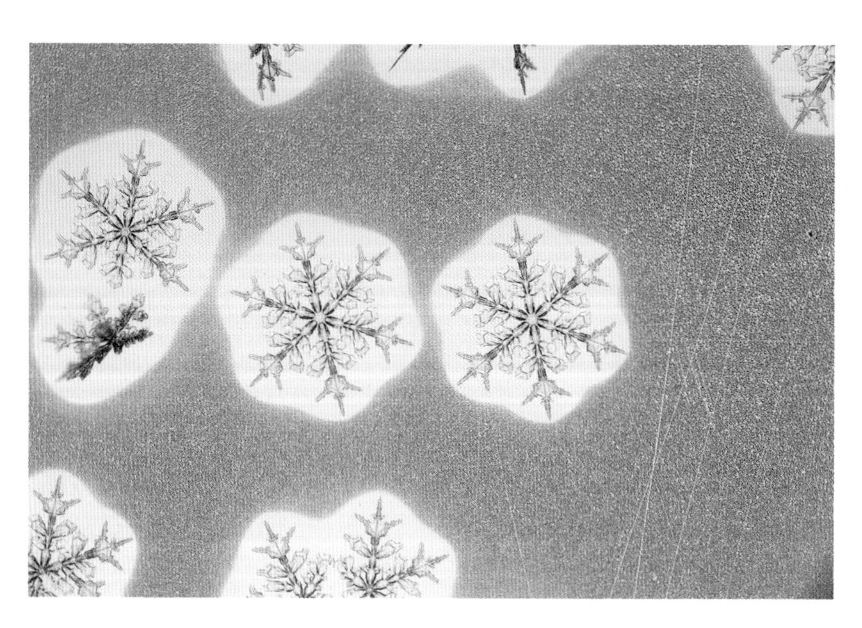

FIGURE 9.29. The unretouched, uncropped, straight-out-of-the-camera original version of the image shown in Figure 9.28.

branching, faceting, and other effects at various times. A key trick for growing identical-twin snow crystals is to make a continuous series of large, abrupt changes in growth conditions. Rapidly changing the environment by a substantial amount causes a correspondingly large and abrupt change in the growth behavior. A series of such events creates an exceptionally complex morphology, which tends to accentuate the similarities between the two crystals. Smaller, gradual changes typically yield less-perfect symmetry, even among the six branches of a single crystal. There are inevitably some weak temperature gradients in the growth region, and neighboring crystals also perturb the local environment to some extent. Making large, abrupt changes tends to mask these weaker effects, thus improving the overall symmetry of the growing crystals.

Another trick is to create a field of water droplets around the two crystals early on and to maintain a well-defined droplet perimeter around both crystals, always with a "barrier" of droplets separating the crystals, as shown in Figure 9.28. The droplets provide a stabilizing influence on the supersaturation, greatly reducing the perturbations that arise from neighboring crystals. This makes sense, because the droplets hold the supersaturation at $\sigma \approx \sigma_{water}$ in their immediate vicinity, and the boundary condition of having a clean droplet perimeter around both crystals tends to improve the overall symmetry of the two crystals.

The final trick is just knowing when to stop. In Figure 9.28, for example, the line of droplets between the crystals will soon evaporate as the crystals grow larger, and then the crystals will begin to interfere with each other's growth. Figure 9.30 shows what happens when the droplet barrier disperses while the crystals continue to grow. The branches growing between the two crystals compete for the available water vapor, thereby stunting their growth relative to the outer branches, which are still supplied by nearby droplets. Viewing a video that shows first the nearly identical growth of nearby crystals, followed by the stunted growth of the branches between them, gives one a good appreciation of how the droplet field affects the overall growth process.

FIGURE 9.30. (Top) The water droplets surrounding a PoP snow crystal tend to isolate it and stabilize its growth. The droplet field thus contributes to providing the same level of supersaturation for each growing branch. (Bottom) The same pair of crystals after the droplet barrier between the crystals evaporated. In this environment, the facing branches grow more slowly, yielding asymmetrical crystals.

Note that the phenomenon of twelve branches growing in synchrony is essentially no different than with six branches. For a natural snow crystal, only its six conjoined branches experience the same growth conditions as a function of time (although twelve-branched crystals can be quite symmetrical also, as seen in Chapter 10). In the PoP apparatus, however, all the crystals are connected by virtue of the substrate. There is no need to stop at twelve in this line of reasoning, and Figure 9.31 shows that larger groups can also grow into similar-looking snow crystals.

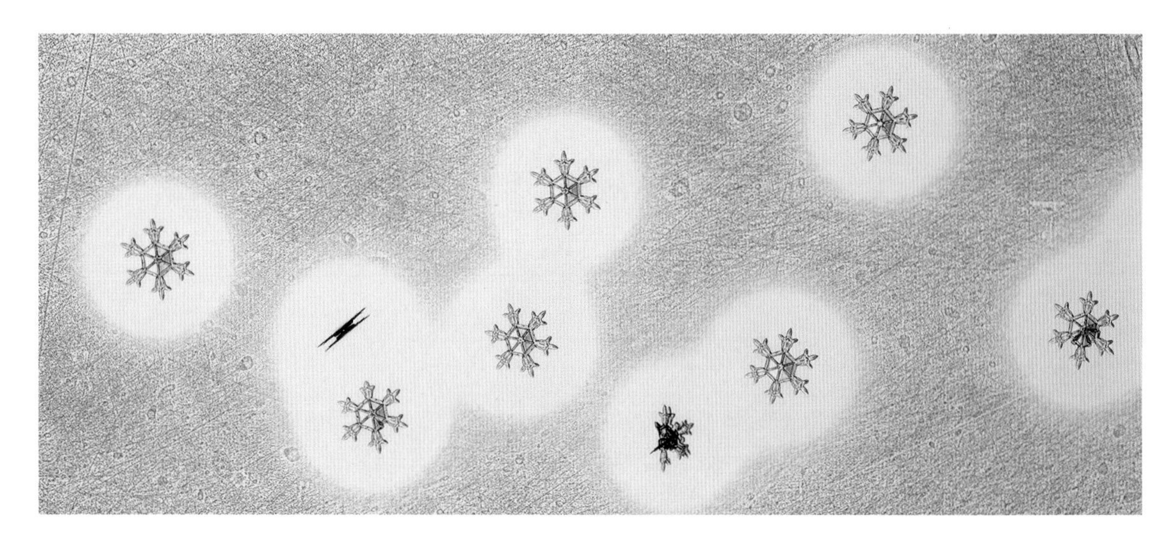

FIGURE 9.31. When any cluster of small PoP crystals exhibits the same growth conditions as a function of time, the individual crystals will all develop into similar morphologies.

POP ART

The remainder of this chapter presents a gallery of synthetic PoP snow crystals that I grew using the apparatus and techniques described in the previous sections. The images that follow show real PoP snow crystals, with essentially no digital modifications of the overall crystal structures. However, I did use a fair bit of artistic license when adjusting brightness, contrast, cropping, sharpness, and a host of color effects. In some images, I also removed droplets, specks of dirt, or other distractions from the background around the growing crystals. My overarching goal in this gallery is to present the growth of synthetic snow crystals as a novel art form, rather than as a tool for scientific discovery. While science and art are normally quite distinct endeavors, they come together beautifully in snow crystal growth. An understanding of the science was needed to engineer the PoP apparatus, making possible the creation of high-resolution images and videos showing details not observable in natural snow crystals. I believe there is much left to explore in this novel form of additive ice sculpture.

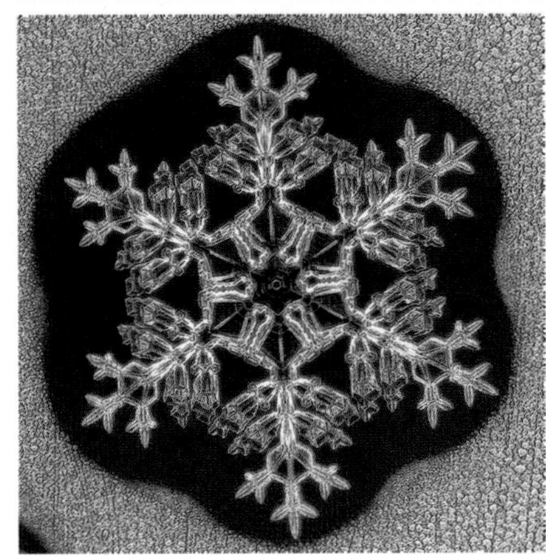

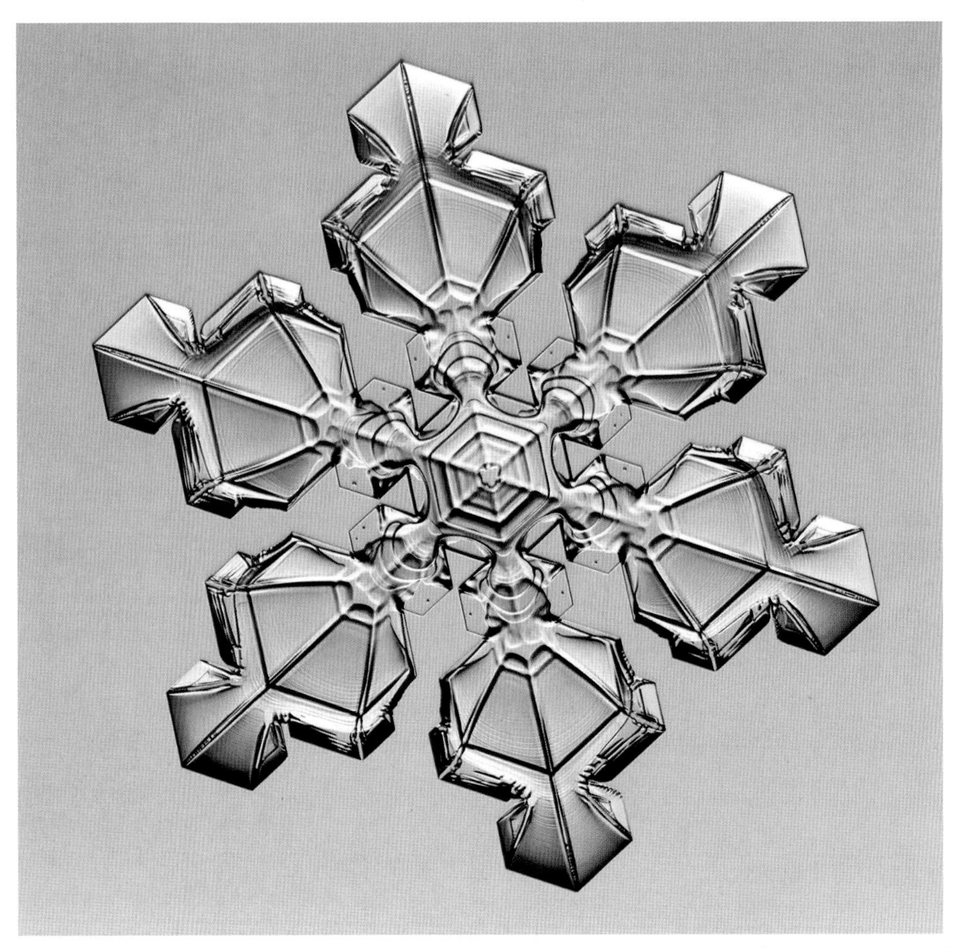

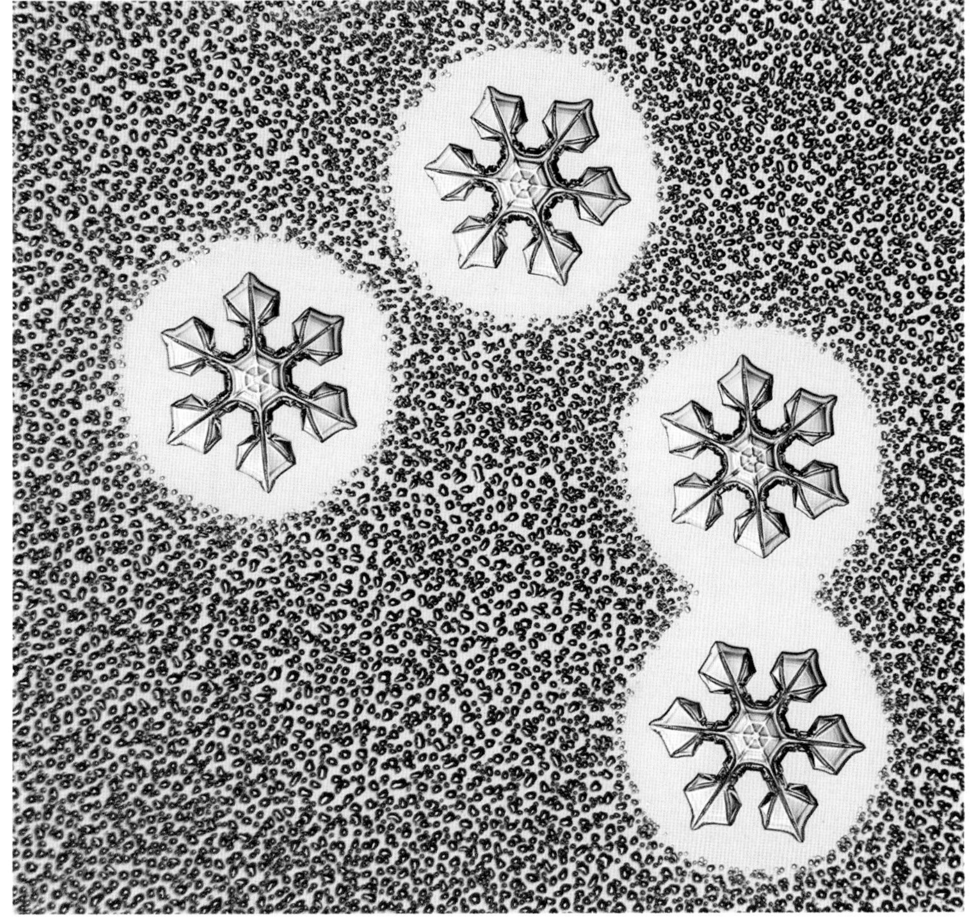

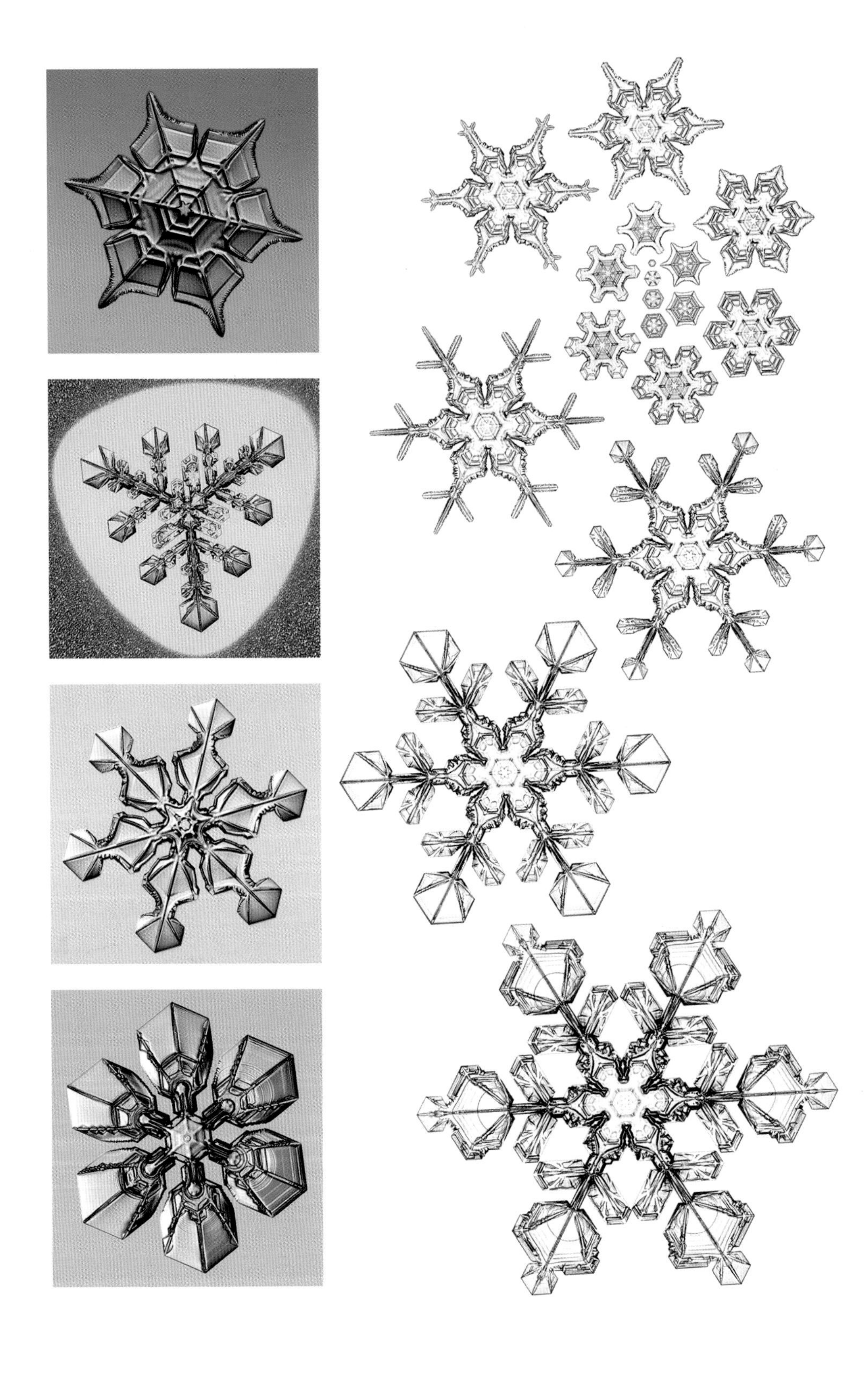

FIGURE 10.1. A large stellar-plate snow crystal with complex surface markings, measuring just over 3 mm from tip to tip, captured by the author in Burlington, Vermont.

TEN

Natural Snowflakes

One cannot fix one's eyes on the commonest natural production without finding food for a rambling fancy.

—JANE AUSTEN, *MANSFIELD PARK*, 1814

Nature provides a marvelous laboratory for examining the morphological diversity of snow crystals. With a simple magnifier and a robust tolerance for cold weather, one can observe a remarkable variety of crystalline forms falling from the winter clouds. The possibilities range from simple plates and prisms to hollow columns, sectored plates, fernlike stellar dendrites, capped columns, and a host of rare and exotic varieties. Each snowfall has its own character, and there is always something new to discover.

In this chapter, I present an extensive catalogue of photographic examples showing different types of snowflakes and snow crystals along with a discussion of their various identifying features and characteristics. The presentation is in the form of a field guide [2006Lib], aimed at assisting snowflake photographers or other readers who want to see for themselves what the winter clouds have to offer. Although such exploration could be done without any guidance, the venture is generally more rewarding when you start out knowing what others have observed and documented over the years. Unless otherwise indicated, I photographed all the snow crystals in this chapter during snowfalls in northern Ontario, Alaska, Michigan, Vermont, northern Sweden, and other locations.

As an occasionally avid bird watcher, I like to think of snowflake watching as an entirely analogous activity. It can be entertaining, educational, and a surprisingly enjoyable recreation. Keeping an eye out for interesting crystals is a worthwhile pursuit any time you happen to be outside during a light snowfall. You could be riding the chair lift at your local ski area, taking a stroll through the winter woods, or just waiting in your car somewhere. If the snow is falling all around you, why not have a look from time to time to see what you can find?

Growing up on a farm in North Dakota, I experienced a lot of snow, and I saw my share of birds. But I never really noticed either until someone showed me what to look for, which happened long after my childhood days. Looking back on this particular aspect of my youth, these were lost opportunities. We all live in nature, but it takes a bit of effort to develop an awareness and appreciation of the natural world around us. If you happen to live in a cold climate, I heartily recommend that you think about snowflakes occasionally, and perhaps go outside to have a look for yourself. You never know what works of art the clouds might be sending down.

SNOWFLAKE WATCHING

Ukichiro Nakaya described snow crystals as "hieroglyphs from the sky" [1954Nak], because the form of each crystal can be interpreted, at least in principle, to reveal the atmospheric conditions it experienced as it grew and developed. There is some truth to this supposition, because the process of snow crystal growth is largely deterministic, meaning that two crystals experiencing the same conditions as a function of time will grow into nearly identical shapes. The "identical-twin" snow crystals presented in Chapter 9 provide a direct confirmation of this deterministic behavior.

A primary tool for deciphering the shape of a specific snow crystal is the Nakaya diagram illustrated in Figure 10.2, as this morphological overview connects the seemingly disparate observations of falling snow into a generally coherent picture of what is happening up in the clouds. Of course, there are limitations to how much one can say about the growth history of a snowflake by examining a single photograph of its final state; but usually one can envision a plausible scenario to explain an observed morphology. Another use of the Nakaya diagram is to predict what types of crystals will appear in different weather conditions. For example, if one wishes to find large, well-formed stellar snow crystals (a popular photographic goal), it is useful to know that such

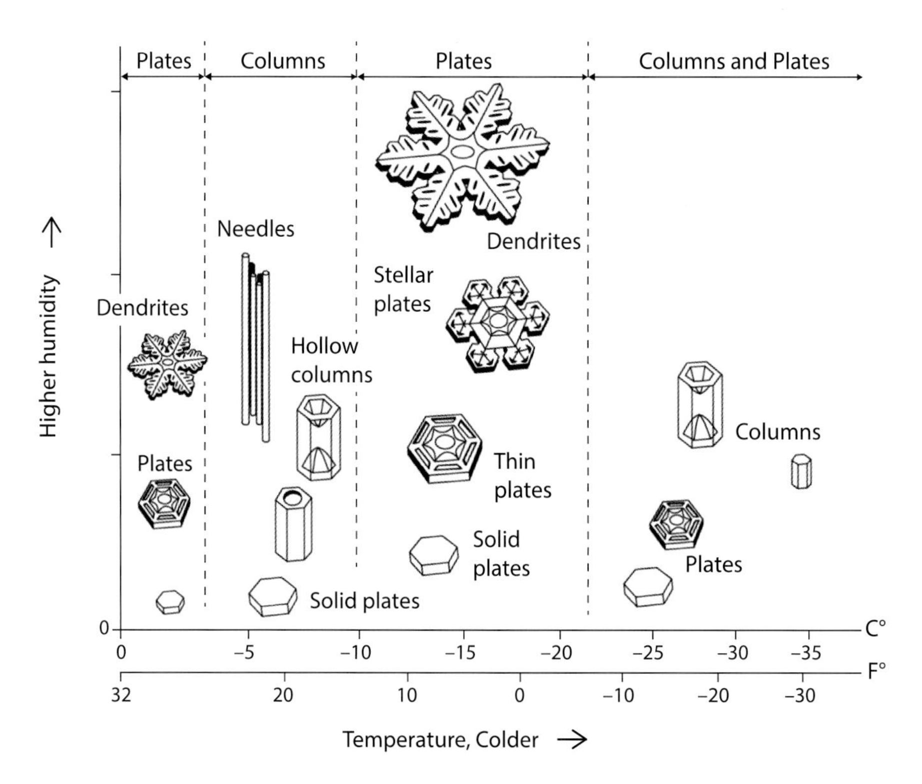

FIGURE 10.2. The Nakaya diagram illustrates what types of crystals form in air as a function of temperature and humidity level. This chart provides just a rough approximation of the different morphologies, however, plus it applies only if the growth conditions are constant in time.

crystals can only be found when the cloud temperatures are around −15°C.

Snow Crystal Classification

We name snowflakes for the same reason we name most things—so we can talk about them. Certain morphologies are commonly found in nature and have a distinctive appearance, and those have historically well-defined names. Stellar plates, stellar dendrites, fernlike stellar dendrites, hollow columns, and capped columns have all been part of the snowflake vernacular for some time. But there is no absolute classification system for snow crystals, and there never will be, because we have no definitive way to divide snow crystals into distinct, nonoverlapping categories.

Some things are better suited to classification than others. Atoms, for example, are neatly classified by how many protons they contain (in the periodic table) and by how many neutrons they contain (in isotope tables). Naming chemical species works quite well also, as each name refers to a specific chemical formula (and perhaps a specific isomer). Biological species cannot easily interbreed, so they too mostly form well-defined, nonoverlapping categories (although there are many exceptions, such as the mule). In these cases, naming conventions can be quite precise, making it reasonably straightforward to identify the named group that a given individual fits into.

Other groups of items are not so easily categorized. We can talk about different types of bread, cheese, cookies, breeds of dogs, types of hobbies, or musical instruments, but the names are generally human constructs with few natural partitions. People organize and catalog all these items, but different people have different lists, and the names often change over time. As a specific example, skiers have many names for different types of snow on the ground, but again the categories are a bit arbitrary. Snow scientists have done the same for falling snow crystals.

Asking "what kind of snowflake is that?" is not an especially good question, because there may not be a well-defined answer. The shape of a snow crystal depends on its entire growth history, which is somewhat analogous to saying that the breed of a dog depends on its entire ancestry. If a specific snow crystal had an unusual history, then it may not fit well into any category, no matter how many categories one defines. (And the same is true for dogs.) There is no way to avoid these ambiguities, so classifying snow crystals is a practice with somewhat limited usefulness. Nevertheless, some taxonomy is useful and necessary to guide the conversation, so classification systems have been devised for this purpose.

Nakaya first recognized the need for nomenclature and constructed a classification system containing seven primary categories that branch out into a total of 41 snowflake types [1954Nak]. This system was later expanded by Magono and Lee to 80 categories [1966Mag] and recently expanded again by Kikuchi and Kajikawa to include 121 distinct snowflake types [2011Kik , 2013Kik], as shown in Figure 10.3.

For the purposes of everyday snowflake watching, I prefer the somewhat simpler chart shown in Figure 10.4. This is essentially a modernized version of Nakaya's original list, placing a greater focus on the physical processes that underlie the different morphological types. With 35 different named categories, this chart includes the common designations that have evolved over the years, and it describes most of what can readily be found in the wild. I have found this chart quite useful for identifying and describing natural snow crystals, so I continue to promote its use. There is no definite, ideal method for classifying snow crystals, but Figure 10.4 is the chart I find most useful in the field.

Biased Sampling

In the spirit of full disclosure, I like to point out that well-formed snow crystals like those illustrated in classification charts are not the norm, and most of the snowflake types shown are quite rare. You may not realize this from photos you have seen, because photographers invariably present a heavily biased sample of what falls

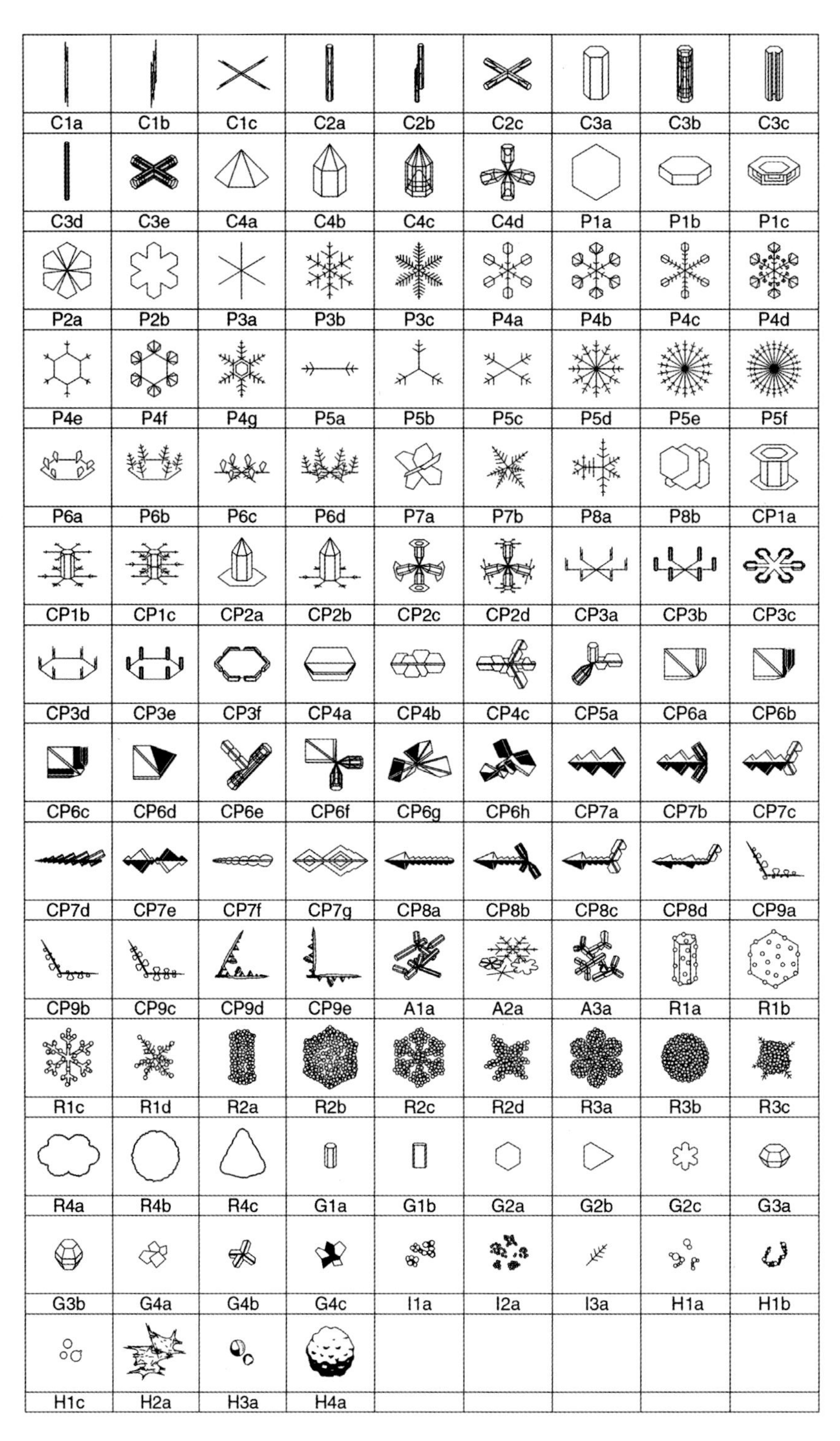

FIGURE 10.3. This large snow crystal classification system contains 121 separate categories [2011Kik, 2013Kik]. Image courtesy of Takao Kameda.

FIGURE 10.4. I devised this simplified snow crystal classification system for everyday observing [2006Lib]. While there is no definitive method for dividing snow crystals into precise categories, these 35 types provide a reasonable overview of the morphological diversity found in natural snow crystals.

FIGURE 10.5. Although this chapter focuses on different types of well-formed snow crystals, clumps of small "irregular" crystals like those shown in these two photos, are far more common.

from the clouds. We stand out in the bitter cold for hours on end, searching for especially photogenic examples that exhibit well-formed, strikingly symmetrical features. Exceptionally beautiful snow crystals are a delight to behold, so we work hard to find them. And because most people are not eager to buy a book or read an article showing unattractive snowflakes, those photos do not get published.

To witness an unbiased snowflake sample, you need only go outside during a light snowfall and have a look. Every snowfall has a different character, and certain weather conditions are conducive to producing photogenic crystals (see Chapter 11). But essentially all snowfalls produce many, many examples from the "Irregulars" category in Figure 10.4, and Figure 10.5 provides a representative sample. These small, somewhat malformed platelike crystals are extremely common, and some snowfalls deliver little else. I sometimes call this "granular snow," because the crystals look a lot like icy grains of sand.

Another common occurrence is when growing snow crystals collide with cloud droplets, creating what is called *rime*—basically collections of tiny, frozen drop-

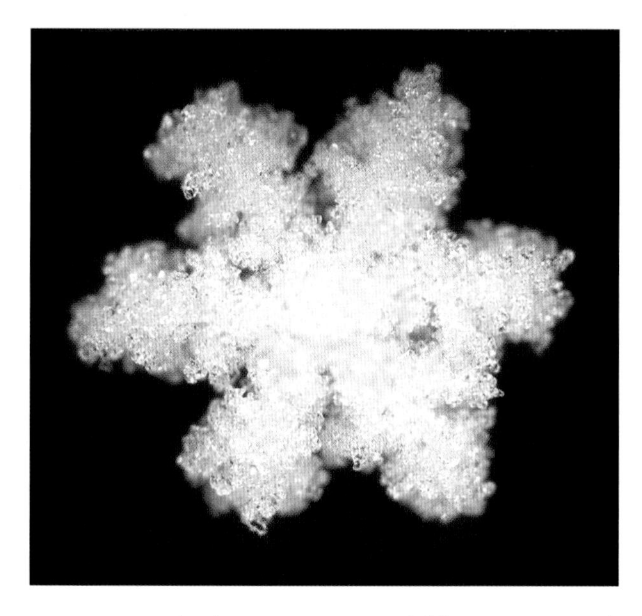

FIGURE 10.6. As they are growing and falling, snow crystals often collide with water droplets from the surrounding clouds. The supercooled droplets immediately freeze onto the ice surface, and this example shows a thick coating of the resultant *rime* on a stellar crystal.

lets. Figure 10.6 shows an example where a stellar snow crystal first developed normally in a region relatively free of cloud droplets, and then it moved into a dense cloud and accumulated a thick coating of rime. If the rime coating becomes so thick that essentially the entire structure is an agglomeration of frozen droplets, then it is called *graupel*, or *soft hail*. Again, some snowfalls deliver mostly rimed crystals.

An unfortunate truth when snowflake watching is that granular snow and rimed crystals are especially prevalent when the temperature is near 0°C, which includes a lot of snowfalls. Warm conditions tend to produce lower quality specimens overall (Figure 10.7), while the most spectacular stellar crystals appear when the cloud temperatures are near −15°C. Because population centers tend to form in moderate climates, and −15°C (5° F) is considered bitter cold by typical standards, the laws of probability suggest that most people will rarely observe exceptionally beautiful snow crystals where they live, even when snow is fairly common. I discuss this and

FIGURE 10.7. Many snow crystals exhibit a somewhat "travel-worn" appearance, especially when the temperature is warm. In this example that I captured in California, the branch tips are rimed, and nearly all the crystal edges are rounded from sublimation.

related problems further when considering snowflake photography in Chapter 11.

The goal and tenor of the guide to snowflakes in this chapter is much like what you find in a guide to mineral crystals. Mineral books tend to focus on beautiful, single-crystal specimens, as these represent the basic mineral types. But if you go hiking up in the mountains to look for yourself, all you will likely encounter is rather ordinary rocks. Small mineral crystals can be seen if you look closely at many rocks, as any geologist will quickly point out. But large single-crystal mineral specimens are exceedingly rare.

The good news for snowflake watchers is that finding high-quality snow crystals is much easier than finding high-quality minerals. Rock hounds have already removed nearly all the nice specimens that were easily retrievable, leaving few behind to discover. You can find large mineral crystals in museums, and for purchase, but not so much in the wild. New mining operations are among the best places to find quality mineral specimens, as they expose unexplored material.

In contrast, snow crystals are all made anew for every snowfall, so you have a good chance of finding some outstanding examples. If the temperature is somewhere between −10°C and −20°C, and you know what to look for, you will almost certainly find some noteworthy crystals if you are a determined observer. Not every snowfall brings exquisite snow crystal gems; but occasionally one can witness beautiful crystals falling to earth in large numbers. Patience and persistence are often needed, but those few magical snowfalls bringing exquisite crystals make up for all the granular and rimed crystals.

The remainder of this chapter presents many examples to illustrate the different snow crystal morphologies presented in Figure 10.4, along with some discussion of the physical processes involved in their creation. If you go outside to look at the falling snow, magnifier in hand, you may find this chart useful for observing and identifying different crystals. Human nature being what it is, you are more likely to spot a triangular crystal, a bullet rosette, or a double plate if you know that they might be out there.

SIMPLE PRISMS

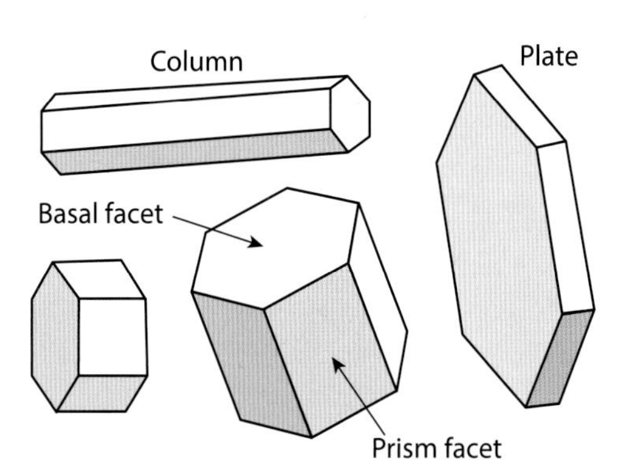

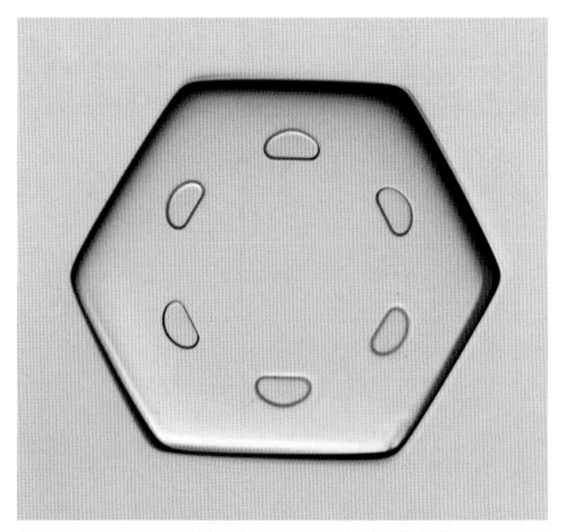

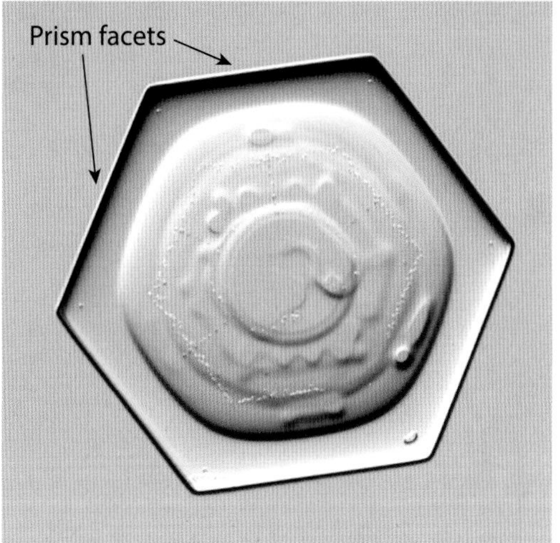

Simple prisms are small, faceted snow crystals that range from platelike to columnar in form. They have relatively plain shapes, with minor patterning and no branching. These minimalist snowflakes are common and can be found in most snowfalls, regardless of temperature. However, most simple prisms are so tiny that you need a microscope to see them clearly.

Every snowflake has its beginning, and these small crystals are essentially young snowflakes that have not had time to grow into larger and more elaborate shapes. The examples shown on this page are roughly 0.3 mm in size, about as large as the period at the end of this sentence.

Faceting is a dominant force in the development of simple prisms, because they are still small. The transition to branching has simply not yet had a chance to occur. A rough rule of thumb is that branching begins when a crystal grows to more than half a millimeter in size, although this rule is only approximate.

Occasionally you can observe these small crystals on bitter cold days when the sun is out, so their mirror-like facets sparkle brightly as they tumble through the air. With that image in mind, you can see why these are also called *diamond dust* snow crystals.

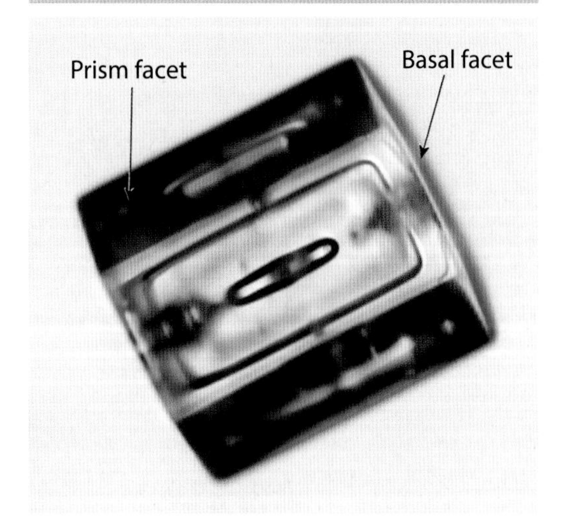

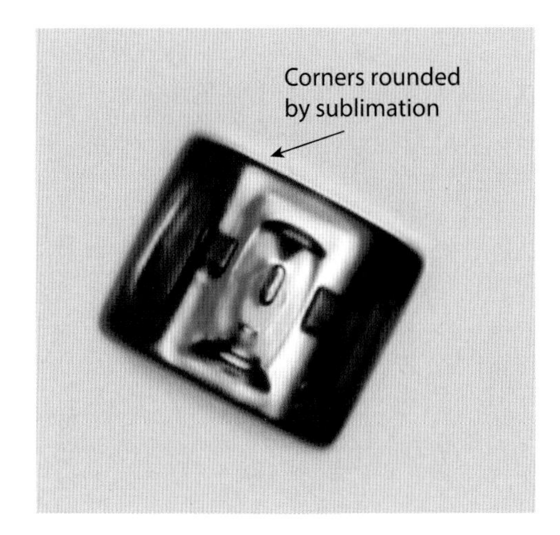

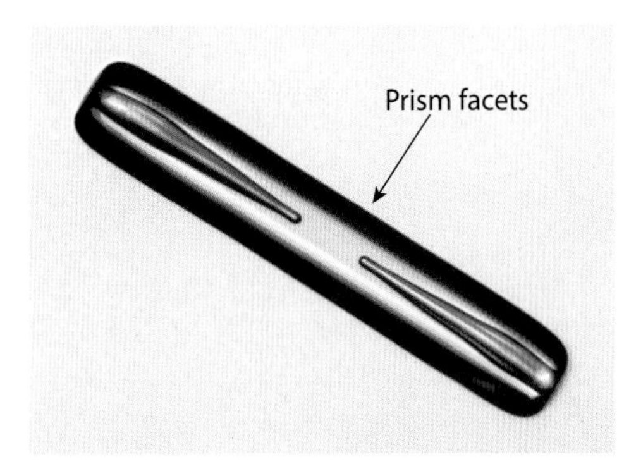

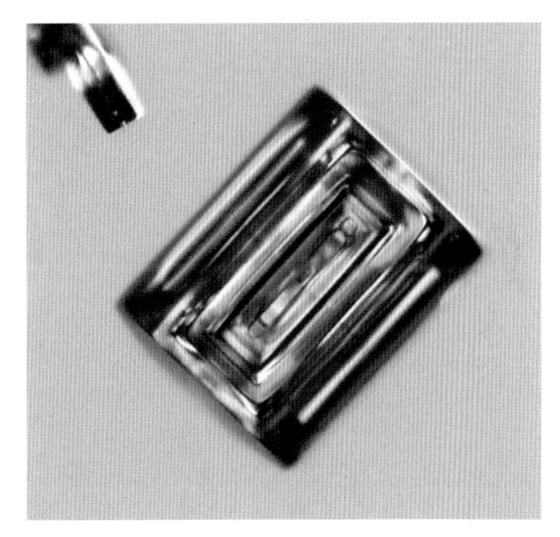

Well-formed crystal facets have razor-sharp corners during growth, but this is not always what you see in the pictures. Sublimation will often round the edges, as you can see with the small prism shown above. The rounding of sharp features is especially noticeable on smaller specimens and when the temperature is warm. Sublimation is always an unknown factor when snowflake watching, because you do not know what conditions the different crystals have been through after forming. By the time it reaches the ground, a crystal may look quite different than it did when it was growing up in the clouds.

When photographing snow crystals, I usually illuminate them from behind with colored lights, giving my photos a bright background. In the photo below/right, Canadian photographer Don Komarechka [2013Kom] used a ring flash to capture these bright, glasslike crystals on a dark background.

Hollow Faces. (right) For this photograph, I focused my camera on one face of a diamond dust prism, about 0.3 mm in size. I caught this crystal quickly on an especially cold day, so sublimation has not yet taken its toll; the corners are still distinct and sharp. This picture also illustrates the hollowing sometimes seen in prism facets.

During growth, diffusion gives the corners of the crystal a greater supply of water vapor. The facet centers receive less, so they accumulate material more slowly. Over time, the facet centers lag behind the growth of the edges, as shown in the accompanying sketch. This is a common growth behavior, and it is a first step in the transition to branching.

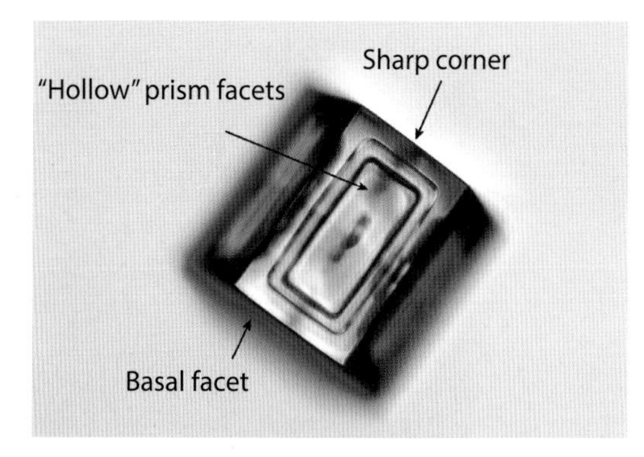

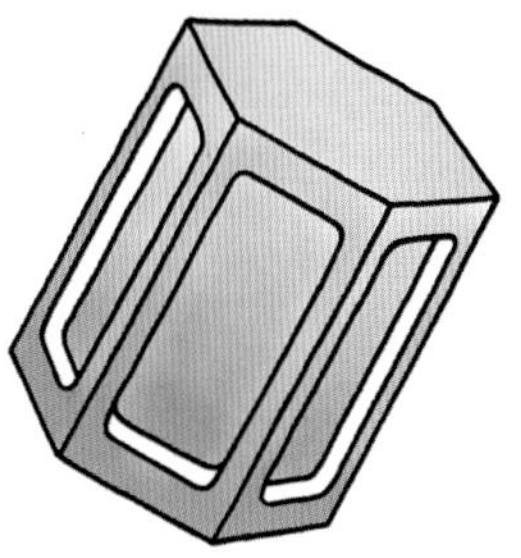

Antarctic Snow Crystals. (below) These tiny crystals were photographed at the South Pole [1990Tap]. In the dry, bitter-cold arctic conditions, snow crystals often grow into simple, sharply faceted prisms like these.

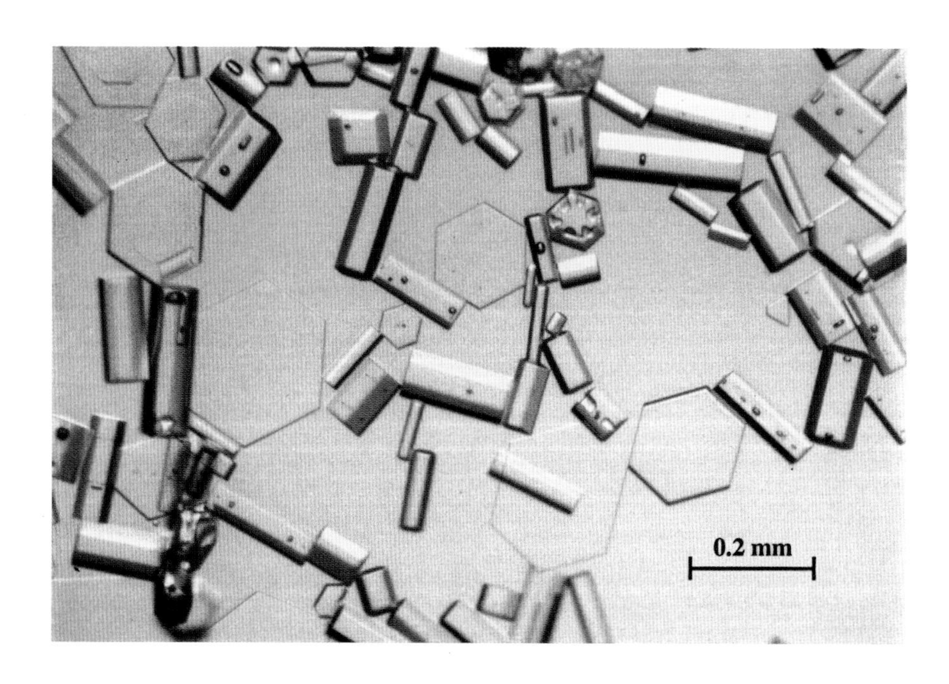

Atmospheric Halos. Simple-prism snow crystals are responsible for a variety of atmospheric halo phenomena. The photo above shows a spectacular halo display captured at the South Pole [1990Tap]. The lower-left image illustrates a simpler light pillar phenomenon that includes the author's thumb blocking the sun, photographed in Cochrane, Ontario. The lower-right photo shows the sun flanked by a pair of sundogs captured in Fargo, North Dakota [2009Gop]. Much has been written about halo phenomena [1980Gre, 1990Tap], and complex reflection/refraction models are needed to explain how falling ice crystals can create such complex patterns of light. For many halo features, including sundogs, the falling crystals must be aligned relative to vertical by aerodynamic forces, as described in Chapter 3.

STELLAR PLATES

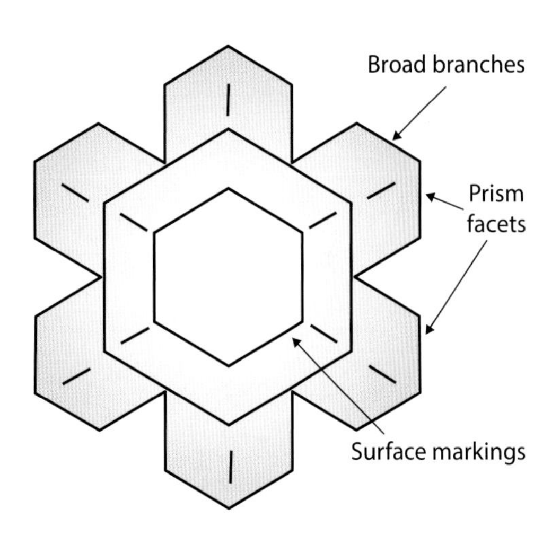

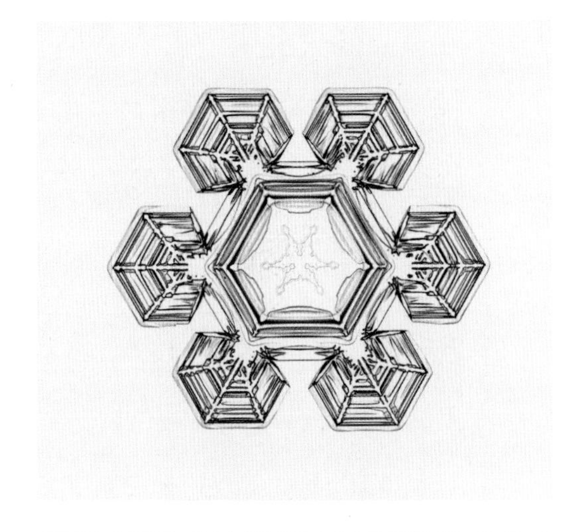

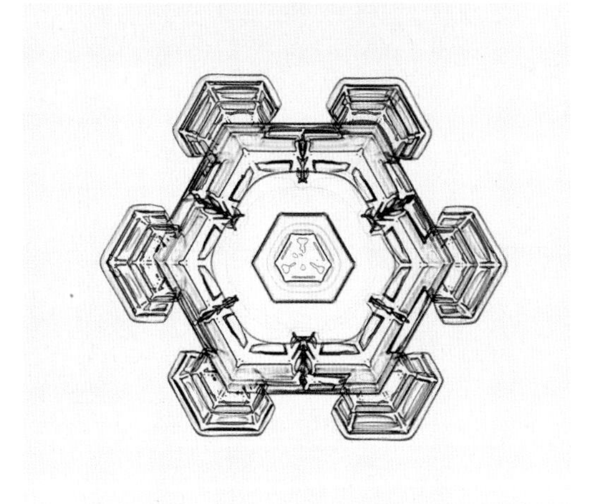

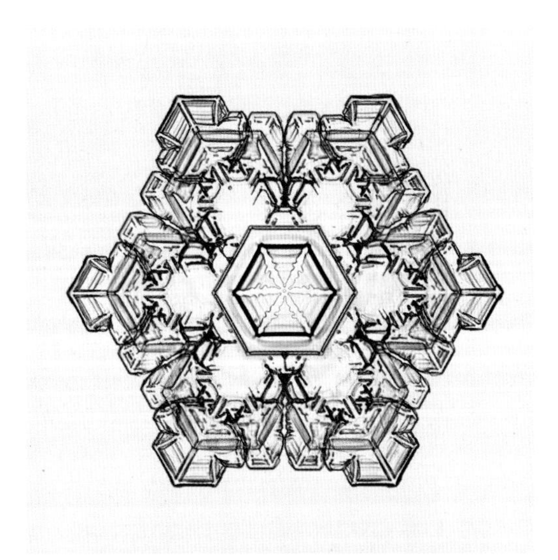

Stellar plates are thin, flat crystals of medium size with an overall sixfold symmetry. They are typically broad branched with little or no sidebranching and a profusion of complex surface markings. Stellar plates can be abundant when conditions are right. The best specimens are found during light snowfalls at fairly low temperatures.

The sparkle you see in falling snow often comes from stellar plates, as their flat basal surfaces reflect incident light like tiny mirrors. These crystals are large enough that a simple magnifier gives you a nice view of their overall structure, as a good-sized specimen might be 2 mm in diameter. A microscope reveals a whole new realm of observing, however, allowing a detailed look at the intricate patterning on each crystal.

Stellar plates form over a narrow range in temperature, so are not present in all snowfalls. The Nakaya diagram tells us that large, platelike crystals will grow when the clouds are near either −15°C or −2°C. At the higher of these temperatures, however, one does not generally find well-formed crystals, because of sublimation and other factors. Thus, large stellar crystals mainly appear when the temperature is within several degrees of −15°C. If you want to find some beautiful stellar plates, you must wait for just the right conditions.

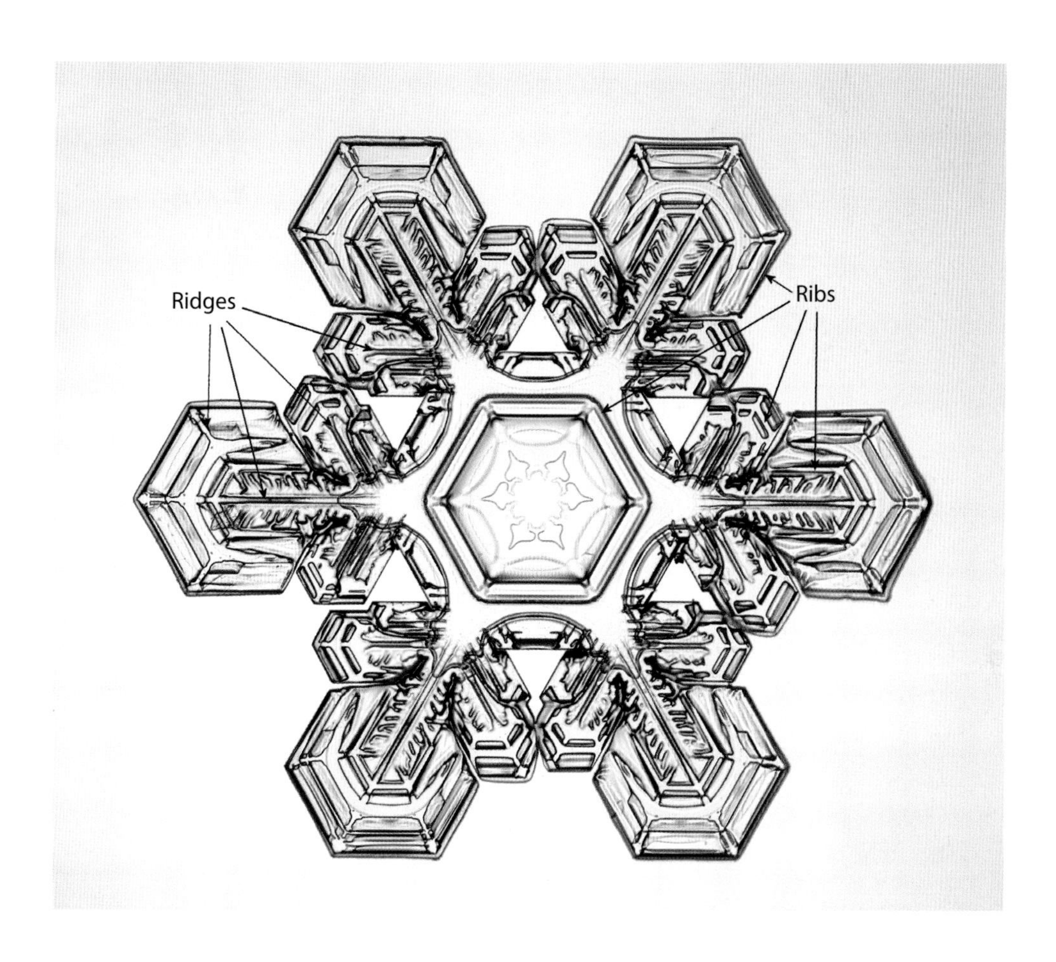

Ribs, Ridges, and Rings. The ribs, ridges, and inwardly propagating rings described in Chapters 3 and 9 are often prominent surface features on stellar-plate snow crystals. The complexity of the patterning reflects the ever-changing conditions experienced by each crystal, brought about by the convoluted path it followed through the atmosphere. The possible permutations are endless, and some crystals develop highly complex surface markings. Laboratory-grown snow crystals allow these different features to be analyzed in isolation, as described in Chapter 9.

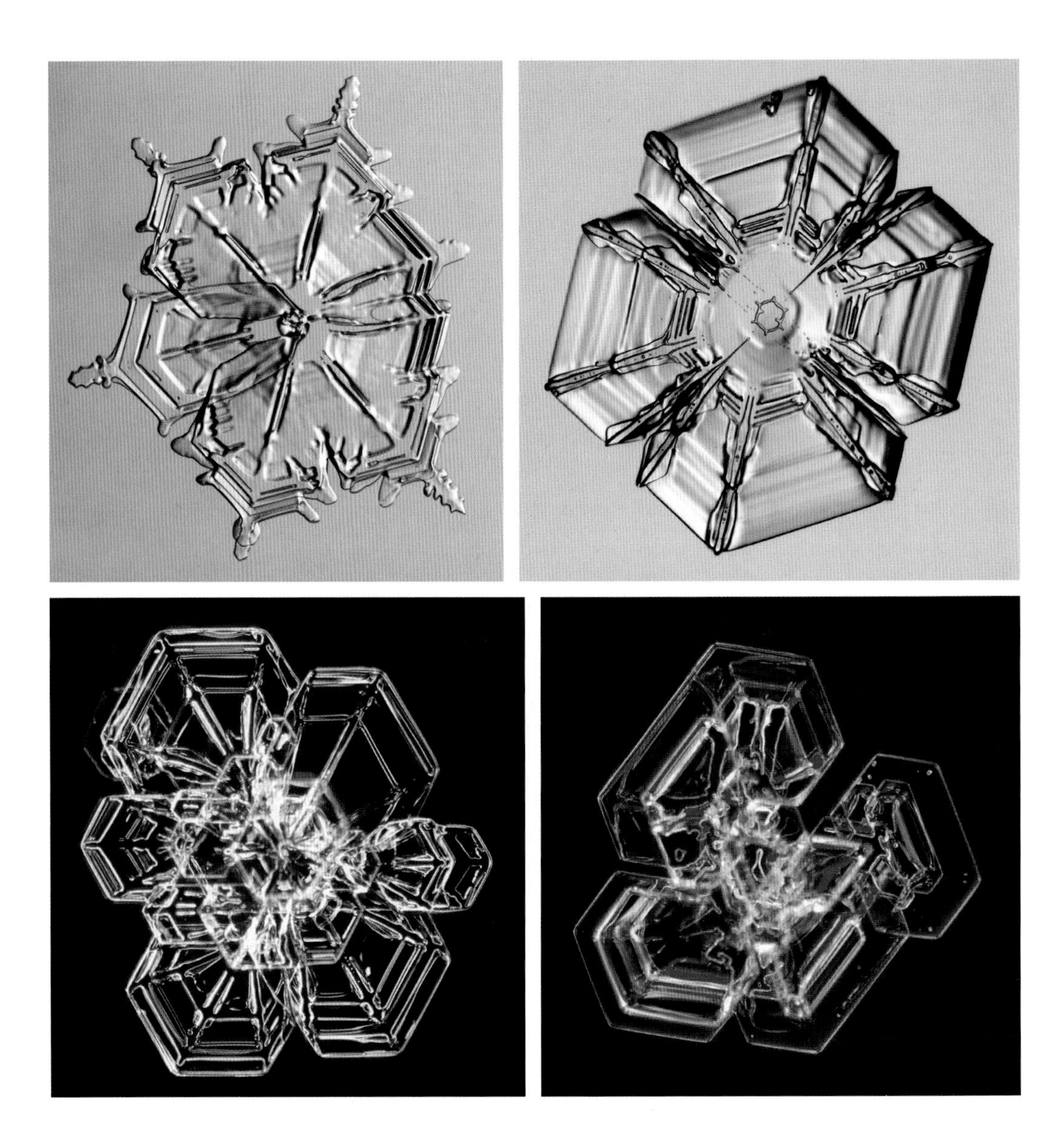

Malformed Plates. Most stellar plates are not beautifully formed and flawlessly symmetrical, as you can verify by spending 10 minutes with a magnifying glass in any snowfall. The quintessential, well-formed snow crystal is actually quite rare. The above pictures provide several examples of imperfect, somewhat malformed stellar plates. These are all single crystals of ice, as you can ascertain from the relative alignment of the various facets on each crystal. The facets reveal the underlying molecular order, and the aligned facets indicate that this order is preserved throughout each plate. The odd shapes of these crystals came about because their growth was perturbed in some way. Perhaps they experienced some lattice defects during growth or suffered collisions with rime particles or other falling crystals. There are many potential problems that can interfere with symmetrical growth.

Surface Patterns. Some stellar plates exhibit remarkably complex and symmetrical surface markings, especially in their central regions. The structures are so small that a microscope is usually needed to see them.

Perpetual Variety. (above plus facing page) Stellar-plate snow crystals exhibit an endless diversity of complex surface patterns.

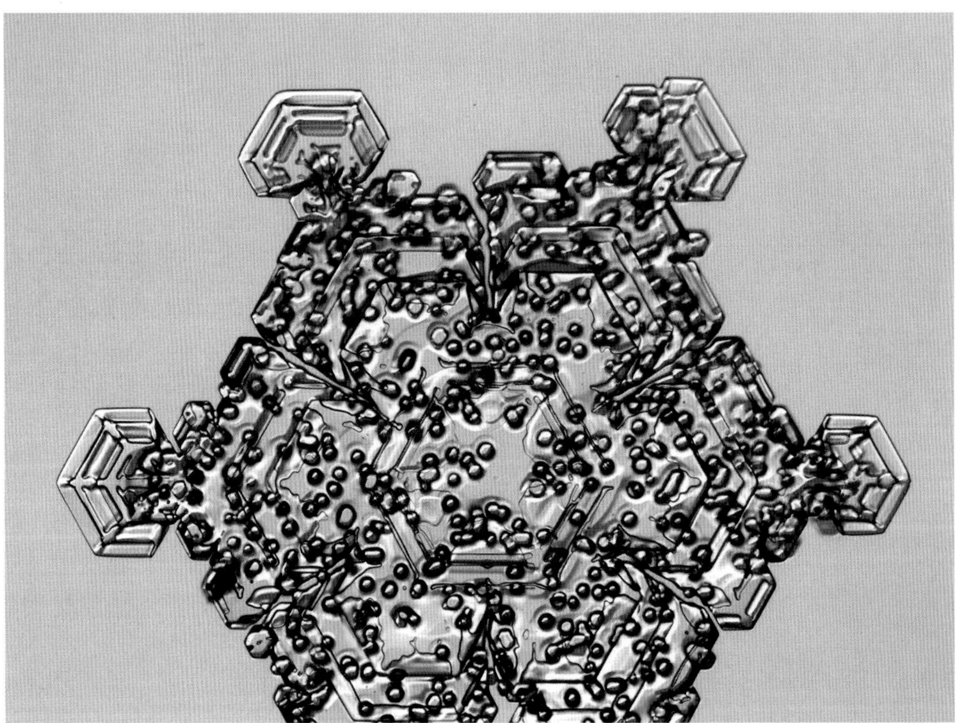

Rimed Crystals. Snow crystals are often decorated with rime particles, ranging in number anywhere from one to thousands. A typical droplet has a size of roughly 30 microns, which is half the diameter of a human hair. Large crystals can be especially prone to rime, as the high humidity necessary for their growth requires a high density of cloud droplets. Aerodynamic forces often deposit rime particles on the edges of large plates, as seen in the upper-left image. The lower crystal is unusual in that it picked up quite a bit of rime and then it moved to a region with fewer cloud droplets, where the crystal tips grew out relatively unperturbed by the rime.

Epitaxial Growth. After growing into a small hexagonal plate, the above crystal wandered into a dense region of a cloud and picked up a good dusting of rime droplets. Each droplet froze upon contact with the ice, and if you look closely, you can see that the facets on the frozen droplets are mostly aligned with the facets of the plate. This is an example of *epitaxial growth*, as the plate ice served as a template to guide the molecular orientation of the freezing liquid. The crystal on the right apparently acquired a single rime droplet when it was smaller, which froze epitaxially and then stimulated the growth of an errant branch that grew differently from its siblings.

SECTORED PLATES

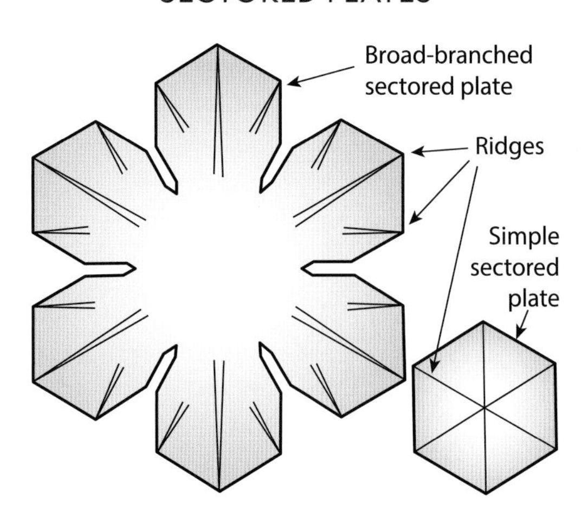

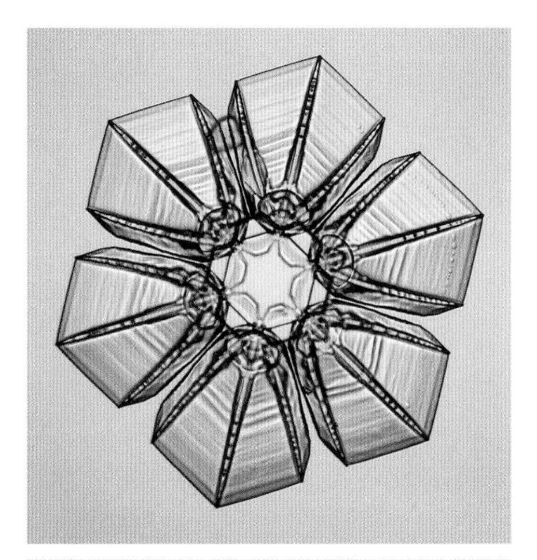

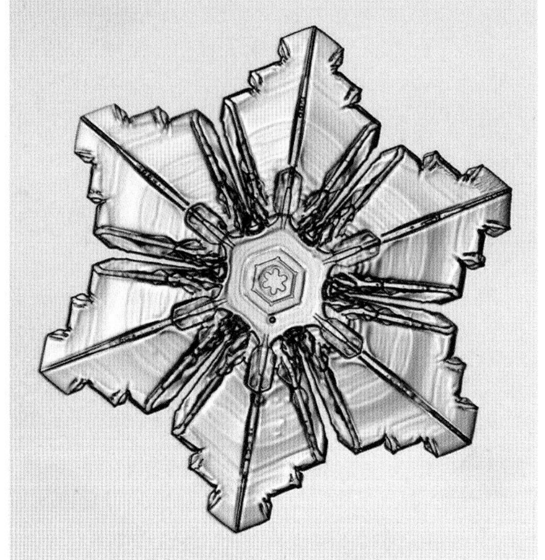

Sectored plates are flat, broad-branched crystals decorated with pronounced radiating patterns of ridges. They get their name from the way the ridges seem to neatly divide the plates into sectors. At times, these surface markings look like veins on a leaf, giving some snowflakes an almost plantlike appearance.

The simplest sectored plate is a hexagonal shape divided into six sectors, like the example shown below right (see also Chapter 9). Broad-branched crystals with sectored-plate extensions are more common, as illustrated in the other photos. Sectored plates can be considered a subclass of stellar plates, and there is no sharp dividing line between the two categories. If a stellar plate shows especially prominent ridging and few other surface markings, then we tend to call it a sectored plate.

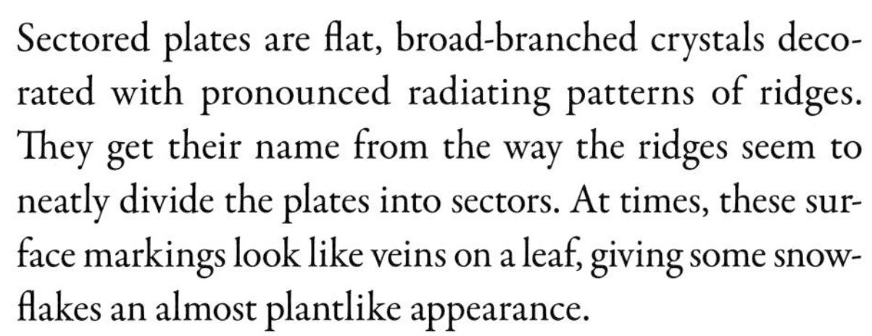

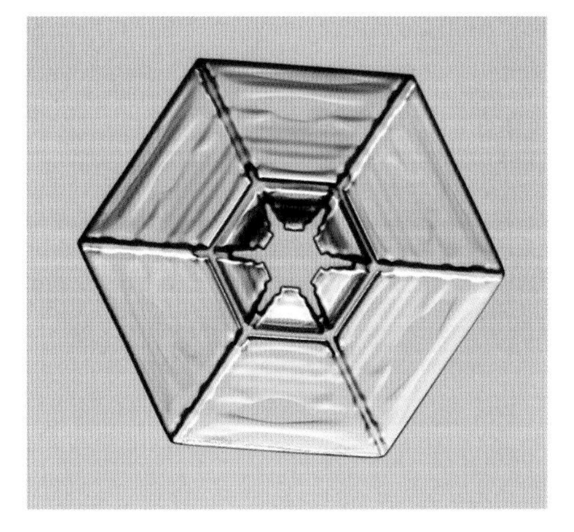

Sectored-Plate Extensions. The relative simplicity of their surface markings indicates that sectored plates form in relatively constant conditions, without large swings in temperature or humidity. Thus, the platelike branches are generally flat and smooth (aside from the ridges), and the prism facets tend to be well formed and large. Then they look a bit like duck's feet.

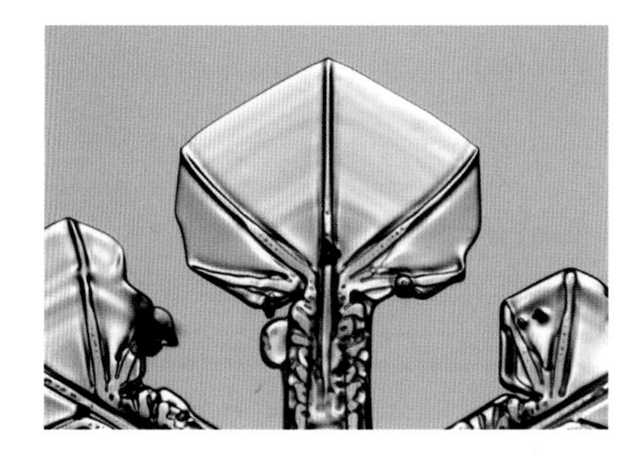

Prominent Ridges. Sectored plates merit a separate name, because ridging is such a robust feature in snow crystal growth. As described in Chapters 4 and 9, ridges form on slightly convex basal surfaces, owing to a 2D version of the branching instability. The pictorial Nakaya diagram in Chapter 8 shows that ridging occurs over a broad range of temperatures and supersaturations, making it one of the most prevalent snow crystal features. If you look carefully, you can usually find some evidence of ridge formation on most thin-plate crystals.

Contemporaneous Crystals. It is not atypical to see a cluster of comparable crystals all falling in a brief period, as these did one day. When the conditions are right to form a particular crystal type, the clouds can release them in large numbers. As usual, some are well formed, but many are not.

STELLAR DENDRITES

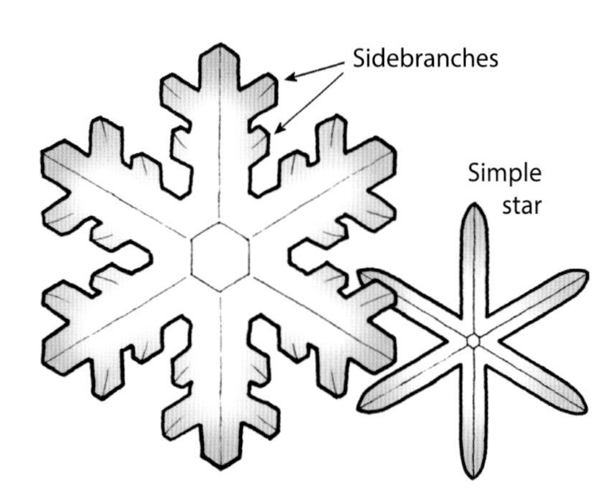

Stellar dendrites are platelike crystals with narrow branches decorated with numerous sidebranches. They tend to be larger than stellar plates, with generally less prominent faceting and more complex shapes. These crystals can be readily found with the naked eye, and considerable detail can be seen with a simple magnifier. Stellar dendrites are common in many snowfalls, often arriving in great numbers.

The word "dendrite" means "treelike," which is an apt description of these extravagant crystals. They form around −15°C when the humidity is quite high. The ample water vapor supply drives the branching instability to produce numerous sidebranches. Stellar dendrites are often conspicuous, as a generously sized specimen might measure 3 mm from tip-to-tip. They are also quite thin and flat. Their ornate shapes with outstanding symmetry make stellar dendrites the much-celebrated canonical holiday snowflakes.

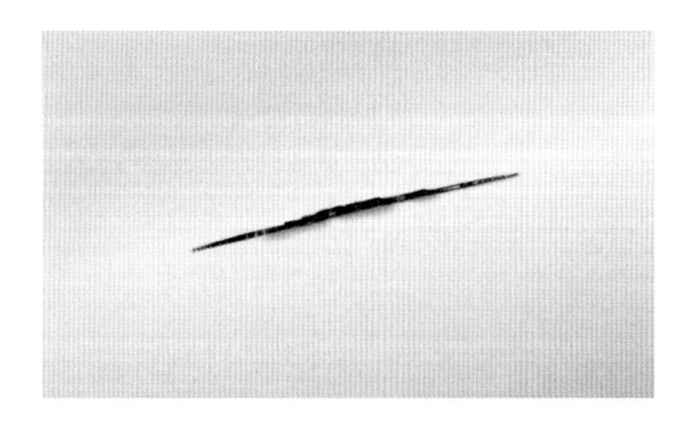

Thin Plates. (left) Stellar plates and dendrites are usually remarkably thin and flat, which can be appreciated by viewing them from the side. This overall shape is why we call them snow***flakes***.

Induced Sidebranching. Symmetrical sidebranching on a stellar dendrite is typically brought about by induced sidebranching, as described in Chapters 4 and 9. This is the only known mechanism that will cause sidebranches to sprout synchronously from all six primary branches on a large stellar crystal. It requires a carefully orchestrated series of events to produce several sets of well-formed symmetrical sidebranches, which is why good examples are rare and difficult to find in nature. They are substantially easier to create in the lab under controlled conditions.

Backward Branches. On rare occasions, one can find sidebranches that appear to be growing 60 degrees off from the usual forward direction, so I call them "backward" sidebranches. Their formation mechanism can be seen in the crystal on the right. When a platelike crystal grows on the end of a long thin branch, all five available plate corners are susceptible to sidebranching.

Simple Stars. These minimal stellar crystals appear when the supersaturation is high enough to produce narrow branched growth, but too low to create much sidebranching. As seen in the pictorial Nakaya diagram in Chapter 8, these crystals grow only over a narrow range of parameter space, so they are not especially common. They also tend to be small with simple features, making them easy to overlook.

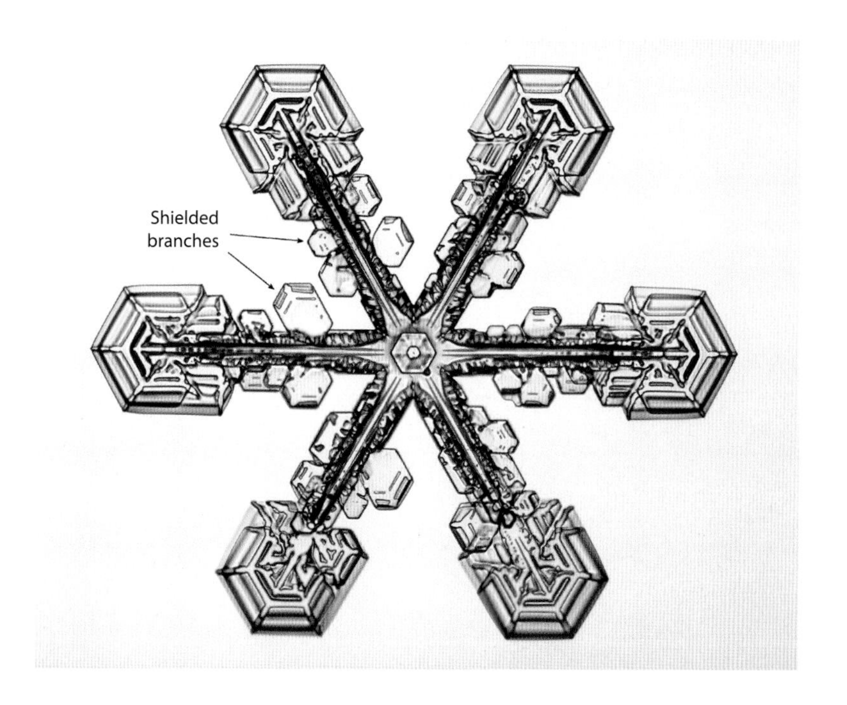

Shielded Branches. Snow crystals grow fastest at their outer edges, which have the greatest supply of water vapor. But sometimes the interior branches will grow substantially even after the outer branches have grown out and left them behind. Because the interior branches are shielded by the outer branches, they receive a reduced supply of water vapor. Such conditions often yield thin, rather featureless platelike structures that are rather asymmetrically placed, as seen in these examples.

Branched Beauty. One does not easily grow bored examining the endless variety of stellar-dendrite snow crystals.

FERNLIKE STELLAR DENDRITES

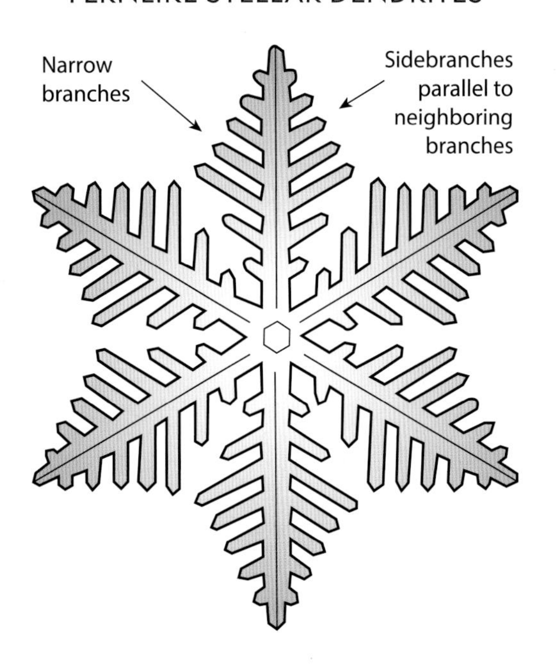

Fernlike stellar dendrites are large, thin plates with narrow branches and sidebranches that look like a fern. Sidebranches typically form at 60-degree angles relative to their primary branches. These crystals are common, and their exceptionally large size makes them easy to spot.

Fernlike stellar dendrites are the largest snow crystals, on rare occasions measuring more than 10 mm in diameter. Their thickness may be 100 times less than this, however, making them extremely thin, flat, platelike crystals. They only form near −15°C when the humidity is exceptionally high, which drives their rapid growth with copious sidebranching.

The well-defined 60 degree angles between the branches and sidebranches of fernlike stellar dendrites indicate that they are single crystals of ice. Despite their complex shapes, the molecules in the ice lattice are all lined up from one tip to the other.

Chaotic Branching. Fernlike stellar dendrites grow rapidly in a humid environment, resulting in copious sidebranching. As soon as the above crystal was born, the abundance of water vapor drove the branching instability hard, so the transition from faceted to branched growth occurred early. As a result, at its center there is little visible remnant of the crystal's initial faceted stage. Once the six principal arms were established, the high humidity resulted in narrow, closely spaced sidebranches with no prism faceting. The absence of faceting means that there were no induced sidebranching events, and thus no sixfold symmetry in the placement of the sidebranches. Even the sidebranches on opposite sides of a single primary branch are uncorrelated. In a sense, the growth of this crystal was too fast to be orchestrated. This is a medium-sized dendritic specimen, just over 2 mm from tip to tip, but it is also quite thin and flat. Basal faceting, with some assistance from the ESI (see Chapter 4), mainly restricted its growth to two dimensions. Because it stayed thin and light, the crystal made a slow descent through the clouds, never falling faster than about half a meter per second.

Monster Snowflake. To my knowledge, this is the largest single snow crystal ever photographed—a fernlike stellar dendrite measuring just over 10 mm from tip to tip. Each branch holds first-generation sidebranches along with second-, third-, and even fourth-generation sidebranches. Extensive higher-order sidebranching like this is rare in snow crystals. I have witnessed such large crystals only twice, both times in Cochrane, Ontario, and both times for just a few minutes. Weather conditions need to be quite finely tuned to create such large crystals, with the temperature close to −15°C, the humidity exceptionally high in a dense cloud, not too many crystals forming at once, and absolutely no wind to break up the delicate branched structure.

Snowflake Sizes. This composite image shows several snow crystals compared to a penny (which is 19 mm in diameter), all at the same magnification. A large stellar dendrite might be 5 mm from tip to tip, while many simple plates are no more than 1 mm in size. The monster snowflake on the preceding page is about as large as Lincoln's head.

Powder Snow. When conditions are right for the formation of fernlike stellar dendrites, they can fall in abundance. The top picture above shows a close-up view of the windshield of my car after a snowfall that dropped almost entirely large stellar snowflakes. You can see how the barbed branches locked together to form an exceptionally light, fluffy blanket of ice. On the ground, this kind of snow is called "fresh powder," and the airy structure is so soft that a skier might sink waist-deep into it, skies and all. After being exposed to the sun and wind for a day or two, however, the snow packs down into a denser, less yielding composition.

Variations. Once again, there is no sharp dividing line between the stellar-dendrite and fernlike stellar-dendrite categories. Many crystals display aspects of both types, so they rightfully belong somewhere between these two classifications.

HOLLOW COLUMNS

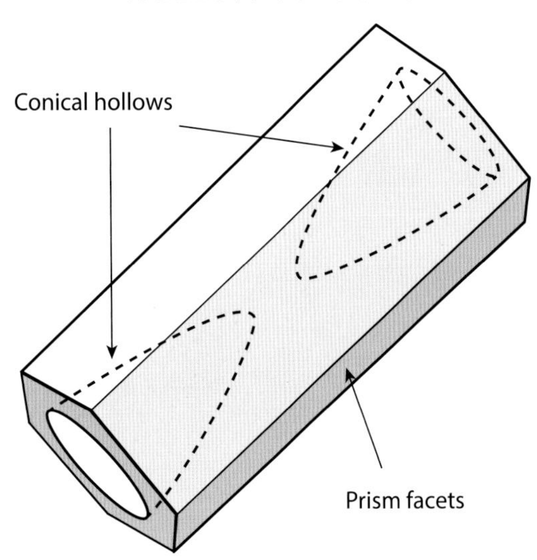

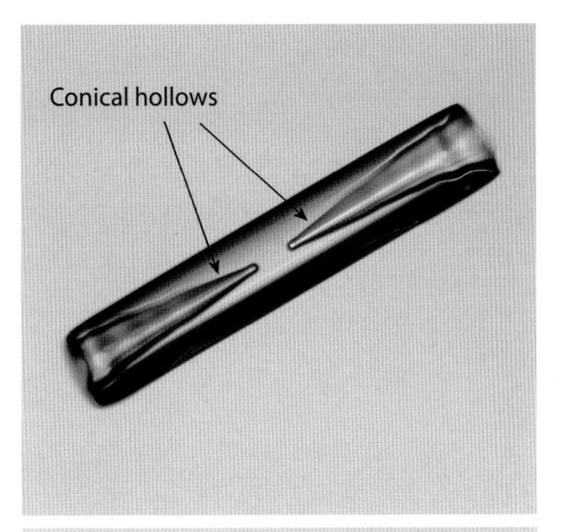

Hollow columns are simple hexagonal ice prisms with conical voids extending down from their ends. The recesses typically appear in a symmetric pair running along the central axis of a crystal, with the tips nearly touching at the waist. Hollow columns are small and easy to overlook with the naked eye, being about 1 mm in length, and their internal structure is best viewed with a microscope. They are a relatively common columnar morphology and can frequently be found in warmer snowfalls.

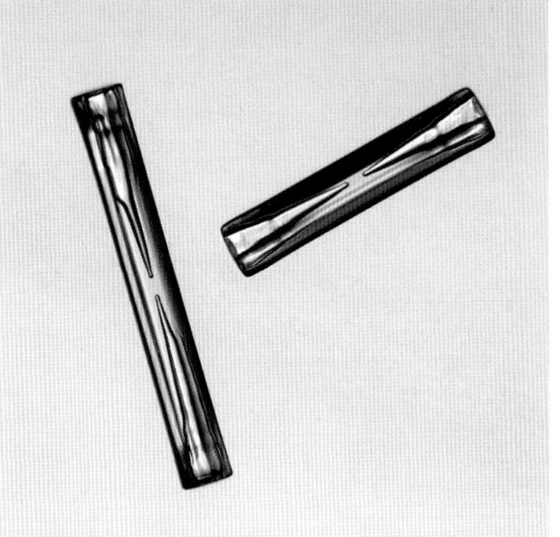

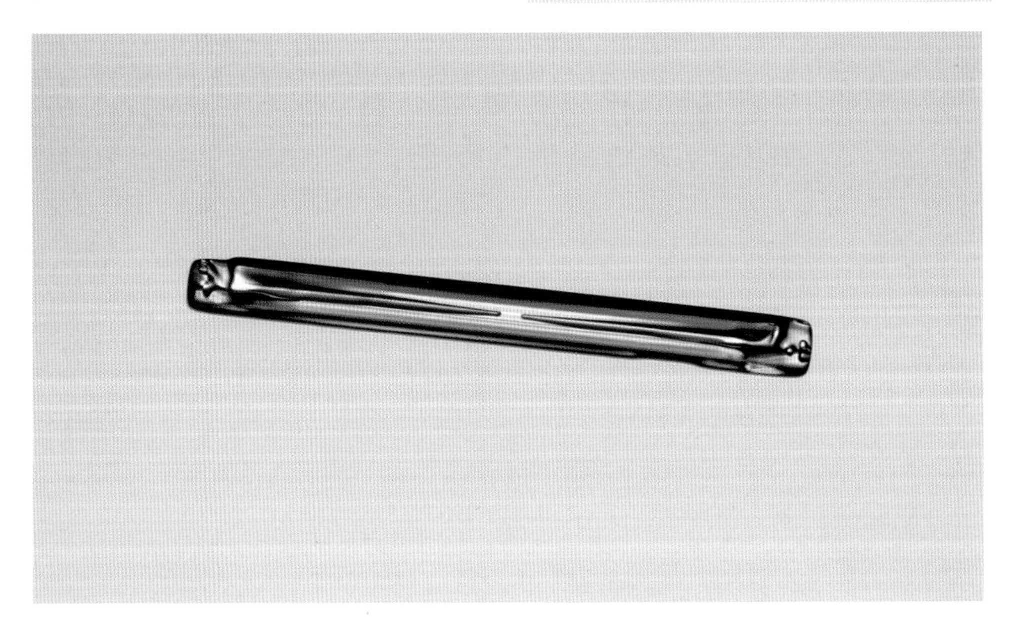

Hollow columns are most likely found when the temperature is near −5°C, as indicated in the Nakaya diagram. The overall hexagonal columnar structure is often not apparent, because the prism corners have been rounded by sublimation, which is especially rapid at these warmer temperatures. Thus, hollow columns may look more like round cylinders than like hexagonal columns.

The best way to find and view these crystals is to let some snow fall onto several glass slides and then view the slides under a microscope. When the temperature is high and granular snow is the norm, one can often find a few well-formed hollow columns in the mix. As is true with most snow crystal types, however, finding and photographing well-formed examples can be a challenge.

The formation of a hollow column is a manifestation of the familiar branching instability (see Chapter 3). The crystal starts out as a tiny ice prism, but soon the basal edges grow faster than the basal centers, resulting in the columnar hollowing. This mechanism predicts that there can never be a fully hollow column (hollow like a pipe),

and none has ever been observed. The initial seed crystal will always leave behind a solid central core.

Columnar Bubbles. In some hollow columns, including the two examples below, the hollows close up when the supersaturation becomes low, leaving pairs of columnar bubbles in the ice.

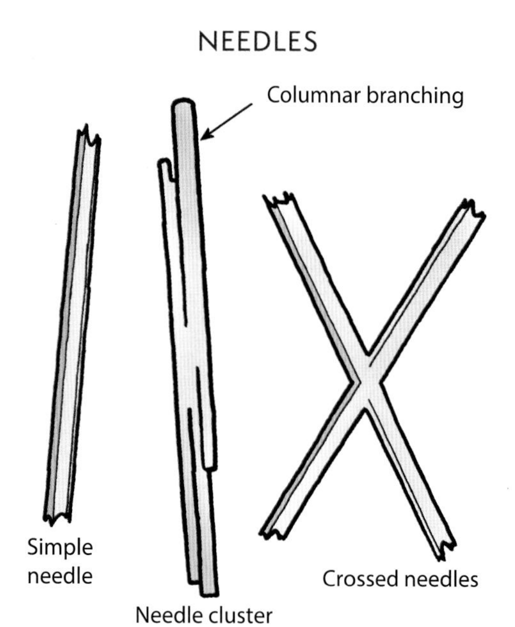

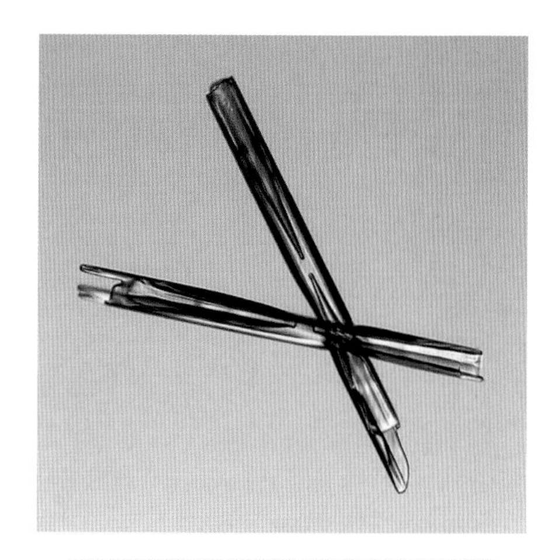

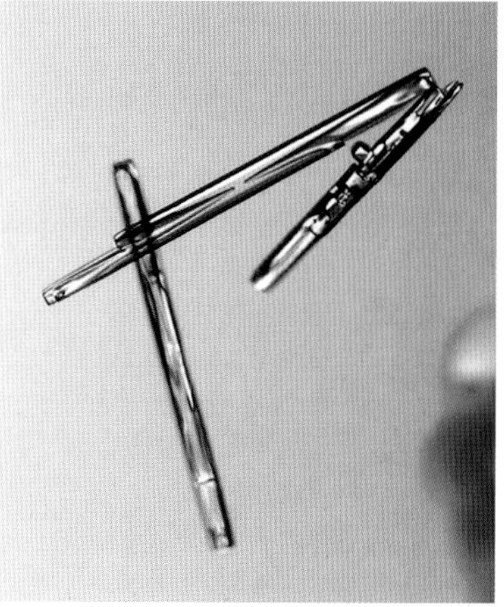

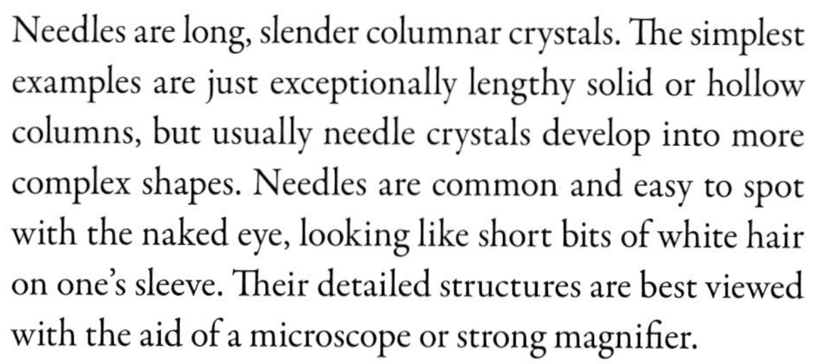

Needles are long, slender columnar crystals. The simplest examples are just exceptionally lengthy solid or hollow columns, but usually needle crystals develop into more complex shapes. Needles are common and easy to spot with the naked eye, looking like short bits of white hair on one's sleeve. Their detailed structures are best viewed with the aid of a microscope or strong magnifier.

Needle crystals are the product of warm, wet snowfalls, forming when the temperature is close to −5°C and the humidity is high. With lengths often up to 3 mm, needles are the longest of the columnar snow crystals. A crossed needle arises either from a polycrystalline seed crystal or from the mid-air collision of two simple needles. Needle clusters are another result of the branching instability, as secondary needles sprout from the corners of a primary needle end.

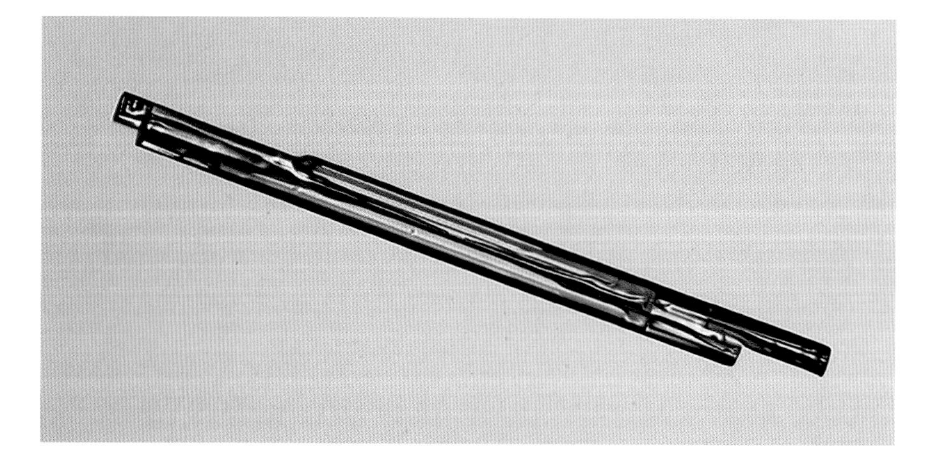

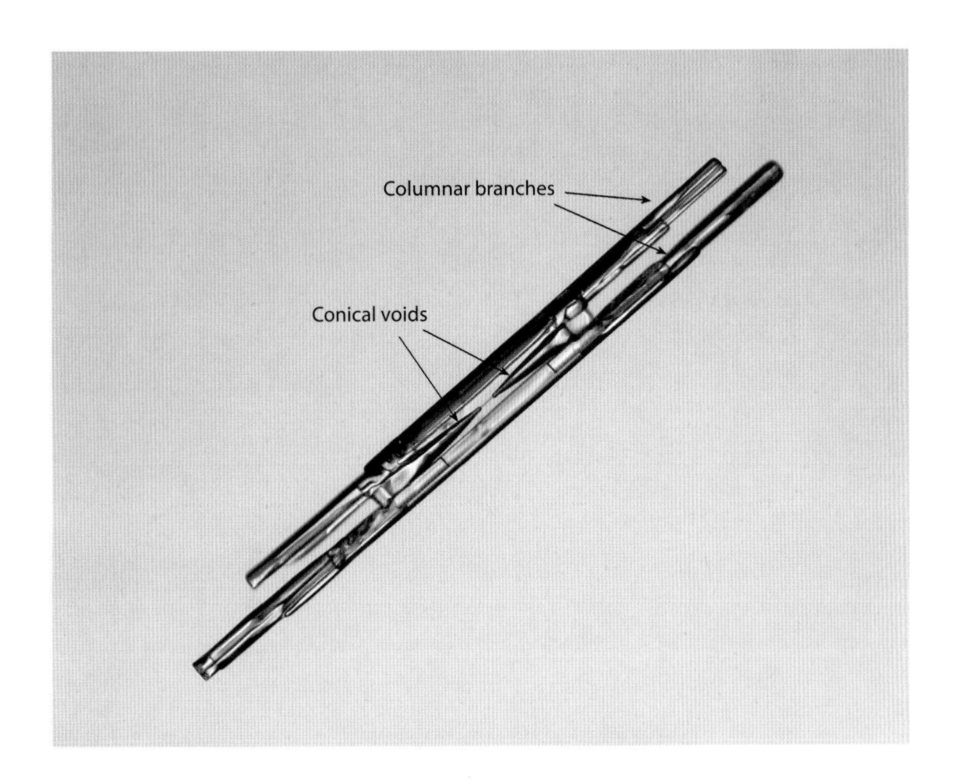

Hollow Column with Needle Extensions. The complex needle crystal above began as a hollow column, as evidenced by the conical voids seen deep inside the structure. As the hollow column grew larger, needlelike branches sprouted from the corners of the columnar ends.

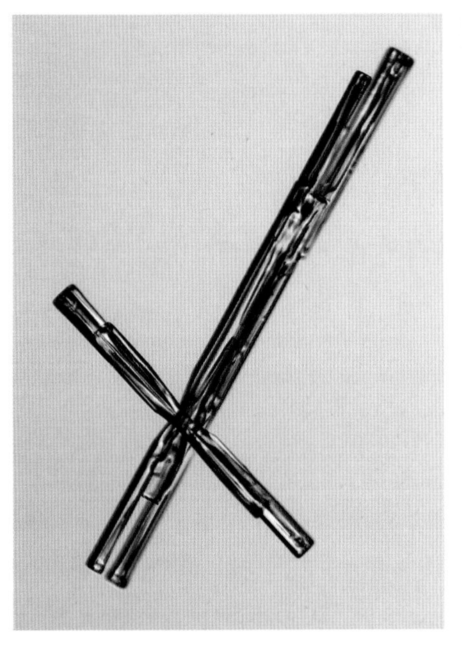

CAPPED COLUMNS

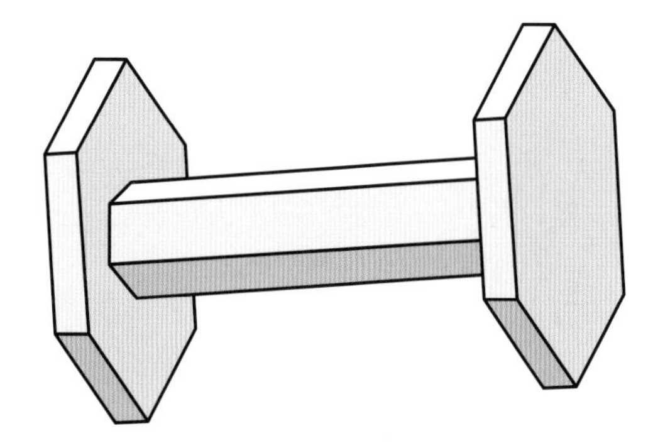

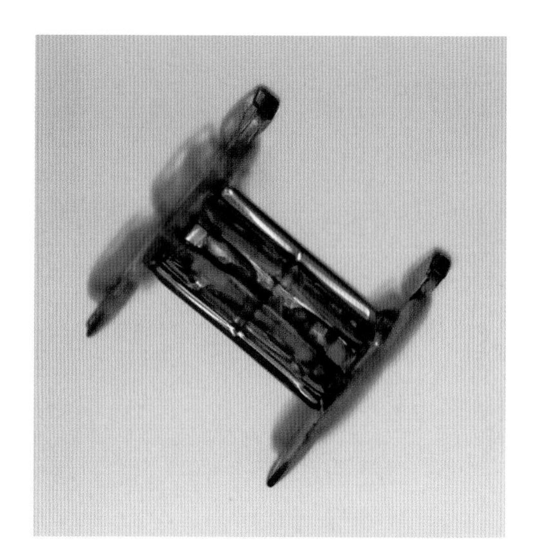

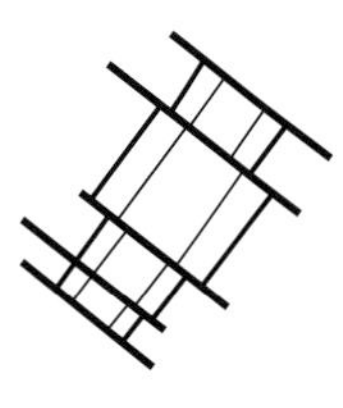

Capped columns are columnar crystals with stellar plates on their ends. A typical specimen looks like a stubby axle flanked by two hexagonal wheels, but multiply capped columns have more than two plates. Although these crystals are not especially common, a trained eye can often find a few mixed in with simple columnar crystals in warmer snowfalls. Capped columns are just large enough to be spotted with the naked eye, and their distinctive shape makes them easy to identify.

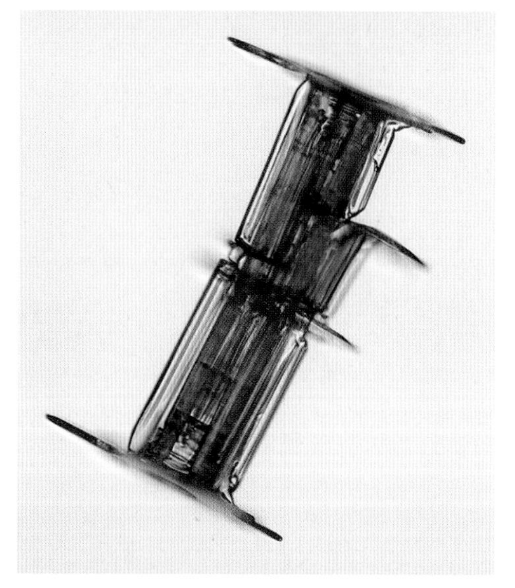

A capped column forms when a snow crystal experiences its own style of midlife crisis, abruptly changing its growth behavior from columnar to platelike. This can happen when a large mass of moist air is pushed upward by a passing storm front. The air cools as it rises, carrying its suspended cloud droplets along with it. When the temperature falls to about −6°C, some of the droplets freeze and begin growing into columns. If the air continues to rise, the temperature may drop to about −12°C, promoting platelike growth on the columnar ends and yielding capped columns.

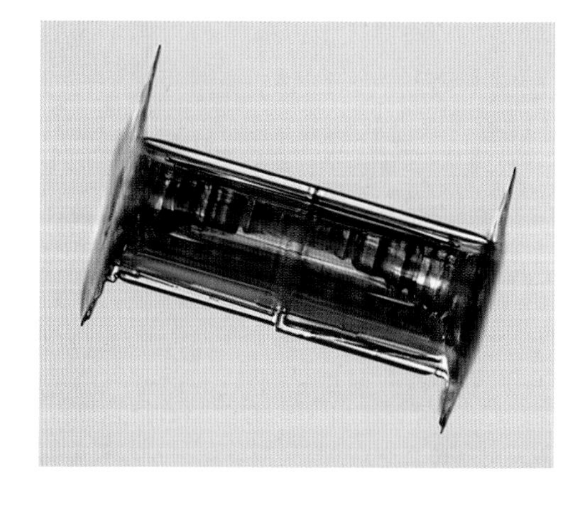

A common feature of capped columns is that the transition from columnar to platelike growth is usually quite abrupt, owing to the ESI (see Chapter 4). This same physical effect allows the formation of PoP crystals described in Chapter 9. In addition to two primary end plates, additional side plates may sprout from the exposed ledges in needle clusters or other features when columns are not simple and smooth.

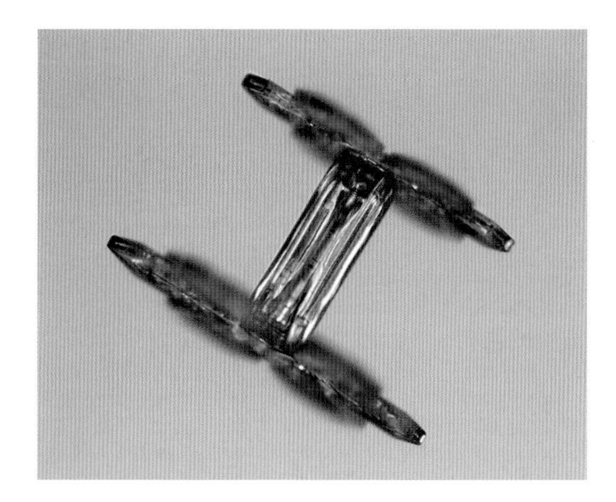

Capped Column Close-Up. The three photographs at left are all of the same crystal, but with two different orientations and with different focal planes. The top picture shows the crystal in the orientation I found it, after it had fallen onto a glass slide. This shows a nice side view of the column on which the plates formed. Some hollowing is present, so at one point this crystal must have looked like a simple hollow column. After photographing the crystal as it had fallen, I then used a fine paintbrush to flip it onto one face. Focusing my microscope on the smaller upper plate yielded the second picture, which looks like a typical stellar plate. The symmetry is subtly imperfect, and you can see a central dark spot where the column attaches. Without moving the crystal, I then refocused on the lower plate to produce the third picture. In this photo, the blurry upper plate now obscures the lower plate to some degree. The lower plate looks a lot like the upper one, as you would expect, because the two formed under nearly identical conditions.

The crystal on the right is a capped column with especially distinctive stellar plates on both ends. The image was captured in Moscow by Russian photographer Alexey Kljatov.

Capped Needles. These two remarkable specimens are essentially capped columns, but they might be more appropriately called "capped needle clusters," as each has multiple plates growing from the ends of sizable needle clusters. Both are about 1.7 mm long, and the various plates (seen edge-on) are all amazingly thin, with razor-sharp edges. Moreover, the column-to-plate transitions are especially abrupt. Here again these crystals provide excellent demonstrations of the ESI in action. I have encountered large capped needles like this only once, on one extraordinary day in the Michigan Upper Peninsula. These two crystals fell within a few minutes of each other, and I spotted several others like them as well. When the conditions are just right, rare snow crystals can fall in abundance.

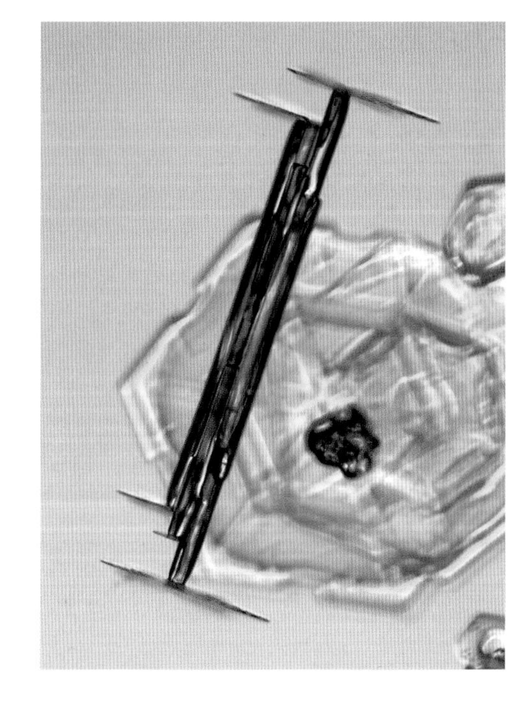

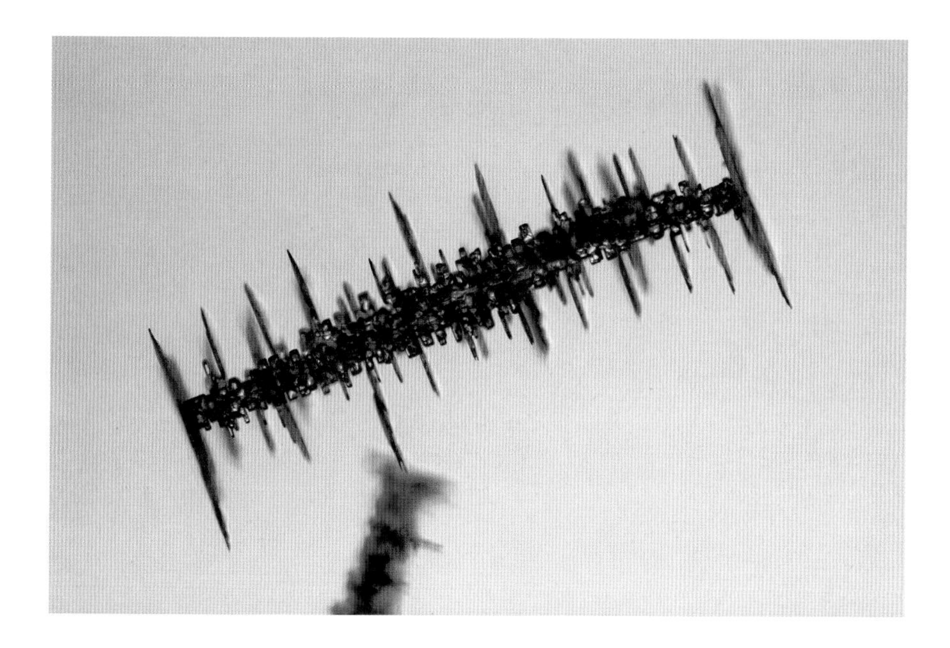

Plates from Rime. The world of multiply capped columns is inhabited by some exotic beasts, such as the two ice caterpillars above and below left. Both are relatively simple needle crystals festooned with copious side plates (the plates are viewed edge-on in the photos). Each of these crystals started out as a simple needle, which then became coated with rime. Next the temperature dropped, and plates sprouted from many of the rime droplets. Note that the rime froze with the same lattice orientation as the underlying needles, so the side plates are all parallel to one another. Thus, each of these seemingly disordered structures is a single crystal of ice, with the water molecules aligned throughout.

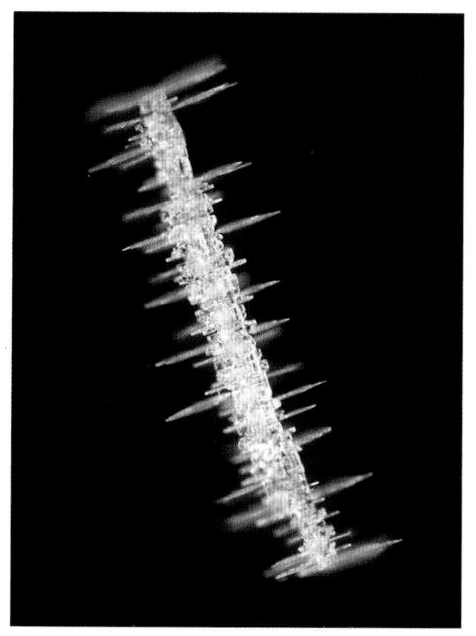

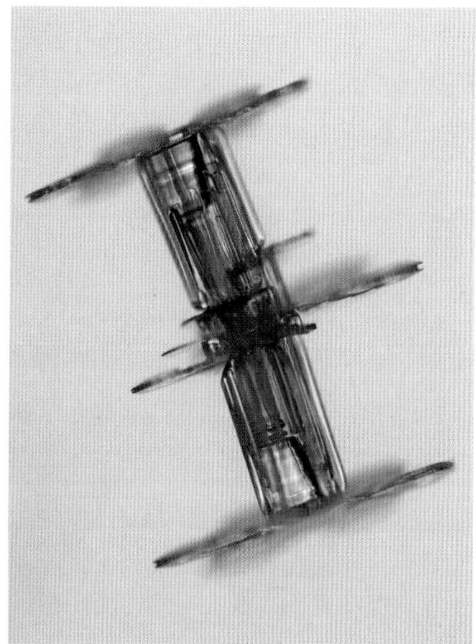

DOUBLE PLATES

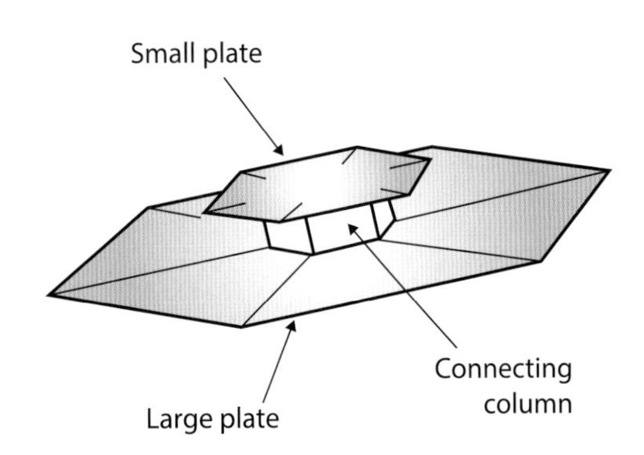

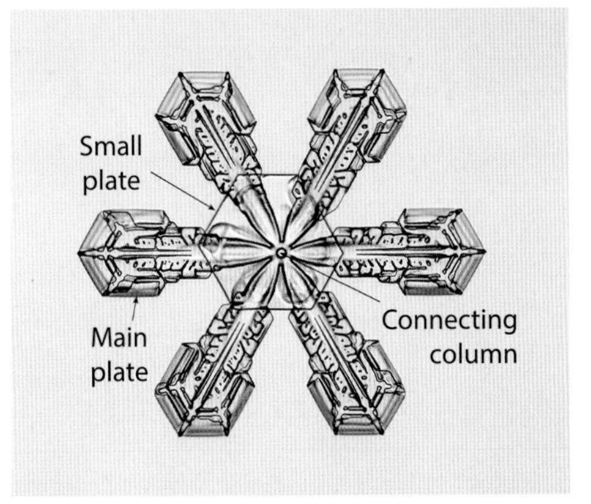

Double plates are pairs of thin, platelike crystals held together by a small connecting column. Often one side is a large stellar plate while the other is a smaller hexagon, although many other variations are possible. This phenomenon is relatively common, and many stellar crystals are actually double plates if you look closely.

Double plates are basically extreme versions of capped columns that result in two closely spaced plates. The two plates compete for water vapor, leading to a growth instability: any slight perturbation can cause one plate to overshadow the other, yielding one dominant and one recessive plate. The two photos below show the same rimed crystal with separate focus on the top and bottom plates.

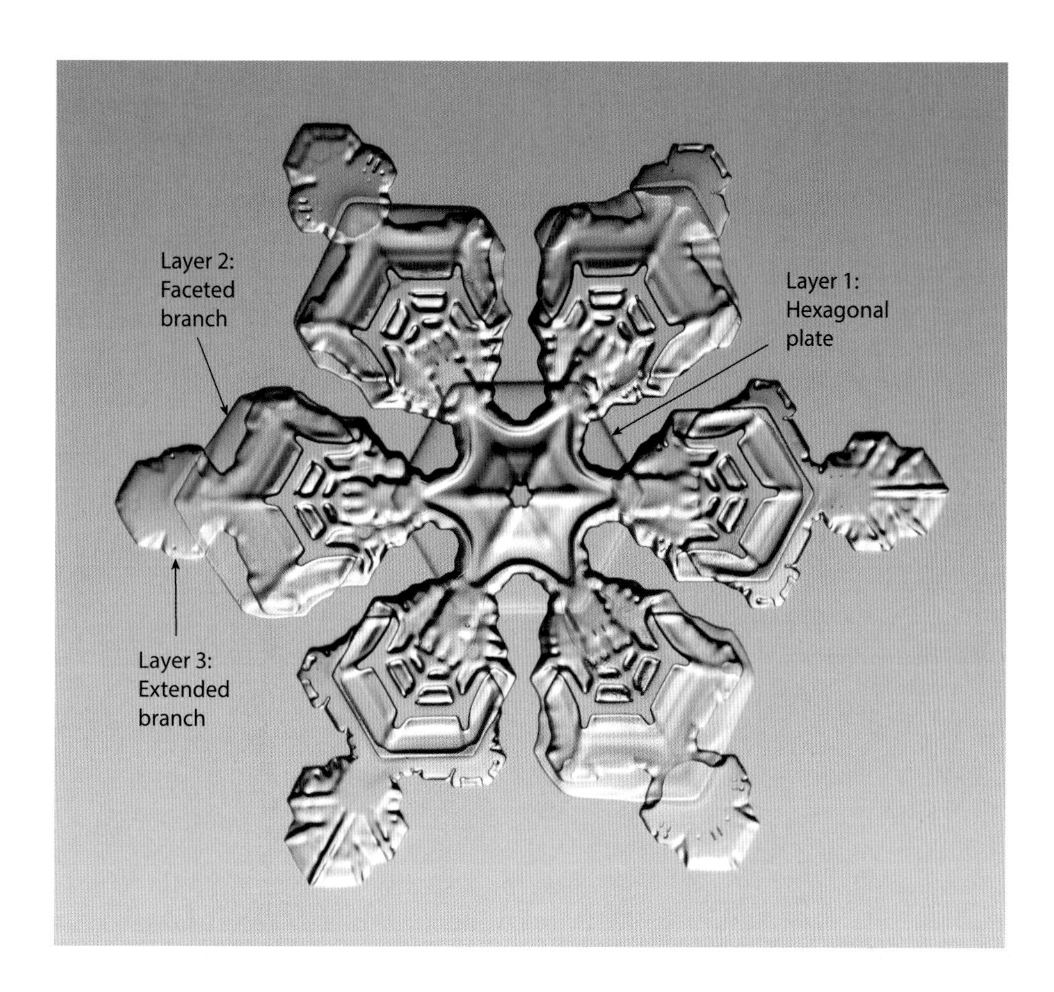

A Multi-Layered Plate. At first glance, this snow crystal may look like an ordinary stellar plate, but a closer inspection reveals three distinct layers, as shown in cross-section in the sketch. Note first the nicely formed hexagonal plate near the center of the crystal (layer 1), which is slightly out of focus in this picture. This hexagon was one half of a double plate when the crystal was small. The other half grew out faster and branched, and in doing so, it deprived the hexagon of water vapor. Because it grew relatively slowly, the hexagon remained smaller and faceted. One often sees double plates where the larger sheet is branched and the smaller one is faceted for this reason. When the crystal was about half its final size, it ran into low humidity and the branches grew thicker. Later the humidity picked up again, and the branches became double plates

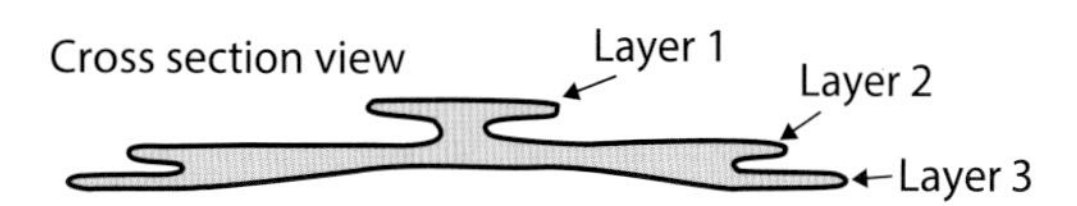

of their own (layers 2 and 3). Here again, one plate was left behind growing slowly (layer 2) while the other grew out more quickly and became branched (layer 3). If you look carefully, many stellar crystals show multiple layers like this one.

SPLIT PLATES AND SPLIT STARS

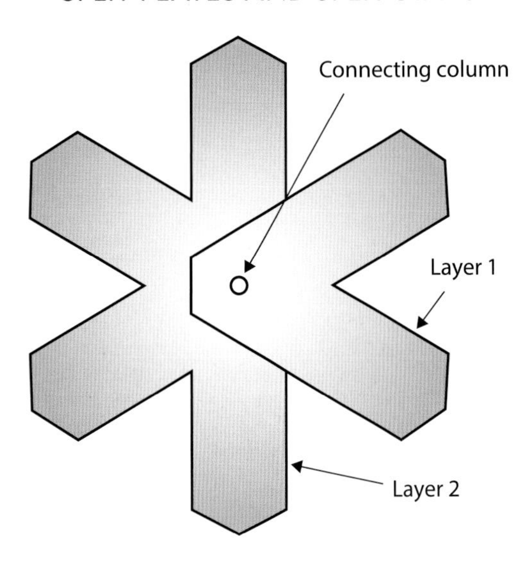

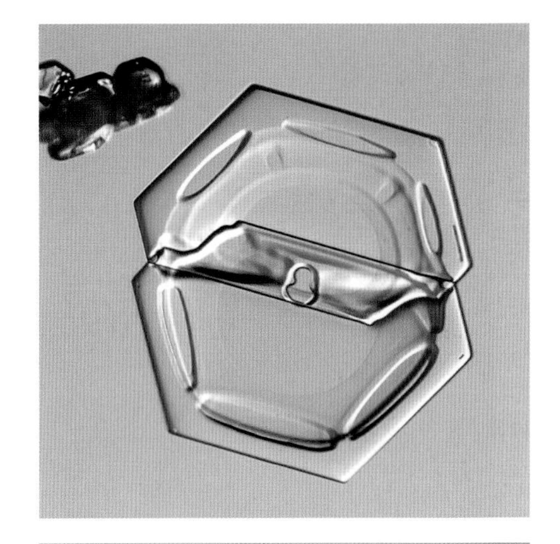

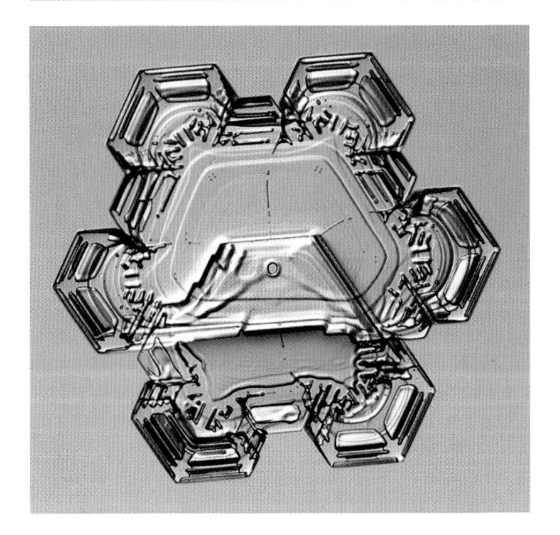

Split crystals are essentially double plates that have experienced asymmetrical growth. A surprising number of stellar dendrites are split stars when you look carefully at their central construction. These snowflakes are commonly mixed in with normal stellar plates, and sometimes one finds isolated partial plates after the two parts of a split crystal have separated during flight.

The formation of a split crystal is driven by a growth competition between the two members of a double plate. The pair starts out symmetrical, looking much like a short capped column. But if one branch or corner happens to edge ahead of its nearby sibling, then the growth of the latter is soon stunted from overshadowing. If one entire plate dominates over the other, then the result is a double plate. But if parts of both plates prevail, then the crystal will develop into a split plate or split star. If the split occurs early, the six dominant branches may grow into a surprisingly symmetrical stellar crystal.

Split Categories. There are eight possible topologies for making a split plate or star, as illustrated in the sketches on the left. The crystal above is one of the three possible 4 + 2 variants, photographed by Patricia Rasmussen in Wisconsin [2003Lib2]. Here the two parts of the crystal broke apart during handling, giving a nice look at a "disassembled" split star. The photo below shows one piece from another 4 + 2 variant.

HOLLOW PLATES

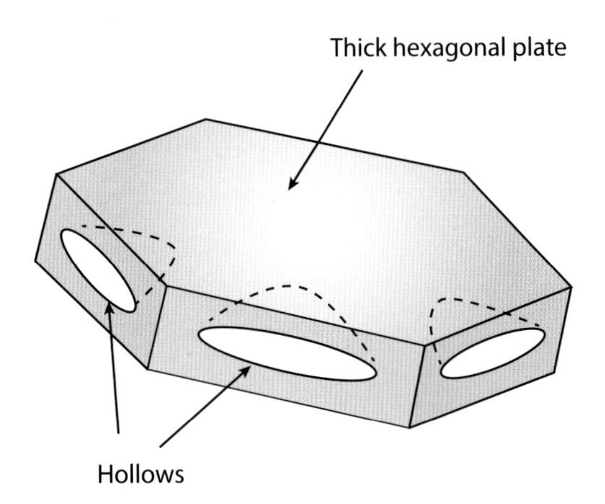

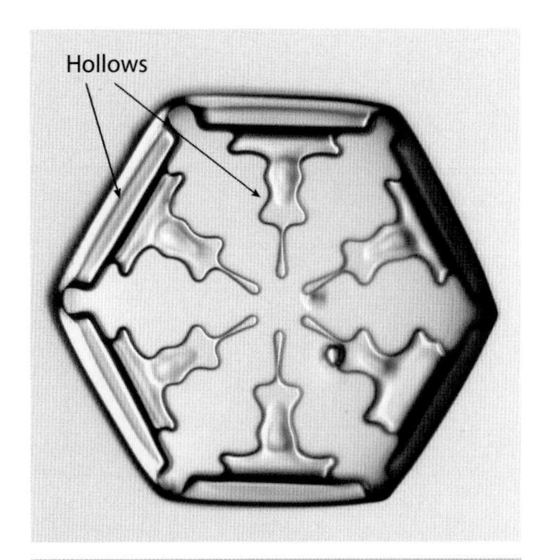

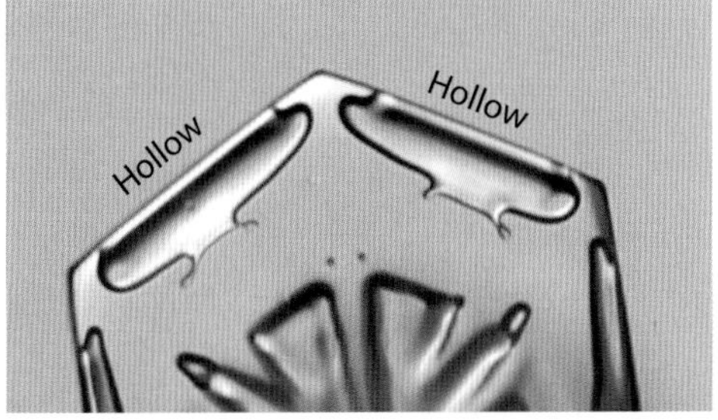

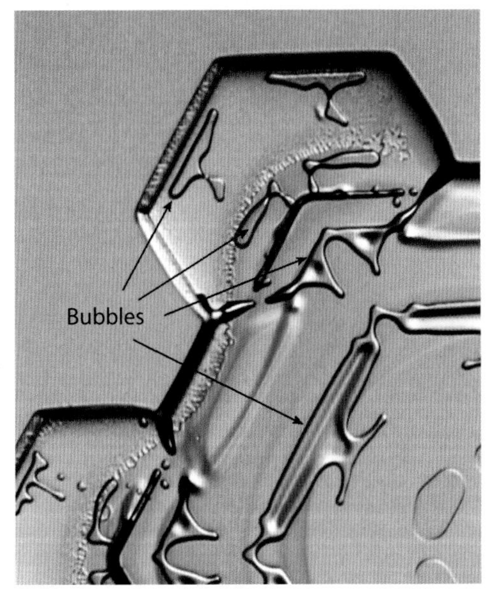

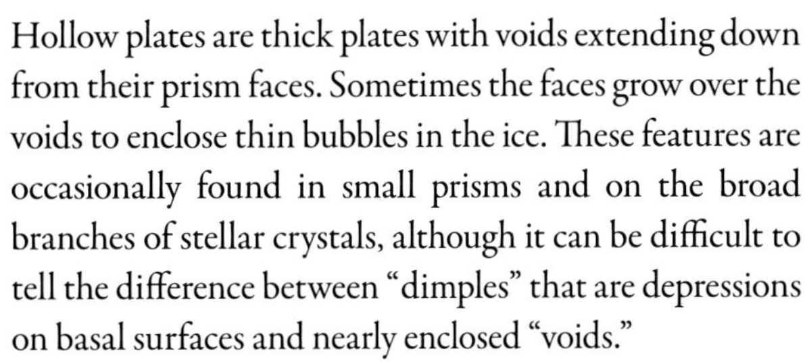

Hollow plates are thick plates with voids extending down from their prism faces. Sometimes the faces grow over the voids to enclose thin bubbles in the ice. These features are occasionally found in small prisms and on the broad branches of stellar crystals, although it can be difficult to tell the difference between "dimples" that are depressions on basal surfaces and nearly enclosed "voids."

Hollow plates are essentially the platelike analog of hollow columns. One starts with a thick-plate crystal, and then the facet edges grow faster than the centers, eventually leaving behind hollows in the prism faces. The sketch above shows a hexagonal hollow-plate crystal, but the phenomenon is more of a structural feature than a snow crystal type. Like ridges and ribs, small hollows are fairly common features in broad-branched stellar plates and other thick-plate crystals.

Hollow plates are most likely to grow when the temperature is either just above or just below −15°C, and fluctuations in temperature and humidity can yield rather oddly shaped voids, although broad, wide voids are more typical.

When you begin to look closely, hollows and bubbles can be found on many platelike snow crystals, as in these two examples.

SKELETAL FORMS

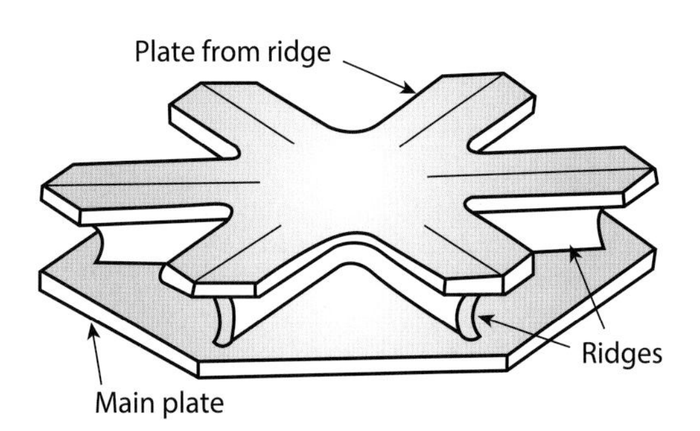

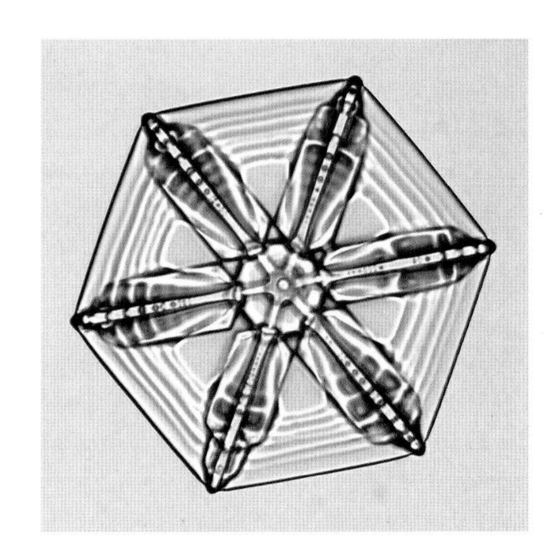

This category refers to the formation of thick ridges on a basal surface followed by secondary plates growing out from the ridges. The distinctive "I-beam" structures that result are a fairly common morphological feature that can be found in many stellar snow crystals.

While thin ridges are the defining feature of sectored-plate snow crystals, thick ridges are the basis for skeletal forms. In both cases, the underlying physical phenomenon is the spontaneous appearance of ridge structures on convex basal surfaces, which is described in Chapters 3 and 8. Both thin-ridge and thick-ridge phenomena are clearly seen over a broad range of conditions in the pictorial Nakaya diagram in Chapter 8, showing that they grow under constant growth conditions (in contrast to capped columns, for example, which cannot form in constant conditions). When circumstances are right, plate-from-ridge skeletal forms can be remarkably common.

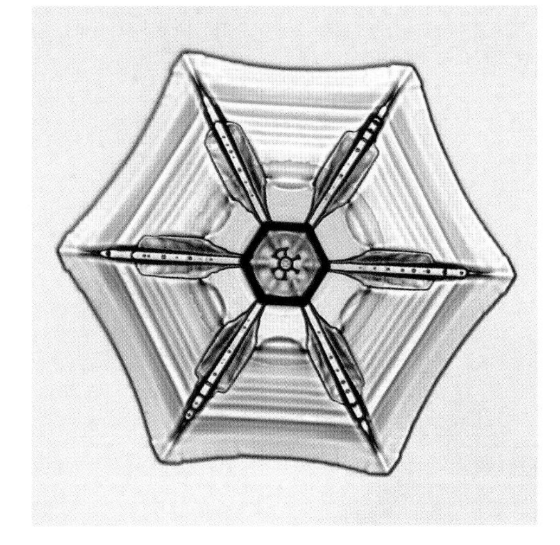

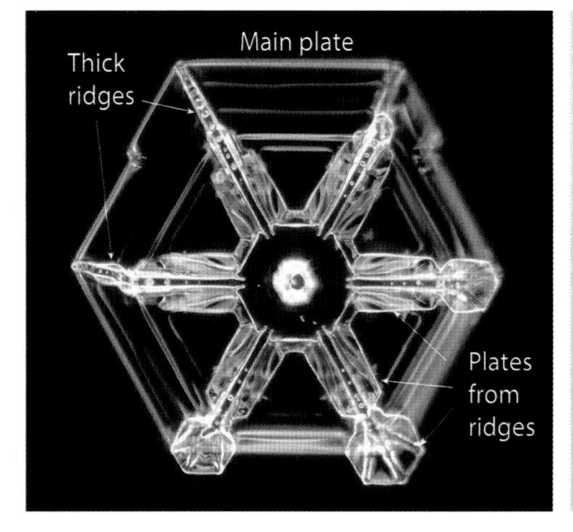

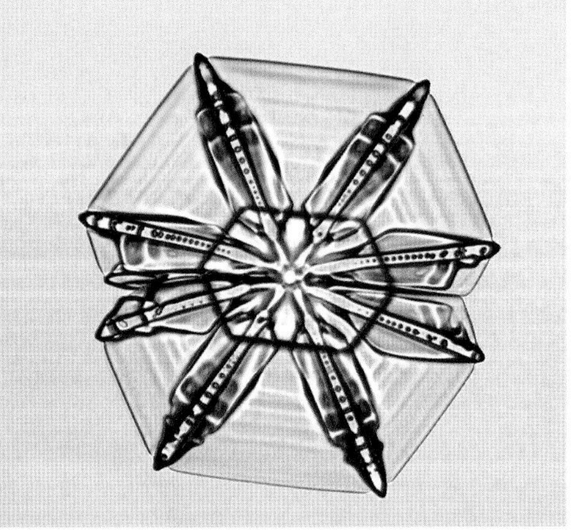

COLUMNS ON PLATES

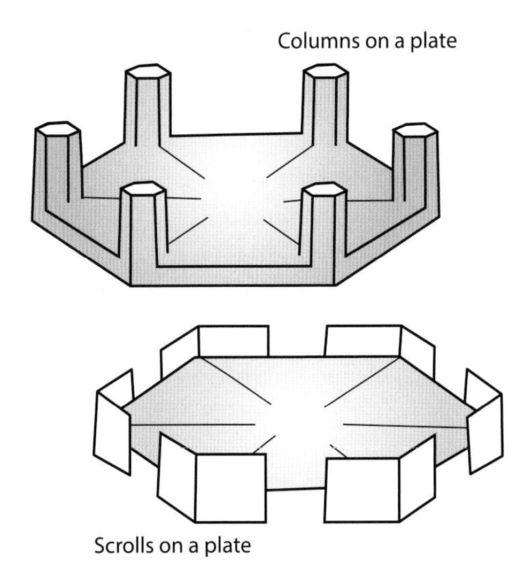

This category includes crystals for which platelike growth was followed by columnar growth, which is essentially the opposite of capped columns. As shown in the sketch, the columnar growth can take the form of simple columns (top), or sections of hollow columns called "scrolls" (bottom). Unlike capped columns, simple examples of these forms are exceedingly rare.

A typical cooling cloud may transition through −2°C (not cold enough to freeze droplets) to −6°C (droplets start to freeze, columns form), to −15°C (plates form), and this common behavior can yield capped columns. Weather scenarios that produce columns after plates are unusual, so any kind of column-on-plate growth behavior is quite rare.

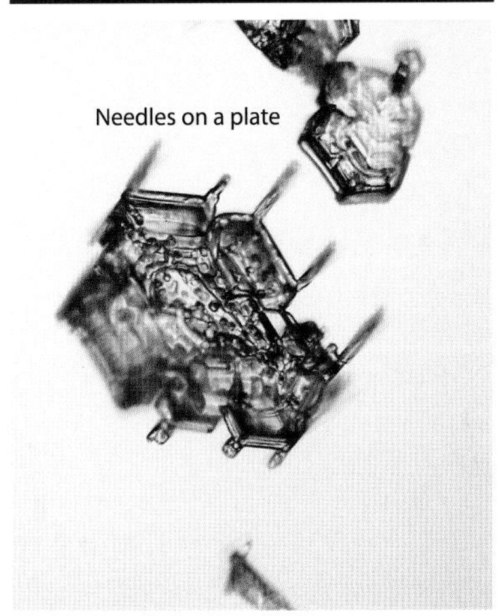

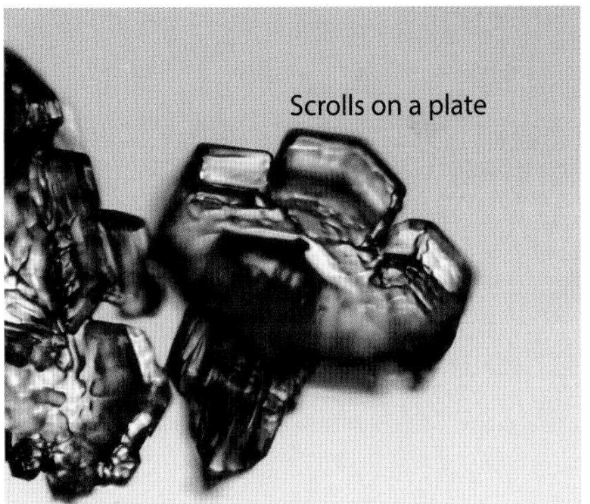

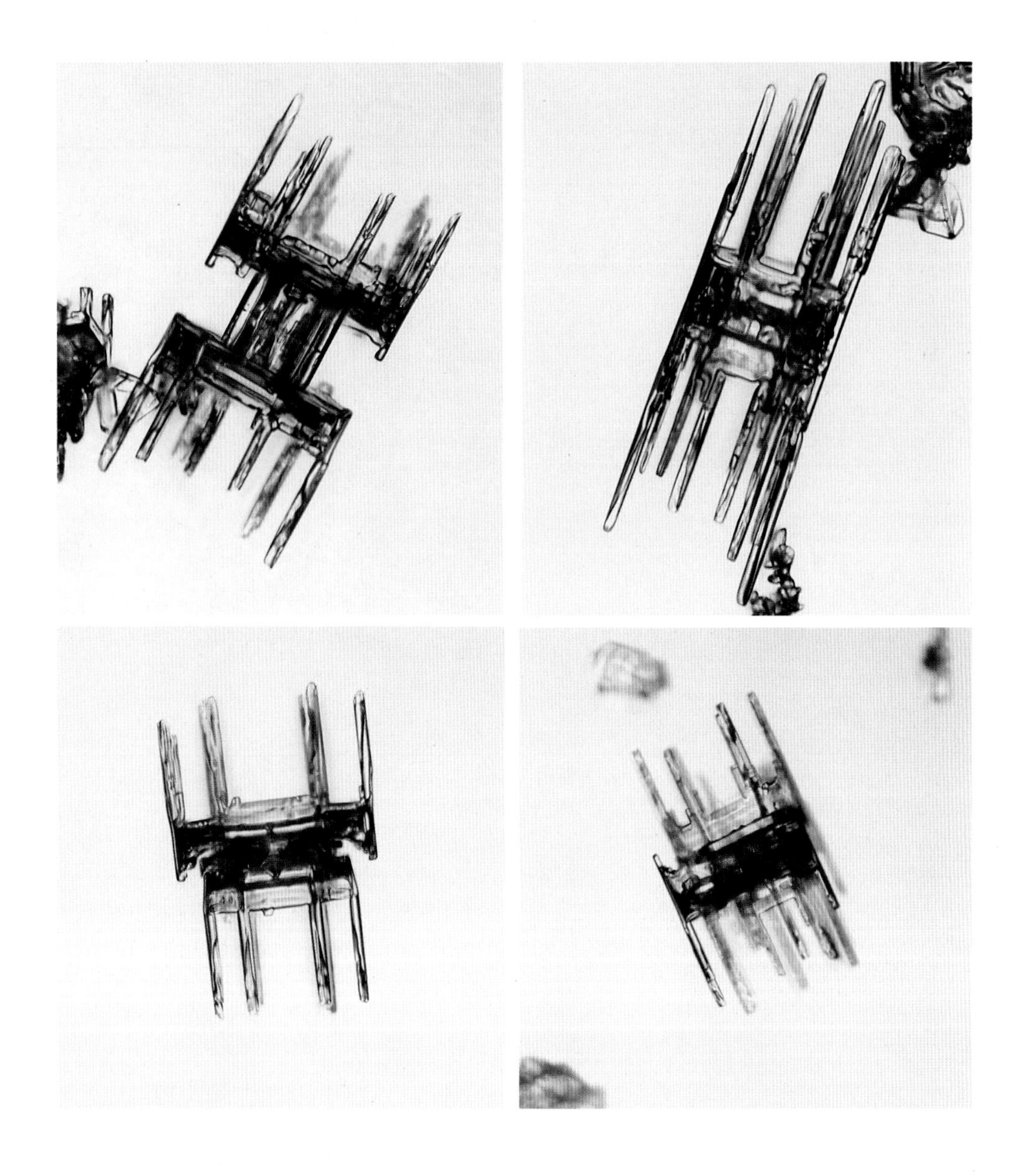

Locally Abundant. In birdwatching, rare but "locally abundant" birds are generally hard to find except when a whole flock of them shows up where you happen to be. Snow crystals can behave in the same way. When a snowfall produces just the right weather conditions, ordinarily rare crystals can be quite abundant, at least for a short while. I photographed all the peculiar column-on-plate crystals on this page during a 20-minute period in Fairbanks, Alaska, when the temperature was near −5°C. One never knows what oddities the clouds may deliver, so you just have to be watching when they appear.

TRIANGULAR CRYSTALS

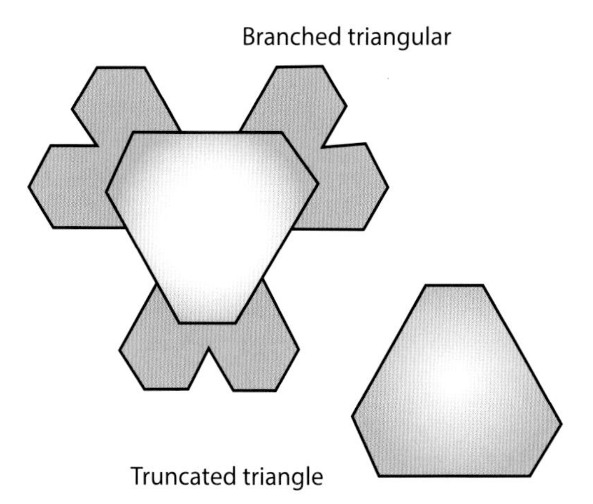

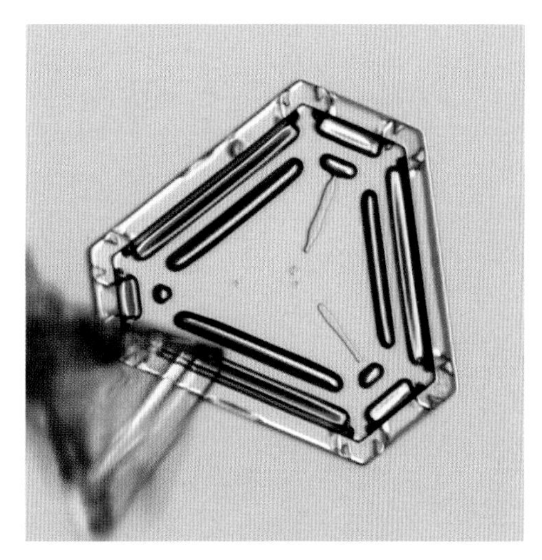

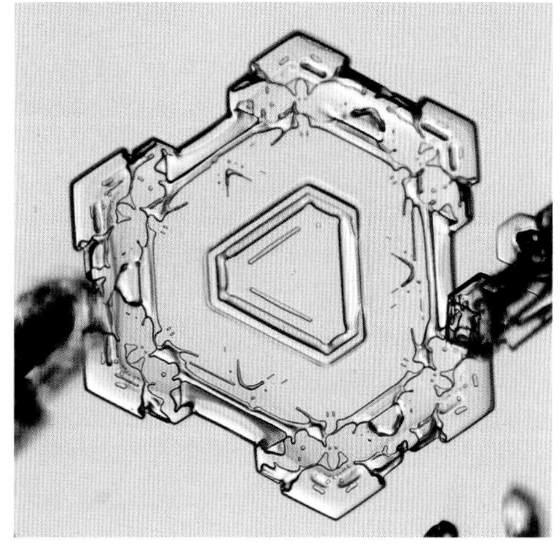

Triangular snow crystals display an overall threefold symmetry rather than the usual sixfold symmetry. The most common shape is a truncated triangular plate, sometimes with branching. Triangular crystals are relatively rare and usually small. They are most likely to be found in warmer snowfalls, mixed in with other small plates.

As described in Chapter 3, there is a weak growth instability that can cause a hexagonal plate to transform into a triangular shape. A slight perturbation in that direction will be amplified by diffusion-limited growth, and once begun, the transition from hexagonal to triangular is irreversible. Exactly how and when this instability is triggered is not yet well understood.

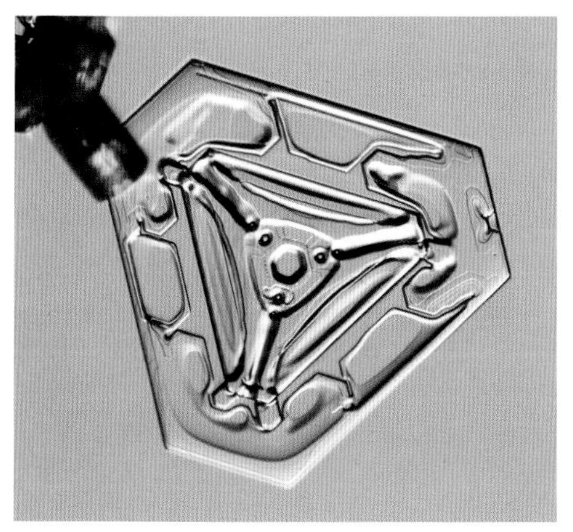

BULLET ROSETTES

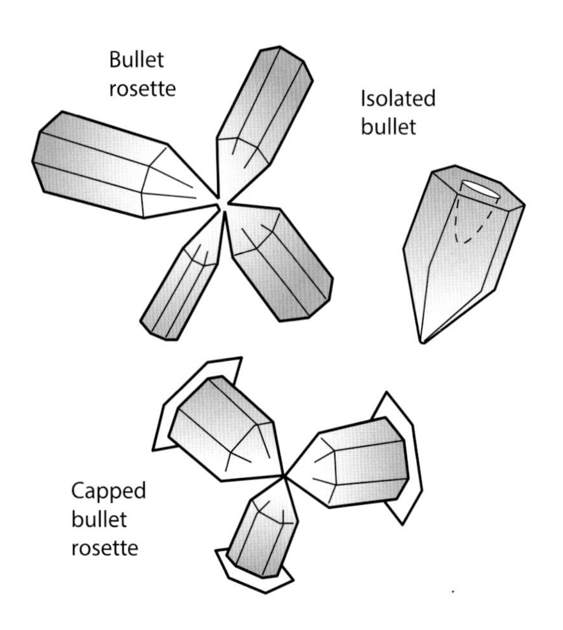

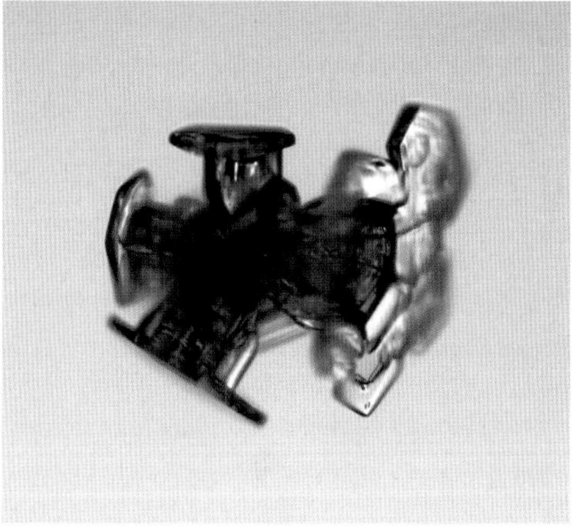

Bullet rosettes are collection of columnar crystals that form together around a single nucleus. Competition for water vapor inhibits growth near the center, giving each column a bullet-like shape. Individual bullets come from the breakup of bullet rosettes. These snowflakes are typically found mixed with columnar crystals in warmer snowfalls. Bullet rosettes are polycrystalline forms, which means that the entire structure is made of several individual crystals that grew out from an initially polycrystalline seed.

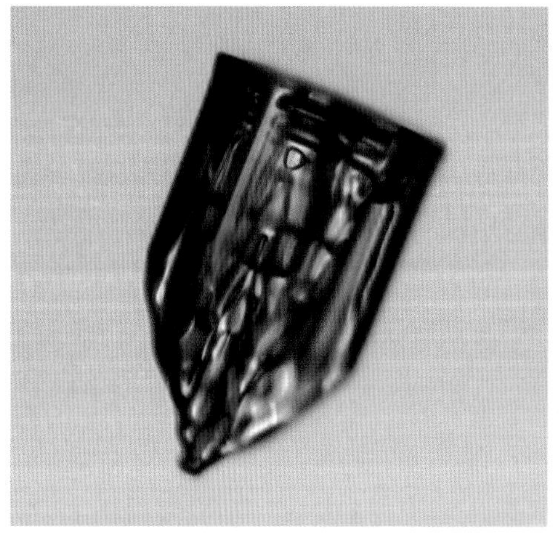

RADIATING PLATES AND DENDRITES

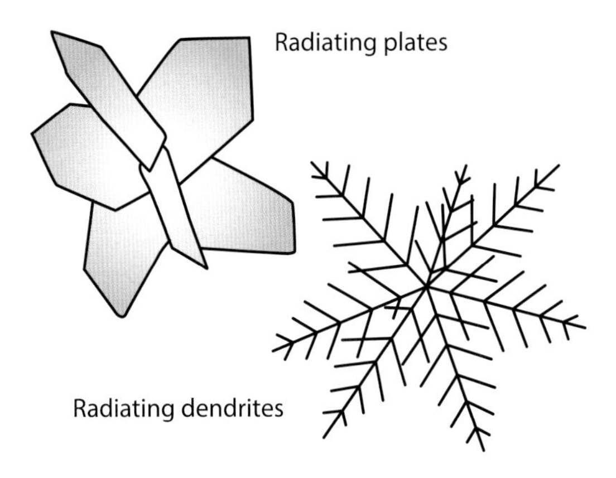

Radiating plates and dendrites are polycrystalline forms much like bullet rosettes, except with a collection of platelike crystals instead of columns. Typically, the different segments grow out from a common center, and their structure can be anything from simple faceted plates to fernlike dendrites. These composite structures are common and typically found mixed in with other platelike crystals.

Whether a cloud droplet freezes into a single ice crystal or a polycrystal depends on many factors. Larger droplets are more likely to become polycrystalline, as are highly supercooled droplets. Polycrystalline forms are generally much more common at lower temperatures than at high temperatures. A polycrystal can also form when a rime particle collides and sticks to a single-crystal form. The crystal at right probably picked up a rime droplet that froze with some random crystal orientation. This nucleated the formation of the additional branches you see growing out of the plane of the photograph.

SHEATHS AND CUPS

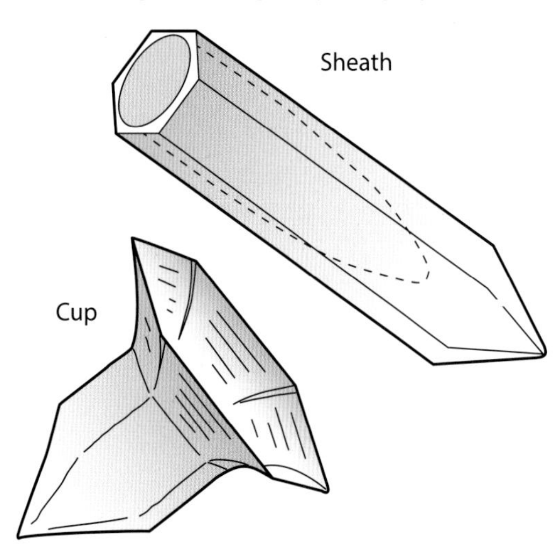

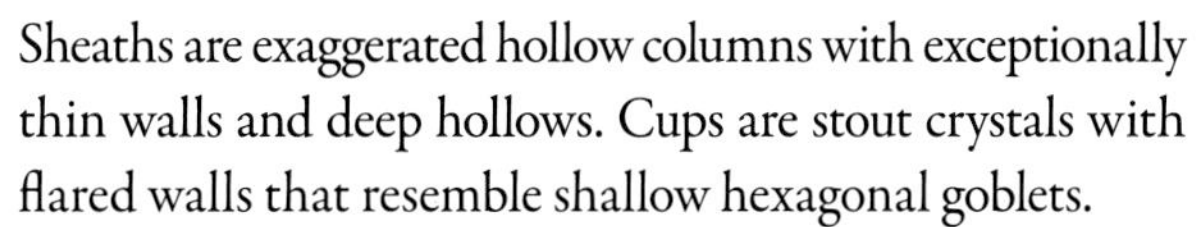

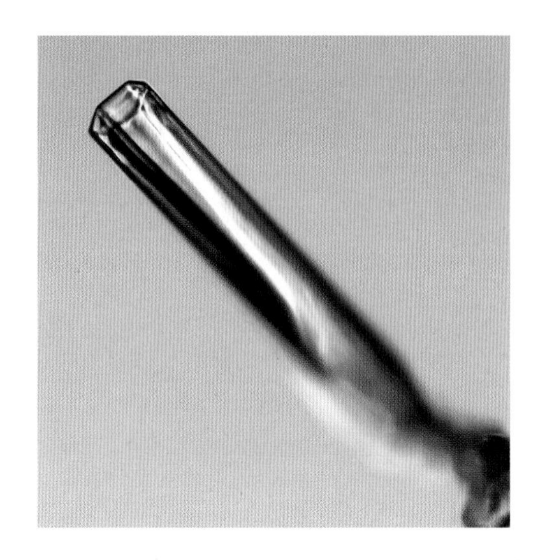

Sheaths are exaggerated hollow columns with exceptionally thin walls and deep hollows. Cups are stout crystals with flared walls that resemble shallow hexagonal goblets.

In terms of growth mechanisms, these crystals could be included in the hollow-column and capped-column categories; but both can be quite distinctive in appearance, so they have picked up their own names over the years. These crystals are generally small and rare, so they are easily overlooked. I found these by scanning over collections of small irregular crystals that had landed on glass slides.

CRYSTAL TWINS

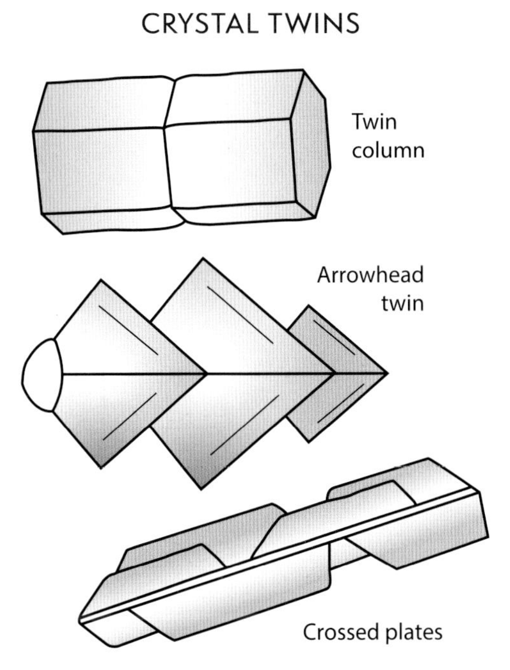

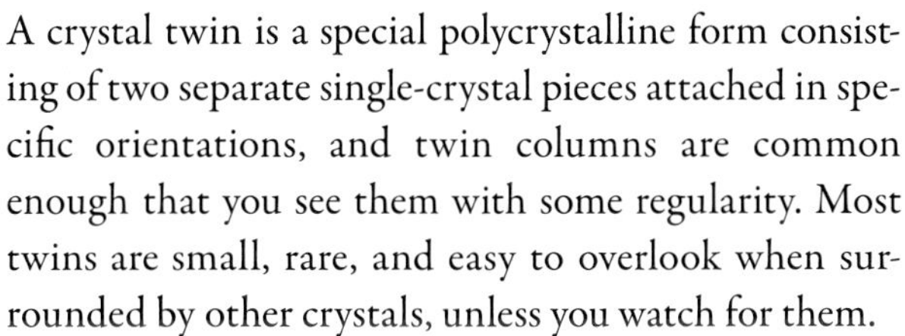

A crystal twin is a special polycrystalline form consisting of two separate single-crystal pieces attached in specific orientations, and twin columns are common enough that you see them with some regularity. Most twins are small, rare, and easy to overlook when surrounded by other crystals, unless you watch for them.

Crystal twinning is a common mineralogical phenomenon involving an initial molecular lattice mismatch that develops into a pair of co-growing crystals. The alignment of the pieces indicates the lattice construction of a twin crystal, and I described the known possibilities for snow crystals in detail in Chapter 2. Crossed plates and arrowhead twins are both quite rare in the wild, although some variants are fairly easy to produce in the lab. Twin columns can be quite common; they look almost exactly the same as normal columns, but often one can see a distinct "evaporation groove" around the column's waist, indicating the weaker molecular bonding in that plane.

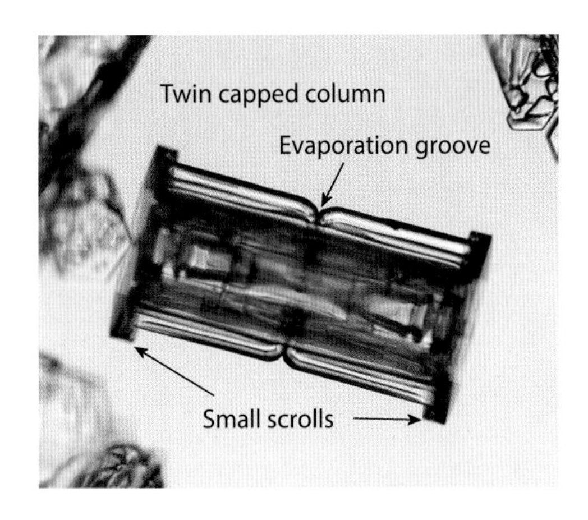

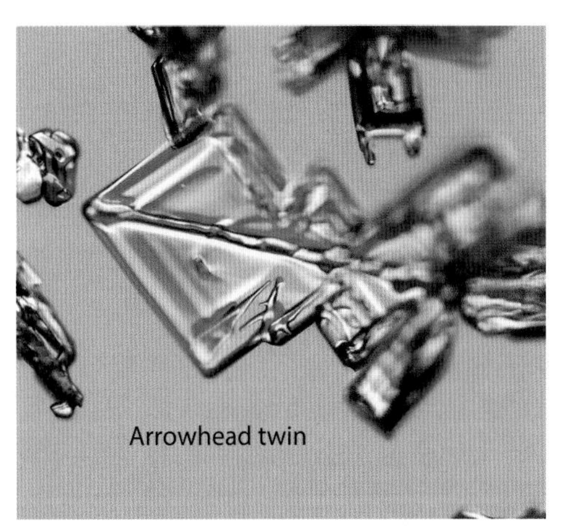

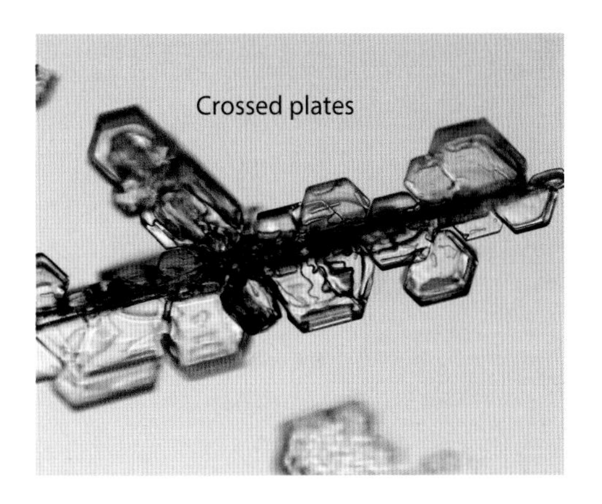

TWELVE-BRANCHED SNOWFLAKES

A twelve-branched snowflake is essentially a matched pair of six-branched stellar crystals attached at their centers, with one rotated 30 degrees relative to the other. Twelve-branched snowflakes are not common, but they can be quite large and distinctive in appearance. Some snowfalls bring more than others, mixed in with normal stellar crystals.

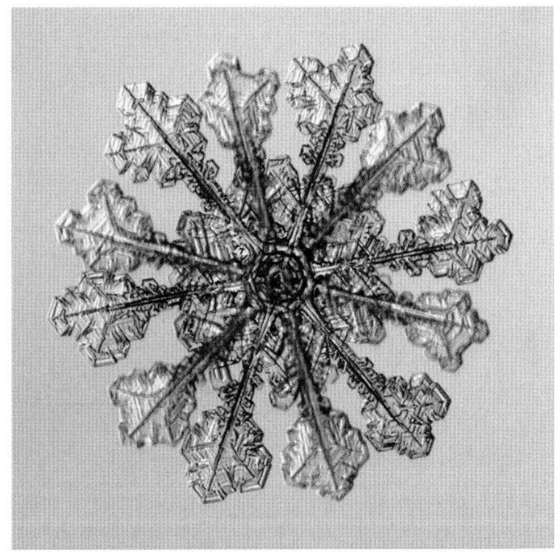

The evidence to date suggests that a twelve-branched snowflake is nothing more than two six-branched crystals that collided and stuck together when they were small. The near-perfect twelvefold symmetry in some examples appears to arise from a selection effect: if two tiny prisms experience a collision that bonds their basal faces together with close to a 30-degree rotation between them, then the pair will develop into a well-developed and easily spotted twelve-branched crystal. However, if the collision is less ideal (which is far more likely), then the pair will develop into an inconspicuous radiating dendrite. A key to this model is that your brain is quite adept at noticing symmetrical snowflakes in the midst of a great deal of malformed clutter. The fact that many twelve-branched crystals are not quite aligned, either in position or angle, supports this selection-bias hypothesis [1975Kob, 1990Uye].

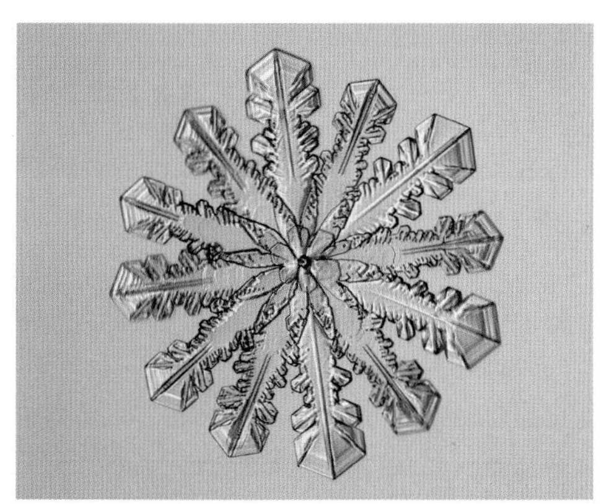

FIGURE 11.1. Freshly fallen snow crystals perched on a branch of eastern hemlock in Vermont. Photo courtesy of Martha Macy.

ELEVEN

Snowflake Photography

It is extremely improbable that anyone has as yet found, or, indeed, ever will find, the one preeminently beautiful and symmetrical snow crystal that nature has probably fashioned when in her most artistic mood.

—WILSON BENTLEY, *THE VERMONTER*, 1922

Snowflake photography has much in common with other forms of nature photography; it requires an artistic eye, some suitable optical gear, and a compelling desire to just go out there and take some pictures. The activity presents its own unique challenges as well, in that snow crystals are small, somewhat fragile, prone to evaporation and melting, and—as if that were not enough—they need to be handled outside in the cold. As a semi-professional snowflake photographer for many years, I have managed to pick up a few tricks and techniques regarding lighting, handling, equipment, and other considerations that matter out in the field. Moreover, I have studied the subject extensively and have learned from other prominent snowflake photographers as well. When you take a deep dive into the subject, there are a substantial number of rather subtle issues involved in capturing quality images of these tiny slivers of ice. In this chapter, I attempt to document what I have learned about snowflake photography, in the hope that others can continue developing this fascinating craft.

In my experience, three factors are of primary importance in snowflake photography: finding suitable subjects, using quality equipment, and developing an effective technique (especially regarding lighting). If any one of these factors is insufficiently developed, the quality of the resulting photographs will suffer. Patience is a virtue as well, along with an artistic eye and a willingness to try different approaches. And, as with all types of nature photography, success sometimes requires just being in the right place at the right time.

FINDING SNOWFLAKES

Perhaps the most common difficulty one encounters in snowflake photography is simply a dearth of quality subjects. One cannot control what the clouds are producing, and not every snowfall brings superb crystals. As described in Chapter 10, the most common bits of frozen precipitation are best classified as "irregular" or "rimed," and these are undoubtedly the least photogenic of all snow crystal types. Crystals from more desirable categories can be quite difficult to find, and they are usually mixed in with a sizable number of irregular specimens.

The first step in snowflake photography, therefore, is learning how to find nice specimens.

To begin, proper snowflake photography can only be done with freshly fallen crystals. Once the flakes hit the ground, they will stick together and soon transform into clumps of crystals with much changed morphologies. The character of ground snow is of considerable interest to skiers and people studying avalanches, but that takes us outside the scope of this book. Moreover, hoarfrost crystals and many other icy phenomena can also be amazingly beautiful, and they too make worthy photographic subjects. But if you want to photograph snow crystals in all their glory, you must catch them before they hit the ground.

The optimal strategy for photographing snowflakes will depend on what kinds of crystals are falling. When it begins to snow, a good first step is to leave the camera behind and just go outside to have a look. My preferred tools at this point are a sheet of dark-blue foam-core cardboard and a small magnifier like the one shown in Figure 11.2. The foam core provides a smooth matte surface that makes it easier to spot nice crystals, and the magnifier is handy for evaluating the quality of the crystals.

It is not unusual to observe a lot of small, grainy, gloppy, rimed, and generally undesirable crystals at this point; my generic name for them is "granular" snow, because the crystals look essentially like small icy grains of sand. As described in Chapter 10, this type of snow offers little appeal for snowflake photography. If there is nothing falling from the clouds but granular snow, then one's best option is probably just to go back inside and try again later. Wishing that the clouds would drop something better to photograph is not especially helpful.

FIGURE 11.2. An inexpensive fold-up magnifier, or *loupe*, is a convenient tool for appraising the overall quality of falling snow crystals. A magnification of about 5X is good for most circumstances.

However, I find it important not to give up too quickly. Even when there is a lot of granular snow all around, some interesting crystals might be in the mix. Moreover, some of the rare and quite captivating crystal types are usually quite small, and I like to photograph those almost as much as the canonical stellar variety. Capturing the full menagerie of snow crystal types is a worthwhile and often fascinating activity in its own right.

One useful trick I have learned is to hold the foam core out under a bright light, perhaps a streetlight or a yard light that is essentially a single point of bright illumination from a distance. By moving the foam core around under such sharp lighting, even small-faceted ice surfaces will sparkle clearly, making them easier to spot when surrounded by unfaceted granular snow. As a general rule of thumb, if you can see some sparkle on the board, then there is a reasonable chance that some interesting crystals are present.

If the snow has been falling for a while, and it happens to be dark outside, another trick is to just look out a window and view the reflection of a bright streetlight off a nearby snowbank. Pure granular snow, especially heavily rimed snow, has little or no sparkle, and this gives a snowbank a flat, white appearance. In contrast, a snowbank that shows some sparkle suggests that some nice, faceted crystals are falling.

It is also important to keep watch on the crystals throughout a snowfall, even when it appears that there will be little of interest to see. The character of the falling crystals can change dramatically with time, and you can miss some great pictures if you are not sufficiently diligent. There have been times when, as I was outside photographing, some exceptional crystals appeared only briefly, for perhaps 10–20 minutes. Granular snow does not usually change to great snow that quickly, so I typically check the crystals every 30 minutes or so. I have witnessed many snowfalls that started out as granular glop, then improved somewhat, then improved more, and then delivered some excellent photographic subjects for an hour or two, only to go back to granular snow as the snowfall waned. Like all other aspects of weather, snow crystal production can be highly variable and quite unpredictable.

In addition, snowflake photography is often best done at night. Partly that is simply because the nights are long in the winter, especially at high latitudes. As one ventures farther north in the dead of winter, working in the dark becomes a matter of statistical necessity, as the daylight hours are so short. Also, the temperature is typically lower at night, and lower temperatures are desirable in most locations. Thus, a dedicated snowflake photographer can expect to spend long hours outside, alone, in the cold and dark. Treating this as a serious hobby is not for everyone, but it does satisfy one's hermit-like tendencies.

Weather and Climate

In principle, the Nakaya diagram could be used to predict what kinds of snow crystals fall in different weather conditions, and this works to a limited extent. For example, like most snowflake photographers, I am always keen to find large stellar crystals, and these occur almost exclusively when the temperature is near −15°C. More precisely, because the temperature is usually slightly higher on the ground than up in the clouds, around −13°C is close to an ideal ground temperature for finding good specimens. However, the weather is not nearly as predictable or precise as that makes it sound. In practice, well-formed stellar crystals might be found anywhere from −10° to −20°C. But the probability falls off substantially outside that temperature range.

Warmer snowfalls often bring a great variety of snow crystal types, including columnar crystals near −5°C, or perhaps capped columns and other exotic forms. These crystals tend to be on the small side, however, and they are nearly always accompanied by lots of granular snow. As I describe later in this chapter, working a warmer snowfall is best done with high magnification and a different collection strategy compared with stellar crystals. But small can be beautiful, and I have captured many excellent photographs of unusual snow crystal types in relatively warm (above −10°C) conditions.

Although temperature is the most important parameter for predicting snow crystal types, many other factors influence quality. For example, wind can be quite detrimental, as the crystals can get beaten up by mid-air collisions. A heavy snowfall is not ideal for the same reason. In my experience, the best specimens can usually be found during calm, cold, light snowfalls, providing just a steady dusting of crystals drifting slowly downward.

Another meteorological phenomenon I have come to appreciate is low-hanging clouds. When the clouds are high in the sky, a kilometer or more above the ground, that usually yields what I call "travel-worn" snowflakes. The problem is that the crystals stop growing once they leave their cloudy nurseries, and they can experience quite a bit of sublimation as they slowly descend. Sublimation rounds the faceted corners and yields somewhat shabby-looking crystals. When I see snow falling from especially high clouds, I know that finding extraordinary specimens will be unlikely.

However, although it can be relatively easy to predict low-quality snow crystals from observing the weather conditions, predicting high-quality crystals is almost impossible. I have experienced some snowfalls that checked off all the boxes for great crystals yet brought nothing but granular snow for hours on end. Moreover, often granular snow gives way to beautiful stellar crystals, or vice versa, with no obvious change in weather conditions. It is certainly true that hollow columns and needles generally form only around −5°C, and large stellar plates are restricted to around −15°C; but there is not much one can reliably say beyond that. The atmosphere is not a precisely controlled laboratory environment, so it is impossible to predict exactly what kinds of crystals will appear, at least not with any real accuracy. For the snowflake photographer, waiting and watching are simply part of the process.

I have been especially attentive when it comes to finding the best conditions for observing large stellar crystals, as these are such a delight to photograph. As a concrete example, Figure 11.3 shows a photo I took during a "perfect storm" that lasted about eight hours and gave me some of my best snowflake photographs. Looking back on this day, I noted several beneficial characteristics of the weather:

1) The temperature had hovered around −13°C all day, which is the ideal temperature for finding stellar crystals.
2) It snowed lightly all day, so the crystals did not much interfere with one another in the clouds or on my collection board.
3) There was essentially no wind all day.
4) The clouds were hanging low in the sky, barely above ground level, so the crystals continued growing during most of their descent, yielding sharply faceted crystals.
5) The clouds were thin and patchy, so the varying conditions resulted in a good deal of morphological diversity in the falling crystals.

Even at a good location, one might encounter a high-quality snowfall like this maybe a few times during a winter season. As I already mentioned, some degree of patience is essential in snowflake photography.

Location Matters

Quality snowflakes can appear anywhere, as long as the temperature and other weather conditions are favorable. Location is a factor only because the probability of experiencing such conditions varies from place to place. Being a snowflake photographer who happens to live in southern California, I have tried to find locations that maximize the probability of finding high-quality snow crystals, especially large stellar dendrites, and I have studied this problem quite a bit over the years.

Temperature is the most important factor, as I mentioned above. One of my favorite locations is the small town of Cochrane, Ontario, where the average January temperature is −18°C and the average daily high is −12°C. This means that the probability of finding stellar crystals is reasonably high on average, notably in the daytime, when being outside is most pleasant. The average January precipitation is a respectable 3.5 cm (water equivalent), and this arrives in

FIGURE 11.3. The scene during a near-perfect storm for snow crystal photography, taken by the author in Cochrane, Ontario. On rare occasions, the meteorological conditions seem to conspire to create the most beautiful snow crystal forms.

FIGURE 11.4. Roads packed with accumulated snow often indicate a good location for snow crystal photography. In Kiruna, Sweden, shown here, some exercise-conscious residents use sleds for their grocery shopping, using the snow-packed roads to good advantage.

frequent, light snowfalls. Wind speeds are generally low as well (7 mph), and I have found Cochrane to be an excellent location for photographing snowflakes overall. Many time zones away, the town of Kiruna, Sweden, has comparable average conditions, and I have found some excellent crystals there also. One noteworthy characteristic of both Cochrane and Kiruna is that residential roads in January are almost always covered with packed snow, as seen in Figure 11.4. This is a good sign for snowflake photography, as it tends to indicate consistent low temperatures (as the snow does not melt) and plenty of falling snow.

My hometown of Fargo, North Dakota, provides a good example where a low average temperature is not the only parameter to consider. The average January temperature in Fargo is −13°C, which sounds good; but the average precipitation is only 0.7 cm, and the brisk winds can be quite incessant. Snowfalls are somewhat infrequent, and much of the winter precipitation comes in the form of intense blizzards. Although Fargo has a favorable average temperature, it is not an ideal location for snowflake photography.

Ukichiro Nakaya lived in Sapporo, Japan, where the January average is a balmy −4°C, although conditions are better in nearby Asahikawa at −8°C. The January precipitation weighs in at an impressive 10 cm water equivalent, with typically calm winds, so there is certainly no shortage of snow. Central Hokkaido is also well known as an excellent location for snowflake photography, as evidenced by Nakaya, Katsuhiro Kikuchi, Yoshinori Furukawa, and others from that region.

Another good case study is Barrie, Ontario, which is home to noted snowflake photographer Don Komarechka. The average January temperature in Barrie is −8°C, and the average low is −12°C, so perhaps this location is a bit on the warm side. Nevertheless, Don has taken some of the world's best snowflake photographs in Barrie, so the site is obviously working for him. It helps that it snows a lot, bringing 4 cm on average in January, and the average wind is not too bad (9 mph). Moscow is worthy of consideration also, as this location is the home of Alexey Kljatov, another renowned snowflake photographer. Here the average January temperature is −8°C with an average snowfall of 4 cm, and Moscow can boast a remarkably low average wind speed (3 mph). Wilson Bentley, the founding father of snowflake photography, made his home in Jericho, Vermont, where the January average is −7°C, although this number was a bit lower in the 1880s. The precipitation and wind speeds are also both suitable, and Vermont remains a prime location for snowflake photography.

The climate data indicate that Barrie, Moscow, and Jericho are all quite similar in average January conditions, so certainly that says something regarding the availability of quality snow crystals. Personally, I would rate Cochrane as slightly higher, with its colder average temperatures, but the statistics are thin all around; even the best locations deliver exceptionally well-formed crystals only rarely. Note also that the population density drops off rapidly with the average winter temperature. I suspect that slightly warmer conditions yield more snowflake photographers in part because fewer people live in colder regions. Details notwithstanding, snowflake photography is a craft best practiced near the cold edges of human civilization.

Other than average weather conditions, there does not seem to be anything exceptional about any of these locations. No magic lurks behind producing quality snowflakes, other than the fact that favorable weather conditions are more likely in some places than in others. I have never been especially fond of mountain locations because of generally high winds, and most highly populated areas are simply too warm. But quality snow crystals can occur whenever and wherever the clouds conspire to make them.

If you happen to live in a place that experiences sufficiently cold winters with plenty of snow, then you will likely find some excellent snow crystals if you go out looking for them. The best way to find out is just to go outside with a simple magnifier to have a look for yourself, preferably sampling multiple snowfalls at multiple times, as not every storm brings exceptionally photogenic crystals. If you like what you see, and you are willing to spend some time outside in the cold, then you might well enjoy snowflake photography.

Handling Snowflakes

When I am photographing snowflakes, how I handle them depends on whether the interesting crystals are larger or smaller than about 2 mm. When larger crystals are falling, I let them fall onto a foam-core collection board, with the result looking something like what is shown in Figure 11.5. Especially photogenic specimens are rare even on the best days, so a large foam-core collection board gives one a lot of crystals to look over, and the eye is remarkably adept at noticing especially nice crystals in a field of mostly granular snow. In this figure, I would say that more than 1 percent of the crystals are reasonably well formed, which is considered quite a good yield. The average yield is much lower than 1 percent, as many snowfalls bring nothing but granular glop. With maybe a thousand snowflakes on the board (which is just a modest 33×33 array of crystals), one can usually search and find the best of the bunch in a minute or two, thus delivering a one-in-a-thousand snow crystal to photograph. With a quick brush of one's sleeve, the board is cleared for another round. Scanning over a board like this every few minutes, before long, one can capture some exceptionally photogenic, one-in-a-million specimens.

When I spot what looks like a promising candidate on my collection board, I pick it up using a small paintbrush and transfer it from the board to a glass microscope slide. This works surprisingly well, as the fine bristles will lift a typical stellar snow crystal with hardly any damage (most of the time, anyway). Best is to gently roll the bristles under the crystal, lifting it in the process. Of course, some breakage in handling is to be expected, and Figure 11.6 shows one example. A worse problem, in my experience, is carefully lifting a highly promising specimen onto the brush and then poof, a slight gust of wind sends it flying off, gone forever.

A glass microscope slide is certainly not the only destination for placing a snow crystal, and one might want to frame a photo in any number of ways. Regardless of how you want to proceed, a large foam-core surface and a small paintbrush can be used to scan through many

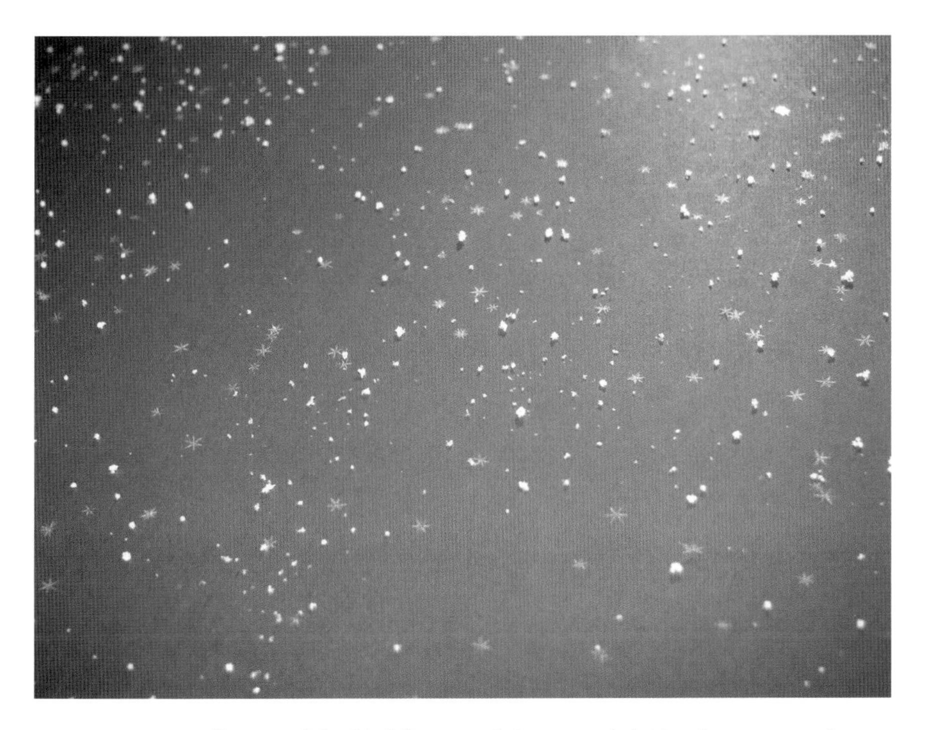

FIGURE 11.5. A collection of freshly fallen snowflakes on a dark-blue foam-core collection board. The glow in the upper right comes from a bright lamp shining down on the crystals, producing strong reflections from smooth faceted surfaces.

FIGURE 11.6. A fine paintbrush works quite well for picking up and placing snow crystals, but damage is not uncommon. I broke this snow crystal when I tried to pick it up and move it onto a glass slide, losing a platelike branch in the process.

snowflakes quickly, thus allowing one to choose the nicest specimens. If you want to photograph rare crystals, including large, well-formed stellar crystals, then it is essential to examine as much falling snow as possible.

Another secret to photographing exquisite crystals is to move fast. A good snowfall will not last forever, so it pays to capture as many crystals as possible while the clouds are being generous. Scan the collection board, find a worthy subject, pick it up, place in on a slide, put it under the microscope, adjust the lighting, and take the shot . . . and then repeat. On a good day, I can do a crystal every minute or two this way, thus achieving a fairly large throughput. I have never met a photographer who managed to get terrific pictures with every single shot; taking lots of pictures is essential for yielding a much smaller number of outstanding photos.

Another good reason to hurry is to avoid sublimation. Figure 11.7 shows a nice example of a small stellar

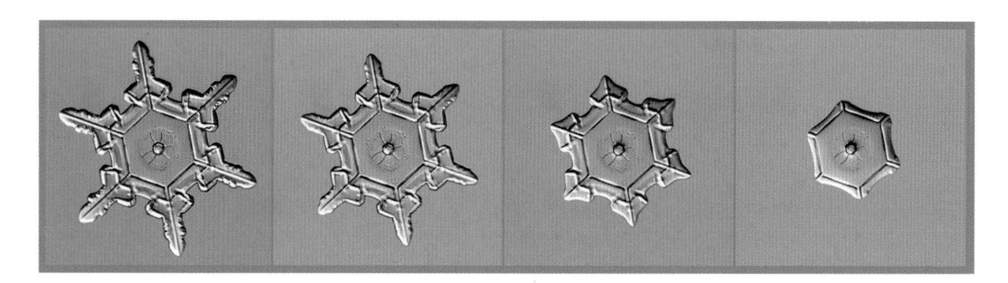

FIGURE 11.7. This snow crystal experienced quite a lot of sublimation during the 2 minutes that elapsed between the first photo and the last. You can see how the finer structural features on the crystal extremities are the usually first to disappear during sublimation.

crystal that slowly evaporated away as it sat in the bright lights of my microscope. Figure 11.8 shows a further example of a snow crystal that melted as it was being photographed. Melting is mostly a problem only when the temperature is close to 0°C, and sublimation is not a huge issue at temperatures near −15°C. Nevertheless, speed is a virtue when photographing snowflakes.

Although I am partial to a foam-core collection board and glass slides, this is by no means the only approach to snow crystal photography. Many practitioners prefer to let the crystals fall onto a dark-colored wooly fabric for direct point-and-shoot photography, as illustrated in Figure 11.9. Nice-looking specimens are often supported by a single cloth fiber, and the tangle of neighboring fibers provides an interesting backdrop for the photo. No collection board is needed when the crystals fall directly on a piece of fabric, but searching through large numbers of specimens will be somewhat slower using this technique.

As I describe further below, my foam-core and glass-slide approach appeals to my science side, as it gives exceptional clarity and resolution, revealing fine details in the crystals. But Alexey Kljatov's point-and-shoot technique is wonderfully pleasing from the artistic side, yielding a more natural view of these tiny slivers of ice. There are many ways to photograph a snow crystal.

Granular Gems

Although photographing large, well-formed stellar snow crystals can yield some spectacular results, I have obtained many excellent pictures while focusing on small

FIGURE 11.8. This series of photos shows a snow crystal melting, and only 27 seconds elapsed between the first and last image. The temperature was just below 0°C during this series, illustrating how challenging it can be to photograph snow crystals in such warm temperatures.

FIGURE 11.9. A snow crystal is supported by a single fiber of a dark fabric in this photograph taken by Alexey Kljatov in Moscow. Note how the out-of-focus fibers provide a pleasing background that adds a sense of depth and scale to the photo.

specimens, typically around 1–2 mm in size or even smaller. Many of the exotic snow crystal varieties described in Chapter 10 are invariably quite small, and these tiny gems are worth pursuing. Their small size adds a new challenge, so different techniques are needed to find and photograph falling snow in this regime.

One big change when working with small crystals is that it is no longer possible to scan over a collection board to pick out promising specimens with the naked eye, at least not effectively. The crystals usually must be placed under a microscope just to see what you have. Although this sounds like something of a painstaking process, it is actually quite simple and enjoyable, and it can yield some remarkably interesting photos even when the clouds seem to be delivering little more than granular snow.

The technique I like best is to lay out a set of glass microscope slides to catch the falling snow, and then just pick one up and look it over under my photomicroscope. If a crystal looks worthy of a photograph, then I focus, adjust the lighting, and take the shot. If I find nothing worthwhile on the slide, then I clean it off, set it back out to catch more snow, and pick up another slide to scan. Cycling through a half-dozen slides usually works quite well, allowing each slide to accumulate a new dusting of snow while looking at the others. I support the slides on a pair of knife-edge "rails" (made from tape) to keep them elevated. This keeps the bottoms of the slides clean and free of snow.

If the clouds are being unkind, nearly all the falling snow may consist of gloppy, granular, or rimed crystals; at such times, there is little one can do but try again later. Small hexagonal plates can usually be found even in quite wretched conditions, but there are only so many photos one can take of those forms, as they all look pretty much the same. Surprisingly often, however, if one has some patience, interesting crystals are waiting to be found in the mix, at least from time to time. Most of the photos of the smaller exotic crystals described near the end of Chapter 10 were taken by scanning over hundreds of glass slides when no large stellar plates were to be had.

OPTICS AND LENSES

Many equipment options are available for photographing snowflakes, depending on the image quality you seek and how much money you are willing to spend. At the low end, a smartphone with a $10 clip-on macro lens can yield some reasonably nice snowflake photos; not super-sharp, but good enough to capture the overall shape of stellar crystals, including some surface detail. Many people have been experimenting with this simplest form of snowflake photography, and a quick web search will yield numerous examples. This is a fine approach for getting started, just to see what kinds of crystals nature has to offer in your part of the world.

A next step up, if you already own some camera equipment, is to use a "reversed lens" at the end of an extension tube to make a relatively inexpensive macro lens. This technique is discussed in considerable detail on various photo blogs and websites, so again, a web search will provide much more information than I care to write down here. By my estimation, a reversed-lens macro system can achieve an optical resolution of perhaps 10–20 microns or even better if done with care using a high-quality lens. This resolution is sufficient to take some excellent snowflake photos, and no one has demonstrated this better than Alexey Kljatov, who has captured many stunning snow crystals using a reversed-lens system. With quality crystals, an artistic eye, and some patience and effort, this technique can yield outstanding photos without spending a lot of money on fancy optical gear.

Given the scientific nature of this book, my primary focus here is on achieving exceptionally high optical resolution, with the overarching goal of revealing the finest details in snow crystal structure, especially with smaller specimens. Obtaining resolutions of 2–5 microns is not an inexpensive undertaking, but the exceptional photos that result take one to a whole new level in snowflake photography. This kind of professional-grade hobby is clearly not for everyone, but it can reveal a delightful world of nature's frozen artistry.

To begin, it has been my experience that the choice of camera sensor is not especially important in snowflake photography. Many high-quality, reasonably priced camera bodies are on the market with sensors in the 20- to 40-megapixel range, and most would work well in this application. Sensors with larger physical dimensions tend to be better than smaller sensors, other things being equal, and the lens requirements are somewhat relaxed with a larger sensor as well; but this detail is probably not terribly important. As long as you have a reasonably modern digital camera, the imaging sensor will likely not be the limiting factor in obtaining quality photographs.

It is necessary to have at least two pixels for each real resolution element, so the optimal camera field-of-view is one that is matched to the optical resolution of the lens (described below). For example, if one wants to achieve a 2-μm optical resolution, the camera field-of-view should have about 1 μm/pixel. Put another way, a 2-micron feature on the object snow crystal should image onto two sensor pixels, regardless of the actual physical size of the sensor pixels. Oversampling the image beyond this just wastes camera real estate, while undersampling will compromise the optical resolution. A 20-megapixel sensor imaged to produce 1 μm/pixel will give about a 4×5 mm field of view, and already this is larger than most snow crystals. If the physical size of the sensor pixels is 5 μm, which is typical for many cameras, then the lens should provide 5× magnification.

This exercise shows that a 100-megapixel sensor would not yield substantially better snowflake photos than a 20-megapixel sensor. A larger field-of-view is of little use, as very few snowflakes would fill it, and more pixels per micron would not help either, because the optical resolution of the lens is what usually limits the photo, as described below. Occasionally one encounters a really huge crystal, and a field of view larger than 4×5 mm would be handy. But those situations are rare, and it is straightforward to just take two or more photos and stitch them together digitally in postprocessing. When you consider the full parameter space, the camera is usually not the limiting factor for achieving high-quality snowflake photos.

While the choice of camera is not so important, the choice of lens is quite critical, especially when the goal is to obtain the highest possible resolution. Because resolution is of central importance in this chapter, I will use resolution as a starting point in the discussion of lens options. The usual definition of optical resolution (a.k.a. resolving power) is about what you would expect—the distance between two pointlike objects that can just barely be resolved in an image. In practice, a resolution of 2 μm will allow you to clearly distinguish features that are separated by at least 4 μm. Features that are 2 μm apart would be "barely resolved," which usually means they are almost completely blurred together and so not easily distinguished. Of course, one can provide a proper mathematical definition of resolution, but this rule-of-thumb is adequate for the present discussion.

In the case of snow crystals, the smallest structural features are about 1 micron in size, and this is a real physical limit imposed by surface tension and the Gibbs-Thomson effect (see Chapter 2). Any significant surface structure (like a rib, ridge, or sharp edge) that is substantially smaller than 1 micron would have such a high vapor pressure that it would soon sublimate away unless under extreme environmental conditions. Thus, unlike with most solid objects, one does not observe ever more detail in snow crystal structure by observing with ever higher resolution. There just is not much to see beyond a resolution of about 1 μm. As a result, electron microscope images do not reveal much more structural details than do optical images, as discussed in Chapter 6.

In my personal experience photographing snow crystals, I have found that using a lens with 2-micron resolution yields noticeably better photos than a lens with 4-micron resolution; the edges are crisper, and the image has a sharper appearance overall. This can be seen quite easily in side-by-side comparisons of a single snow crystal. Put into typical photographic language, I would say that a 4-micron-resolution lens yields a noticeably "softer" snowflake image than a 2-micron-resolution lens. It may sound counterintuitive, but using a 1-micron-resolution lens generally yields lower-quality images than a 2-micron-resolution lens, at least when looking at snowflakes. Depth of focus is the main reason for this (discussed below), but another reason is simply that there are few additional structures to be seen in snow crystals at super-high resolution. The takeaway message is that one gains little by going beyond using a 2-micron-resolution lens when photographing snow crystals. That has been my experience, anyway.

Macro and Micro Lenses

In terms of overall resolution, snowflake photography falls roughly between the usual regimes of macro photography and full-blown microscopy. Macro lenses tend to yield resolutions in the 5- to 20-micron range, and a few exceptional lenses can do a bit better. Unfortunately, the optical resolution of most macro lenses is not listed in their specification sheets, even though high resolution is pretty much the main reason one purchases a macro lens. I have limited experience with the broad range of macro lenses that are available, but my experience has been that a 5-micron-resolution lens is quite good, and photography reviewers will speak of its bitingly sharp images. And it will cost a lot. A 10-micron lens will likely cost less, but it will be reviewed as only okay, yielding somewhat "softer" images in high-resolution tests and lacking in their finer details.

For example, I have done some testing with the Canon MP-E 65 mm lens, which is something of a high-resolution macro stalwart, well reviewed by many macro photographers. When set to its highest-resolution setting (5×), I measured an overall resolution of about 4 microns using this lens (discussed below). I imagine others have made similar measurements, but I have not found much resolution data online, either from the manufacturer or from lens reviewers. The world of macro photography is often not a very quantitative place, which can make it difficult to know what you are buying. It is not always obvious what to make of adjectives like "soft" or "bitingly sharp."

In contrast, microscope objectives invariably list resolving power (resolution), numerical aperture, working distance, and depth of focus as part of their specs. These numbers can still be deceptive, as they refer only to on-axis viewing, and inexpensive microscope objectives can have dreadful off-axis optical quality even with supposedly good specs. However, for most reputable manufacturers (such as Zeiss, Olympus, and Mitutoyo), the specifications provide a reliable assessment of the quality of their objectives. In this respect, it is generally easier to purchase a quality microscope objective of known performance than a high-resolution macro lens, and microscope objectives are usually somewhat cheaper as well.

While most people equate microscope objectives with full-blown (and expensive) microscopes, Figure 11.10 shows how a simple microscope objective can be turned into a DIY photomicroscope. This configuration is identical to the usual reversed-lens setup, just replacing the reversed lens with a higher quality microscope objective. The biggest drawback with this layout is scattered light, which sends unwanted light onto the camera sensor. Fortunately, this problem can often be ameliorated by carefully covering the inside of the extension tube with highly absorbing black flocking paper. A field stop in the object plane is also useful, as this prevents otherwise unused light from entering the objective and rattling around inside the extension tube.

The simple optics and fixed extension tube in Figure 11.10 means that focusing involves either moving the camera or moving the subject, usually with some kind of mechanical translation stage. This focusing method is the norm for both microscope objectives and high-resolution macro lenses. Moving lens elements in a lens (using a focusing ring) is generally not practical at high resolution, nor is in-lens autofocus. Personally, I tend to favor microscope objectives over macro lenses for several reasons:

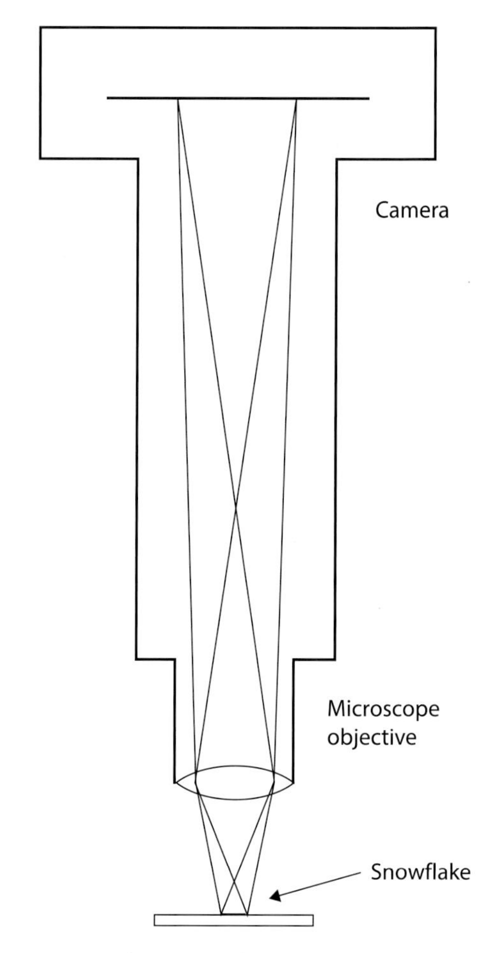

FIGURE 11.10. A basic DIY photomicroscope consists of little more than a microscope objective, an extension tube, and a camera body. Viewing is done through the camera, for example, displaying the image on a TV monitor via the live-feed camera output. The field of view of the camera can be set by choosing an appropriate length for the extension tube, which is the same as for reversed-lens setups.

1) A larger selection of microscope objectives is available, at generally higher quality (in my opinion).
2) Microscope objectives have clear specifications that include their optical resolution, unlike macro lenses.
3) Microscope objectives are typically somewhat cheaper than macro lenses for a given resolution.
4) Microscope objectives are far more compact than macro lenses, making them much easier to incorporate into snow crystal growth chambers.
5) Microscope objectives are easily adaptable for use with different camera bodies.

The Diffraction Limit

In the microscopy world, optics are nearly always diffraction limited, meaning that the wavelength of light is ultimately what limits the image resolution.

This is not true with normal photography, but the diffraction limit will play a role in macro photography at the highest resolutions. As a rule, if the overall image resolution is smaller than about 10λ, where $\lambda \approx 0.5$ µm is the wavelength of visible light, then the diffraction limit will be an important consideration. Because I am mainly concerned with high-resolution imaging in this book, I assume that diffraction is one of the main factors limiting the overall optical resolution.

Because a 2-µm optical resolution is substantially larger than λ, the diffraction limit takes on a relatively simple mathematical form. Using the terms defined in Figure 11.11, and assuming an index-of-refraction of unity for imaging in air, we can assume a small-angle approximation with $\sin\theta \approx \tan\theta \approx \theta$, where θ is measured in radians. For $\theta = 0.15$, $\sin\theta = 0.149$ and $\tan\theta = 0.151$, so this is an excellent assumption.

Microscope objectives are typically specified by a *numerical aperture*, N_A, and in our small-angle approximation, this is given by

$$N_A \approx \theta. \tag{11.1}$$

In photography, a lens is specified by its *f-number*, $f_\#$, which can be adjusted by changing the aperture of

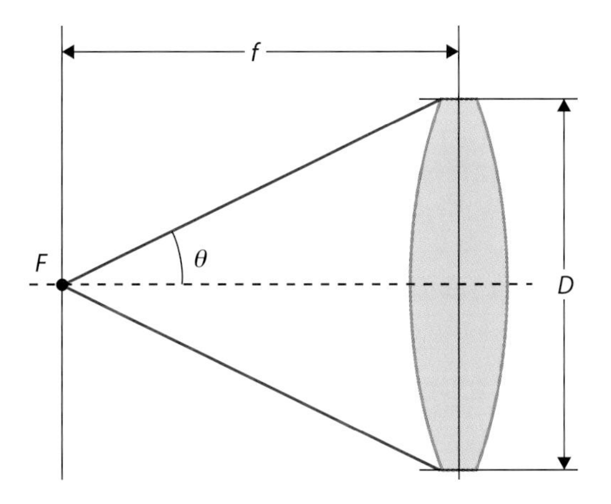

FIGURE 11.11. This sketch defines the focal length *f*, the lens aperture *D*, and the half-angle θ for a simple microscope objective.

the lens, and it is given by (in the small-angle approximation)

$$f_\# \approx \frac{1}{\theta} \approx \frac{1}{N_A}. \tag{11.12}$$

Photographers and microscopists tend to use different nomenclatures, but the underlying optical physics is the same. In the diffraction limit, the optical resolution is given by

$$R_{xy} \approx \frac{\lambda}{2N_A}, \tag{11.3}$$

and I will typically assume $\lambda \approx 0.5$ µm. Additionally, there is a corresponding resolution perpendicular to the image plane, R_z, which is given by

$$R_z \approx \frac{\lambda}{2N_A^2}. \tag{11.4}$$

This is usually called the *depth of focus* or depth of field; parts of the object that are within $\pm R_z$ of the focus position will be essentially in focus, while parts outside this range will be considerably out of focus.

For example, if we want an optical resolution of $R_{xy} = 2$ µm, then we need a microscope objective with a numerical aperture of at least $N_A \approx 0.125$, and this means that the depth of focus will be a scant $R_z \approx 16$ µm. This latter number can be problematic, because most snow crystals are thicker than 16 µm, even platelike crystals. This problem is inescapable in snowflake photography—one cannot have both high resolution and a large depth of focus simultaneously. In normal photography, one closes down the aperture to increase the depth of focus, but that no longer works when the resolution is diffraction limited.

As another example, the Canon MP-E 65 mm macro lens has an f/2.8 aperture, and the above equations give a corresponding numerical aperture of $N_A = 0.36$ with a theoretical resolution of $R_{xy} \approx 0.7$ µm, which would be awesome. However, this lens is not diffraction limited when used at its maximum resolution, and the measured resolution (see Table 11.1) is about 4 µm. Most traditional camera lenses are not diffraction limited, even high-quality macro lenses. Any good microscope objective will be diffraction limited, or close to it, so the above equations are quite useful for evaluating the performance of quality objectives. Inexpensive objectives may have serious deficiencies, but quality objectives should always meet spec, at least on axis. With camera lenses, however, it is often not possible to know the optical resolution unless you measure it yourself.

One straightforward way to estimate the optical resolution of a lens is to image a calibrated resolution target, as demonstrated in Figure 11.12 for several example lenses. In all cases, the camera sensor was not a limiting factor in determining the quality of the images. Although making an absolute measurement of R_{xy} is difficult, comparisons between lenses are straightforward. By my reckoning, these images reveal that the 5X Mitutoyo objective seems to meet its spec of having a 2-μm resolution, and from this, I obtained the measurements shown in Table 11.1.

The microscope objectives mostly met spec, except for the Compact 3X, whose specified $R_{xy} = 2.5$ μm simply does not agree with the measured resolution of about 4 μm. However, the specified numerical aperture (0.07) gives a theoretical resolution of $R_{xy} \approx 3.6$ μm, which agrees reasonably well with the measurement. As far as I can tell, this is a specification error by Mitutoyo, which is unusual for this company. The Canon lens has no resolution specification, and the 4-μm number is the best I could get using this lens. The Canon resolution-target images at 5X with f/2.8 and f/4.0 were similar, and the resolution rapidly deteriorated at lower magnification or higher f-numbers, as one would expect.

Another useful method for examining resolution is to sprinkle 10-μm beads onto a glass substrate, as shown in Figure 11.13. A quality microscope objective will produce a good focus across a wide field of view, whereas cheaper lenses often focus well only near the center of the field. Even in this example, the Mitutoyo 5X produced some noticeable variation in focus across the field, but I corrected this with focus stacking (see the next section).

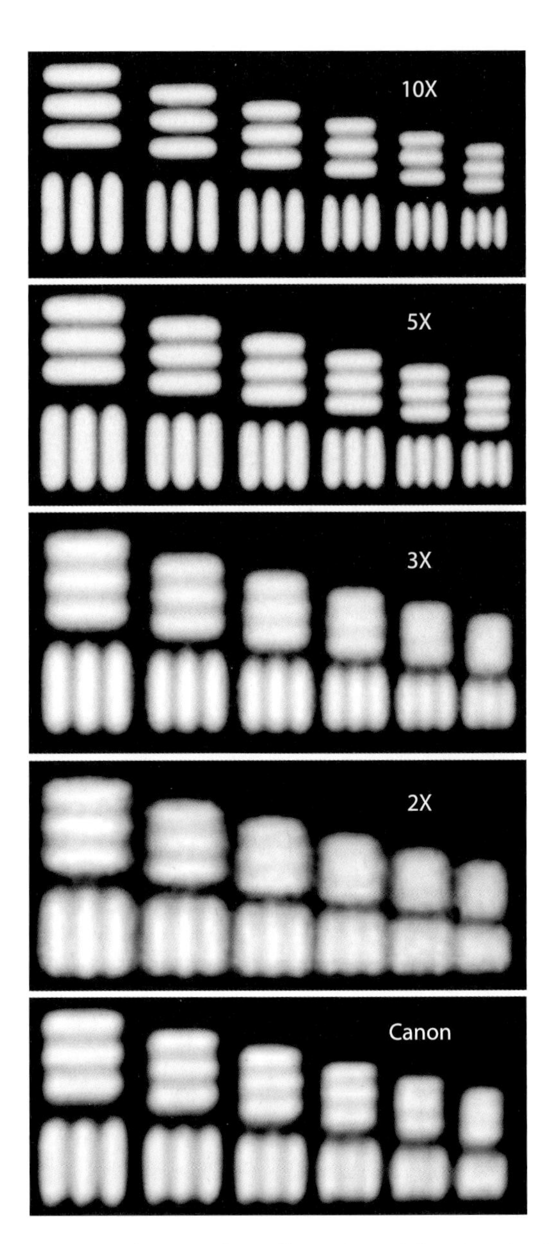

FIGURE 11.12. Images of a calibrated resolution target using three Mitutoyo Plan APO Objectives (10X, 5X, and 2X), a Mitutoyo 3X Compact Objective (3X), and the Canon MP-E 65mm 1-5X Macro Lens set at 5X/f2.8 (f4.0 is similar). The Mitutoyo 2X Compact Objective (not shown here) yields a resolution-target image that is quite similar to the 3X objective. The spacing between the bars is 7.8, 7.0, 6.2, 5.6, 4.9, and 4.4 μm.

TABLE 11.1

Measurements and specifications of several microscope objectives and the Canon MP-E macro lens

Lens/Objective	NA spec	Rxy (μm) spec	Rz (μm) spec	Rxy (μm) meas	Rz (μm) calc	Rxy (μm) calc	Working distance (mm)	Price (US$)
Mitutoyo Plan APO 10X	0.28	1	3.5	1.5	3.2	0.9	34	880
Mitutoyo Plan APO 5X	0.14	2	14	2	13	1.8	34	700
Mitutoyo Plan APO 2X	0.055	5	91	5	83	4.5	34	930
Mitutoyo Compact 3X	0.07	(2.5)	(23)	4	51	3.6	78	500
Mitutoyo Compact 2X	0.06	4.6	76	4.5	70	4.2	92	460
Canon MP-E 65mm @5X			50	4			42	950

Note: For the Compact 3X objective, my measurements suggest that the manufacturer's online specified numerical aperture is accurate, but the specified resolution (shown in parentheses) is incorrect.

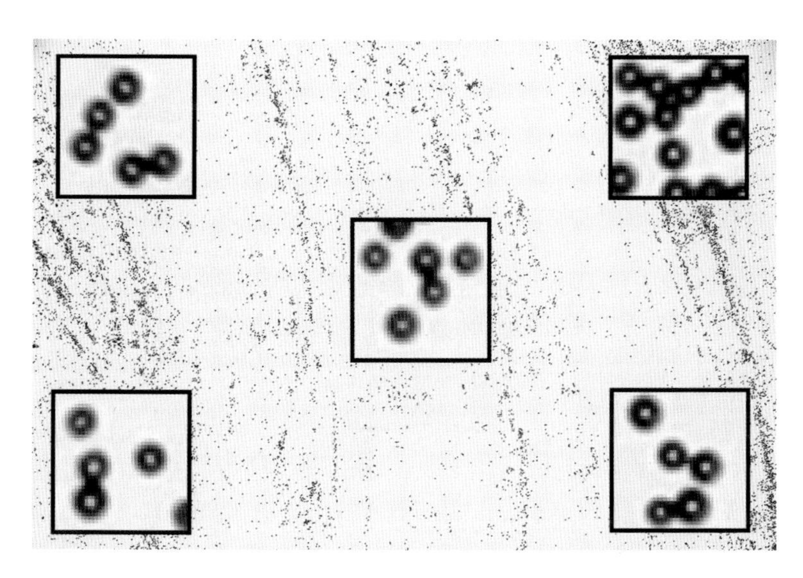

FIGURE 11.13. This photograph shows a glass substrate sprinkled with 10-μm glass beads, imaged with a bare Mitutoyo Plan APO 5X objective. The total field of view is about 5 × 4 mm, and the inset images show magnified snippets of the main image taken from the center and the four corners. Adding a secondary lens to compensate for the infinity-corrected objective gave better off-axis focus, and combining several images by using focus stacking yielded a sharp focus across the entire image.

Every objective has its foibles, so care is needed to deliver the highest resolution. Imaging an array of small beads makes it easy to evaluate the focus and resolution across an entire field of view, which is useful for optimizing an optical system.

Focus Stacking

Snowflake photography always involves a trade-off between resolution and depth of focus. When the resolution is high, the depth of focus is low, so only a thin plane is brought into sharp focus on the sensor. If the snow crystal is tilted with respect to that plane, or if the crystal is not thin and flat, then not all parts of the crystal can be brought into focus at the same time. This is a fundamental feature of diffusion-limited optics, so there is no optical means to avoid this trade-off.

However, focus stacking is an effective workaround that allows one to photograph complex snow crystals at high resolution over their entire structure. The basic idea is to take several pictures at different focus settings, with each photo bringing a different part of the crystal into focus. The images can then be combined digitally in postprocessing to stitch together the in-focus pieces of each of the individual photos, thus creating a single image that appears to be in focus throughout. Several software packages are available to do the image reconstruction (for example, Helicon Focus), and much information about focus stacking can be found online. There are even hardware systems (such as from StackShot) that will automatically move the camera focus in programmable steps using a translation stage to acquire the desired series of images.

Nearly all serious snowflake photographers use focus stacking to some extent, as this is a straightforward technique for effectively increasing the depth of focus while maintaining a high optical resolution. Large stellar snow crystals are intrinsically thin and flat, so photographing these crystals face-on usually requires minimal focus stacking even at high resolution. It is hard to avoid some tilt of the crystal relative to the image plane, however, so I often take two or three pictures while adjusting the focus to make sure all the branch tips are nicely in focus. This kind of minimal focus stacking is easy to apply and nearly always yields good results. Moreover, it is often desirable to tilt a flat crystal over quite large angles to obtain specular light reflections (see the section on Specular Reflection illumination later in this chapter), and in this case, a great deal of focus stacking is needed at high resolution. Don Komarechka is the undisputed focus-stacking champion in snow crystal photography, often combining 30–50 individual shots to obtain a single in-focus image, as I describe below [2013Kom].

Like any photographic tool, focus stacking can be employed or not, depending on what is being photographed and what kinds of optical effects are desired. If one opts for a lower overall optical resolution, then perhaps a single image is sufficient. Moreover, having the extremities of a crystal appearing slightly out of focus often gives an image a pleasing sense of depth, and this type of optical illusion is often used by photographers (a version of *bokeh*). But if super-high resolution is desired over an entire crystal, then some focus stacking (and perhaps a lot of it) is usually required. Focus stacking is a nice trick that is both easy and inexpensive to use, so it has become a valuable addition to any snow crystal photographer's toolkit.

Point-and-Shoot versus Stable Mounting

In most circumstances, photomicroscopy is not performed in a point-and-shoot fashion using hand-held optics. Microscopes tend to be rigid structures where the camera, the optics, and the object being viewed are all solidly mounted. The reason is that photomicroscopy subjects are so tiny that it is nearly impossible to hold everything steady enough by hand to get good pictures. In contrast, macro photography is often a point-and-shoot affair, as a hand-held camera plus lens gives the photographer plenty of freedom to move around an object to get just the right angle for an artistic shot. Snowflake photography is somewhere between these two, as a resolution of 5–10 microns is quite low by microscopy standards but is quite high for macro photography. Given this intermediate position, some snowflake photographers use the point-and-shoot method, while others go with rigidly mounted hardware. Both can be made to work, but there are trade-offs for each.

For low-resolution imaging at 10–20 microns, point-and-shoot is relatively easy, inexpensive, and effective. Not as easy as normal photography, but doable. You let the snowflakes fall where they may, and then simply photograph them as you would anything else. It requires a steady hand, because the crystals are small, and a bright flash is useful to freeze any remaining camera motion. The point-and-shoot method is especially convenient as there is no additional investment in mounting hardware. Even at relatively low resolutions, focusing is not accomplished by rotating a lens ring, but simply by moving the camera+lens in and out.

For high-resolution imaging, point-and-shoot becomes substantially more challenging, and the savings gained by not having to buy a lot of mounting hardware tends to be lost in the need for high-performance camera equipment and a great deal of image processing. A bright flash becomes an absolute necessary at high resolution, as it is practically impossible to hold a camera steady enough by hand unless you use a superfast shutter speed. In addition, you probably want to take a lot of photos quickly, because the camera is moving around somewhat and changing the focus, which means that both your camera and flash need to be capable of taking several pictures per second for best results. Don Komarechka has described his point-and-shoot methods [2013Kom], and they involve some high-end camera gear.

Because I like to achieve the highest possible resolution with relatively simple gear, I prefer to use a stable mounting platform, essentially like a traditional microscope. There are many options for mounting the hardware, but a tripod is one of the worst, as tripods are generally too unwieldy and unstable for microscopy. If money is no object, then a commercial microscope with a camera attachment would work fine, provided it has a field of view up to 3–5 mm to fit large snow crystals. On a tighter budget, it is possible to build a quality DIY mounting system. Focusing is typically the most difficult and expensive part of a rigid mounting system, as focusing requires a mount that is both stable and movable. Linear positioning stages and focusing rings are reasonable options, and one can move either the lens, the subject, or even the camera. Motorized stages are available as well, such as from StackShot, and these are designed specifically for focus stacking.

One of the biggest advantages of a rigid mounting system is ease of use. You drop a snowflake onto a glass slide (for example), place it under the microscope objective, and there it sits, solid as a rock. You can move it around in the field of view, adjust the focus, adjust the lighting, and take the shot when everything looks right. No flash is needed, and slow shutter speeds are not a problem. Focus stacking is easy as well, because the crystal does not move laterally when you tweak the focus, at least not if you use a quality linear positioning stage. I clearly prefer the up-front costs of stable mounting hardware over the constant trials and tribulations of point-and-shoot photography out in the cold, but that is a matter of personal taste. Overall, rigid mounting tends to win over point-and-shoot at the highest resolutions.

ILLUMINATION MATTERS

One thing that separates snow crystal photography from other types of photography is that ice is transparent, like glass. With opaque materials, one can simply shine some light on an object and expect to get a reasonably good picture. Of course, lighting is important for taking excellent photos in any situation, but the type of illumination one uses is particularly critical in producing high-quality snow crystal photographs.

To understand why the type of illumination matters so much when photographing snow crystals, it is necessary to examine how transparent objects scatter and refract light. For example, a bank of snow looks white even though the ice crystals it contains are all quite transparent. When you look at an individual sliver of ice up close, it looks clear, like glass. But a large pile of clear objects always looks white. When light shines onto the surface of a snowbank, some light reflects off every air/ice interface. Only a small fraction of the light is reflected from each surface, while the rest is transmitted, and very little light is absorbed in the process. After encountering thousands of air/ice surfaces in the snow, the light is mostly scattered this way and that, until it makes its way back out of the snowbank. The net result is that light striking the snow is scattered in all directions with little absorption, which is exactly what being "white" means. Any pile of transparent grains appears white, as shown in Figure 11.14. Paper is white for the same reason, because it is made from layers of tiny, transparent cellulose fibers. The whitest paper comes from using extremely transparent fibers. Similarly, a snowbank is bright white because its constituent ice particles are especially clear. This answers that amusing question: when a snowbank melts, where does all the white go?

A pane of glass looks transparent because it contains only two smooth, planar surfaces. When you look at the

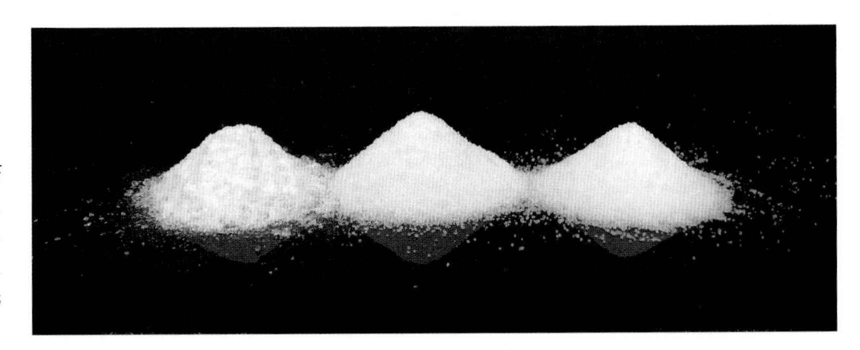

FIGURE 11.14. This photo shows piles of crushed glass (left), sugar crystals (center), and salt grains (right). In all cases, the individual particles are essentially transparent. The piles look white, because light scatters repeatedly off the countless small surfaces.

pane from certain angles, you can see the light reflecting like a weak mirror. But otherwise, light transmits through the glass and it looks transparent. If you scratch up the surfaces with sandpaper, however, then the light scatters in all directions, and the surface takes on a whitish appearance. It is not bright white because some light incident on the front surface of the scratched pane makes it out the back surface. With a pile of ground glass, the light mostly keeps scattering around until it comes back out the front; the probability that the light makes it through to the back of the pile is extremely low.

In addition to reflecting light from its surfaces, a transparent object also bends light via refraction. This is how lenses work, and a lens-shaped piece of clear ice would behave similarly. The fact that ice is clear—transmitting, reflecting, scattering, and refracting light—brings an added dimension of lighting effects to snow crystal photography. Even interference effects can be important in some circumstances.

In many ways, a snowflake can be thought of as a complex lens that refracts light through various angles as it is transmitted through the clear ice. In other circumstances, the snowflake can be thought of as a small sliver of scratched-up glass that scatters light from its highly structured surface. And a heavily rimed snowflake begins to look like a small pile of crushed glass, because its surface is covered with a dense layer of frozen droplets.

One aspect of light transmission that is completely negligible in snowflakes is color dispersion. A beam of light transmitted through a glass prism will be dispersed into a rainbow of colors, and this happens with ice as well. But it would have to be an exceptionally large block of ice for this to be even remotely noticeable. One does not normally observe color dispersion from glass bowls, pitchers, cups, or other glass objects. Likewise, color dispersion in tiny snow crystals is completely negligible.

Perhaps the easiest way to understand the different ways illumination affects snowflake photography is by example. People have been experimenting with different types of lighting for many years, and it is straightforward to categorize different photographs by the type of lighting used. The sections that follow focus on these lighting categories.

Side Illumination

What I am calling "side illumination" could also be called "ambient light illumination." The basic idea is illustrated in Figure 11.15. This is essentially the type of lighting you get when you simply photograph a snowflake resting on an opaque surface with a point-and-shoot camera. Light shines down on the crystal from all around, and some of that light enters the camera lens and is focused onto the sensor.

If you supply your own lighting, then an infinite number of variations of the side lighting method are possible. For example, you might shine a bright light in from one side only. Or you might shine blue light in from one direction and red light in from another. The creative possibilities for using colored light in side illumination

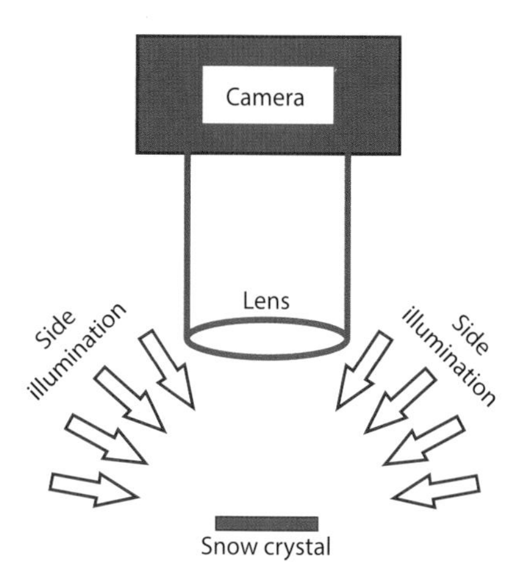

FIGURE 11.15. Side illumination. In this straightforward method, light shines in from the side to illuminate the snow crystal, and scattered light enters the lens and is focused onto the camera sensor.

have received little attention, and I believe there are some interesting opportunities for branching out into new photographic directions here.

One aspect of side illumination that I want to focus on is that only scattered light can contribute to a snowflake photograph taken using this method. Refraction is irrelevant, because the transmitted light strikes the opaque substrate and is absorbed. As a result, side illumination tends to accentuate the crystal edges and surface structures.

Another important aspect of side illumination is that a flat ice plate that lies perpendicular to the viewing angle will not reflect any light directly into the camera lens. When a flat plate rests perpendicular to the camera, as shown in Figure 11.15, any direct reflection of the side illumination will not enter the camera lens. Because no light shines down on the crystal from above, no directly reflected light from the plate surface can enter the camera. Thus, flat plates appear somewhat invisible, like a pane of glass, when using side illumination.

Alexey Kljatov is a master of side-illumination snowflake photography, and several examples from Alexey illustrate many features of this illumination method. Several shared features can be seen in many of Alexey's side-illumination photos, including:

1) Thin, platelike regions scatter little light, making them appear almost invisible in some side-illuminated photos (Figures 11.16 and 11.17) This gives the accurate impression that snow crystals are clear, which is good, although many people expect snow crystals to be white.
2) Crystal edges are generally quite bright, as they strongly scatter light into the camera. Surfaces with

FIGURE 11.16. With side illumination, the flat, platelike parts of a crystal scatter almost no light into the camera lens, so they appear invisible, like small panes of glass. Edges tend to scatter a lot of light, however, so they look bright white. This crystal is supported by a few fibers from the underlying piece of cloth. Photo by Alexey Kljatov.

a lot of structural detail (Figure 11.18) and thicker crystals like capped columns (Figure 11.19) tend to appear quite white also.

3) Rimed structures appear bright white, like a pile of crushed glass.

FIGURE 11.17. Holes in snowflakes? This snow crystal looks like it has a central hole, but holes do not spontaneously form from small seed crystals. In fact, there is a thin ice sheet at the center, so clear that it looks just like a hole. Photo by Alexey Kljatov.

Although I have long been enamored with Alexey's beautiful photos, I have found that taking high-quality snow crystal photos with side illumination is a difficult skill to master, and it does not work well with all types of snow crystals. The bright edges of thin, platelike crystals are easy to oversaturate, giving the photo a garish look. Thicker crystals, however, like capped columns, provide more varied light scattering and often photograph well using side illumination.

Specular Reflection

Figure 11.20 illustrates another variation of side illumination in which a platelike snow crystal is tilted slightly compared with the face-on arrangement shown in Figure 11.15. Because of this tilt, flat basal surfaces will produce a mirror-like (specular) reflection of some of the side-illumination light into the camera lens. When this reflection is especially strong, the face of the crystal has a much brighter appearance compared to the side-illumination examples described above. This added specular reflection solves, or greatly reduces, the main problem associated with using side-illumination for

FIGURE 11.18. In this wonderfully creative photo, Alexey dropped a bit of snow onto a plastic surface and then used the snow to balance a stellar crystal on its edge. Note the beautiful combination of white-light side illumination with colored back illumination. Photo by Alexey Kljatov.

FIGURE 11.19. Side illumination yields especially pleasing photos of thick crystals like this capped column. The body of the crystal scatters a good amount of light, and the out-of-focus regions give the picture an overall sense of depth. Photo by Alexey Kljatov.

platelike crystals, namely, that bright edges dominate the photo while flat plates are nearly invisible.

In Figure 11.21, for example, the main body of this flat stellar crystal is quite bright, so the image is not dom-

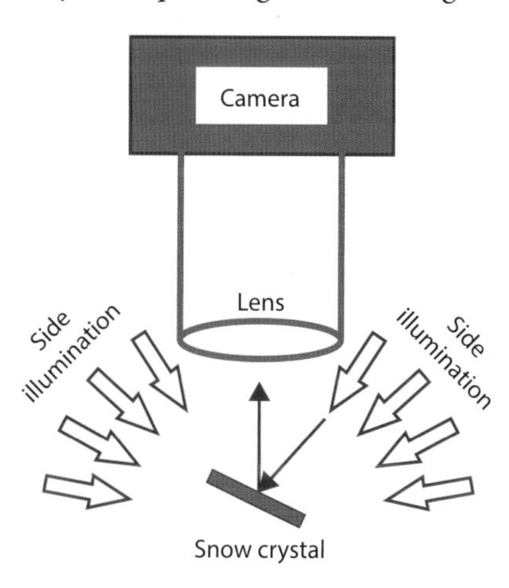

FIGURE 11.20. Specular reflection. In this variation of side illumination, a platelike snow crystal is tilted so that the flat face of the crystal reflects light into the camera lens. The direct reflection makes thin plates appear much brighter than for the face-on version of side illumination.

inated by high-contrast edges. Moreover, this gives the crystal an overall white appearance, which satisfies many viewers' desire that snowflakes ought to look white. What is somewhat lost in the process, however, is the glassy look that gives one the (correct) impression that a snow crystal is not intrinsically white but is rather made from a sliver of transparent ice. Reducing the intensity of the specular reflection can alter this effect, and the right balance is in the eye of the beholder. Specular-reflection illumination is especially popular for point-and-shoot snowflake photographers working at moderate (10–20 micron) resolution. Finer structural details are diminished at this resolution, but there are endless opportunities for artistically placed crystals on colorful, textured backgrounds.

Perhaps the biggest drawback associated with specular-reflection illumination is that a tilted crystal does not lie in the image plane of the camera. From Equation 11.4, we see that the depth of focus is inversely proportional to the square of the resolution, so this issue is not so bad with a low-resolution lens. Perhaps only the

FIGURE 11.21. Specular reflection gives this platelike snow crystal a bright-white appearance. Although the ice plate is still quite transparent, it reflects the lamp much like a pane of flat glass. A red cloth supports the delicate crystal and provides an excellent textured background. Photo by Katy Turk.

FIGURE 11.22. Specular reflection works especially well at modest optical resolution, as this avoids depth-of-focus issues at high resolution. This illumination method is well suited to point-and-shoot snowflake photograph, affording many possibilities for artistic compositions. Photo by Delena-Jane Lane.

edges of a platelike crystal will be out of focus, and this can even be beneficial, as it gives the image a sense of depth (Figure 11.22). But things get worse quickly as one goes to higher-resolution lenses. At exceptionally high resolution (<5 microns), just a small portion of a tilted, platelike crystal can be in focus in a single photo, as illustrated in Figure 11.23.

Don Komarechka has been a pioneering proponent of high-resolution snow crystal photography using specular-reflection illumination, managing the depth-of-focus problem by using an unprecedented amount of focus stacking. Don typically captures 200–300 images in quick succession using a hand-held camera with a high-speed ring flash, and he then selects the best 30–50

FIGURE 11.23. Don Komarechka took these images using the Canon MP-E 65 mm macro lens at 5X, giving an optical resolution of $R_{xy} \approx 4$ μm and a depth-of-focus of $R_z \approx 50$ μm (see Table 11.1). With these parameters and specular-reflection illumination, only a small sliver of the tilted snow crystal is in focus in a single shot (left image). Don manages this problem by focus stacking 30–50 images into a single composite photo (right image) [2013Kom].

to combine into a single composite image in postprocessing [2013Kom].

Don's technique is perhaps the most demanding in terms of hardware and software capabilities, computing power, and overall effort expended. Shooting hundreds of images in burst mode at six frames per second or faster requires a high-end camera, a rapid-refresh ring flash, and a high-resolution lens, none of which is cheap. Moreover, a high-speed computer running first-rate software is needed when focus stacking so many images, and Don estimates that he spends up to 4 hours processing a single composite image [2013Kom]. This amount of expense and effort is not for everyone, but Don has captured many world-class snow crystal photographs using this innovative technique, as illustrated in Figure 11.24.

While experimenting with specular-reflection illumination, Don discovered the colorful appearance of internal bubbles and thin hollows in platelike snow crystals, as illustrated in Figure 11.25. The colors arise from light-interference effects when specular reflections from the top and bottom faceted surfaces of a hollow region interfere with one another, which is analogous to similar interference effects seen in soap films or oil films on water. Figure 11.26 shows a calculation of the interference color as a function of the thickness of the hollow cavity in a snow crystal. The details of the calculation depend on several factors relating to the physics of optical interference, such as the spectrum of the incident light source and the RGB sensitivity of the camera sensor. These details notwithstanding, Don's images suggest that the most colorful hollow regions and voids are roughly 0.5–1.5 microns in thickness. The structure and stability of such remarkably thin hollows and bubbles is described briefly in Chapter 3, but their formation is not well understood at present.

Front Illumination

Figure 11.27 shows a different type of specular-reflection illumination that avoids having to tilt platelike snow crystals. With this geometry, the flake lies flat with respect to the image plane, so an entire thin-plate crystal

FIGURE 11.24. Don Komarechka employs extensive focus stacking to capture exquisite high-resolution images using specular-reflection illumination of tilted snow crystals. This small-sized reproduction does not do the photo justice, but many similar images can be viewed at high resolution online. Don clearly has a strong preference for the Bentley-esque style of a jet-black background, so he often digitally removes background cloth fibers that support the crystals, so they seem to float through the night [2013Kom].

FIGURE 11.25. Specular reflection sometimes yields intense colors caused by interference effects between closely spaced faceted surfaces defining hollow regions within thick-plate crystals. Photos by Don Komarechka.

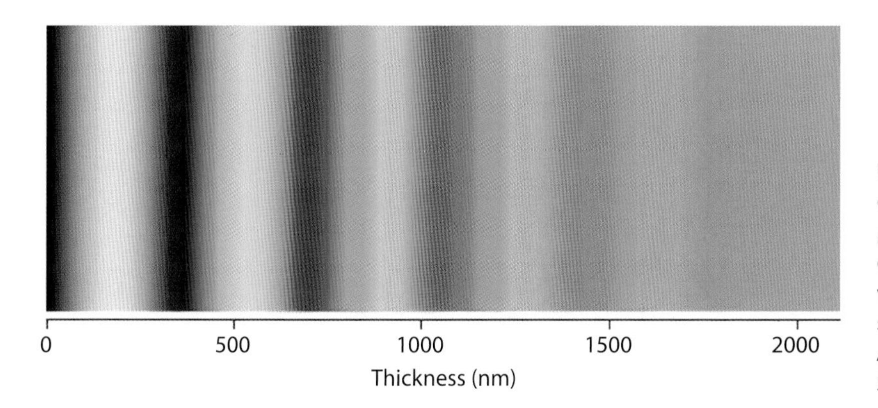

FIGURE 11.26. Estimated interference colors for white light reflecting off the two parallel surfaces of a hollow void in an ice crystal. The calculation will change somewhat depending on the camera color response and the spectrum of the light source. Adapted from an image by Björn Böttcher, Seifenblasenmann.de, with permission.

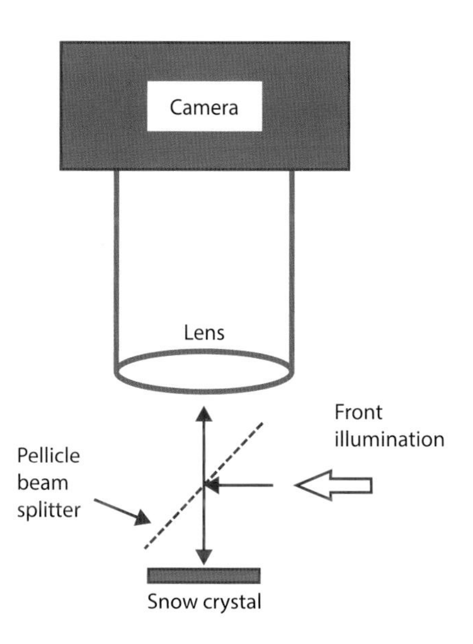

FIGURE 11.27. Front illumination. This largely unexplored variation of specular-reflection illumination uses a pellicle beamsplitter to direct light onto a platelike crystal. In contrast to Figure 11.20, the face-on crystal lies nicely in the image plane, avoiding depth-of-focus issues.

can be brought into focus in a single photo. This avoids the need for extensive focus stacking while still providing a direct specular reflection off the face of the crystal. From a scientific perspective, this technique has the additional advantage that a face-on view provides a more accurate depiction of the hexagonal geometry of platelike crystals. With a tilted crystal, the overall hexagonal shape in the image depends on viewing angle, so the measured angles between prism facets is no longer 120 degrees. A face-on view would make it possible to measure these and other angles accurately.

A pellicle beamsplitter is shown in Figure 11.27, because it is generally ill advised to image through a glass-plate or cube beamsplitter. Because the pellicle is only a few microns in thickness, it introduces minimal image distortion compared to other beamsplitter options. Front illumination like this would be difficult to achieve using a hand-held setup, so it is probably necessary to use a rigid mounting system for the various components. This method of front illumination appears to have some significant advantages over tilted-crystal specular-reflection illumination, and it should be straightforward to implement. However, this technique appears to be essentially untried in snow crystal photography, so I have no example images to show here.

Back Illumination

Back illumination is another straightforward and adaptable method for photographing snow crystals. As illustrated in Figure 11.28, a specimen is placed on a transparent substrate and photographed using light that is transmitted through the clear ice. Unlike the previously discussed illumination methods, refraction of the light by the ice now plays a major role in defining the overall appearance of the image, while specular reflection is relatively unimportant.

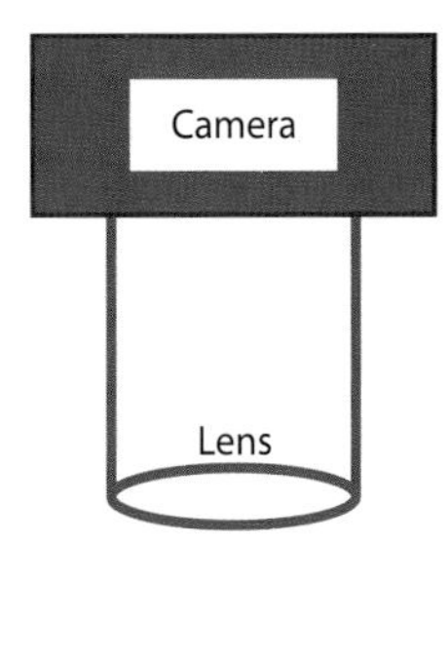

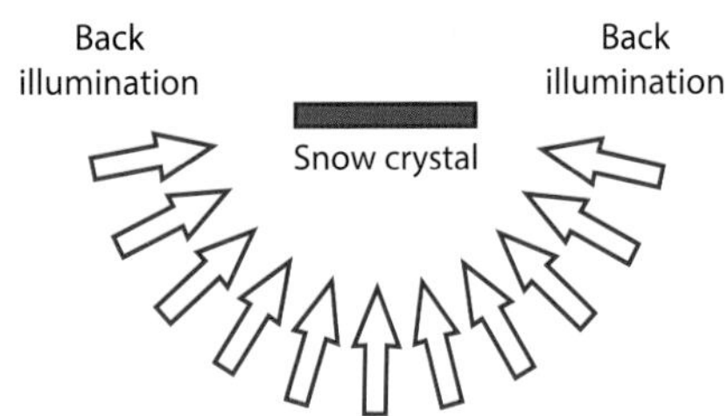

FIGURE 11.28. Back illumination. With this method, a snow crystal is photographed using light transmitted through the clear ice. The method requires placing the crystal on a transparent substrate, typically a glass microscope slide, and it is best implemented using a rigid mounting system. Many illumination effects can be achieved by modifying the direction, intensity, and color of the illumination.

In the absence of a snow crystal, quite a lot of light passes through the substrate and strikes the camera sensor, yielding a featureless bright background, as illustrated in Figure 11.29. The background color is determined simply by the color of the illumination source. Refraction by the edges of the crystal deflects light that would otherwise have struck the camera, so the edges are typically quite dark, while clear ice plates transmit the background without reducing its brightness. Overall, the crystal acts like a complex ice lens that can refract the light in many interesting ways.

Front illumination tends to reveal the interior snow crystal structure quite clearly, providing a detailed look at surface patterns and other morphological characteristics. The edges are not oversaturated (as they can be with side illumination), and the broad illumination fills the lens to provide the highest possible image resolution. A downside, however, is that this technique yields a dark snowflake on a bright background, which does not always invoke a feeling of an icy crystal.

Dark-Field Illumination

Figure 11.30 shows a special variation of back illumination that can be used to good effect in snow crystal photography. Light rays from the side proceed undeflected and do not enter the camera lens if no snow crystal is present, giving a completely black background. But a snow crystal placed in the field will deflect some of the rays into the lens, thereby yielding a bright snow crystal, as

FIGURE 11.29. With back illumination, refraction by the snow crystal deflects light that would have struck the camera sensor, so the snow crystal is always darker than the background.

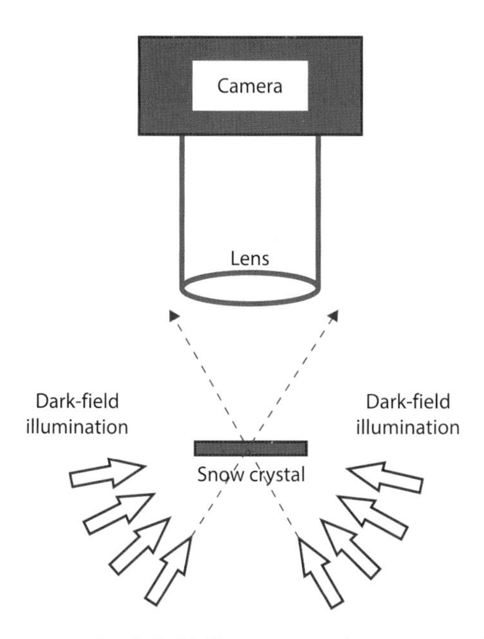

FIGURE 11.30. Dark-field illumination. This variation of back illumination uses only light that does not enter the lens in the absence of a snow crystal. If not deflected, the light rays shown in the sketch all pass outside the lens. A snow crystal will refract some of those rays, however, sending them into the lens and resulting in a bright snow crystal on a dark background.

illustrated in Figure 11.31. The overall look is much like that obtained using side illumination on a featureless black background, and the problems are similar as well. In particular, the crystal edges often appear bright to the point of being oversaturated, while thin platelike regions contribute little to the photograph.

Rheinberg Illumination

Rheinberg illumination is another variation of back illumination that adds a highly flexible method for achieving a variety of colorful effects on a uniform background. The overall optical layout is illustrated in Figure 11.32, and the key new addition is a *field lens* (so named because it is near the image field) that images a color filter onto the optical pupil inside the microscope objective. Adding this additional lens complicates the optics, and it requires a rigidly mounted optical system, but it provides the ability to rapidly experiment with all different types of back illumination simply by using different color filters. This capability is especially useful because different types of snow crystals tend to photograph better under different types of lighting. With Rheinberg illumination, switching the lighting around can be done simply by swapping in different color filters.

Figure 11.33 shows a variety of different color filters that I have used with my Rheinberg illumination setup. These were made by photographing my computer screen using 2 × 2 color slide film, as the mounted slides are a

FIGURE 11.31. Dark-field illumination gives results that are like side illumination in the absence of a background surface. In both cases, the lighting accentuates the edges of the crystal while thin plates are essentially invisible.

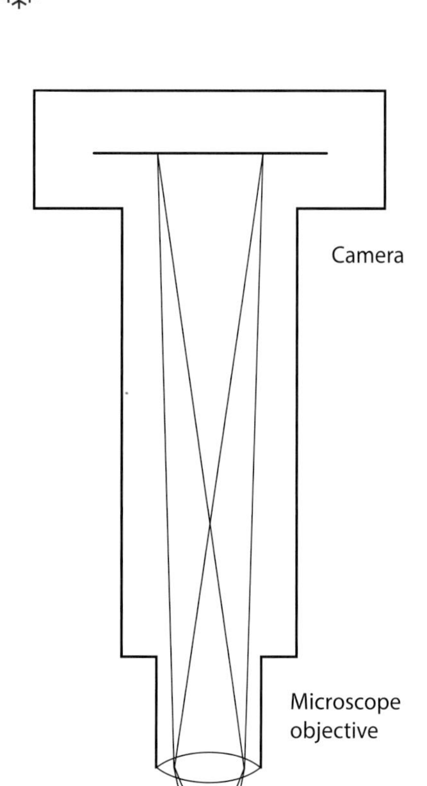

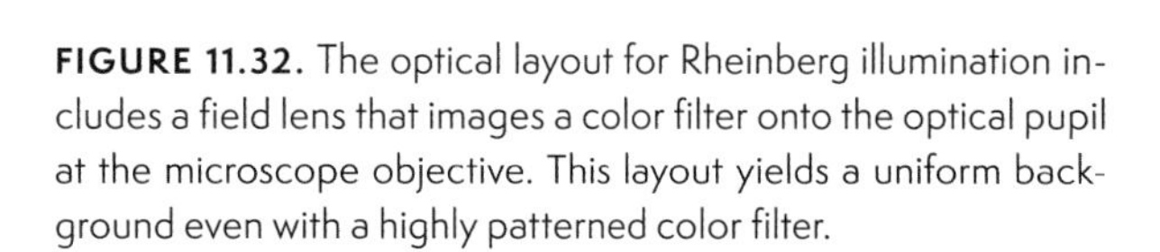

FIGURE 11.32. The optical layout for Rheinberg illumination includes a field lens that images a color filter onto the optical pupil at the microscope objective. This layout yields a uniform background even with a highly patterned color filter.

convenient size for my Rheinberg setup. Filters could also be made from any colored plastic sheet. Looking at the fourth filter in this set, following the rays in Figure 11.32 shows how the red side of the filter produces a set of nearly parallel red light rays impinging onto the snow

FIGURE 11.33. Nine different color filters I have found useful with Rheinberg illumination, yielding a variety of colorful effects.

crystal from the left, while the blue side of the filter produces rays impinging from the right. Both sets of rays fill the entire image plane uniformly, and these rays combine to yield an overall red+blue background color. The result is angle-dependent back illumination, uniform across the image plane, with the colors of the different rays determined by the pattern on the filter. By imaging the filter onto the optical pupil, the overall background color is uniform regardless of the pattern in the filter. Figure 11.29 was taken using this filter, and I obtained most of the photos in Chapter 10 using Rheinberg illumination.

Figure 11.34 illustrates some of the different effects one can obtain using different color filters in a Rheinberg illumination system. For the first image in this set, the plain filter produced uniform white light impinging on the crystal from a broad range of angles. This is like covering a flashlight with waxed paper and placing it directly behind the crystal. As can be seen in the image, the uniform lighting tends to wash out the structural details in the crystal, producing a rather flat, bland image.

Better results were obtained in the second image by using a patterned red/white filter. Now the snow crystal acts like a complex lens that refracts the light through different angles. The refraction depends on the shape of the ice, so the resulting image shows a variety of subtle red

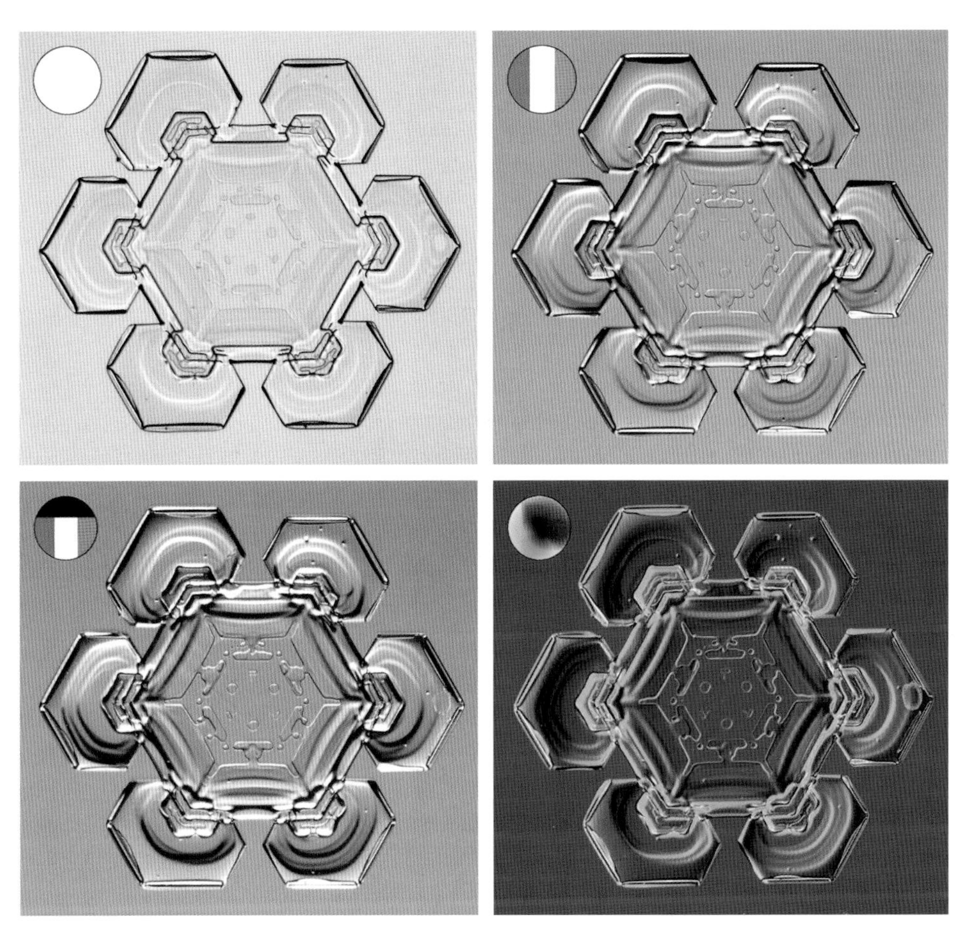

FIGURE 11.34. This snow crystal was photographed swapping in several different Rheinberg filters in quick succession. The insets show the filter used for each corresponding photo.

and blue highlights that reveal more structural details and give the image a pleasing sense of depth. Meanwhile the background has a uniform color and brightness, because the back illumination varied only by angle, not by position at the object plane. The third image is similar, but there is greater variation in the illumination as a function of angle, resulting in stronger shading and higher contrast overall. Note how the inward-propagating rings (see Chapter 4) on the outer branches are barely visible in first photo, quite clear in second, and highly visible in the third. Finally, the fourth photo used a "rainbow" filter that shines many different colors in from different angles around the crystal, yielding a somewhat psychedelic effect. I would not use this as my go-to illumination method every day, but I like to photograph snowflakes using a full range of lighting techniques, just for variety.

Figure 11.35 shows a side-by-side comparison of Rheinberg back illumination with dark-field illumination. The Rheinberg image shows off subtle curvature effects like the faint ribbing in the platelike sections, but it gives the crystal a somewhat "plastic" look that may be unappealing to some. Dark-field illumination produces a white-on-black image, but the sense of depth is reduced. The Rheinberg method can also be used to add some color to dark-field illumination, as illustrated in Figure 11.36. The basic idea is to use a filter with a dark

FIGURE 11.35. The left image shows a photo taken using Rheinberg back illumination, while the right image shows the same crystal using dark-field illumination. Surface features and edges refract light through large angles, so these features appear dark in the left image and bright in the right image. Note the central dark "hole" in the dark-field image, where the plate is thin and featureless.

FIGURE 11.36. Rheinberg illumination can be used to add some color to dark-field illumination, yielding the colorful effects shown in these photos. Because blue is associated with cold, the fourth image has an especially frigid look.

spot at its center (see Figure 11.33) and various colors at its extremities.

The SnowMaster 9000

For many years I have traveled to various cold locations to photograph snowflakes using the apparatus shown in Figure 11.37, constructed in 2003, which I have come to call the *SnowMaster 9000*. The overall optical layout is basically that depicted in Figure 11.32, using Rheinberg illumination. This hardware has become a real workhorse for me, and I have used it to take more than 10,000 snowflake photos. Starting at the top, the camera is an older Canon EOS single-lens reflex (SLR) model with a 20-megapixel full-frame (36 × 24 mm) sensor, outfitted with a right-angle eyepiece for through-the-lens focusing. I put the camera in a styrofoam lined box and included a low-power heater to keep the temperature

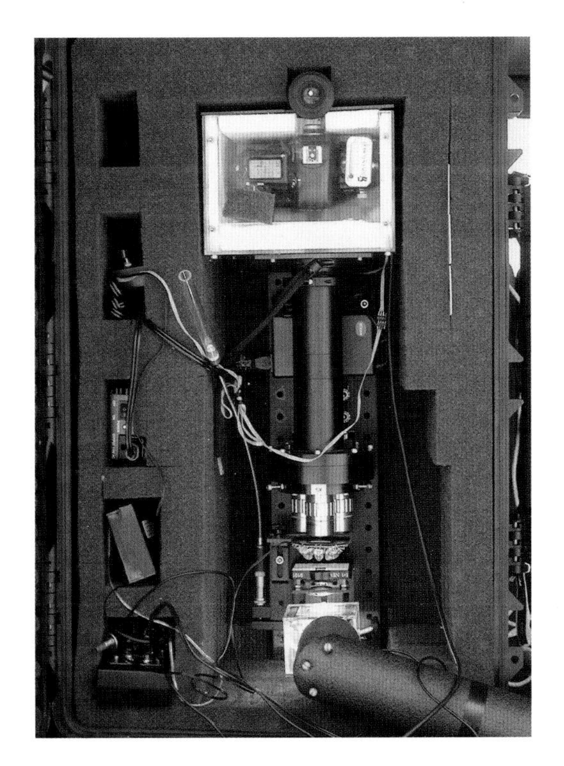

FIGURE 11.37. The author's traveling snow crystal photomicroscope, the *Snowmaster 9000*.

above freezing, as the camera specifications showed it rated down only to 0°C. I have since learned that this spec is highly conservative, and most digital cameras have no trouble working at temperatures down to −20°C, although the battery capacity tends to be reduced (temporarily) when the temperature is low.

Below the camera is a length of extension tube from Thorlabs, attached to the camera using an appropriate adaptor. Below this is a home-built turret, complete with the Mitutoyo Plan APO 2X, 5X, and 10X objectives (see Table 11.1). In hindsight, the turret was largely unnecessary, as I have used the 5X objective about 90 percent of the time. However, the "monster" fernlike stellar dendrite described in Chapter 10 was so big that it required four separate photos at 2X, so I was happy to have the ability to switch objectives quickly that day.

Directly below the microscope objectives there is an array of LEDs I used for experimenting with side illumination. The LEDs point down toward a spot on the center of the image plane, where a snow crystal rests on a microscope slide. I was never very happy with the results I obtained using the LED light source, however, so I mainly used Rheinberg illumination, which allowed for a much greater variety of illumination effects by changing color filters, as described above. The microscope slide rests on a 90-degree angle plate that is attached to a linear translation stage, both from Thorlabs. The micrometer on the stage is attached to a flexible cable that includes a large plastic handle at its end, used to adjust the focus. This thin plastic tube makes it easier to change the focus using ungloved hands in the cold.

The field lens sits directly below the snow crystal, mounted to the bottom surface of the angle plate. The field lens is just a basic one-inch-diameter achromatic lens, as there is no need to focus the color filter onto the pupil with great precision. An incandescent light source is contained in the glowing box in the photo, although this was later replaced with an LED bulb to reduce the amount of heat generated. Because this system provides plenty of light, I usually set my camera to ISO 100, as this reduces noise in the image relative to

FIGURE 11.38. The author photographing snowflakes using the SnowMaster 9000 in the frozen north outside Cochran, Ontario. Note the blue foam-core collection board mounted at a convenient height using a tripod, the small paintbrush in one hand, and a glass slide in the other. Fingerless gloves are used to retain enough dexterity to be able to handle small snow crystals.

FIGURE 11.39. Rheinberg illumination was used to produce red highlights in this somewhat asymmetrical stellar snow crystal.

higher ISO settings. The shutter speed was quite slow, of order 1/100 sec, but this is plenty fast when using a rigid mounting system.

Although this hardware setup may look imposing, the overall optical layout is quite simple, as illustrated in Figure 11.32. If you remove the unnecessary heated camera box, get rid of the hard shell case by putting the setup in an unheated garage or shed, and forgo the turret, then what remains is not particularly difficult to build. You will not find microscope objectives or translation stages at your local camera store, but these tools are not difficult to work with, and they are no more expensive than traditional macro photography gear.

Figure 11.38 shows the author working with the SnowMaster 9000 under typical winter conditions. Figure 11.39 illustrates a photo taken with it in northern Ontario. Once the system is set up, the workflow is straightforward: search for suitable crystals on the foam-core collection board by eye, pick up a promising specimen using a fine paintbrush, set it carefully on a glass microscope slide, place the slide on the observing plate, adjust the focus, adjust the color filter, and take the shot. Take several shots for focus stacking as needed. If the crystal looks especially nice, try a few additional color filters for different visual effects. Then clean the slide and repeat.

APPENDIX

List of Variables and Physical Constants

a—Size of a water molecule:

$a \approx c_{ice}^{-1/3} \approx 0.32$ nm

a_0—Ice lattice parameter

A, A_{basal}, A_{prism}—Nucleation parameters in attachment kinetics

B—Cylinder/parabolic diffusion parameter:

$B = \log(r_{far}/R)$

c—Water vapor number density

c_0—Ice lattice parameter

c_{ice}—Ice molecular number density:

$c_{ice} = \rho_{ice}/m_{mol}$

$c_{ice} \approx 3.1 \times 10^{28}\ \mathrm{m}^{-3}$

$c_{p,air}$—Heat capacity of air:

$c_{p,air} \approx 1.0$ kJ/kg-K

$c_{p,s}$—Heat capacity of ice:

$c_{p,s} \approx 2.1$ kJ/kg-K

$c_{p,l}$—Heat capacity of water:

$c_{p,l} \approx 4.2$ kJ/kg-K

$c_{p,v}$—Heat capacity of water vapor:

$c_{p,v} \approx 2.0$ kJ/kg-K

c_{sat}—Saturated water vapor number density

$c_{sat,water}$—Saturated water vapor density of supercooled water

c_{surf}—Water vapor number density at the surface

C_{diff}—Second derivative of $c_{sat}(T)$:

$$C_{diff} = \frac{1}{2}\frac{1}{c_{sat}}\frac{d^2 c_{sat}}{dT^2}$$

D—Diffusion constant

D_{air}—Particle diffusion constant in air:

$D_{air} \approx 2 \times 10^{-5}\ \mathrm{m}^2/\mathrm{sec}$

D_{therm}—Thermal diffusion constant in air:

$D_{therm} \approx 2 \times 10^{-5}\ \mathrm{m}^2/\mathrm{sec}$

D_{water}—Thermal diffusion constant in water:

$D_{water} \approx 1.4 \times 10^{-7}\ \mathrm{m}^2/\mathrm{sec}$

d_{sv}—Gibbs-Thomson length:

$d_{sv} = \gamma_{sv}/c_{ice}kT$

≈ 1 nm

F—Flux

$f_{\#}$—Lens f-number

k—Boltzmann constant:

$k \approx 1.38 \times 10^{-23}\ \mathrm{m}^2\ \mathrm{kg}\ \mathrm{sec}^{-2}\ \mathrm{K}^{-1}$

$kT_{-15} \approx 3.6 \times 10^{-21}$ Joules

K_T—Kinetic coefficient for ice growth from liquid water

ℓ—Latent heat per molecule, or . . .

ℓ—A generic length

$\ell_{vicinal}$—Step spacing on a vicinal surface

L—A generic length

L_{sl}—Latent heat of melting (solid/liquid):

$L_{sl} \approx 0.33 \times 10^6$ J/kg

L_{lv}—Latent heat of evaporation (liquid/vapor):

$L_{sl} \approx 2.5 \times 10^6$ J/kg

L_{sv}—Latent heat of sublimation (solid/vapor)

$L_{sv} \approx 2.8 \times 10^6$ J/kg

m_{mol}—Mass of water molecule:

$m_{mol} \approx 3.0 \times 10^{-26}$ kg

N_A—Lens numerical aperture

P_{ice}—Ice equilibrium vapor pressure

P_{water}—Water equilibrium vapor pressure

R—A generic radius

R_{tip}—Tip radius of parabolic dendrite

R_{xy}—Optical resolution (resolving power)

R_z—Optical depth of focus

s_0—Solvability parameter

Δt—A small time step

T—Temperature

T_K—Temperature in Kelvin

T_0—Ice/water nucleation parameter

T_{-15}—258 K (−15°C)

T_∞—Temperature at infinity

T_m—Ice/water melting point (0°C)

T_{sl}—Solid/liquid transition temperature

T_{surf}—Surface temperature

ΔT_{surf}—Surface supercooling

u_{term}—Terminal velocity

v—Crystal growth velocity

v_n—Crystal growth velocity normal to the surface

v_{kin}—Kinetic velocity:

$$v_{kin} = \frac{c_{sat}}{c_{ice}} \sqrt{\frac{kT}{2\pi m_{mol}}}$$

v_{mol}—Water vapor molecular velocity

v_{tip}—Growth velocity of dendrite tip

w—Width of a top facet terrace

X_0—Characteristic diffusion length:

$$X_0 = \frac{c_{sat}}{c_{ice}} \frac{D}{v_{kin}}$$

In air at −15°C, $X_0 \approx 0.145\ \mu$m

x_{basal}—Spacing between basal layers

x_{prism}—Spacing between prism layers

x_{diff}—Diffusion length for a water molecule on an ice surface

α—Kinetics attachment coefficient

α_{basal}—Attachment coefficient on a basal faceted surface

α_{diff}—Attachment coefficient for diffusion-limited growth of a sphere:

$\alpha_{diff} = X_0/R$

$\alpha_{diff,heat}$—Attachment coefficient including heat diffusion

$\alpha_{diffcyl}$—Attachment coefficient for diffusion-limited growth of a cylinder

α_{facet}—Attachment coefficient on a generic faceted surface

α_{prism}—Attachment coefficient on a prism faceted surface

α_{therm}—Attachment coefficient for heat flow to substrate

β—Step energy on an ice/vapor surface

β_0—Fictitious "rigid terrace" step energy:

$\beta_0 \equiv a\gamma_{sv}$

$\approx 3 \times 10^{-11}$ J/m

γ—Surface energy

γ_0—Ice solid/vapor surface energy of rough, unfaceted surface

γ_{lv}—Ice liquid/vapor surface energy:

$\gamma_{lv} \approx 76$ mJ/m^2

γ_{sl}—Ice solid/liquid surface energy:

$\gamma_{sl} \approx 30 \pm 5$ mJ/m^2

γ_{sv}—Ice solid/vapor surface energy:

$\gamma_{sv} \approx 106 \pm 15$ mJ/m^2

δ—Step density (number per unit length) on a faceted surface

η—Logarithmic change in c_{sat} with T:

$$\eta = \left(\frac{1}{c_{sat}}\right)\left(\frac{dc_{sat}}{dT}\right), \text{ or} \ldots$$

η—Part of the (η, ξ, φ) parabolic coordinate system

η_{eff}—Dynamical viscosity for liquid water

θ_{surf}—Surface normal relative to a faceted surface

κ—Surface curvature, or . . .

κ—Thermal conductivity coefficient

κ_{air}—Thermal conductivity of air:

$\kappa_{air} \approx 0.025$ W m^{-1} K^{1}

κ_{ice}—Thermal conductivity of ice:

$\kappa_{ice} \approx 2.3$ W m^{-1} K^{1}

κ_{water}—Thermal conductivity of water:

$\kappa_{water} \approx 0.6$ W m^{-1} K^{1}

κ_{wv}—Thermal conductivity of water vapor:

$\kappa_{wv} \approx 0.02$ W m^{-1} K^{1}

λ—Wavelength of light

μ—Dynamical viscosity of air:

$\mu \approx 1.8 \times 10^{-5}$ kg/(m-sec)

$\nu_{kinematic}$—Kinematic viscosity of air:

$\nu_{kinematic} = \mu/\rho_{air} \approx 1.4 \times 10^{-5}$ m^2/sec at 1 atm pressure

ξ—Ice filling fraction on substrate

ρ_{air}—Mass density of air:

$\rho_{air} \approx 1.2$ kg/m^3

ρ_{ice}—Mass density of ice:

$\rho_{ice} \approx 917$ kg/m^3

ρ_{water}—Mass density of water:

$\rho_{water} \approx 1000$ kg/m^3

σ—Supersaturation

σ_0—Nucleation parameter

σ_∞—Supersaturation at infinity

σ_{subst}—Supersaturation at a substrate

σ_{surf}—Supersaturation at ice surface

σ_{water}—The supersaturation of supercooled liquid water relative to ice:

$$\sigma_{water} = \frac{c_{sat,water} - c_{sat,ice}}{c_{sat,ice}}$$

χ_0—Thermal parameter (dimensionless):

$$\chi_0 = \frac{\eta D L_{sv} \rho_{ice}}{\kappa_{air}} \frac{c_{sat}}{c_{ice}}$$

BIBLIOGRAPHY

[1611Kep] Johannes Kepler, *De Nive Sexangular*, 1611; translated as *The Six-Cornered Snowflake*, by Colin Hardie, Oxford University Press, Oxford, 1966; also by Jacques Bromberg, Paul Dry Books, 2010.

[1637Des] Réne Descartes, *Les Météores, in Discours de la Methode*, Leiden, 1637.

[1665Hoo] Robert Hooke, *Micrographia*, Royal Society, London, 1665.

[1820Sco] William Scoresby, *An Account of the Arctic Regions with a History and Description of the Northern Whale-Fishery*, Edinburgh, 1820.

[1859Far] Michael Faraday, *Experimental Researches in Chemistry and Physics*, Taylor and Frances, London, 1859.

[1871Tho] W. Thomson, On the equilibrium vapor at a curved surface of liquid, *Phil. Mag.* 42, 448–452, 1871.

[1880Twa] Mark Twain [Samuel Clemens], *A Tramp Abroad*, American Publishing Company, 1880.

[1882Her] H. Hertz, *Ann. Phys. Lpz.* 17, 193, 1882.

[1900Wil] H. A. Wilson, On the velocity of solidification and viscosity of supercooled liquids, *Phil. Mag.* 50, 238, 1900.

[1904Koc] H. von Koch, Sur une courbe continue sans tangente, obtenue par une construction géométrique élémentaire, *Ark. Math. Astron. Fys.* 1, 681–702, 1904.

[1910Lin] F. A. Lindemann, The calculation of molecular natural frequencies, *Z. Phys.* 11, 609–612, 1910.

[1915Knu] M. Knudsen, Die maximale verdampfungsgeschwindigkeit des quecksilbers, *Ann. Phys. Lpz.* 47, 697, 1915.

[1917Tho] D'Arcy Wentworth Thompson, *On Growth and Form*, Cambridge University Press, Cambridge, 1917.

[1928Str] I. N. Stranski, *Z. Phys. Chem.* 136, 259, 1928.

[1929Bar] William Howard Barnes, The crystal structure of ice between 0 C and −183 C, *Proc. R. Soc. Lond.* A, 1929.

[1931Ben] Wilson Bentley and William Humphreys, *Snow Crystals*, McGraw-Hill, New York, 1931.

[1932Fre] J. Frenkel, *Phys. Z. Sowjetunion* 1, 498, 1932.

[1933Ber] J. D. Bernal and R. H. Fowler, A theory of water and ionic solution, with particular reference to hydrogen and hydroxyl ions, *J. Chem. Phys.* 1, 515, 1933.

[1934Str] I. N. Stranski and R. Kaischew, *Z. phys. Chem.* B 26, 31, 1934.

[1935Bec] R. Becker and W. Döring, *Ann. Phys. Lpz.* 24, 719, 1935.

[1938Ber] W. F. Berg, Crystal growth from solution, *Proc. Roy. Soc. London A* 164, 79–95, 1938.

[1939Vol] M. Volmer, *Kinetik der Phasenbildung*, Steinkopff, Dresden and Leipzig, 1939.

[1947Iva] G. P. Ivantsov, *Dokl. Akad. Nauk SSSR* 58, 1113, 1947.

[1948Von] B. Vonnegut, G. E. Res. Lab., Project Cirrus Occas. Rep. 5, Schenectady, NY, 1948.

[1949Sch] V. J. Schaefer, The formation of ice crystals in the laboratory and the atmosphere, *Chem. Rev.* 44, 291, 1949.

[1950Gri] L. J. Griffin, Observation of unimolecular growth steps on crystal surfaces, *Phil. Mag.* 41, 196–199, 1950.

[1951Bur] W. K. Burton, N. Cabrera, and F. C. Frank, The growth of crystals and the equilibrium structure of their surfaces, *Phil. Trans. Royal Soc. London* 243, 299–358, 1951.

[1951Ver] Ajit Ram Verma, Spiral growth on carborundum crystal faces, *Nature* 167, 939–940, 1951.

[1952Sch] Vincent J. Schaefer, Formation of ice crystals in ordinary and nuclei-free air, *Indust. Engr. Chem.* 44, 1300–1304, 1952.

[1954Nak] Ukichiro Nakaya, *Snow Crystals, Natural and Artificial*, Harvard University Press, Cambridge, MA, 1954.

[1958Hal] J. Hallett and B. J. Mason, The influence of temperature and supersaturation on the habit of ice crystals grown from the vapour, *Proc. Roy. Soc. A* 247, 440–453, 1958.

[1958Hil] W. B. Hillig, The kinetics of freezing of ice in the direction perpendicular to the basal plane, in *Growth and Perfection of Crystals* (ed. R. H. Doremus, B. W. Roberts, and D. Turnbull), 350–360, 1958.

[1958Nak] U. Nakaya, M. Hanajima, and J. Mugurama, Physical investigations of the growth of snow crystals, *J. Fac. Sci. Hokkaido Univ.* ser. II-5, 87, 1958.

[1959Heu] A. P. van den Heuvel and B. J. Mason, Habit of ice crystals grown in hydrogen, carbon dioxide and air at reduced pressure, *Nature* 184, 519–520, 1959.

[1961Kob] T. Kobayashi, The growth of snow crystals at low supersaturations, *Phil. Mag.* 6, 1363–1370, 1961.

[1961Tho] D'Arcy Wentworth Thompson, *On Growth and Form* (abridged edition, ed. John Tyler Bonner), Cambridge University Press, Cambridge, 1961.

[1963Bar] J. T. Bartlett, A. P. van den Heuvel, and B. J. Mason, Growth of ice crystals in an electric field, *Zeit. fur Ange. Math. Phys.* 14, 599–610, 1963.

[1963Mas] B. J. Mason, G. W. Bryant, and A. P. van den Heuvel, The growth habits and surface structure of ice crystals, *Phil. Mag.* 8, 505–526, 1963.

[1963Mul] W. W. Mullins and R. F. Sekerka, Morphological stability of a particle growing by diffusion or heat flow, *J. Appl. Phys.* 34, 323–329, 1963.

[1964Fey] Richard Feynman, *The Character of Physical Law*, MIT Press, Cambridge, MA, 1964.

[1964Mul] W. W. Mullins and R. F. Sekerka, Stability of a planar interface during solidification of dilute binary alloy, *J. Appl. Phys.* 35, 444–451, 1964.

[1965Kni] Charles A. Knight and Nancy C. Knight, "Negative" Crystals in Ice: A Method for Growth, *Science* 150, 1819–1821, 1965.

[1965Rei] F. Reif, *Fundamentals of Statistical and Thermal Physics*, McGraw-Hill, New York, 1965.

[1966Mag] C. Magono and C. Lee, Meteorological classification of natural snow crystals, *J. Fac. Sci. Hokkaido* 2, 321–355, 1966.

[1966Mic] A. S. Michaels, P. L. T. Brian, and P. R. Sperry, Impurity effects on the basal plane solidification kinetics of supercooled water, *J. Appl. Phys.* 37, 4649–4661, 1966.

[1969And] B. J. Anderson, J. D. Sutkoff, and J. Hallett, Influence of methyl 2-cyanoacrylate monomer on habit of ice crystals grown from vapor, *J. Atmos. Sci.* 26, 673–674, 1969.

[1969Fuk] Norihiko Fukuta, Experimental studies on the growth of small ice crystals, *J. Atmos. Sci.* 26, 522–530, 1969.

[1970Fle] N. H. Fletcher, *The Chemical Physics of Ice*, Cambridge University Press, Cambridge, 1970.

[1971Iwa] Kunimoto Iwai, Note on Snow Crystals of Spatial Type, *J. Meteor. Soc. Japan* 49, 516–520, 1971.

[1971Mas] B. J. Mason, *The Physics of Clouds*, Clarendon Press, Oxford, 1971.

[1972Lam] D. Lamb and W. D. Scott, Linear growth rates of ice crystals grown from the vapor phase, *J. Cryst. Growth* 12, 21–31, 1972.

[1973Bil] J. H. Bilgram, Perfect highly doped ice crystals by the Czochralski method, in *Physics and Chemistry of Ice* (ed. E. Whalley, S. Jones, and L. Gold), Royal Society of Canada, Ottawa, 1973.

[1973Yam] Akira Yamashita, On the trigonal growth of ice crystals, *J. Meteor. Soc. Japan* 51, 307–317, 1973.

[1974Hob] Peter V. Hobbs, *Ice Physics*, Oxford University Press, Oxford, 1974.

[1974Kob] T. Kobayashi and T. Ohtake, Hexagonal twin prisms of ice, *J. Atmos. Sci.* 31, 1377, 1974.

[1975Kob] T. Kobayashi and Y. Furukawa, On twelve-branched snow crystals, *J. Cryst. Growth* 28, 21–28, 1975.

[1976Gli] M. E. Glicksman, R. J. Schaefer, and J. D Ayers, Dendritic growth—test of theory, *Metal. Trans. A* 7, 1747–1759, 1976.

[1976Gon] Takehiko Gonda, The growth of small ice crystal in gases of high and low pressures, *J. Meteor. Soc. Japan* 54, 233–240, 1976.

[1976Rya] B. F. Ryan, E. R. Wishart, and D. E. Shaw, The growth rates and densities of ice crystals between −3°C and −21°C, *J. Atmos. Sci.* 22, 123–133, 1976.

[1977Gol] I. Golecki and C. Jaccard, The surface of ice near 0°C studied by 100 keV proton channeling, *Phys. Lett. A* 63, 374–376, 1977.

[1978Fur] Y. Furukawa and T. Kobayashi, On the growth mechanism of polycrystalline snow crystals with a specific grain boundary, *J. Cryst. Growth* 45, 57–65, 1978.

[1978Lan] J. S. Langer, R. F. Sekerka, and T. Fujioka, Evidence for a universal law of dendritic growth rates, *J. Cryst. Growth* 44, 414–418, 1978.

[1979Kik] Katsuhiro Kikuchi and Austin W. Hogan, Properties of diamond dust type ice crystals observed in summer season at Amundsen-Scott South Pole Station, Antarctica, *J. Meteor. Soc. Japan* 57, 180–190, 1979.

[1980Bea] D. Beaglehole and D. Nason, Transition layer on the surface on ice, *Surf. Sci.* 96, 357–363, 1980.

[1980Gre] Robert Greenler, *Rainbows, Halos, and Glories*, Cambridge University Press, Cambridge, 1980.

[1980Lan] J. S. Langer, Instabilities and pattern formation in crystal growth, *Rev. Mod. Phys.* 52, 1–28, 1980.

[1981Hua] S.-C. Huang and M. E. Glicksman, Overview 12: Fundamentals of dendritic solidification—II Development of sidebranch structure, *Acta Metall.* 29, 717–734, 1981.

[1981Sch] Vincent J. Schaefer and John A. Day, *A Field Guide to the Atmosphere*, Houghton Mifflin, Boston, 1981.

[1981Wit] T. Witten and L. Sander, Diffusion-limited aggregation, a kinetic critical phenomenon, *Phys. Rev. Lett.* 47, 1400–1403, 1981.

[1982Bec] W. Beckmann and R. Lacmann, Interface kinetics of the growth and evaporation of ice single crystal from the vapor phase—Part II: Measurements in a pure water vapour environment, *J. Cryst. Growth* 58, 433–442, 1982.

[1982Fra] F. C. Frank, Snow crystals, *Contemp. Phys.* 23, 3–22, 1982.

[1982Gon] T. Gonda and T. Koike, Growth rates and growth forms of ice crystal grown from the vapour phase, *J. Cryst. Growth* 56, 259–264, 1982.

[1982Kel] V. W. Keller and J. Hallett, Influence of air velocity on the habit of ice crystal growth from the vapor, *J. Cryst. Growth* 60, 91–106, 1982.

[1982Kur] T. Kuroda and R. Lacmann, Growth kinetics of ice from the vapour phase and its growth forms, *J. Cryst. Growth* 56, 189–205, 1982.

[1983Bec] W. Beckmann, R. Lacmann, and A. Blerfreund, Growth rates and habits of ice crystals grown from the vapor phase, *J. Phys. Chem.* 87, 4142–4146, 1983.

[1984Kur] Toshio Kuroda, Rate determining processes of growth of ice crystals from the vapour phase—Part I: Theoretical considerations, *J. Meteor. Soc. Japan* 62, 1–11, 1984.

[1984Kur1] Toshio Kuroda and Takehiko Gonda, Rate determining processes of growth of ice crystals from the vapour phase—Part II: Investigation of surface kinetic processes, *J. Meteor. Soc. Japan* 62, 563–572, 1984.

[1985Col] S. C. Colbeck, Temperature dependence of the equilibrium shape of ice, *J. Cryst. Growth* 72, 726–732, 1985.

[1986Pac] N. H. Packard, Lattice models for solidification and aggregation, in *Science on Form: Proceedings of the First International Symposium for Science on Form* (ed. S. Ishizaka et al.), Scientific Publishers, Tokyo, 1986.

[1987Ban] H. L. Bank and K. G. M. Brockbank, Basic principles of cryobiology, *J. Cardiac Surgery* 2, 137–143, 1987.

[1987Kar] O. A. Karim and A. D. J. Haymet, The ice water interface, *Chem. Phys. Lett.* 138, 531–534, 1987.

[1987Kob] T. Kobayashi and T. Kuroda, *Snow Crystals: Morphology of Crystals—Part B*, Terra Scientific, Tokyo, 1987.

[1988Kar] O. A. Karim and A. D. J. Haymet, The ice water interface—A molecular-dynamics study, *J. Chem. Phys.* 89, 6889–6896, 1988.

[1988Kes] D. A. Kessler, J. Koplik, and H. Levine, Pattern selection in fingered growth phenomena, *Adv. Phys.* 37, 255–339, 1988.

[1988Sai] Y. Saito, G. Goldbeck-Wood, and H. Muller-Krumbhaar, Numerical simulation of dendritic growth, *Phys. Rev. A* 38, 2148–2157, 1988.

[1988Tak] Tsuneya Takahashi and Norihiko Fukuta, Supercooled cloud tunnel studies on the growth of snow crystals between −4°C and −20°C, *Meteor. Soc. Japan* 66, 841–855, 1988.

[1988Xia] R.-F. Xiao, J. I. D. Alexander, and F. Rosenberger, Morphological evolution of growing crystals: A Monte Carlo simulation, *Phys. Rev. A* 38, 2447–2456, 1988.

[1989Lan] J. S. Langer, Dendrites, viscous fingers, and the theory of pattern formation, *Science* 243, 1150–1156, 1989.

[1989Sei] T. Sei and T. Gonda, The growth mechanism and the habit change of ice crystals growing from the vapor phase, *J. Cryst. Growth* 94, 697–707, 1989.

[1990Gon] T. Gonda, S. Nakahara, and T. Sei, The formation of side branches of dendritic ice crystals growing from vapor and solution, *J. Cryst. Growth* 90, 183–187, 1990.

[1990Gon1] T. Gonda, H. Kakiuchi, and K. Moriya, In situ observation of internal structure in growing ice crystals by laser scattering tomography, *J. Cryst. Growth* 102, 167–174, 1990.

[1990Tap] Walter Tape, *Atmospheric Halos*, American Geophysical Union, Washington, DC, 1990.

[1990Uye] Hiroshi Uyeda and Katsuhiro Kikuchi, Formation mechanisms of twelve-branched snow crystals, *J. Meteor. Soc. Japan* 68, 549, 1990.

[1990Yok] Etsuro Yokoyama and Toshio Kuroda, Pattern formation in growth of snow crystals occurring in the surface kinetic process and the diffusion process, *Phys. Rev. A* 41, 2038–2049, 1990.

[1991Bre] E. A. Brener and V. I. Melnikov, Pattern selection in 2-dimensional dendritic growth, *Adv. Phys.* 40, 53–97, 1991.

[1991Elb] M. Elbaum, Roughening transition observed on the prism facet of ice, *Phys. Rev. Lett.* 67, 2982–2985, 1991.

[1991Tak] T. Takahashi et al., Vapor diffusional growth of free-falling snow crystals between −3°C and −23°C, *J. Meteor. Soc. Japan* 69, 15–30, 1991.

[1992Don] Yayi Dong and John Hallett, Charge separation by ice and water drops during growth and evaporation, *J. Geophys. Res. Atmos.* 97, 20361–20371, 1992.

[1992Mus] M. Muschol, D. Liu, and H. Z. Cummins, Surface-tension-anisotropy measurements of succinonitrile and pivalic acid: Comparison with microscopic solvability theory, *Phys. Rev. A* 46, 1038–1050, 1992.

[1993Cro] M. C. Cross and P. C. Hohenburg, Pattern formation outside of equilibrium, *Rev. Mod. Phys.* 65, 851–1112, 1993.

[1993Elb] M. Elbaum, S. G. Lipson, and J. G. Dash, Optical study of surface melting on ice, *J. Cryst. Growth* 129, 491–505, 1993.

[1993Fur] Yoshinori Furukawa and Shigetsugu Kohata, Temperature dependence of the growth form of negative crystal in an ice single crystal and evaporation kinetics for its surfaces, *J. Cryst. Growth* 129, 571–581, 1993.

[1993Furl] Yoshinori Furukawa and Wataru Shimada, Three-dimensional pattern formation during growth of ice dendrites—its relation to universal law of dendritic growth, *J. Cryst. Growth* 128, 234–239, 1993.

[1994Gon] T. Gonda, Y. Matsuura, and T. Sei, In situ observation of vapor-grown ice crystals by laser two-beam interferometry, *J. Cryst. Growth* 142, 171–176, 1994.

[1994Ven] J. A. Venables, Atomic processes in crystal growth, *Surf. Sci.* 299, 798–817, 1994.

[1995Dos] H. Dosch, A. Lied, and J. H. Bilgram, Glancing-angle X-ray scattering studies of the premelting of ice surfaces, *Surf. Sci.* 327, 145–164, 1995.

[1995Hua] Jinfan Huang and Lawrence S. Bartell, Kinetics of homogeneous nucleation in the freezing of large water clusters, *J. Phys. Chem.* 99, 3924–3931, 1995.

[1995Wer] William P. Wergin, Albert Rango, and Eric F. Erbe, Observations of snow crystals using low-temperature scanning electron microscopy, *Scanning* 17, 41–49, 1995.

[1996Kar] A. Karma and W. J. Rappel, Phase-field method for computationally efficient modeling of solidification with arbitrary interface kinetics, *Phys. Rev. E* 53, R3017–R3020, 2006.

[1996Kni] Charles A. Knight, A simple technique for growing large, optically perfect ice crystals, *J. Glaciology* 42, 585–587, 1996.

[1996Nad] H. Nada and Y. Furukawa, Anisotropic growth kinetics of ice crystals from water studied by molecular dynamics simulation, *J. Cryst. Growth* 169, 587–597, 1996.

[1996Nel] Jon Nelson and Charles A. Knight, A new technique for growing crystals from the vapor, *J. Cryst. Growth* 169, 795–797, 1996.

[1996Ran] A. Rango, W. P. Wergin, and E. F. Erbe. Snow crystal imaging using scanning electron microscopy: I. Precipitated snow, *Hydrological Sci. J.* 41, 219–233, 1996.

[1996Sai] Y. Saito, *Statistical Physics of Crystal Growth*, World Scientific Books, Singapore, 1996.

[1996Sch] A. Schmidt, Computation of three-dimensional dendrites with finite elements, *J. Comp. Phys.* 125, 293–312, 1996.

[1997Fur] Yoshinori Furukawa and Hiroki Nada, Anisotropic surface melting of an ice crystal and its relationship to growth forms, *J. Phys. Chem. B* 101, 6167–6170, 1997.

[1997Gon] T. Gonda and S. Nakahara, Dendritic ice crystals with faceted tip growing from the vapor phase, *J. Cryst. Growth* 173, 189–193, 1997.

[1997Kar] Alain Karma and Wouter-Jan Rappel, Phase-field simulation of three-dimensional dendrites: Is microscopic solvability theory correct?, *J. Cryst. Growth* 174, 1–4, 54–64, 1997.

[1997Pet] V. F. Petrenko, Study of the surface of ice, ice/solid, and ice/liquid interfaces with
scanning force microscopy, *J. Phys. Chem. B* 101, 6276–6281, 1997.

[1997Pru] H. R. Pruppacher and J. D. Klett, *Microphysics of Clouds and Precipitation*, Kluwer Academic Publishers, Dordrecht, 1997.

[1998Dop] A. Doppenschmidt, M. Kappl, and H. J. Butt, Surface properties of ice studied by atomic force microscopy. *J. Phys. Chem. B* 102, 7813–7819, 1998.

[1998Kar] A. Karma and W. J. Rappel, Quantitative phase-field modeling of dendritic growth in two and three dimensions, *Phys. Rev. E* 57, 4323–4349, 1998.

[1998Lib] K. G. Libbrecht and V. M. Tanusheva, Electrically induced morphological instabilities in free dendrite growth, *Phys. Rev. Lett.* 81, 176–179, 1998.

[1998Nel] Jon Nelson and Charles Knight, Snow crystal habit changes explained by layer nucleation, *J. Atmos. Sci.* 55, 1452–1465, 1998.

[1998Pit] B. Pittenger et al., Investigation of ice-solid interfaces by force microscopy: Plastic flow and adhesive forces, *J. Vac. Sci. Technol. A* 16, 1832–1837, 1998.

[1999Bre] E. A. Brener and H. Muller-Krumbhaar, Comment on "Electrically induced morphological instabilities in free dendrite growth," *Phys. Rev. Lett.* 83, 1698, 1999.

[1999Fuk] Norihiko Fukuta and Tsuneya Takahashi, The growth of atmospheric ice crystals: A summary of findings in vertical supercooled cloud tunnel studies, *Am. Meteor. Soc.* 56, 1963–1979, 1999.

[1999Lib] K. G. Libbrecht and V. M. Tanusheva, Comment on "Electrically induced morphological instabilities in free dendrite growth"—Libbrecht and Tanusheva reply, *Phys. Rev. Lett.* 83, 1699, 1999.

[1999Lib1] Kenneth G. Libbrecht and Victoria M. Tanusheva, Cloud chambers and crystal growth: Effects of electrically enhanced diffusion on dendrite formation from neutral molecules, *Phys. Rev. E* 59, 3253–3261, 1999.

[1999Pet] Victor F. Petrenko and Robert W. Whitworth, *Physics of Ice*, Oxford University Press, Oxford, 1999.

[1999Pim] Alberto Pimpinelli and Jacques Villain, *Physics of Crystal Growth*, Cambridge University Press, Cambridge, 1999.

[1999Swa] B. D. Swanson et al., Electrodynamic trapping and manipulation of ice crystals, *Quart. J. Roy. Meteor. Soc.* 125, 1039–1058, 1999.

[2000Bol] Kim Bolton and Jan B. C. Pettersson, A molecular dynamics study of the long-time ice Ih surface dynamics, *J. Phys. Chem. B* 104, 1590–1595, 2000.

[2000Wan] W. Wang, Lyophilization and development of solid protein pharmaceuticals, *Int. J. Pharmaceutics* 203, 1–60, 2000.

[2001Bon] H. P. Bonzel, Equilibrium crystal shapes: Towards absolute energies, *Prog. Surf. Sci.* 67, 45–58, 2001.

[2001Gir] C. Girardet and C. Toubin, Molecular atmospheric pollutant adsorption on ice: A theoretical survey, *Surf. Sci. Rep.* 44, 159–238, 2001.

[2001Kni] C. A. Knight and A. Wierzbicki, Adsorption of biomolecules to ice and their effects on ice growth. 2. A discussion of the basic mechanism of antifreeze phenomena, *Cryst. Growth & Design* 1, 439–446, 2001.

[2001Nel] J. Nelson, Growth mechanisms to explain the primary and secondary habits of snow crystals, *Phil. Mag.* 81, 2337–2373, 2001.

[2001Shi] A. A. Shibkov et al., Kinetics and morphology of nonequilibrium growth of ice in supercooled water, *Crystallography Rep.* 46, 496–502, 2001.

[2001Woo] Stephen E. Wood and Marcia B. Baker, New model for vapor growth of hexagonal ice crystals in the atmosphere, *J. Geophys. Res.* 106, 4845–4870, 2001.

[2001Zep] S. Zepeda, Y. Yeh, and C. A. Orme, Atomic force microscope chamber for in situ studies of ice, *Rev. Sci. Instr.* 72, 4159–4163, 2001.

[2002Blu] H. Bluhm et al., The premelting of ice studied with photoelectron spectroscopy, *J. Phys. Cond. Matter* 14, L227–L233, 2002.

[2002Boe] W. J. Boettinger et al., Phase-field simulation of solidification, *Ann. Rev. Mat. Sci.* 32, 163–194, 2002.

[2002Bra] J. Bragard et al., Linking phase-field and atomistic simulations to model dendritic solidification in highly undercooled melts, *Interface Sci.* 10, 121–136, 2002.

[2002Lib] K. G. Libbrecht, T. Crosby, and M. Swanson, Electrically enhanced free dendrite growth in polar and non-polar systems, *J. Cryst. Growth* 240, 241–254, 2002.

[2002Mat] M. Matsumoto, S. Saito, and I. Ohmine, Molecular dynamics simulation of the ice nucleation and growth process leading to water freezing, *Nature* 416, 409–413, 2002.

[2002Mut] Boyan Mutaftschiev, *The Atomistic Nature of Crystal Growth*, Springer-Verlag, Berlin, 2002.

[2002Pou] F. Pouleta and J. N. Cuzzi, The composition of Saturn's rings, *Icarus* 160, 350–358, 2002.

[2002Sad] V. Sadtchenko, and G. E. Ewing, Interfacial melting of thin ice films: An infrared study, *J. Chem. Phys.* 116, 4686–4697, 2002.

[2002Wan] Pao K. Wang, Shape and microdynamics of ice particles and their effects in cirrus clouds, *Adv. Geophys.* 45, 1–258, 2002.

[2002Wei] X. Wei et al., Sum-frequency spectroscopic studies of ice interfaces, *Phys. Rev. B* 66, 085401, 2002.

[2002Wer] William P. Wergin, Irregular snow crystals: Structural features as revealed by low
temperature scanning electron microscopy, *Scanning* 24, 247–256, 2002.

[2003Abb] J. P. D. Abbatt, Interactions of atmospheric trace gases with ice surfaces: Adsorption and reaction, *Chem. Rev.* 103, 4783–4800, 2003.

[2003Bac] N. Bacon, M. Baker, and B. Swanson, Initial stages in the morphological evolution of vapour-grown ice crystals: A laboratory investigation, *Q. J. R. Meteorol. Soc.* 129, 1903–1927, 2003.

[2003Du] Ning Du, Xiang Y. Liu, and Choy Leong Hew, Ice nucleation inhibition mechanism of antifreeze by antifreeze protein, *J. Bio. Chem.* 278, 36000–36004, 2003.

[2003Lib] K. G. Libbrecht, Growth rates of the principal facets of ice between −10C and −40C, *J. Cryst. Growth* 247, 530–535, 2003.

[2003Lib1] K. G. Libbrecht, Explaining the formation of thin ice-crystal plates with structure-dependent attachment kinetics, *J. Cryst. Growth* 258, 168–175, 2003.

[2003Lib2] Kenneth Libbrecht, with photographs by Patricia Rasmussen, *The Snowflake: Winter's Secret Beauty*, Voyageur Press, Stillwater, MN, 2003.

[2003Shi] A. A. Shibkov et al., Morphology diagram of nonequilibrium patterns of ice crystals growing in supercooled water, *Physica A* 319, 65–79, 2003.

[2004Bai] M. Bailey and J. Hallett, Growth rates and habits of ice crystals between −20C and −70C, *J. Atmos. Sci.* 61, 514–544, 2004.

[2004Czi] D. J. Cziczo et al., Observations of organic species and atmospheric ice formation, *Geophys. Res. Lett.* 31, L12116, doi:10.1029/2004GL019822, 2004.

[2004Gli] M. E. Glicksman and A. O. Lupulescu, Dendritic crystal growth in pure materials, *J. Crystal Growth* 264, 541–549, 2004.

[2004Ike] T. Ikeda-Fukazawa and K. Kawamura, Molecular-dynamics studies of ice Ih, *J. Chem. Phys.* 120, 1395–1401, 2004.

[2004Mar] Ivan V. Markov, *Crystal Growth for Beginners: Fundamentals of Nucleation, Crystal Growth, and Epitaxy*, World Scientific, Singapore, 2004.

[2004Nad] H. Nada, J. P. van der Eerden, and Y. Furukawa, A clear observation of crystal growth of ice from water in a molecular dynamics simulation with a six-site potential model of H_2O, *J. Cryst. Growth* 266, 297–302, 2004.

[2004Shi] A. A. Shibkov et al., Effect of surface kinetics on the dendritic growth of ice in supercooled water, *Crystallography Rep.* 46, 1056–1063, 2004.

[2005Bre] Gary A. Breaux et al., Melting, premelting, and structural transitions in size-selected aluminum clusters with around 55 atoms, *Phys. Rev. Lett.* 94, 173401, 2005.

[2005Car] M. A. Carignano, P. B. Shepson, and I. Szleifer, Molecular dynamics simulations of ice growth from supercooled water, *Mol. Phys.* 103, 2957–2967, 2005.

[2005Joh] G. P. Johari, Water's size-dependent freezing to cubic ice, *J. Chem. Phys.* 122, 194504, 2005.

[2005Lib] K. G. Libbrecht, The physics of snow crystals, *Rep. Prog. Phys.* 68, 855–895, 2005.

[2005Mar] Minoru Maruyama, Roughening transition of prism faces of ice crystals grown from melt under pressure, *J. Cryst. Growth* 275, 598–605, 2005.

[2005Rei] C. A. Reiter, A local cellular model for snow crystal growth, *Chaos, Solitons, and Fractals* 23, 1111–1119, 2005.

[2005Shi] A. A. Shibkov et al., Crossover from diffusion-limited to kinetics-limited growth of ice crystals, *J. Cryst. Growth* 285, 215–227, 2005.

[2006Das] J. G. Dash, A. W. Rempel, and J. S. Wettlaufer, The physics of premelted ice and its geophysical consequences, *Rev. Mod. Phys.* 78, 695–741, 2006.

[2006Fer] R. G. Fernandez, J. L. F. Abascal, and C. Vega, The melting point of ice Ih for common water models calculated from direct coexistence of the solid-liquid interface, *J. Chem. Phys.* 124, 1.2183308, 2006.

[2006Gra] Janko Gravner and David Griffeath, Modeling snow crystal growth I: Rigorous results for Packard's digital snowflakes, *Expt. Math.* 15, 421–444, 2006.

[2006Gra1] Laszlo Granasy et al., Phase field theory of crystal nucleation and polycrystalline growth: A review, *J. Mater. Res.* 21, 309–319, 2006.

[2006Lib] Kenneth Libbrecht, *Ken Libbrecht's Field Guide to Snowflakes*, Voyageur Press, Stillwater, MN, 2006.

[2006Mag] Nathan Magee, Alfred M. Moyle, and Dennis Lamb, Experimental determination of the deposition coefficient of small cirrus-like ice crystals near −50 C, *Geophys. Res. Lett.* 33, L17813, doi:10.1029/2006GL026665, 2006.

[2006Tap] Walter Tape and Jarmo Moilanen, *Atmospheric Halos and the Search for Angle x*, American Geophysical Union, Washington, DC, 2006.

[2007Li] Yimin Li and Gabor A. Somorjai, Surface premelting of ice, *J. Phys. Chem.* 111, 9631–9637, 2007.

[2007Mic] Angelos Michaelides and Karina Morgenstern, Ice nanoclusters at hydrophobic metal surfaces, *Nature Materials* 6, 597–601, 2007.

[2007Nin] Chen Ning and Clifford A. Reiter, A cellular model for three-dimensional snow crystallization, *Comp. and Graphics* 31, 668–677, 2007.

[2008Con] M. M. Conde, C. Vega, and A. Patrykiejew, The thickness of a liquid layer on the free surface of ice as obtained from computer simulation, *J. Chem. Phys.* 129, 014702, 2008.

[2008Gra] Janko Gravner and David Griffeath, Modeling snow crystal growth II: A mesoscopic lattice map with plausible dynamics, *Physica D* 237, 385–404, 2008.

[2008Lib] Kenneth G. Libbrecht, Physically derived rules for simulating faceted crystal growth using cellular automata, arXiv:0807.2616, 2008.

[2008Lib1] Kenneth G. Libbrecht, Helen C. Morrison, and Benjamin Faber, Measurements of snow crystal growth dynamics in a free-fall convection chamber, arXiv:0811.2994, 2008.

[2008Lib2] Kenneth G. Libbrecht and Helen C. Morrison, A convection chamber for measuring ice crystal growth dynamics, arXiv:0809.4869, 2008.

[2009Bai] Matthew Bailey and John Hallett, A comprehensive habit diagram for atmospheric ice crystals: Confirmation from the laboratory, AIRS II, and other field studies, *J. Atmos. Sci.* 66, 2888–2899, 2009.

[2009Goe] M. Goertz, X.-Y Zhu, and J. Houston, Exploring the liquid-like layer on the ice surface, *Langmuir* 25, 6905–6908, 2009.

[2009Gop] Photo by Gopherboy6956, taken February 18, 2009, in Fargo, ND; Wikimedia Commons.

[2009Gra] Janko Gravner and David Griffeath, Modeling snow-crystal growth: A three-dimensional mesoscopic approach, *Phys. Rev. E* 79, 011601, 2009.

[2009Kne] T. N. Knepp, T. L. Renkens, and P. B. Shepson, Gas phase acetic acid and its qualitative effects on snow crystal morphology and the quasi-liquid layer, *Atmos. Chem. Phys.* 9, 7679–7690, 2009.

[2009Lib] K. G. Libbrecht and H. M. Arnold, Measurements of ice crystal growth rates in air at −5 C and −10 C, arXiv:0912.2518, 2009.

[2009Lib1] Kenneth G. Libbrecht, Identification of a novel "fishbone" structure in the dendritic growth of columnar ice crystals, arXiv:0912.2522, 2009.

[2009Lib2] K. G. Libbrecht, Aerodynamical effects in snow crystal growth, arXiv:0911.4733, 2009.

[2009Lib3] K. G. Libbrecht and H. M. Arnold, Aerodynamic stability and the growth of triangular snow crystals, arXiv:0911.4267, 2009; also published in *The Microscope* 57, 4, 157–163, 2009.

[2009Nes] S. Neshyba et al., Molecular dynamics study of ice-vapor interactions via the quasi-liquid layer, *J. Phys. Chem.* 113, 4597–4604, 2009.

[2009Pae] Francesco Paesani and Gregory A. Voth, The properties of water: Insights from quantum simulations, *J. Phys. Chem. B* 113, 5702–5719, 2009.

[2009Pat] Sajal M. Patel, Chandan Bhugra, and Michael J. Pikal, Reduced pressure ice fog technique for controlled ice nucleation during freeze-drying, *AAPS Pharm. Sci. Tech.* 10, 1406, 2009.

[2009Pet] Guillermo Petzold and José M. Aguilera, Ice morphology: Fundamentals and technological applications in foods, *Food Biophys.* 4, 378–396, 2009.

[2010Bai] M. Bailey and J. Hallett, Laboratory measured ice crystal capacitances and mass dimensional relations. 13th Conference on Cloud Physics, Portland, American Meteorological Society, P1.30, https://ams.confex.com/ams/pdfpapers /171204.pdf, 2010.

[2010Bar] J. W. Barrett, Harald Garcke, and Robert Nürnberg, On stable parametric finite element methods for the Stefan problem and the Mullins-Sekerak problem with applications to dendritic growth, *J. Comp. Phys.* 229, 6270–6299, 2010.

[2010Dad] Ruzica Dadic, Bonnie Light, and Stephen G. Warren, Migration of air bubbles in ice under a temperature gradient, with application to "Snowball Earth," *J. Geophys. Res.* 115, D18125, 2010.

[2010Pfa] W. C. Pfalzgraff, R. M. Hulscher, and S. P. Neshyba, Scanning electron microscopy and molecular dynamics of surfaces of growing and ablating hexagonal ice crystals, *Atmos. Chem. Phys.* 10, 2927–2935, 2010.

[2010Saz] G. Sazaki et al., Elementary steps at the surface of ice crystals visualized by advanced optical microscopy, *PNAS* 107, 19702–19707, 2010.

[2010Zha] Jing Zhang, Feng Huang, and Zhang Lin, Progress of nanocrystalline growth kinetics based on oriented attachment, *Nanoscale* 2, 18–34, 2010.

[2011Agu] Andres Aguado and Martin F. Jarrold, Melting and freezing of metal clusters, *Annu. Rev. Phys. Chem.* 62, 151–172, 2011.

[2011Kik] Katsuhiro Kikuchi and Masahiro Kajikawa, *Picture Book of Natural Snow Crystals*, Hokkaido Shinbun, Sapporo, Japan, 2011.

[2011Lib] K. G. Libbrecht and R. Bell, Chemical influences on ice crystal growth from vapor, arXiv:1101.0127, 2011.

[2011Mil] Matthias Militzer, Phase field modeling of microstructure evolution in steels, *Curr. Opin. Solid State Mater. Sci.* 15, 3, 106–115, 2011.

[2011Par] P. Parent et al., HCl adsorption on ice at low temperature: A combined X-ray absorption, photoemission and infrared study, *Phys. Chem. Chem. Phys.* 13, 7111–7177, 2011.

[2011Roz] Dmitri Rozmanov and Peter G. Kusalik, Temperature dependence of crystal growth of hexagonal ice (Ih), *Phys. Chem. Chem. Phys.* 13, 15501–15511, 2011.

[2012Bai] Matthew Bailey and John Hallett, Ice crystal linear growth rates from −20 to −70C: Confirmation from wave cloud studies, *J. Atmos. Sci.* 69, 390–402, 2012.

[2012Bar] John Barrett, Harald Garcke, and Robert Nürnberg, Numerical computations of faceted pattern formation in snow crystal growth, *Phys. Rev. E* 86, 011604, 2012.

[2012Fro] T. Frolov and M. Asta, Step free energies at faceted solid-liquid interfaces from equilibrium molecular dynamics simulations, *J. Chem. Phys.* 137, 214108, 2012.

[2012Gar] T. J. Garrett et al., Fallspeed measurement and high-resolution multi-angle photography of hydrometeors in freefall, *Atmos. Meas. Tech. Discuss.* 5, 4827–4850, 2012.

[2012Hoo] C. Hoose and O. Mohler, Heterogeneous ice nucleation on atmospheric aerosols: A review of results from laboratory experiments, *Atmos. Chem. Phys.* 12, 9817–9854, 2012.

[2012Kni] Charles A. Knight, Ice growth from the vapor at −5°C, *J. Atmos. Sci.* 69, 2031–2040, 2012.

[2012Lib] Kenneth G. Libbrecht, Managing systematic errors in ice crystal growth experiments, arXiv:1208.5064, 2012.

[2012Lib1] K. G. Libbrecht, On the equilibrium shape of an ice crystal, arXiv:1205.1452, 2012.

[2012Mel] A. J. Melendez and C. Beckermann, Measurements of dendrite tip growth and sidebranching in succinonitrile-acetone alloys, *J. Crystal Growth* 340, 175–189, 2012.

[2012Roz] Dmitri Rozmanov and Peter G. Kusalik, Anisotropy in the crystal growth of hexagonal ice I-h, *J. Chem. Phys.* 137, 094702, 2012.

[2012Seo] M. Seo et al., Understanding anisotropic growth behavior of hexagonal ice on a molecular scale: A molecular dynamics simulation study, *J. Chem. Phys.* 137, 154503, 2012.

[2013Dev] S. Deville, Ice-templating, freeze casting: Beyond materials processing, *J. Mat. Res.* 28, 2202–2219, 2013.

[2013Kel] James G. Kelly and Everett C. Boyer, Physical improvements to a mesoscopic cellular automaton model for three-dimensional snow crystal growth, arXiv:1308.4910, 2013.

[2013Kik] Katsuhiro Kikuchia, Takao Kameda, Keiji Higuchi, and Akira Yamashita, A global classification of snow crystals, ice crystals, and solid precipitation based on observations from middle latitudes to polar regions, *Atmos. Res.* 132–133, 460–472, 2013.

[2013Kom] Don Komarechka, *Sky Crystals: Unraveling the Mysteries of Snowflakes*, Self-published, 2013.

[2013Lib] Kenneth G. Libbrecht and Mark E. Rickerby, Measurements of surface attachment kinetics for faceted ice crystal growth, *J. Crystal Growth* 377, 1–8, 2013. Preprint at arXiv: 1208.5982.

[2013Lib1] Kenneth G. Libbrecht, Quantitative modeling of faceted ice crystal growth from water vapor using cellular automata, *J. Comp. Meth. Phys.*, ID-174806, 2013. Preprint at arXiv:0807.2616, 2008.

[2013Men] Fei Meng et al., Screw dislocation driven growth of nanomaterials, *Acc. Chem. Res.* 46, 7, 1616–1626, 2013.

[2013Thu] Konrad Thürmer and Shu Nie, Formation of hexagonal and cubic ice during low-temperature growth, *PNAS* 110, 11757–11762, 2013.

[2014Asa] H. Asakawa et al., Roles of surface/volume diffusion in the growth kinetics of elementary spiral steps on ice basal facets grown from water vapor, *Cryst. Growth Des.* 14, 3210–3220, 2014.

[2014Bar] J. W. Barrett, H. Garcke, and R. Nürnberg, Stable phase field approximations of anisotropic solidification, *IMA J. Numerical Anal.* 34, 1289–1237, 2014.

[2014Bar1] John W. Barrett, Harald Garcke, and Robert Nürnberg, Phase field models versus parametric front tracking methods: Are they accurate and computationally efficient? *Commun. Comput. Phys.* 15, 506–555, 2014.

[2014Kel] J. G. Kelly and E. C. Boyer, Physical improvements to a mesoscopic cellular automaton model for three-dimensional snow crystal growth, *Cryst. Growth Des.* 14, 1392–1405, 2014. Preprint at arXiv:1308.4910.

[2014Lib] Kenneth G. Libbrecht, Toward a comprehensive model of snow crystal growth dynamics: 3. The correspondence between ice growth from water vapor and ice growth from liquid water, arXiv:1407.0740, 2014.

[2014Lib1] Kenneth G. Libbrecht, A dual diffusion chamber for observing ice crystal growth on c-axis ice needles, arXiv:1405.1053, 2014.

[2014Lim] David T. Limmer and David Chandler, Premelting, fluctuations, and coarse-graining of water-ice interfaces, *J. Chem. Phys.* 141, 18C505, 2014.

[2014Mag] N. B. Magee, A. Miller, M. Amaral, and A. Cumiskey, Mesoscopic surface roughness of ice crystals pervasive across a wide range of ice crystal conditions, *Atmos. Chem. Phys.* 14, 12357–12371, 2014.

[2014Shu] Mary Jane Shultz, Patrick J. Bisson, and Alexandra Brumberg, Best face forward: Crystal-face competition at the ice–water interface, *J. Phys. Chem. B* 118, 7972–7980, 2014.

[2015Abd] A. Abdelmonem, J. Lützenkirchen, and T. Leisner, Probing ice-nucleation processes on the molecular level using second harmonic generation spectroscopy, *Atmos. Meas. Tech.* 8, 3519–3526, 2015.

[2015Asa] Harutoshi Asakawa et al., Prism and other high-index faces of ice crystals exhibit two types of quasi-liquid layers, *Cryst. Growth Des.* 15, 3339–3344, 2015.

[2015Ick] Luisa Ickes et al., Classical nucleation theory of homogeneous freezing of water: Thermodynamic and kinetic properties, *Phys. Chem. Chem. Phys.* 17, 5514–5537, 2015.

[2015Lib] Kenneth G. Libbrecht, The surface diffusion length of water molecules on faceted ice: A reanalysis of "Roles of surface/volume diffusion in the growth kinetics of elementary spiral steps on ice basal faces grown from water vapor" by Asakawa et al., arXiv:1509.06609, 2015.

[2015Lib1] Kenneth G. Libbrecht, Incorporating surface diffusion into a cellular automata model of ice growth from water vapor, arXiv:1509.08543, 2015.

[2015Lib2] Kenneth Libbrecht et al., Toward a comprehensive model of snow crystal growth dynamics: 4. Measurements of diffusion-limited growth at −15 C, arXiv:1512.03389, 2015.

[2015Lib3] Kenneth G. Libbrecht, An experimental apparatus for observing deterministic structure formation in plate-on-pedestal ice crystal growth, arXiv: 1503.01019, 2015.

[2016Asa] Harutoshi Asakawaa et al., Two types of quasi-liquid layers on ice crystals are formed kinetically, *PNAS* 113, 1749–1753, 2016.

[2016Ben] Jorge Benet et al., Premelting-induced smoothening of the ice-vapor interface, *Phys. Rev. Lett.* 117, 096101, 2016.

[2016Bis] P. Bisson et al., High yield, single crystal ice via the Bridgman method. *Rev. Sci. Instrum.* 87:034103, 2016.

[2016Har] Alexander Harrison et al., Levitation diffusion chamber measurements of the mass growth of small ice crystals from vapor, *J. Atmos. Sci.* 73, 2743–2758, 2016.

[2016Kar] Alain Karma and Damien Tourret, Atomistic to continuum modeling of solidification microstructures, *Curr. Opin. Solid State Mater. Sci.* 20, 25, 2016.

[2016Li] Jessica Li and Laura P. Schaposnik, Interface control and snow crystal growth, *Phys. Rev. E* 93, 023302, 2016.

[2016Lib] Kenneth G. Libbrecht, Measurements of cylindrical ice crystal growth limited by combined particle and heat diffusion, arXiv:1602.02683, 2016.

[2016Lib1] Kenneth G. Libbrecht, Toward a comprehensive model of snow crystal growth dynamics: 5. Measurements of changes in attachment kinetics from background gas interactions at −5°C, arXiv:1602.08528, 2016.

[2016Mai] S. Maier, B. A. Lechner, G. A. Somorjai, and M. Salmeron, Growth and structure of the first layers of ice on Ru(0001) and Pt(111). *J. Am. Chem. Soc.* 138, 3145–3151, 2016.

[2016Mur] K. Murata et al., Thermodynamic origin of surface melting on ice crystals, *PNAS* E6741–E6748, 2016.

[2016Sei] John H. Seinfeld and Spyros N. Pandis, *Atmospheric Chemistry and Physics: From Air Pollution to Climate Change*, Wiley, Hoboken, NJ, 2016.

[2016Shi] W.ataru Shimada and Kazuki Ohtake, Three-dimensional morphology of natural snow crystals, *Cryst. Growth Des.* 16, 5603–5605, 2016.

[2017Bru] Alexandra Brunberg et al., Single-crystal Ih ice surfaces unveil connection between macroscopic and molecular structure, *PNAS* 114, 5349–5354, 2017.

[2017Dem] G. Demange et al., Growth kinetics and morphology of snowflakes in supersaturated atmosphere using a three-dimensional phase-field model, *Phys. Rev. E* 96, 022803, 2017.

[2017Dem1] Gilles Demange et al., A phase field model for snow crystal growth in three dimensions, *Comput. Mater.* 3, 15, 2017.

[2017Jaa] Mohamad Ali Jaafar et al., A review of dendritic growth during solidification: Mathematical modeling and numerical simulations, *Renew. Sust. Energ. Rev.* 74, 1064–1079, 2017.

[2017Lib] Kenneth G. Libbrecht, Physical dynamics of ice crystal growth, *Annu. Rev. Mater. Res.* 47, 271–295, 2017.

[2017Lup] Laura Lupi et al., Role of stacking disorder in ice nucleation, *Nature* 551, 218 2017.

[2017San] M. Alejandra Sánchez et al., Experimental and theoretical evidence for bilayer-by-bilayer surface melting of crystalline ice, *PNAS* 114, 227–232, 2017.

[2018Che] Phase-field crystal simulation facet and branch crystal growth, *App. Phys. A* 124, 385, 2018.

[2018Cmg] Cmglee, creativecommons.org, from https://en.wikipedia.org/wiki/Ice#/media/File:Phase_diagram_of_water.svg.

[2018Con] J. G. Constantin et al., The quasi-liquid layer of ice revisited: The role of temperature gradients and tip chemistry in AFM studies, *Atmos. Chem. Phys.* 18, 14965–14978, 2018.

[2018Ino] Masahiro Inomata et al., Temperature dependence of the growth kinetics of elementary spiral steps on ice basal faces grown from water vapor, *Cryst. Growth Des.* 18, 786–793, 2018.

[2018Lib] Kenneth G. Libbrecht and Eric D. Black, Improved microparticle electrodynamic ion traps for physics teaching, *Am. J. Phys.* 86, 539–549, 2018.

[2018Lou] Patrick B. Louden and J. Daniel Gezelter, Why is ice slippery? Simulations of shear viscosity of the quasiliquid layer on ice, *J. Phys. Chem. Lett.* 9, 3686–3691, 2018.

[2018Moh] Ali Mohandesi and Peter G. Kusalik, Probing ice growth from vapor phase: A molecular dynamics simulation approach, *J. Cryst. Growth* 483, 156–163, 2018.

[2018Mur] Ken-ichiro Murata, Ken Nagashima, and Gen Sazaki, In situ observations of spiral growth on ice crystal surfaces, *Phys. Rev. Mater.* 2, 093402, 2018.

[2018Qui] Yuqing Qiu and Valeria Molinero, Why is it so difficult to identify the onset of ice premelting? *J. Phys. Chem. Lett.* 9, 17, 5179–5182, 2018.

[2019Ben] Jorge Benet et al., Structure and fluctuations of the premelted liquid film of ice at the triple point, *Molecular Phys.* 0026-8978, 2019.

[2019Har] J. Y. Harrington et al., On calculating deposition coefficients and aspect-ratio evolution in approximate models of ice crystal vapor growth, *J. Atmos. Sci.* 76, 1609, 2019.

[2019Lib] Kenneth G. Libbrecht, A quantitative physical model of the snow crystal morphology diagram, arXiv:1910.09067, 2019.

[2019Lib1] Kenneth G. Libbrecht, Toward a comprehensive model of snow crystal growth: 6. Ice attachment kinetics near −5°C, arXiv:1912.03230, 2019.

[2019Lib2] Kenneth G. Libbrecht, A versatile apparatus for measuring the growth rates of small ice prisms from the vapor phase, arXiv:1912.09440, 2019.

[2019Liu] S. Liu et al., Free dendritic growth model based on non-isothermal interface and microscopic solvability theory, *Trans. Nonferrous Metals Soc. of China* 29, 3, 601–607, 2019.

[2019Llo] P. Llombart, R. M. Bergua, E. G. Noya, and L. G. MacDowell, Structure and water attachment rates of ice in the atmosphere: Role of nitrogen, *Phys. Chem. Chem. Phys.* doi: 10.1039/C9CP03728D, 2019.

[2019Nag] Yuki Nagata et al., The surface of ice under equilibrium and nonequilibrium conditions, *Acc. Chem. Res.* 52, 1006–1015, 2019.

[2019Nel] Jon Nelson and Brian Swanson, Air pockets and secondary habits in ice from lateral-type growth, *Atmos. Chem. Phys. Discuss.*, https://doi.org/10.5194/acp-2019-280, 2019.

[2019Nel1] Jon Nelson and Brian D. Swanson, Lateral facet growth of ice and snow—Part 1: Observations and applications to secondary habits, *Atmos. Chem. Phys.* 19, 15285–15320, 2019.

[2019Sla] Ben Slater and Angelos Michaelides, Surface premelting of water ice, *Nature Rev. Chem.* 3, 172–188, 2019.

[2020Lib] Kenneth G. Libbrecht, Toward a comprehensive model of snow crystal growth: 7. Ice attachment kinetics near −2°C, arXiv:2004.06212, 2020.

[2020Llo] Pablo Llombart, Eva G. Noya, and Luis G. MacDowell, Surface phase transitions and crystal growth rates of ice in the atmosphere, *Sci. Adv.* 6, 21, eaay9322, doi: 10.1126/sciadv.aay9322, 2020. Also arXiv:2004.10465, 2020.

[2020Ma] Runze Ma et al., Atomic imaging of the edge structure and growth of a two-dimensional hexagonal ice, *Nature* 577, 60–63, doi.org/10.1038/s41586-019-1853-4, 2020.

INDEX

Page numbers in italics refer to figures and tables.